NATURAL POLYMERS, BIOPOLYMERS, BIOMATERIALS, AND THEIR COMPOSITES, BLENDS, AND IPNS

Advances in Materials Science
Volume 2

NATURAL POLYMERS, BIOPOLYMERS, BIOMATERIALS, AND THEIR COMPOSITES, BLENDS, AND IPNS

Edited By
**Sabu Thomas, PhD, Neethu Ninan, Sneha Mohan
and Elizabeth Francis**

Apple Academic Press

TORONTO   NEW JERSEY

© 2013 by
Apple Academic Press Inc.
3333 Mistwell Crescent
Oakville, ON L6L 0A2
Canada

Apple Academic Press Inc.
1613 Beaver Dam Road, Suite # 104
Point Pleasant, NJ 08742
USA

First issued in paperback 2021

Exclusive worldwide distribution by CRC Press, a Taylor & Francis Group

ISBN 13: 978-1-77463-213-0 (pbk)
ISBN 13: 978-1-926895-16-1 (hbk)

Library of Congress Control Number: 2012935654

Library and Archives Canada Cataloguing in Publication

Natural polymers, biopolymers, biomaterials, and their composites, blends, and IPNs/edited by Sabu Thomas ... [et al.].
(Recent advances in materials science; 2)

Includes bibliographical references and index.
ISBN 978-1-926895-16-1
1. Biopolymers. 2. Polymeric composites. I. Thomas, Sabu
II. Series: Recent advances in materials science (Toronto); 2

QP801.B69N38 2012   572'.33   C2012-900042-6

Advances in Materials Science

Series Editors-in-Chief

Sabu Thomas, PhD

Dr. Sabu Thomas is the Director of the School of Chemical Sciences, Mahatma Gandhi University, Kottayam, India. He is also a full professor of polymer science and engineering and Director of the Centre for nanoscience and nanotechnology of the same university. He is a fellow of many professional bodies. Professor Thomas has authored or co-authored many papers in international peer-reviewed journals in the area of polymer processing. He has organized several international conferences and has more than 420 publications, 11 books and two patents to his credit. He has been involved in a number of books both as author and editor. He is a reviewer to many international journals and has received many awards for his excellent work in polymer processing. His h Index is 42. Professor Thomas is listed as the 5th position in the list of Most Productive Researchers in India, in 2008.

Mathew Sebastian, MD

Dr. Mathew Sebastian has a degree in surgery (1976) with specialization in Ayurveda. He holds several diplomas in acupuncture, neural therapy (pain therapy), manual therapy and vascular diseases. He was a missionary doctor in Mugana Hospital, Bukoba in Tansania, Africa (1976-1978) and underwent surgical training in different hospitals in Austria, Germany, and India for more than 10 years. Since 2000 he is the doctor in charge of the Ayurveda and Vein Clinic in Klagenfurt, Austria. At present he is a Consultant Surgeon at Privatclinic Maria Hilf, Klagenfurt. He is a member of the scientific advisory committee of the European Academy for Ayurveda, Birstein, Germany, and the TAM advisory committee (Traditional Asian Medicine, Sector Ayurveda) of the Austrian Ministry for Health, Vienna. He conducted an International Ayurveda Congress in Klagenfurt, Austria, in 2010. He has several publications to his name.

Anne George, MD

Anne George, MD, is the Director of the Institute for Holistic Medical Sciences, Kottayam, Kerala, India. She did her MBBS (Bachelor of Medicine, Bachelor of Surgery) at Trivandrum Medical College, University of Kerala, India. She acquired a DGO (Diploma in Obstetrics and Gynaecology) from the University of Vienna, Austria; Diploma Acupuncture from the University of Vienna; and an MD from Kottayam Medical College, Mahatma Gandhi University, Kerala, India. She has organized several international conferences, is a fellow of the American Medical Society, and is a member of many international organizations. She has five publications to her name and has presented 25 papers.

Dr. Yang Weimin

Dr. Yang Weimin is the Taishan Scholar Professor of Quingdao University of Science and Technology in China. He is a full professor at the Beijing University of Chemical Technology and a fellow of many professional organizations. Professor Weimin has authored many papers in international peer-reviewed journals in the area of polymer processing. He has been contributed to a number of books as author and editor and acts as a reviewer to many international journals. In addition, he is a consultant to many polymer equipment manufacturers. He has also received numerous award for his work in polymer processing.

Contents

List of Contributors

Ibrahim Abdullah
Polymer Research Center (PORCE), School of Chemical Sciences & Food Technology.
Faculty of Science & Technology, Universiti Kebangsan Malaysia (UKM), 43600 Bangi, Selangor, Malaysia.

Ishak Ahmad
Polymer Research Center (PORCE), School of Chemical Sciences & Food Technology.
Faculty of Science & Technology, Universiti Kebangsan Malaysia (UKM), 43600 Bangi, Selangor, Malaysia.

Mansur Ahmed
Department of Materials and Metallurgical Engineering, Bangladesh University of Engineering and Technology (BUET), Dhaka-1000, Bangladesh.

Qumrul Ahsan
Department of Materials and Metallurgical Engineering, Bangladesh University of Engineering and Technology (BUET), Dhaka-1000, Bangladesh.

Z. Ansar Ali
Research scholar, Bharathiyar University, Coimbatore-641 046, Tamil Nadu, India.

D. Saravana Bavan
Department of Mechanical Engineering, National Institute of Technology Karnataka
Surathkal, Mangalore-575025, India.

P.V. Brevern
Faculty of Engineering and Technology, Multimedia University, Jalan Ayer Keroh Lama, 75450, Melaka, Malaysia.

M. L. Botelhoand
Instituto Tecnológico e Nuclear, Estrada Nacional nº 10, Sacavém, Portugal.

C. C. Castro
Instituto Tecnológico e Nuclear, Estrada Nacional nº 10, Sacavém, Portugal.

Frédéric Chivrac
CREAGIF Biopolymères, 18, avenue de la voie aux coqs, 14760 Bretteville–sur–Odon, France.

K. Cotton
Institute for Biomedical Engineering, Louisiana Tech University, Ruston, Louisiana, USA 71270.

A. Cyganiuk
Faculty of Chemistry, Nicholas Copernicus University, ul. Gagarina 11, 87-100 Toruń, Poland.

Mark A. DeCoster
Institute for Micromanufacturing, Biomedical Engineering, Louisiana Tech University, Ruston, Louisiana, USA 71270.
Corresponding author: Mark A. DeCoster, Biomedical Engineering, 818 Nelson Avenue, Louisiana Tech University, Ruston, LA, USA 71270.

Brigitte Deschrevel
Laboratoire "Polymères, Biopolymères, Surfaces", UMR 6270 CNRS–Université de Rouen Institut Normand de Chimie Moléculaire, Médicinale et Macromoléculaire, FR 3038 CNRS.

Gokulnath Dillibabu
Materials Science and Engineering Division, Department of Mechanical Engineering, Anna University, Chennai 600 025, India.

K. Evans
Institute for Mathematics and Statistics, Louisiana Tech University, Ruston, Louisiana, USA 71270.

Chester A. Faunce
The University of Salford, Joule Physics Laboratory, Faculty of Science, Engineering and Environment, Manchester M5 4WT, United Kingdom.

Susana C. M. Fernandes
Department of Chemistry, CICECO, Campus Universitário de Santiago, University of Aveiro, 3810-193 Aveiro, Portugal.

Souza Jr., Fernando Gomes
Instituto de Macromoléculas, Universidade Federal do Rio de Janeiro, Rio de Janeiro, Brasil.

Nadège Follain
Université de Rouen, Laboratoire «Polymères, Biopolymères, Surfaces», UMR 6270 CNRS & FR3038 INC3M, 76821 Mont-Saint-Aignan Cedex, France.

Carmen S. R. Freire
Department of Chemistry, CICECO, Campus Universitário de Santiago, University of Aveiro, 3810–193 Aveiro, Portugal.

C. Frias
Instituto de Engenharia Mecânica e Gestão Industrial Faculdade de Engenharia, Universidade do Porto (FEUP), Rua Roberto Frias s/n, 4200-465 Porto, Portugal.

Alessandro Gandini
Department of Chemistry, CICECO, Campus Universitário de Santiago, University of Aveiro, 3810–193 Aveiro, Portugal.

Swarupini Ganesan
Materials Science and Engineering Division, Department of Mechanical Engineering, Anna University, Chennai 600 025, India.

N. S. Gangurde
P.G. Department of Microbiology, Shri S. I. Patil Arts, G. B. Patel Science & S.T.S.K.V. S. Commerce College, Shahada, Dist. Nandurbar, (MS) 425409 India.

Florent Girard
CREAGIF Biopolymères, 18, avenue de la voie aux coqs, 14760 Bretteville–sur–Odon, France.

Nina Graupner
Hochschule Bremen–University of Applied Sciences, Faculty 5–Biomimetics, Biological Materials, Neustadtswall 30, D-28199 Bremen

Dustin Green
Institute for Biology, Louisiana Tech University, Ruston, Louisiana, USA 71270.

A. Gnanamani
Microbiology Division, Central Leather Research Institute (Council of Scientific and Industrial Research), Adyar, Chennai-600020, Tamil Nadu, India.

Jamil Hashim
Faculty of Engineering and Technology, Multimedia University, Jalan Ayer Keroh Lama, 75450, Melaka, Malaysia.

My Hedhammar
Department of Anatomy, Physiology and Biochemistry, Swedish University of Agricultural Sciences, the Biomedical Centre, 751 23 Uppsala, Sweden.
Department of Neurobiology, Care Sciences and Society (NVS), Karolinska Institutet, Novum 5th floor, 141 86 Stockholm, Sweden.
Spiber Technologies AB, Doppingv 12, 756 51 Uppsala, Sweden.

R. Hemalatha
Research scholar, Manonmonium Sundaranar University, Thirunelveli, Tamilnadu, India.

R. Idowu
Institute for Mathematics and Statistics, Louisiana Tech University, Ruston, Louisiana, USA 71270.

MdSaiful Islam
Department of Materials and Metallurgical Engineering, Bangladesh University of Engineering and Technology (BUET), Dhaka-1000, Bangladesh.

C. Jeyasankar
Institute for Biomedical Engineering, Louisiana Tech University, Ruston, Louisiana, USA 71270.

Hanieh Kargarzadeh
Polymer Research Center (PORCE), School of Chemical Sciences & Food Technology.
Faculty of Science & Technology, UniversitiKebangsan Malaysia (UKM), 43600 Bangi, Selangor, Malaysia.

Seiichi Kawahara
Department of Materials Science and Technology, Faculty of Engineering, Nagaoka University of Technology, Nagaoka, Niigata 940-2188, Japan
Center for Green-Tech Development in Asia, Nagaoka University of Technology, Nagaoka, Niigata 940-2188, Japan.
Department of Chemical Science and Engineering, Tokyo National College of Technology, 1220-2 Kunugidamachi, Hachioji, Tokyo 193-0997, Japan.
JST-JICA SATREPS.

R. Klimkiewicz
Institute of Low Temperature and Structure Research PAN, ul.Okólna 2, 50-422 Wrocław, Poland.

Annaleena Kokko
VTT Technical Research Centre of Finland, P.O. Box 1000, FI-02044 VTT, Finland.

A. Kucinska
Faculty of Chemistry, Nicholas Copernicus University, ul.Gagarina 11, 87-100 Toruń, Poland.

Mohan Kumar. G. C
Department of Mechanical Engineering, National Institute of Technology Karnataka
Surathkal, Mangalore-575025, India.

Sandeep Kumar
Centre for Polymer Science and Engineering, Indian Institute of Technology Delhi,
Hauz Khas, New Delhi-110016, India.

Vesa Kunnari
VTT Technical Research Centre of Finland, P.O. Box 1000, FI-02044 VTT, Finland.

J. P. Lukaszewicz
Faculty of Chemistry, Nicholas Copernicus University, ul.Gagarina 11, 87-100 Toruń, Poland.

Y. Lvov
Institute for Micromanufacturing, Louisiana Tech University, Ruston, Louisiana, USA 71270.

Lopes, Magnolvaldo Carvalho
Instituto de Macromoléculas, Universidade Federal do Rio de Janeiro, Rio de Janeiro, Brasil.

Kausik Majumder
Post-Graduate Department of Botany, Darjeeling Government College, Darjeeling-734101
West Bengal, India.

S. K. Malhotra
Professor, Dept of Mechanical Engineering, RMK Engineering College, RSM Nagar, Kavaraipettai-601206
(TN), India.

A B Mandal
Microbiology Division, Central Leather Research Institute (Council of Scientific and Industrial Research),
Adyar, Chennai-600020, Tamil Nadu, India.

Stéphane Marais
Université de Rouen, Laboratoire «Polymères, Biopolymères, Surfaces», UMR 6270 CNRS & FR3038
INC3M, 76821 Mont-Saint-Aignan Cedex, France.

A. T. Marques
Instituto de Engenharia Mecânica e Gestão Industrial Faculdade de Engenharia, Universidade do Porto
(FEUP), Rua Roberto Frias s/n, 4200-465 Porto, Portugal.

J. McNamara
Institute for Biomedical Engineering, Louisiana Tech University, Ruston, Louisiana, USA 71270.

Tapas Mitra
Microbiology Division, Central Leather Research Institute (Council of Scientific and Industrial Research),
Adyar, Chennai-600020, Tamil Nadu, India.

Mohammad S. Mubarak
Department of Chemistry, the University of Jordan, Amman 11942, Jordan.

JörgMüssig
Hochschule Bremen–University of Applied Sciences, Faculty 5–Biomimetics, Biological Materials, Neustadtswall 30, D-28199 Bremen.

Umar Nirmal
Faculty of Engineering and Technology, Multimedia University, Jalan Ayer Keroh Lama, 75450, Melaka,
Malaysia.

Carlos Pascoal Neto
Department of Chemistry, CICECO, Campus Universitário de Santiago, University of Aveiro, 3810–193
Aveiro, Portugal

A. Olejniczak
Faculty of Chemistry, Nicholas Copernicus University, ul.Gagarina 11, 87-100 Toruń, Poland.

Geiza Esperandio Oliveira
DQUI, CCE, Universidade Federal do Espírito Santo, Vitória, Brasil.

Henrich H. Paradies
The University of Salford, Joule Physics Laboratory, Faculty of Science, Engineering and Environment,
Manchester M5 4WT, United Kingdom.

Krystyna Pietrucha
Technical University of Lodz, Zeromskiego 116, 90-924 Lodz, Poland.

J. Potes
ICAAM, Departamento de Medicina Veterinária, Universidade de Évora,Pólo da Mitra, Ap. 94,7002-554
Évora, Portugal.

Hariharan Raja
Materials Science and Engineering Division, Department of Mechanical Engineering, Anna University, Chennai 600 025, India.

Perumal Ramasamy
Department of Physics, Anna University, Chennai-600 025, India.

R. Ramya
Research Scholar, Manonmaniam Sundaranar University, Palayamkottai-627 002, Tamil Nadu, India.

Hendrik Reichelt
The University of Salford, Joule Physics Laboratory, Faculty of Science, Engineering and Environment, Manchester M5 4WT, United Kingdom.

J. Reis
Departamento de Medicina Veterinária, Universidade de Évora,Pólo da Mitra, Ap. 94,7002-554 Évora, Portugal.

Dirk Rilling
Faculty of Engineering and Technology, Multimedia University, Jalan Ayer Keroh Lama, 75450, Melaka, Malaysia.

Anna Rising
Department of Anatomy, Physiology and Biochemistry, Swedish University of Agricultural Sciences, the Biomedical Centre, 751 23 Uppsala, Sweden.
Department of Neurobiology, Care Sciences and Society (NVS), Karolinska Institutet, Novum 5th floor, 141 86 Stockholm, Sweden.
Spiber Technologies AB, Doppingv 12, 756 51 Uppsala, Sweden.

G. Sailakshmi
Microbiology Division, Central Leather Research Institute (Council of Scientific and Industrial Research), Adyar, Chennai-600020, Tamil Nadu, India.

R. Z. Sayyed
P.G. Department of Microbiology, Shri S. I. Patil Arts, G. B. Patel Science & S.T.S.K.V. S. Commerce College, SHAHADA, Dist. Nandurbar, (MS) 425409 India.

A. Shanmugapriya
Research scholar, Bharathiyar University, Coimbatore-641 046, Tamilnadu, India.

F. Silva
ICAAM, Departamento de Biologia, Universidade de Évora,Pólo da Mitra, Ap. 94,7002- 554 Évora, Portugal.

Armando J. D. Silvestre
Department of Chemistry, CICECO, Campus Universitário de Santiago, University of Aveiro, 3810-193 Aveiro, Portugal.

J. A. Simões
Departamento de Mecânica, Universidade de Aveiro,Campus de Universitário de Santiago, s/n, 3810-193 Aveiro, Portugal.

Ravisankar Subburayalu
Materials Science and Engineering Division, Department of Mechanical Engineering, Anna University, Chennai 600 025, India.

P. N. Sudha
Department of Chemistry, DKM College, Thiruvalluvar University, Vellore-632 001, India.

Jari Vartiainen
VTT Technical Research Centre of Finland, P.O. Box 1000, FI-02044 VTT, Finland.

Mona Widhe
Spiber Technologies AB, Doppingv 12, 756 51 Uppsala, Sweden.

Q. Xing
Institute for Micromanufacturing, Louisiana Tech University, Ruston, Louisiana, USA 71270.

Yoshimasa Yamamoto
Department of Chemical Science and Engineering, Tokyo National College of Technology, 1220-2 Kunugi-da-machi, Hachioji, Tokyo 193-0997, Japan.
JST-JICA SATREPS.

F. Yousif
Faculty of Engineering and Surveying, University of Southern Queensland, Toowoomba, West St., 4350 QLD, Australia.

Siti Yasmine, Z. Z.
Polymer Research Center (PORCE), School of Chemical Sciences & Food Technology.
Faculty of Science & Technology, UniversitiKebangsan Malaysia (UKM), 43600 Bangi, Selangor, Malaysia.

Hiba M. Zalloum
Hamdi Mango Center for Scientific Research, the University of Jordan, Amman 11942, Jordan.

Kurt Zimmermann
The SymbioHerborn Group Ltd., Auf den Lüppen 8, Herborn D-35745, Germany.

List of Abbreviations

AA	Alginic acid
ADP	Adenosine diphosphate
AMP	Adenosine monophosphate
ASC	Acid soluble collagen
ATP	Adenosine triphosphate
BBB	Blood brain barrier
BET	Brummer-emmet-teller
bFGF	Basic fibroblast growth factor
BOD	Block-On-Disc
BSA	Bovine serum albumin
BT-HAase	Bovine testicular HAase
CAM	Cationic antimicrobial
CCdBE	Crosslinked chitosan crown ether
CGC	Chemically modified chitosans
CMC	Carboxymethylcellulose
CMO	Chloromethyloxirane
CNSL	Cashew nut shell liquid
CNTs	Carbon nanotubes
CS	Chondroitin sulfate
CSIS	Chitosan-isatin schiff's base
CSMO	Chitosan–diacetylmonoxime
CT	Chloromethyl thiirane
CTPP	Chitosan-tripolyphosphate
CTS	Chitosan
CTS-B15	Chitosan-benzo-15-crown-5
CTS-B18	Chitosan-benzo-18-crown-6
CTS-DMTD	Chitosan hydrogel with 2,5-dimercapto-1,3,4-thiodiazole
DCP	Dicumyl peroxide
DG	Degree of grafting
DHT	Dehydrothermal
DI	Deionoized
DoE	Design technique
D-R	Dubinin-radushkevitch
dRG	Debranched RG
DRS	Dielectric relaxation spectroscopy

DSC	Differential scanning calorimetry
DTG	Derivative thermogravimetric
DTPA	Diethylenetriaminepentaacetic acid
ECH	Epichlorohydrin
ECM	Extra-cellular matrix
E-DPNR	Enzymatic deproteinization
EDTA	Ethylenediaminetetraacetic acid
EDTriA	Ethylenediamine-N,N,N'-triacetate
EGF	Epidermal growth factor
ELP	Elastin like peptide
ESPI	Electronic speckle pattern interferometry
FBS	Fetal bovine serum
FESEM	Field emission scanning electron microscope
FITC	Fluorescein isothiocyanate
FRP	Fibre reinforced plastics
FTIR	Fourier transforms infrared
GAB	Guggenheim-Anderson-de-Boer
GABA	Gamma-aminobutyric acid
GAG	Glycosaminoglycan
GD	Grafting degree
GMA	Glycidyl methacrylate
HA	Hyaluronan
HAases	Hyaluronidases
HAL	Hessian untreated/polyester
HFIP	Hexafluoroisopropanol
HGs	Homogalacturonans
HN	Hyaluronectin
HPMC	Hydroxypropylmethylcellulose
HS	Horse serum
HSAL	Hessian starch alkali treated/polyester
HSALMC	Hessian starch alkali treated with mechanical constraint/polyester
HSTHER	Hessian starch thermal treated/polyester
HSUNTR	Hessian starch untreated/polyester
HUNTR	Hessian untreated/polyester
HV	Hydroxyvalerate
ICP-MS	Inductively coupled plasma-mass spectrometry
IDT	Initial decomposition temperature
IPN	Interpenetrating polymer network
LDI	Lysine-based diisocyanate

L-F	Langmuir-freundlich
LLDPE	Linear low density polyethylene
LPS	Lipopolysaccharides
LS	Light scattering
LYS	Lysozyme
MA	Maleic acid
MAPP	Maleic anhydride grafted polypropylene
MaSp	Major ampullate spidroin
MC	Methylcellulose
MCL	Medium chain lengths
MD	Machine direction
MEKP	Methyl Ethyl Ketone Peroxide
MEKP	Methyl ethyl ketone peroxide
MIN	Minimal dimensions
MPEGs	Monomethoxy poly(ethylene glycol)
MW	Molecular weight
MWS	Maxwell-wagner-sillars
NCM	Neuronal culture media
NR	Natural rubber
OPP	Oriented polypropylene
OXP	Oxalate-soluble pectic
PBS	Phosphate buffered saline
PCB	Polychlorinated biphenyl
PCNA	Proliferating cell nuclear antigen
PDF	Probability density function
PE	Polyethylene
PEA	polyesteramide
PEGs	Poly(ethylene glycols)
PET	Poly(ethylene terephthalate)
PHAs	Polyhydroxyalkanoates
PHB	Poly-β-hydroxybutyrate
PHBHx	Poly(hydroxybutyrate-co-hydroxyhexanoate)
PHBO	Poly(hydroxybutyrate-co-hydroxyoctanoate)
PHBOd	Poly(hydroxybutyrate-co-hydroxy octadecanoate)
PHBV	Poly(hydroxyl butyrate-co-hydroxyvalerate)
PHH	Polyhydroxyhexanoate
PHV	Polyhydroxyvalerate
PI	Isoelectric point
PLA	Poly(lactic acid)

PLGA	Poly(lactic-co-glycolic acid)
PLL	poly-L-lysine
PLLA	Poly-L-lactic acid
PMMA	Poly (methyl methacrylate)
PP	Polypropylene
PSD	Pore size distribution
PTFE	Polytetrafluroethylene
PVA	Poly(vinyl alcohol)
PVDF	Polyvinylidene Fluoride
Ra	Roughness averages
RASC	Reconstituted acid soluble collagen
RBMVECs	Rat brain microvascular endothelial cells
RG	Rhamnogalacturonan
ROP	Ring-opening polymerization
RTM	Resin transfer molding
SCL	Short chain lengths
SDBF	Short delignified bamboo-fiber
SEM	Scanning Electron Microscope
SF	Silk fibroin
SFPT	Single fiber pullout tests
2D	Two dimensional
3D	Three dimensional
T-BFRP	Treated betelnut fiber reinforced polyester
Tcp	Tetrahedrally close-packed
TEM	Transmission Electron Microscope
TG	Thermo gravimetric
TGA	Termogravimetric analysis
TNF	Tumor necrosis factor
TPP	Triphosphate
TPS	Thermoplastic starch
UF	Urea-formaldehyde
UT-BFRP	Untreated betelnut fiber reinforced polyester
UTM	Universal testing machine
UTS	Ultimate tensile strength
VARTM	Vacuum assisted resin transfer molding
VEGF	Vascular endothelial cell growth factor
WHO	Word Heath Organization
WSP	Water-soluble pectic
XRD	X-ray diffraction

Preface

Science and technology of polymeric biomaterials as a whole have seen extraordinary development, research interest and investment by industry in recent decades. Within this broad field, natural polymers have in particular witnessed major studies. Indeed biopolymers have virtually moulded the modern world and transformed the quality of life in innumerable areas of human activity. They have added new dimensions to standards of life and inexpensive product development. From transportation to communications, entertainment to health care, the world of biopolymers has touched them all.

Natural polymers include RNA and DNA that are so important in genes and life processes. In fact, messenger RNA is what makes possible proteins, peptides, and enzymes. Enzymes do the chemistry inside living organisms and peptides make up some of the most interesting structural components of skin, hair, and even the horns of rhinos. Other natural polymers include polysaccharides (sugar polymers) and polypeptides like silk, keratin, and hair. Natural rubber is a natural polymer, made from just carbon and hydrogen.

Biopolymers are polymers produced by living organisms. They contain monomeric units that are covalently bonded to form larger structures. There are three main classes of biopolymers based on the differing monomeric units used and the structure of the biopolymer formed. Polynucleotides are long polymers which are composed of 13 or more nucleotide monomers, Polypeptides are short polymers of amino acids and polysaccharides are often linear bonded polymeric carbohydrate structures. Cellulose is the most common organic compound and biopolymer on earth. About 33 percent of all plant matter is cellulose. The cellulose content of cotton is 90 percent and that of wood is 50 percent.

Composite materials are solids which contain two or more distinct constituent materials or phases, on a scale larger than the atomic. The term "composite" is usually reserved for those materials in which the distinct phases are separated on a scale larger than the atomic, and in which properties such as the elastic modulus are significantly altered in comparison with those of a homogeneous material. Accordingly, reinforced plastics such as fiberglass as well as natural materials such as bone are viewed as composite materials, but alloys such as brass are not. Foam is a composite in which one phase is empty space. Natural biological materials tend to be composites. Natural composites include bone, wood, dentin, cartilage, and skin. Natural foams include lung, bone, and wood. Natural composites often exhibit hierarchical structures in which particulate, porous, and fibrous structural features are seen on different micro-scales. Composite materials offer a variety of advantages in comparison with homogeneous materials. These include the ability for the scientist or engineer to exercise considerable control over material properties. There is the potential for stiff, strong, lightweight materials as well as for highly resilient and compliant materials. In biomaterials, it is important that each constituent of the composite be biocompatible. Moreover, the interface between constituents should not be degraded by the body environment. Some applications of composites in biomaterial applications are: (1) dental filling composites, (2) reinforced methyl methacrylate bone cement and ultrahigh molecular weight polyethylene, and (3) orthopedic implants with porous surfaces.

A polymer blend or polymer mixture is a member of a class of materials analogous to metal alloys, in which at least two polymers are blended together to create a new material with different physical properties. Polymer blends can be broadly divided into three categories namely immiscible polymer blends, compatible polymer blends, miscible polymer blends. Immiscible polymer blends (heterogeneous polymer blends) are by far the most populous group. If the blend is made of two polymers, two glass transition temperatures will be observed. Compatible polymer blends are immiscible polymer blend that exhibits macroscopically uniform physical properties. The macroscopically uniform properties are usually caused by sufficiently strong interactions between the component polymer. Miscible polymer blends (homogeneous polymer blend) are polymer blends that are single-phase structures. In this case, one glass transition temperature will be observed. Examples of miscible polymer blends are Polyphenylene oxide (PPO) - polystyrene (PS), Polypropylene (PP) - EPDM etc.

An Interpenetrating polymer network (IPN) is a polymer comprising of two or more networks which are at least partially interlaced on a polymer scale but not covalently bonded to each other. The network cannot be separated unless chemical bonds are broken. The two or more networks can be envisioned to be entangled in such a way that they are concatenated and cannot be pulled apart, but not bonded to each other by any chemical bond. Simply mixing two or more polymers does not create an interpenetrating polymer network nor does creating a polymer network out of more than one kind of monomers which are bonded to each other to form one network. There are semi-interpenetrating polymer networks and pseudo-interpenetrating polymer networks.

This book focuses on the recent advances in natural polymers, biopolymers, biomaterials, and their composites, blends and IPNs. Biobased polymer blends and composites occupy a unique position in the dynamic world of new biomaterials. The growing need for lubricious coatings and surfaces in medical devices - an outcome of the move from invasive to noninvasive medicines /procedures - is playing a major role in the advancement of biomaterials technology. Natural polymers have attained their cutting edge technology through various platforms yet there are a lot of novel informations about them which will be discussed in the following chapters. This book covers topics like chitosan composites for biomedical applications and wastewater treatment, coal biotechnology, biomedical and related applications of Second Generation Polyamidoamines, silk fibers, PEG hydrogels, bamboo fibre reinforced PE composites, jute/ polyester composites, magnetic biofoams and many other interesting aspects which may attract readers greatly.

— **Sabu Thomas, PhD**

Chapter 1

Chitosan and Chitosan Derivatives as Chelating Agents

Hiba M. Zalloum and Mohammad S. Mubarak

INTRODUCTION

Chitosan, as shown in Figure 1, is a well-known hetero biopolymer (natural polysac-charide) made of glucosamine and a fraction of acetylglucosamine residues (Krishnapriya and Kandaswamy, 2010). It is a biodegradable, biocompatible, and nontoxic natural polymer with a metal uptake capacity, (Jung et al., 1999; Rabea et al., 2003; Sashiwa et al., 2003; Tsigos et al., 2000; Varma et al., 2004; Xing et al., 2005) that can be obtained from the alkaline deacetylation process of the second most abundant biopolymer, chitin as given in Figure 2, which is found widely in nature and can be extracted from fungi, lobster, shells of shrimp and crab, and in the cuticles of insects. (Brugnerotto et al., 2001a, 2001b; Heux et al., 2000; Kittur et al., 1991; Volesky, 2001). The main characteristics of chitosan are hydrophilicity, harmlessness for living things and biodegradability, easy chemical derivatization, and capability to adsorb a number of metal ions. Therefore, chitosan seems to be a very interesting starting material for chelating resins (Katarina et al., 2006). The degree of polymerization and deacetylation and the distribution of acetyl groups along the polymer chain are of crucial importance for chitosan metal interacting characteristics. Making chemical derivatives is a way to alter the metal interacting characteristics of chitosan (Onsoyen and Skaugrud, 1990).

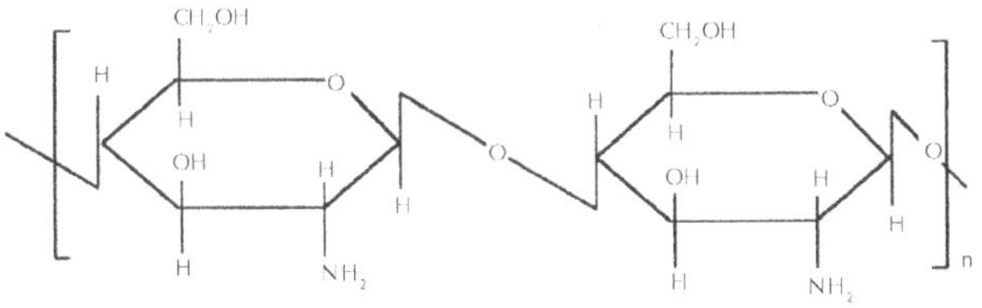

Figure 1. Structure of Chitosan.

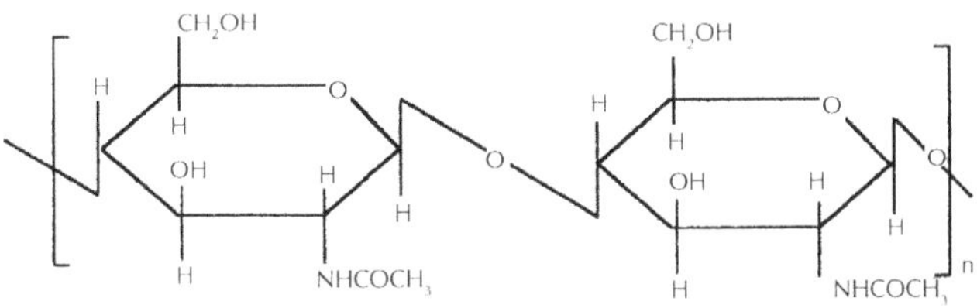

Figure 2. Structure of Chitin.

In recent years, chitosan has been described as a suitable natural polymer and a renewable resource for the collection of metal ions, (Arrascue et al., 2003; Ding et al., 2006; Ruiz et a., 2002; Tabakci and Yilmaz, 2008) since the amine and hydroxyl functional groups on the chitosan chain can act as chelation sites for metal ions. Besides inherent sorption, the adsorption capacity and selectivity of chitosan could also be enhanced by chemical and physical modification on both amine and hydroxyl groups (Ramesh et al., 2008; Ruiz et al., 2003; Wan Ngah et al., 1999). Additionally, it is possible to reinforce the stability of the biopolymer in acidic conditions by cross-linking (Butewicz et al., 2010).

Chitosan was found to have the highest chelating ability in comparison with other natural polymers (Varma et al., 2004). It has been realized, that the mechanism of chitosan–metals complex formation is manifold and probably dominated by different processes taking place simultaneously such as adsorption, ion-exchange and chelation, under different conditions (Butelman, 1991). The presence of amine groups leads to the binding of metal cations by complexation or chelation, and after protonation to the binding of anions by electrostatic attraction or ion exchange (Guibal, 2004; Guibal et al., 2006a).

In addition to medical applications, the most important and significant developments in chitin/chitosan technology have been in the area of environmental applications which include among others the removal of dyes (Chiou and Li, 2003; McKay et al., 1983; Wong et al., 2003, 2004; Yoshida et al., 1993). Other environmental applications include removal of polychlorinated biphenyl (PCB) (Ikeda et al., 1999) and chemical waste detoxification (Wagner and Nicell, 2002). Further developments in the field of water treatment include filtration (Juang and Chiou, 2001), desalination (Arai and Akiya, 1978), and flocculation/coagulation (Eikebrokk and Saltnes, 2002). An interesting area of research, however, has been generated by the ability of chitosan to remove metal ions from wastewaters by the process of adsorption. Chitosan has demonstrated the potential to adsorb significant amounts of metal ions, and this has generated an interest in assessing its feasibility to remove metal ions over a wide range of effluent systems and types.

CHITOSAN AND MODIFIED CHITOSAN AS CHELATING AGENTS FOR METAL IONS

The ability of chitosan to bind transition metal in presence of alkali and alkaline earth metal is well investigated (Deans and Dixon, 1992). The adsorption of Cu^{2+}, Hg^{2+}, Ni^{2+}, and Zn^{2+} on chitosan with various particle sizes and as a function of temperature was studied at neutral pH (McKay et al., 1989). Moreover, metal complexation by chitosan and its derivatives has been reviewed (Gerente et al., 2007; Guibal et al., 2006b; Varma et al., 2004). Recently, chitosan–Zn complex attracted great interest for its potential uses as medicament or nutriment (Paik, 2001; Tang and David, 2001; YoneKura and Suzuki, 2003). It is well known that both of chitosan and metal ions (Zn^{2+}, Zr^{2+}, and Ag^{1+}) have the properties of disinfection and bactericide (Varma et al., 2004). In addition, and with the growing need for new sources of low-cost adsorbent, the increased problems of waste disposal, the increasing cost of synthetic resins undoubtedly make chitosan one of the most attractive materials for wastewater treatment (Babel and Kurniawan, 2003).

Chitosan can easily be modified by chemical or physical processes to prepare chitosan derivatives. These processes may be used for controlling the reactivity of the polymer (improving the affinity of the sorbent for the metal, changing the selectivity series for sorption, changing the pH range for optimum sorption) or enhancing sorption kinetics (controlling diffusion properties, for example) (Guibal, 2004). Eric Guibal et. al. (2006b) has published a review on the use of chitosan for the removal of particulate and dissolved contaminants and concluded that chitosan was very efficient at removing particulate and dissolved contaminants through coagulation-flocculation processes involving several mechanisms such as charge neutralization, precipitative coagulation, bridging, and electrostatic patch. Combining these processes (as a function, for example, of pH conditions) enables the design of competitive procedures for the treatment of wastewaters or pre-treatment of potable water. They have also concluded that chitosan offers a promising alternative to the use of mineral reagents (alum salts, ferric salts) or synthetic polymers. The use of a polymer of biological origin, coming from a renewable resource, biodegradable, and thus less aggressive for final discharge in the environment (sludge landfill, dispersion in the aqueous phase of residues) is an important criterion for future developments. Its attractiveness as an analytical reagent arises from the fact that it can offer simple and inexpensive determinations of various organic and inorganic substances (Cimerman et al., 1997). The insertion of functional groups in the chitosan matrix may improve their capacity for interaction with metallic ions by complexation, thereby increasing their adsorption properties (Hall and Yalpani, 1980; Krishnapriya_and Kandaswamy, 2010; Rodrigues et al., 1998). Also, according to Pearson's hard and soft acid–base theory, polymers containing functional groups with N or S donor atoms should be promising as sorbents of precious metal ions; some published research papers have demonstrated this viewpoint (Aydin et al., 2008; Fujiwara et al., 2007; Hubicki et al., 2007; Kałedkowski and Trochimczuk, 2006).

Heavy metal is a general collective term that applies to the group of metals and metalloids with an atomic density greater than 4 g/cm^3. Heavy metals include lead, cadmium, zinc, mercury, arsenic, silver, chromium, copper, iron, and the platinum group elements (Negm and Ali, 2010). Contamination of aquatic media by heavy metals is a serious environmental problem, mainly due to discard of industrial waste (Prasad et al., 2002; Reddad et al., 2002). Heavy metals are highly toxic at low concentrations and can accumulate in living organisms, causing several disorders and diseases (Bailey et al., 1999; Gotoh et al., 2004; Sag and Aktay, 2002). Environmental pollution is defined as the presence of a pollutant in the environment (air, water, and soil) that may be poisonous or toxic and will cause harm to living things in the polluted environment (Kotrba and Ruml, 2000). Environmental pollution by heavy metals is very prominent in areas of mining and at old mine sites, and pollution decreases with increasing distance away from the mining sites (Peplow, 1999).

This review attempts to analyze and discuss some of the literature that has been published to date on the ability of chitosan, modified chitosan, and chitosan derivatives to adsorb heavy metal ions from solution and some kinetic models studied.

CHEMICALLY MODIFIED CHITOSAN

Metal Ion Uptake of Chitosan-crown Ethers Derivatives

Two kinds of novel chitosan-crown ether resins, Schiff base type chitosan-benzo-15-crown-5 (CTS-B15) and chitosan-benzo-18-crown-6 (CTS-B18), were synthesized through the reaction between-NH_2 in chitosan and -CHO in 4'-formyl benzo-crown ethers as shown in Figure 3. The structures of these derivatives were characterized and the elemental analysis results show that the mass fractions of nitrogen in CTS-B15 and CTS-B18 are much lower than those of chitosan. The adsorption properties of CTS-B15 and CTS-B18 for Pd^{2+}, Cu^{2+} and Hg^{2+} were studied and the experimental results revealed that these adsorbents have good adsorption characteristics and high particular adsorption selectivity for Pd^{2+} when Cu^{2+} and Hg^{2+} are in coexistence. These novel chitosan-crown ether resins of CTS-B15 and CTS-B18 will have wide-range of applications for the separation and concentration of heavy or precious metal ions (Chang-hong et al., 2003).

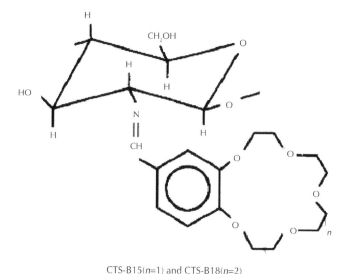

CTS-B15(*n*=1) and CTS-B18(*n*=2)

Figure 3. Structure of (CTS-B15) and (CTS-B18), (Chang-hong et al., 2003).

Varma et al. (2004) discussed the metal complexation of different chitosan crown ethers derivatives in detail (Varma et al., 2004). Newly prepared Schiff base-type chitosan-azacrown ethers were prepared by a reaction of chitosan (CTS) with *N*-(4'-formylphenyl) aza-crown ethers, and were then converted to secondary-amino derivatives by the reduction of CTS-azacrown ethers with sodium borohydride (Malkondu et al.,

2009). The ability of these adsorbents to extract Cu(II) and Ni(II) ions from water by a solid-liquid extraction process was studied. The effects of adsorbent amount, contact time and pH on the adsorption of CTS-azacrown ethers were investigated. The extraction of metal ions by chitosan itself has been mostly found to be pH sensitive (Peng et al., 1998; Roberts, 1992; Tang et al., 2002). Here, the adsorption of Cu(II) and Ni(II) ions on CTS-azacrown ethers was observed at different pH values; results showed that CTS-azacrown ethers had good sorption capacities for Cu(II) ions at $25 \pm 1°C$ and pH = 5.5. These newly prepared chitosan derivatives could be used in environmental analysis and hazardous waste remediation as toxic-metal binding agents in aqueous environments.

Recently, Alsarra and coworkers (Radwan et al., 2010) have employed microwave irradiation in the synthesis of a *N*-Schiff base-type cross-linked chitosan crown ether (CCdBE) *via* the reaction between the $-NH_2$ and $-CHO$ groups in chitosan and 4,4'-diformyldibenzo-18-c-6, respectively; the adsorption capacity of the obtained CCdBE was much higher for Hg^{2+} than that for Pb^{2+}. The reported cross-linking method could retain higher adsorption capacity of CTS and, at the same time, improve the acidic resistance of CTS with a desirable selectivity towards mercury ions over lead ions. The reusability tests for CCdBE for Pb^{2+} adsorption showed that complete recovery of the ion was possible with CCdBE after 10-multiple reuses while CTS had no reusability in acidic solution because of its higher dissolution. The studied features of CCdBE suggested that the material could be considered as a new adsorbent. It is envisaged that the cross-linking of CTS into CCdBE would enhance practicality and effectiveness of adsorption in ion separation and removal procedures.

Metal Ion Adsorption Studies of Magnetic Chitosans

Donia and coworkers (Donia et al., 2008) have prepared magnetic chitosan resin modified by Schiff's base derived from thiourea and glutaraldehyde. The resin obtained was applied for the separation of Hg(II) from aqueous solution and an uptake value of 2.8 mmol/g was reported. The nature of interaction between the metal ion and the resin was found to be dependent upon the acidity of the medium. At pH 1, Hg(II) could be selectively separated from Cu(II), Pb(II), Cd(II), Zn(II), Ca(II), and Mg(II). Kinetic and thermodynamic studies indicated that the adsorption process is pseudo-second-order exothermic spontaneous reaction and proceeds according to Langmuir isotherm. The resin was regenerated effectively using 0.1 M potassium iodide. The studied resin showed an efficient uptake behavior towards Hg(II) relative to the commercial resin Dowex-D3303. Moreover, the same authors have recently reported on the use of different chelating resins with various functionalities for the selective separation of mercury (Atia, 2005; Atia et al., 2003, 2005; Donia et al., 2005). They also reported on the use of magnetic resins in removal of some metals from aqueous solutions. These magnetic resins shown in Figure 4, are easily collected from aqueous media using an external magnetic field and displayed higher uptake capacity compared to the magnetic particles-free resin (Donia et al., 2006a, b). These methods are also cheap and often highly scalable. Moreover, techniques employing magnetism are more amenable to automation (An et al., 2003).

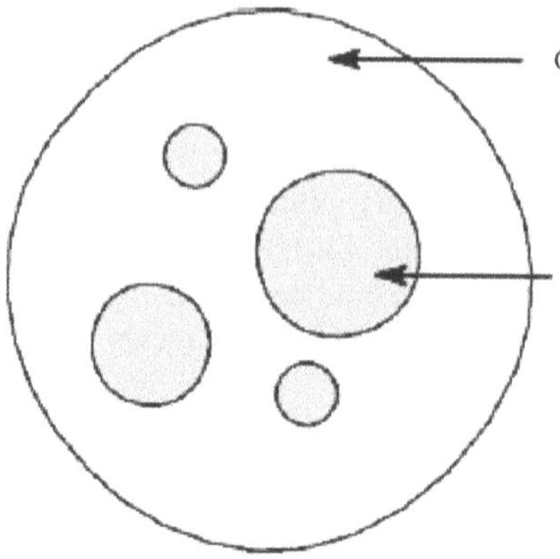

Figure 4. Schematic representation of chitosan resin with embedded magnetic particle, (Donia et al., 2008).

Another cross-linked magnetic chitosan-isatin Schiff's base (CSIS) resin was obtained and characterized (Monier et al., 2010a). The adsorption properties of this resin toward Cu^{2+}, Co^{2+}, and Ni^{2+} ions were investigated. The kinetic parameters were evaluated utilizing the pseudo-first-order and pseudo-second-order approach. The equilibrium data were analyzed using the Langmuir, Freundlich, and Tempkin isotherm models. It was found that the adsorption kinetics followed the mechanism of the pseudo-second-order equation for all systems studied, meaning that the chemical sorption was the rate-limiting step of the adsorption mechanism and not involving a mass transfer in solution. The best interpretation for the equilibrium data was given by Langmuir isotherm, and the maximum adsorption capacities were 103.16, 53.51, and 40.15 mg/g for Cu^{2+}, Co^{2+} and Ni^{2+} ions, respectively. Cross-linked magnetic CSIS displayed higher adsorption capacity for Cu^{2+} in all pH ranges studied. Additionally, it was found that the adsorption capacity of the metal ions decreased with increasing temperature. Regeneration of cross-linked magnetic CSIS obtained was achieved by using 0.01–0.1 M EDTA with efficiency greater than 88%. Feasible improvements in the uptake properties along with the magnetic properties encourage efforts for cross-linked magnetic CSIS to be used in water and wastewater treatment.

Similarly, the same group also prepared and characterized a cross-linked magnetic chitosan–diacetylmonoxime Schiff's base resin (CSMO) to study its metal ions adsorption toward Cu^{2+}, Co^{2+} and Ni^{2+} ions. Various factors affecting the uptake behavior were investigated and results showed greatly improved uptake properties of the resin compared to the unmodified ones, as well as previously reported synthetic ones. The adsorption kinetics followed the pseudo-second-order equation for all systems studied. The equilibrium data were well described by the Langmuir isotherm, and the maximum adsorption capacities were 95 ± 4, 60 ± 1.5, and 47 ± 1.5 mg/g for Cu^{2+}, Co^{2+} and Ni^{2+} ions, respectively. Cross-linked magnetic CSMO displayed higher adsorption capacity for Cu^{2+} in all pH ranges studied and the adsorption capacity of the metal ions decreased with increasing temperature. The metal ion-loaded cross-linked magnetic

CSMO were regenerated with an efficiency of greater than 84% using 0.01–0.1 methylendiamine tetraacetic acid (EDTA). These results could encourage researchers to use the cross-linked magnetic CSMO resin in water and wastewater treatment (Monier et al., 2010b).

Metal Ion Adsorption on Chitosan-Phosphate Derivatives

Tri-polyphosphate has been selected as a possible cross-linking agent, which can be used for the preparation of chitosan gel beads by the coagulation/neutralization effect (Mi et. al., 1999a, 1999b; Monier et al., 2010b). Chitosan-tripolyphosphate (CTPP) beads were synthesized, characterized and used for the adsorption of Pb(II) and Cu(II) ions from aqueous solutions. The experimental data were correlated with the Langmuir, Freundlich and Dubinin-Radushkevich isotherm models. The maximum adsorption capacities of Pb(II) and Cu(II) ions in a single metal system based on the Langmuir isotherm model were 57.33 and 26.06 mg/g, respectively. However, the beads showed higher selectivity towards Cu(II) over Pb(II) ions in the binary metal system. The kinetic data were evaluated based on the pseudo-first and -second order kinetics and intraparticle diffusion models and showed that this adsorption obeyed the pseudo-second order kinetic model, whereby intraparticle diffusion was not the sole rate controlling step. A desorption study was also carried out to show that Pb(II) and Cu(II) ions adsorbed on the beads can be easily and effectively desorbed using 0.1, 0.01, and 0.001 M EDTA solution. Infrared spectra were used to elucidate the mechanism of Pb(II) and Cu(II) ions adsorption onto CTPP beads and revealed that the nitrogen and oxygen atoms found in the beads were the binding sites for the metal ions. This new chitosan derived beads might be used to treat wastewaters containing Pb(II) and Cu(II) ions (Ngah and Fatinathan, 2010). Similarly, Laus R. and his group prepared chitosan (CTS) cross-linked with both epichlorohydrin (ECH) and triphosphate (TPP), by covalent and ionic cross-linking, respectively. The new CTS–ECH–TPP adsorbent was characterized and its adsorption and desorption of Cu(II), Cd(II) and Pb(II) ions in aqueous solution were investigated. It was discovered that the adsorption is dependent on the solution pH, and the optimum pH values for the adsorption were 6.0 for Cu(II), 7.0 for Cd(II) and 5.0 for Pb(II). The kinetics study demonstrated that the adsorption process proceeded according to the pseudo-second order model. The three isotherm models (Langmuir, Freundlich, and Dubinin-Radushkevich) used in the study mentioned above were also employed in the analysis of the adsorption equilibrium data. The Langmuir model resulted in the best fit and the new adsorbent had maximum adsorption capacities for Cu(II), Cd(II), and Pb(II) ions of 130.72, 83.75 and 166.94 mgg^{-1}, respectively. Desorption studies revealed that HNO_3 and HCl were the best eluents for desorption of Cu(II), Cd(II), and Pb(II) ions from the cross-linked chitosan. These results suggested that this new adsorbent could be used in the separation, pre concentration, and Cu(II), Cd(II), or Pb(II) ion uptake from aqueous solutions (Laus et al., 2010).

Metal Ion Adsorption on EDTA- and/or DTPA-Modified Chitosan Derivatives

The adsorption properties of surface modified chitosans with ligands such as ethylenediaminetetraacetic acid (EDTA) or diethylenetriaminepentaacetic acid (DTPA) which were immobilized onto polymer matrices of chitosan in aqueous solutions containing Co(II) and/or Ni(II) ions were investigated (Repo et al., 2010). Metal uptake by EDTA-chitosan was 63.0 mgg^{-1} for Co(II) and 71.0 mgg^{-1} for Ni(II) and by DTPA-chitosan 49.1 mgg^{-1} for Co(II) and 53.1 mgg^{-1} for Ni(II). The adsorption efficiency of the studied adsorbents ranged from 93.6% to 99.5% from 100 mgl^{-1} Co(II) and/or Ni(II) solution, when the adsorbent dose was 2 gl^{-1} and solution pH 2.1. The kinetics of Co(II) and Ni(II) on both of the modified chitosan followed the pseudo-second-order model but the adsorption rate was influenced by intraparticle diffusion. The equilibrium data was best described by the Sips isotherm and its extended form was also well fitted to the two-component data obtained for systems containing different ratios of Co(II) and Ni(II). Nevertheless, the obtained modeling results indicated relatively homogenous system for Co(II) and heterogeneous system for Ni(II) adsorption. The adsorption studies in two-component systems showed that the two new modified chitosan had much better affinity for Ni(II) than for Co(II) suggesting that Ni(II) could be adsorbed selectively from the contaminated water in the presence of Co(II).

Moreover, Katarina and coworkers (Katarina et al., 2008) have reported on a sample-pretreatment method using a chitosan-based chelating resin, ethylenediamine-N,N,N'-triacetate-type chitosan (EDTriA-type chitosan), for the preconcentration of trace metals in seawater and separation of the seawater matrix prior to their determination by inductively coupled plasma–mass spectrometry (ICP-MS). According to those authors, the resin showed very good adsorption for transition metals and rare-earth elements without any interference from alkali and alkaline-earth metals in acidic media and that the adsorption capacity of Cu(II) on the EDTriA-type chitosan resin was 0.12 mmol g^{-1} of the resin. Additionally, Shimizu and colleagues (Shimizu et al., 2008) have prepared chemically modified chitosans with a higher fatty acid glycidyl (CGCs) by the reaction of chitosan with a mixture of 9-octadecenic acid glycidyl and 9,12-octadecanedienic acid glycidyl (CG). The new chitosan modified polymer, CGC$_s$ was further modified through the reaction with ethylenediamine tetraacetic acid dianhydride afforded CGCs (EDTA-CGCs). The same researchers have studied the adsorption behavior of CGCs towards the metal ions Mo^{6+}, Cu^{2+}, Fe^{2+}, Fe^{3+}, and found that Mo^{6+} displayed remarkable adsorption toward the CGCs. In addition, they examined the adsorption of Cu^{2+} on the ethylenediamine tetraacetic acid dianhydride modified CGCs (EDTA-CGCs) and the adsorption of phosphate ions onto the resulting substrate/metal-ion complex was measured. Similarly, Ni and Xu (1996) have synthesized a series of cross-linked chelating resins containing amino and mercapto groups, in addition to chitosan by reacting chitosan with chloromethyl thiirane (CT) using different rations of chitosan to CT. The adsorbing capacities, adsorption rates, and adsorption selectivities of these resins towards Ag(I), Au(III), Pd(II), Pt(IV), Cu(II), Hg(II), and Zn(I1) were investigated. They discovered that these chelating resins containing mercapto and amino groups have remarkable adsorbing capacities and rates for some noble metal ions and can be used to concentrate and retrieve precious metal ions from dilute solutions.

Metal Ion Adsorption on Carbonyl Containing Chitosan Derivatives

A new chitosan derivative was synthesized by the chemical modification of chitosan (CTS) with vanillin-based complexing agent namely 4-hydroxy-3-methoxy-5-[(4-

methylpiperazin-1-yl) methyl] benzaldehyde, (L) (Figure 5) by means of condensa-tion. The new polymer was characterized by various techniques such as elemental, spectral, and structural analysis. Its kinetics of adsorption was evaluated utilizing the pseudo first order and pseudo second order equation models and the equilibrium data were analyzed by Langmuir isotherm model. The CTSL showed good adsorp-tion capacity for the metal ions investigated in the order Cu(II) > Ni(II) > Cd(II) ≥Co ≥Mn(II) > Fe(II) > Pb(II) in all studied pH ranges due to the presence of many coordinating moieties present in it. The binding capacities of the polymer for Mn, Fe, Co, Cu, Ni, Cd, and Pb were found to be 19.8, 18.4, 22.4, 56.5, 34.63, 46.1, and 51.8 mg/g, respectively, and were higher than the values obtained for unmodified chitosan. The higher adsorption capacity toward Cu(II) ions than other metal ions studied, in-dicates that it shows selectivity toward Cu(II) ions in aqueous solution and hence it can be used to extract Cu(II) ions from the industrial waste water (Krishnapriya and Kandaswamy, 2010).

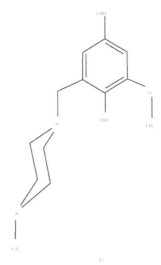

Figure 5. New Vanilin-based Ligand (L) (Krishnapriya and Kandaswamy, 2010).

Recently (Zalloum et al., 2009), the adsorption of Cu(II) ions onto chitosan de-rived Schiff bases obtained from the condensation of chitosan with salicyaldehyde, 2,4-dihydroxybenzaldehyde and with 4-(diethylamino) salicyaldehyde was investi-gated. The isothermal behavior and the kinetics of adsorption of Cu(II) ions on these polymers were investigated. Various factors that affect the adsorption process were also explored. The highest Cu(II) ions uptake was achieved at pH 7.0 and by using so-dium perchlorate as an ionic strength adjuster. Adsorption isothermal equilibrium data could be clearly explained by the Langmuir monolayer model. The experimental data of the adsorption equilibrium from Cu(II) solution correlates well with the Langmuir isotherm equation. Moreover, the adsorption kinetics of Cu(II) ions on these chitosan derived Schiff bases indicated that the kinetic data correlated well with the pseudo-second order model.

Another adsorption study was conducted for the two new chitosan derivatives that have been prepared from the reaction of cinnamoyl chloride (ChitoCin) and cinnamoyl isothiocyanate (ChitoThioCin) with chitosan. The modified chitosans were characterized to confirm their structures. The metal uptake capacity of the two polymers was measured at different pH values as well as under competitive and non-competitive conditions. At pH 5.6, the (ChitoCin and ChitoThioCin) polymers exhibited higher capacity for Cu(II) (0.461–0.572 mmol/g) than the other metal ions used; the capacities of the other metal ions are: Fe(III) (0.235–0.341 mmol/g), Cr(III) (0.078–0.099 mmol/g), Co(II) (0.046–0.057 mmol/g) and Ni(II) (0.041–0.053 mmol/g). Two absorption isotherms were examined for the absorption of metal cations with these two modified chitosan and it was found that the adsorption mechanism fits the Langmuir isotherm better than the Freundlich one.

Chitosan sorbents, cross-linked and grafted with amido or carboxyl groups, were also prepared and their sorption properties for Cu(II) and Cr(VI) uptake were studied (Kyzas et al., 2009). Equilibrium sorption experiments were carried out at different pH values and initial ion concentrations. The equilibrium data were successfully fitted to the Langmuir–Freundlich (L–F) isotherm. The calculated maximum sorption capacity of the carboxyl grafted sorbent for Cu(II) was found to be 318 mg/g at pH 6, while the respective capacity for Cr(VI) uptake onto the amido-grafted sorbent was found to be 935 mg/g at pH 4.

Metal Ion Adsorption of Other Chemically Modified Chitosans (Chitosan Derivatives)

The synthesis and chelating properties of chitosan-based polymers has attracted the attention of several research groups. A number of publications pertaining to the subject have appeared in the literature. Oshita and coworkers (Oshita et al., 2009) have prepared cross-linked chitosan resins with catechol, iminodiacetic acid (IDA-type chitosan), iminodimetylphosphonic acid (IDP-type chitosan), phenylarsonic acid (phenylarsonic acid-type chitosan), or serine (serine-type chitosan) for the collection and concentration of uranium(VI). The adsorption behavior of U(VI) and other ionic species, such as metal ions and oxo-acid ions, on the cross-linked chitosan (base material) and chitosan resins modified with chelating moieties was investigated using a column procedure. They discovered that the adsorption ability for U(VI) was in the order: catechol-type chitosan (type 2) > serine-type chitosan > phenylarsonic acid-type chitosan > the others. They concluded that the catechol-type chitosan was useful for the collection and concentration of uranium(VI). In addition, Katarina et al. (2009) have synthesized a high-capacity chitosan-based chelating resin, N-(2-hydroxyethyl)glycine-type chitosan, using chloromethyloxirane (CMO) as a cross-linker and as coupling arms and hydroxylethylamine and bromoacetic acid as a synthesizer for the N-(2-hydroxyethyl)glycine chelating moiety. The CMO could bind with both of hydroxyl and amino group of the chitosan resin, and then couple with the chelating moiety. They have found that most transition and rare-earth metals could adsorb quantitatively on the resin at wide pH ranges and could be separated from alkaline and alkaline-earth metals. The resin was packed in a mini-column (40 mmlength×2 mm i.d.) which was installed in a Multi-Auto-Pret system. The Multi-Auto-Pret system coupled with ICP-AES was successfully applied to the determination of transition and rare-earth metals in river water samples.

Similarly, Li and his group (Li et al., 2009) synthesized a new chemically modified chitosan hydrogel with 2,5-dimercapto-1,3,4-thiodiazole (CTS-DMTD). A range of static sorption studies were performed on this adsorbent, which showed the selectivity towards cations of the precious metals over other transition metal cations. They have also found that adsorption capacities were significantly affected by the pH of solution, with optimum pH values of 3.0 for Au(III), 2.0 for Pd(II) and Pt(IV). The saturated adsorption capacities were 198.5, 16.2, and 13.8 mg/g for Au(III), Pd(II), and Pt(IV), respectively. Langmuir and Freundlich isotherm adsorption models were applied to analyze the experimental data and the results showed that adsorption isotherms of Pd(II) and Pt(IV) could be well described by the Langmuir equation. The adsorption kinetic investigations indicated that the kinetic data correlated well with the pseudo second-order model. The recovery experimental data showed that CTS-DMTD had a higher affinity toward Au(III), Pd(II), and Pt(IV) in the coexistence system containing Cu(II), Fe(III), Cd(II), Ni(II), Mg(II), and Zn(II). The studies of desorption were carried out using various reagents and the optimum effect was obtained using thiourea. The sorption studies also revealed a considerable capacity for Au(III) ions, which might be useful in the removal of gold from ores. However, the CTS-DMTD was a little bit more difficult to reuse.

Quite recently, Guibal and his research team (Butewicz et al., 2010) have immobilized thiourea onto chitosan; the new polymer was employed for the sorption and recovery of platinum and palladium from acidic solutions (up to 1–2 M HCl concentrations). The kinetics of the sorption process was investigated and the pseudo-second rate equation was used for modeling the uptake kinetics. Similarly, Chanthateyanonth et al. (2010) reported the successful immobilization of vinyl sulfonic acid sodium salt onto dendritic hyper branched chitosan. The new chitosan derivatives displayed improved water solubility as compared to the starting material. In addition, the new material showed better antimicrobial activity and chelating behavior with cadmium(II), copper(II), and nickel(II) than chitosan itself.

Calix[n]arenes are cyclic oligomers composed of phenol units and are very well known as attractive and excellent ionophores because they provide a unique three-dimensional structure with almost unlimited derivatization possibilities (Asfari et al., 2001; Gutsche, 1998; Vicens and Bohmer, 1991). Tabakci and Yilmaz evaluated the sorption properties of a calix[4]arene-based chitosan polymer (C[4]BCP) toward some heavy metal cations (Co^{2+}, Ni^{2+}, Cu^{2+}, Cd^{2+}, Hg^{2+}, and Pb^{2+}) and dichromate anions (Cr$_2$O$_7^{-2}$/HCr$_2$O$_7^{-1}$) as sorbent materials. The results for heavy metal cations showed that C[4]BCP was an excellent sorbent while chitosan exhibits poor sorption than C[4] BCP (Tabakci and Yilmaz, 2008). In the sorption studies of dichromate anions, C[4] BCP was a highly effective sorbent at pH 1.5. The sequence of sorption efficiency of C[4]BCP is Hg^{2+} > Pb^{2+} > Cd^{2+} > Cu^{2+} > Ni^{2+} > Co^{2+}. From the previous works (Memon et al., 2006; Tabakci et al., 2006; Yilmaz et al., 2006), it has been concluded that the amide derivatized calixarenes are effective sorbents for these cations because NHC=O group is preferable to complex the more polarizable transition metal ions especially

Pb, Hg, Cd and Cu due to cation-p interactions. The phenomenon may also reflect the "principle of hard and soft acids and bases" introduced by Pearson (1963). Furthermore, highest sorption for mercury cation by C[4]BCP is in agreement with previous studies (Yilmaz et al., 2006) where calixarenes bearing nitrile functionalities showed much more affinity toward this cation.

ADSORPTION ISOTHERMS MODELS

The most common sorption models used to fit the experimental data are the Langmuir and Freundlich isotherm models (Tan et al., 2007). Readers interested in a detailed discussion of sorption isotherms should refer to the comprehensive reference works by Tien and McKay et al. (Tien, 1994; McKay et al., 2002).

The Langmuir isotherm model assumes that the adsorption occurs at specific homogeneous adsorption sites within the adsorbent. Furthermore, it assumes monolayer adsorption and maximum adsorption occurs when adsorbed molecules on the surface of the adsorbent form a saturated layer. All adsorption sites involved are energetically identical and the intermolecular force decreases as the distance from the adsorption surface increases (Unlu and Ersoz, 2006; Uysal and Ar, 2007; Vasconcelos et al., 2008). The form for the liquid phase sorption system is as follows:

$$Q_e = \frac{bQ_oC_e}{1+bC_e} \tag{1}$$

where, C_e is the equilibrium concentration of the adsorbate (mg/L), Q_e is the amount of adsorbate adsorbed per unit mass of adsorbent (mg/g), Q_o is Langmuir constant related to adsorption capacity (mg/g) and b (L/mg) is a constant related to the affinity between the adsorbent and the adsorbate. The values of Q_o and b can be determined by plotting C_e/Q_e versus C_e.

The linear form of Langmuir model is given as:

$$\frac{C_e}{Q_e} = \frac{1}{bQ_o} + \frac{1}{Q_o}C_e \tag{2}$$

The essential characteristics of the Langmuir isotherm can be expressed in terms of a dimensionless equilibrium parameter (R_L), (Weber and Chakravorti, 1974) which is defined as:

$$R_L = \frac{1}{1+bC_o} \tag{3}$$

where, b is the Langmuir constant and C_o is the highest ion concentration (mg/l). The value of R_L indicates the type of the isotherm to be either favorable ($0<R_L<1$), unfavorable ($R_L>1$), linear ($R_L = 1$) or irreversible ($R_L = 0$).

The widely used empirical Freundlich model expresses adsorption at multilayer and on energetically heterogeneous surface multi-site adsorption isotherm for heterogeneous surfaces (Hasan et al., 2008; Unlu and Ersoz, 2006). It assumes an initial surface adsorption followed by a condensation effect resulting from extremely strong solute-solute interaction. The general form of Freundlich model is as follows:

$$Q_e = K_F(C_e)^{1/n} \qquad (4)$$

where, K_F ((L/mg)$^{1/n}$) is Freundlich isotherm constant and n is the Freundlich isotherm exponent constant. K_F is correlated to the maximum adsorption capacity and n gives an indication of how favorable the adsorption process is (Hosseini et al., 2003). The linear form of this model is expressed as in the following equation:

$$\log Q_e = \log K_F + (1/n)\log C_e \qquad (5)$$

The values of K_F and n can be obtained by plotting $\log Q_e$ versus $\log C_e$.

The Redlich–Peterson equation only differs from the Langmuir–Freundlich equation by the absence of exponent on C_e at the numerator part of the equation (Ng et al, 2002). Meanwhile, the Dubinin-Radushkevitch (D-R) isotherm describes the adsorption on a single type of uniform pores and can be applied to distinguish between physical and chemical adsorption. This isotherm does not assume a homogeneous surface or a constant adsorption potential (Unlu and Ersoz, 2006).

KINETIC MODELS (ADSORPTION KINETICS)

Three kinetic models, pseudo-first-order, pseudo-second-order, and intra-particle diffusion model are used to fit the experimental data. The mathematical description of these models is given below. The conformity between data predicted by any of these models and the experimental data is indicated by the correlation coefficient R^2. The model of higher values of R^2 means that it successfully describes the adsorption kinetics.

The differential form of the pseudo-first-order kinetic model could be expressed by the following equation (Ho and McKay, 1998; Igwe and Abia, 2007)

$$\frac{dQ_t}{dt} = k_1(Q_e - Q_t) \qquad (6)$$

where, t is the time (min) and k_1 is the equilibrium rate constant of the pseudo first order adsorption (min^{-1}). Integrating equation (6) by applying the boundary conditions, $t = 0$ to $t = t$ and $Q_t = 0$ to $Q_t = Q_e$, yields the following integral equation:

$$\log Q_e / (Q_e - Q_t) = 0.4342k_1 t \qquad (7)$$

The value of the model parameters k_1 can be determined by plotting $\log (Q_e - Q_t)$ versus t to give a straight line of slope $0.4342\, k_1$ and intercept of $\log Q_e$.

The differential form of the pseudo-second-order kinetic model is expressed by the following equation (Ho and McKay, 1998):

$$\frac{dQ_t}{dt} = k_1 (Q_e - Q_t)^2 \tag{8}$$

where k_2 is the equilibrium rate constant of the pseudo-second-order kinetic model (g/mg h). Integration of equation (8) between the boundary conditions of equation (6) yields the following equation:

$$t / Q_t = 1 / k_2 (Q_e)^2 + t / Q_e \tag{9}$$

The value of k_2 can be determined by plotting t/Q_t versus t to obtain a straight line of slope $1/Q_e$ and intercept of $1/(k_2 Q_e^2)$.

Since, the above two models cannot give definite mechanisms for the adsorption process, another simplified model that represents the intra-particle diffusion model is tested (Wu et al., 2001). This model, which is based on the theory proposed by Weber and Morris assumes that the intra-particle mass transfer resistance is the rate determining step which means that the adsorption process is pore diffusion controlled (Weber and Morris, 1962). According to this theory, the initial rates of intra-particle diffusion can be obtained from the following equation:

$$Q_t = x_i + k_p t^{1/2} \tag{10}$$

where x_i is a constant proportional to the boundary layer thickness, mg/g, k_p is the intraparticle diffusion rate constant (mg/(g h$^{1/2}$)) (Kavitha and Namasivayam, 2007). For pore diffusion controlled sorption, a plot of Q_t versus $t^{1/2}$ gives a straight line of slope k_p and an intercept of x_i.

CONCLUSION

Chitosan, a biopolymer, has attracted considerable attention due to its many physical and chemical properties and due to its many applications including its usage as a material for medical applications, such as artificial skin and immunosuppressant. In addition, resins derived from chitosan have been synthesized and characterized. The chelation and sorption properties of these chitosan-based polymers towards heavy metal ions in aqueous solutions have been investigated by several research groups. Therefore, the aim of this article will be to shed some light on some of these chitosan-derived resins and their application to sorption of heavy metal ions and some anions.

KEYWORDS

- **Biodegradable**
- **Intraparticle diffusion**
- **Magnetic field**
- **Polychlorinated biphenyl**
- **Thermodynamic**

Chapter 2

Chitosan Composites for Biomedical Applications: An Overview

R. Ramya, Z. Ansar Ali, and P.N. Sudha

INTRODUCTION

Natural polymers received great attention in the biomaterials field; their structural similarities with biological macromolecules make them easily recognized by the bio-environment and therefore easily metabolized into noncytotoxic residues and naturally eliminated. Chitin is one of the most abundant polysaccharides and can be found in the exoskeleton of crustaceans which can be obtained from the shell waste of the crab, shrimp, crawfish processing industries, various invertebrates, and lower plants. Chitosan is one of chitin's derivatives, achieved by N-deacetylation of chitin. It is one of the major components used in vascular surgery, tissue culture, and tissue regeneration as a hemostatic agent (Muzzarelli et al., 1986). Improving the fragile nature of films and membrane permeability are the key challenges that need to be addressed for improving chitosan as a biomaterial. In addition, chitosan is expected to be useful in the development of composite materials such as blends or alloys with other polymers, since chitosan has many functional properties (Japanese Chitin and Chitosan Society, 1995). Chitosan is a biopolymer which has many interesting properties that have been utilized in many pharmaceutical applications (Patel and Amiji, 1996). There have been many studies on the blends of chitosan with various kinds of polymers (Guan et al., 1998; Hasegawa et al., 1992; Kim et al., 1992; Xiao et al., 2000; Yao et al., 1996) in order to obtain some improved properties.

Natural silk fiber is a fiber obtained from silkworms. Silk has been used as a high quality fiber material, because it has characteristics such as high tensile strength, peculiar luster, and excellent dyability. Silk fibroin (SF) is a fibrous protein that is composed of 17 amino acids and its main components are nonpolar ones such as glycine, alanine, and serine. A silk fiber has a structure with two strands of fibroin surrounded by a sericin wall. By preparing in various forms such as a membrane, powder, gel, and aqueous solution, the SF is used in various fields such as food, cosmetics, and medical goods because it has excellent biocompatibility and does not give any adverse effect to surrounding tissues. Silks have been investigated as biomaterials due to the successful use of silk fibers from *Bombyx mori* as suture material for centuries (Moy et al., 1991). Functional differences among silks of different species and within a species are a result of structural differences due to differences in primary amino acid sequence, processing, and the impact of environmental factors (Vollrath and Knight, 2001). Silks represent a unique family of structural proteins that are biocompatible, degradable, and mechanically superior, offer a wide range of properties, are amenable to aqueous or organic solvent processing, and can be chemically modified to suit a wide range of biomedical applications.

BIOMEDICAL APPLICATIONS OF BIOPOLYMER AND ITS COMPOSITES

Chitosan and its Derivatives

Chitosan, poly-β-(1$\rightarrow$4)-2-amino-2-deoxy-D-glucose, is an aminopolysaccharide derived from the N-deacetylation of chitin, which is a structural element in the exoskeleton of crustaceans (crabs, shrimps, etc.) and cell wall of fungi and it is also classified as a natural polymer because of the presence of a degradable enzyme, chitosanase. This polymer possesses hydrogel like properties through a reaction with glutaraldehyde as cross-linking agent. It is known that cross-linked chitosan hydrogel can swell extensively due to the positive charges on the network and in response to changes in the pH of the medium. It has been shown (Nunthanid et al., 2004; Puttipipatkhachorn et al., 2001) that the drug release behavior of chitosan is governed mainly by the swelling property, the dissolution characteristic of the polymer films, the pKa of the drug, and the drug-polymer interaction. Thus chitosan has been studied as a unique vehicle for the sustained delivery of drugs. For example, it was investigated for the delivery of drugs such as prednisolone (Kofuji et al., 2000) and diclofenac sodium (Gupta and Ravi Kumar, 2000). Chitosan displays interesting properties such as biocompatibility, biodegradability, and its degradation products are non-toxic, non-immunogenic, and non-carcinogenic. Moreover, chitosan is metabolized by certain human enzymes, especially lysozyme, and is considered as biodegradable.

The physicochemical and biological properties of chitosan led to the recognition of this polymer as a promising material for biomedical applications. One of these properties is the antimicrobial activity of chitosan which is suggested for use in a variety of different formulations, for example tapes for wound dressing, tooth paste or artificial tears (Felt et al., 2000; Kim et al., 1999). The interaction between positively charged chitosan and negatively charged microbial cell wall will lead to the leakage of intracellular constituents of the microorganism. The binding of chitosan with DNA and inhibition of mRNA synthesis occurs via the penetration of chitosan into the nuclei of the microorganisms and interfering with the synthesis of mRNA and proteins. Wound healing, another major application, is a complex process that can be compromised by a number of factors.

Chitosan has been investigated by many researchers for a long time as possible wound healing accelerators (Muzzarelli et al., 1988; Nishimura et al., 1986; Zikakis, 1984). Chitosan-based wound dressing delivers substances, which are active in wound healing; either by delivery of bioactive compounds or dressings are constructed from materials having endogenous activity. Chitosan has been known as being able to accelerate the healing of wound in human (Kojima et al., 1998). Kifune et al. (1988) recently developed a new wound dressing material, Beschitin W, a commercial product that is composed of chitin nonwoven fabric and that has been found to be beneficial in clinical practice (Kifune et al., 1988). The ability of chitosan to promote neovascularization has been demonstrated by implanting it in the cornea. It is also reported that chitosan does not induce other inflammatory events. Certain cartilaginous tissues could also be repaired in view of the angiogenic action of chitosan.

Chitosan with structural characteristics similar to glycosaminoglycans could be considered for developing a skin replacement. Yannas and Burke (1980) proposed a design for artificial skin applicable to long-term chronic use, with the focus on a non antigenic membrane that performs as a biodegradable template for neodermal tissue (Yannas and Burke, 1980). The functionality of chitosan has also attracted the interest of many researchers to use it in the areas of dentistry and orthopedics (Davidenko et al., 2010; Di Martino et al., 2005; Jeon et al., 2000; Jiang et al., 2008; Kong et al., 2005; Venkatesan and Kim, 2010; Venkatesan et al., in press; Teng et al., 2009). Chitosan has been known to possess many biological activities such as antibacterial activity (Jeon and Kim, 2000; Jeon et al., 2001), anti-diabetic (Liu et al., 2007), immunoenhancing effect (Suzuki et al., 1986), antioxidant (Je et al., 2004), matrix metalloproteinase inhibitor (Kim and Kim, 2006; Rajapakse et al., 2006; Van Ta et al., 2006), anti-HIV (Artan et al., In press), anti-inflammatory (Yang et al., 2010), drug delivery (Liu et al., 2007) heavy metal removal (Sudha and Celine, 2008; Karthik et al., 2009) and so on.

In addition, it has the potential to be used as artificial kidney membrane, absorbable sutures, hypocholesterolemic agents, drug delivery systems, and supports for immobilized enzymes. Chitosan properties allow it to clot blood rapidly and have recently gained approval in the United States and Europe for use in bandages and other hemostatic agents. Chitosan hemostatic products have been shown in testing by the U.S. Marine Corps to quickly stop bleeding and result in 100% survival. Chitosan hemostatic products reduce blood loss in comparison to gauze dressings and increase patient survival (Pusateri et al., 2003). Chitosan is hypoallergenic, and has natural anti-bacterial properties, further supporting its use in field bandages. According to Qi, chitosan nanoparticles could exhibit effective antitumor activities (Qi and Xu, 2006). Chitosan has been combined with a variety of delivery materials such as alginate, hydroxyapatite, hyaluronic acid, calcium phosphate, PMMA, poly-L-lactic acid (PLLA), and growth factors for potential application in orthopedics. Overall, chitosan offers broad possibilities for cell-based tissue engineering (Hu et al., 2006). Besides, it has been claimed to weight reducing process. The aim of this review is to discuss the recent developments on the biopolymer like chitosan composites that are specially designed for the tissue engineering and biomedical applications.

Due to the abundance of hydrophilic functional groups, chitosan is not soluble in most organic solvents. In order to solve this problem, some chemical modifications to introduce hydrophobic nature to chitosan such as phthaloylation (Nishimura et al., 1991), alkylation (Yalpani and Hall, 1984), and acylation (Hirano et al., 1976; Moore and Roberts, 1981; Zong et al., 2000) reactions can be done. Several studies showed that acylated chitosans are very interesting derivatives of chitosan to be used in biomedical applications. Chitosan and its derivatives such as trimethyl chitosan have been used in non-viral gene delivery. Trimethyl chitosan, or quaternised chitosan, has been shown to transfect breast cancer cells; with increased degree of trimethylation increasing the cytotoxicity and at approximately 50% trimethylation the derivative is the most efficient at gene delivery. Oligomeric derivatives (3–6 kDa) are relatively non-toxic and have good gene delivery properties (Kean et al., 2005).

Recent clinical data indicate that modified chitosan with amino acid moieties and substituents at the N-atom play an active role in wound healing. In general, these induce formation of vascularized and non refractive tissues having well-oriented collagen. Trimethyl-chitosan and monocarboxymethyl chitosan has been shown to be effective as intestinal absorption enhancers due to their physiological properties. Chitosan-thioglycolic acid conjugates have been found to be a promising candidate as scaffold material in tissue engineering due to their physicochemical properties. Niamsa et al. (2009) successfully prepared the nanocomposite blend films containing methoxy poly(ethylene glycol)-b-poly(D,L-lactide) nanoparticles with different chitosan/silk fibroin ratios. These biodegradable nanocomposite blend films may have potential for use in drug delivery, wound dressing, and tissue engineering applications (Niamsa et al., 2009). Goy et al. (2009) stated that water soluble chitosan derivatives, which can be attained by chemical introduction of CH_3 in the main chain, enhancing the chitosan applicability in a large pH range and also improve the antimicrobial activity, opening up a broad range of possibilities (Goy et al., 2009). According to Rujiravanit et al. (2003), crosslinked chitosan and its blend films with SF using glutaraldehyde as crosslinking agent exhibit drug release characteristics. The drug release was high in acidic medium because of the protonation of the amino groups on chitosan at acidic pH, resulting in the dissociation of hydrogen bonds between chitosan and SF (Rujiravanit et al., 2003).

Hirano et al. (1980) prepared a series of membranes from chitosan and its derivatives (Hirano, 1978; Hirano et al., 1980), and the membranes showed improved dialysis properties (Arai et al., 2004; Fuchs et al., 2006). They observed that permeability properties of N-acetyl chitosan membranes were similar to those of an Amicon Diaflo membrane UM-10 (Amicon Ltd., England). Chitosan membranes were modified with vinyl monomers using 60-Co γ-ray irradiation, and their physicochemical properties were also studied. The modified membranes showed improved permeability and blood compatibility (Singh and Ray, 1994, 1997). The reactive functional groups present in chitosan (amino group at the C2 position of each deacetylated unit and hydroxyl groups at the C6 and C3 positions) can be readily subjected to chemical derivatization allowing the manipulation of mechanical and solubility properties enlarging its biocompatibility.

Fibrous Protein—Silk

The SF, the typical natural macromolecule spun by *Bombyx mori* silkworm, has been used as textile fiber and suture. The natural silk fibers are one of the strongest and toughest materials mainly because of the dominance of well orientated β-sheet structures of protein chains. Recently, several researchers have investigated SF as one of promising resources of biotechnology and biomedical materials due to its unique properties including good biocompatibility, good oxygen and water vapor permeability, biodegradability, minimal inflammatory reaction (Minoura et al., 1990; Sakabe et al., 1989; Santin et al., 1999), good cell adhesion and growth characteristics, protease susceptibility, and high tensile strength (Altman et al., 2003; Gotoh et al., 2002; Lv and Feng, 2006). Recently, silk and silk-based materials have attracted renewed interest, because of their biological applications. According to early records, silk fibers have been used for wound closure by surgeons for at least 3,000 years.

The biomedical applications of silk protein have been studied since the 1960s. The excellent biocompatibility and functionality of silk has led to the development of various biomedical devices (Altman et al., 2003; Vepari and Kaplan, 2007; Wang et al., 2006). For example, in the early era of silk biomaterials, many researchers developed silk films and sponges as would dressing (Fini et al., 2005; Roh et al., 2006). Recent studies reported the use of silk in oral administration (Oh et al., 2007). The biological applications for silk now include tissue engineering scaffolds (Moreau et al., 2006; Nazarov et al., 2004), nerve conduits (Yang et al., 2007), and artificial ligaments (Fan et al., 2008). The biocompatibility and functionality of silk is similar to collagen, and the physical and mechanical properties of silk make it suitable in biomedical devices. Silks from silkworms and orb-weaving spiders have impressive mechanical properties (Table 1), in addition to environmental stability, biocompatibility, controlled proteolytic biodegradability, morphologic flexibility and the ability for amino acid side change modification to immobilize growth factors (Altman et al., 2003; Arai et al., 2001; Fuchs et al., 2006; Horan et al., 2005; Hu et al., 2006; Karageorgiou et al., 2004; Kim et al., 2005; Li et al., 2006; Minoura et al., 1995; Motta et al., 2004; Vepari and Kaplan, 2006; Wong Po Foo and Kalpan, 2002).

Table 1. Mechanical properties of biodegradable polymeric materials (Arai et al., 2001).

Source of biomaterial	Modulus (GPa)	UTS (MPa)	Strain (%) at break	References
B. mori silk (with sericin)	5–12	500	19	(Perez-Rigueiro et al., 2000)
B. mori silk (without sericin)	15–17	610–690	4–16	(Perez-Rigueiro et al., 2000)
B. mori silk	10	740	20	(Cunniff et al., 1994)
N. clavipes silk	11–13	875–972	17–18	(Cunniff et al., 1994)
Collagen	0.0018–0.046	0.9–7.4	24–68	(Pins et al., 1997)
Cross-linked collagen	0.4–0.8	47–72	12–16	(Pins et al., 1997)
Polylactic acid	1.2–3.0	28–50	2–6	(Engelberg and Kohn, 1991)

Zhen-ding et al. (2009) suggested that combining the advantages of SF and chitosan, the SFCS scaffold should be a prominent candidate for soft tissue engineering, especially liver tissue engineering (Zhen-ding et al., 2009). It has been reported that chitosan could induce the conformational transition of SF from a random coil to a β-sheet structure (Park et al., 1999) and a polymer blend of these two biopolymers could also form a hydrogel having a semi-interpenetrating polymer network by using glutaraldehyde as a crosslinking agent (Kweon et al., 2001).

It has been reported that SF film has good oxygen permeability in its wet state (Minoura et al., 1990), which suggests promising applications of SF as wound dressing, artificial skin, surgical sutures, and biocompatible devices with controlled drug release. However, SF in a dry state is very brittle and unsuitable for practical uses (Freddi et al., 1995). To overcome this limitation, SF has been reported to blend with other synthetic polymers, such as polyacrylamide (Freddi et al., 1999) and poly(vinyl alcohol) (Yamaura et al., 1990), or natural polymers, such as cellulose (Freddi et al., 1995) and sodium alginate (Liang and Hirabayashi, 1992), to improve its mechanical and physical properties.

Alginate

Alginate is a naturally occurring polysaccharide extracted from seaweed (Gombotz and Wee, 1998). Alginates are well established as food additives and as encapsulation agents in biotechnology. Commercial production from harvested brown seaweeds commenced in the early 20th century. Brown seaweeds and only the two Gram-negative bacteria genera *Azotobacter* and *Pseudomonas* are capable of alginate production. Alginates belong to exopolysaccharides and are non-repeating copolymers of b-d-mannuronic acid (M) and a-l-guluronic acid (G) which are linked by 1–4 glycosidic bonds. The comonomer composition and arrangement strongly impact on the alginate material properties, which range in nature from slimy and viscous solutions to pseudo plastic materials. Alginate has the unique ability to form gels and is an excellent membrane material. It is also chemically very stable at pH values between 5 and 10. In addition, a chitosan–alginate polyelectrolyte complex has been used to prepare devices used for the controlled release of drugs (Foldvari, 2000; Mi et al., 2002; Ribeiro et al., 2005; Shu and Zhu, 2002).

Alginates well meet the entire requirement for their use in pharmaceutical and biomedical applications. They have been largely used in wound dressings, dental impression, and formulations for preventing gastric reflux. When alginate comes into contact with wound exudates, ion exchange occurs between the calcium ions of the alginate and the sodium ions in the exudates resulting in the formation of a gel on the surface of the wound. This gel absorbs moisture and maintains an appropriately moist environment that is considered to promote optimal wound healing. For these reasons, SF/alginate-blended sponge is likely to be an effective material that can be used for wound dressing and provide the necessary requirements for recovery. The SF and alginate have been proved to be invaluable natural materials in the field of biomedical engineering. It was revealed that SF/alginate-blended sponge treatment produced the most prominent wound healing effect as compared with either SF or alginate sponge treatment. It was found that the synergic effect of SF/alginate blended sponge is mainly involved in the promotion of re-epithelialization rather than collagen deposition.

Furthermore, chitosan and alginate are also well known for accelerating the healing of wounds in humans (Qin, 2008; Ueno et al., 2001; Wang et al., 2002). These two kinds of polymer form complexes through chemical binding, after lyophilization, which creates the porous structure, the sponge formed to act as a topical applications matrix. Wound dressing based on alginic material is well known, in literature as well as from commercial point of view, in wound management (Paul and Sharma, 2004).

Calcium alginate being a natural haemostat, alginate based dressings is indicated for bleeding wounds. The gel forming property of alginate helps in removing the dressing without much trauma, and reduces the pain experienced by the patient during dressing changes. It provides a moist environment that leads to rapid granulation and re epithelialization. In a controlled clinical trial, significant number of patients dressed with calcium alginate was completely healed at day 10 compared with the members of the paraffin gauze group. Calcium alginate dressings provide a significant improvement in healing split skin graft donor sites (O'Donoghue et al., 1997).

In another study with burn patients, calcium alginate significantly reduced the pain severity and was favored by the nursing personnel because of its ease of care. The combined use of calcium sodium alginate and a bio-occlusive membrane dressing in the management of split-thickness skin graft donor sites eliminated the pain and the problem of seroma formation and leakage seen routinely with the use of a bio-occlusive dressing alone (Disa et al., 2001). A dressing with an optimal combination of chitosan, alginate and poly ethylene glycol containing a synergistic combination of an antibiotic and an analgesic was studied on human subjects with chronic non-healing ulcers. It was observed that this material made the ulcer cleaner and had beneficial effect in the control of infection. Application of chitosan wound dressing made the chronic ulcers heal faster and also the ulcers became sterile than usually by take of local applications. Alginate is currently used for a variety of industrial purposes, and the production was hitherto exclusively based on brown seaweeds. However, the possibility to engineer bacterial production strains capable of producing tailor-made alginates for medical applications especially has become increasingly attractive. In addition, engineering of alginate molecules, by tailor-making their composition and properties or by introducing cell-specific signals, represents an important step forward for future novel application in the biotechnology field.

CONCLUSION

This review highlights the uses of chitosan-based derivatives for various biomedical applications. Chitosan has offered itself as a versatile and promising biodegradable polymer. In addition, chitosan possesses immense potential as an antimicrobial packaging material owing to its antimicrobial activity and non-toxicity. The functional properties of chitosan films can be improved when chitosan films are combined with other film forming materials such as SF, alginate and other biopolymers. All these studies indicate that, in the near future, several commercial biomedical products based on silk fibroin and chitosan will be available in the world market.

KEYWORDS

- **Alginate**
- **Biomedical**
- **Chitin**
- **Chitosan**
- **Silk fibroin**

Chapter 3

Chitosan-based Materials in Waste Water Treatment—Recent Developments

Hemlatha R., Shanmugapriya A., and Sudha P. N.

INTRODUCTION

Contamination of aquatic media by metal ions is a serious environmental problem, mainly due to the discarding of industrial wastes (Bose et al., 2002; Baraka et al., 2007). Heavy metals are highly toxic even at low concentrations and can accumulate in living organisms, causing several disorders and diseases (Aksu, 2005; Crini, 2006; Forgacs et al., 2004). Metals can be distinguished from other toxic pollutants, since they undergo chemical transformations, are non bio-degradable, and have great environmental, economic, and public health impacts (Kozlowski and Walkowiak, 2002; Rio and Delebarre, 2003).

TYPES OF TREATMENTS

Heavy metals are not bio-degradable and tend to accumulate in living organisms, causing diseases and disorders (Vander Oost et al., 2003). The presence of heavy metals in water should be controlled. Different technologies and processes are currently used such as biological treatments (Pearce et al., 2003), membrane processes (Vander Bruggen and Vandecasteele, 2003), advanced oxidation processes (F-Al Momani et al., 2002) and adsorption procedures (Robinson et al., 2002) which are the most widely used technologies for removing metals and organic compounds from industrial effluents. Adsorption is now recognized as an effective, efficient and economic method for decontamination applications and separation analysis.

Adsorption

The conventional methods such as chemical precipitation, oxidation, reduction, filtration, ion exchange, membrane separation, and adsorption (Mohan and Pittman, 2006). Ion exchangers and membrane separation are relatively of very high cost (Nomanbhay and Palanisam, 2005). Adsorption is the most frequently applied technique owing to its advantages such as variety of adsorbents materials and high efficiency at a relatively lower cost (Babel and Kurniawan, 2003). Although, activated carbon is one of the most popular adsorbents for removal of metal ions, it is not supported now owing to its high cost. Current investigations tend towards achieving high removal efficiencies with much cheaper non-conventional materials which are mostly abundant in biological matter.

Activated Carbon as an Adsorbent

The first introduction for heavy metals removal, activated carbon has undoubtedly been the most popular and widely used adsorbent in wastewater treatment applications throughout the world (Babel and Kurniawan, 2003). This capacity is mainly due to its structural characteristics and porous texture, which gives it a large surface area and its chemical nature, which can be easily modified by chemical treatment in order to increase the properties. However, activated carbon, presents several disadvantages. It is non-selective, quite expensive, and higher the quality, greater the cost. The regeneration of saturated carbon by thermal and chemical procedure is also expensive, and results in loss of the adsorbent. This had led many workers to search for more economic and efficient adsorbents. Due to the problems mentioned above, research interest in to the production of alternative sorbents to replace the costly activated carbon has intensified in recent years. Attention has focused on various adsorbents, in particular natural solid supports, which are able to remove pollutants from contaminated water at low cost (Crini, 2005).

Natural-polymers as Adsorbents

Recently numerous approaches have been studied for the development of cheaper and more effective adsorbents containing natural polymers. The removal of metals, compounds and particulates from solution by biological material is recognized as an extension to adsorption and is named as biosorption (Boddu et al., 2003). Many biosorptive agents such as fungi (Acosta et al., 2004), algae (Gupta et al., 2001), seaweeds (Elangovan et al., 2008; Kratochvil et al., 1998), microorganisms (Fan et al., 2008; Sahin and Ozturk, 2005), and several biopolymers (Bailey et al., 1999; Wu et al., 2008) have been utilized in the removal of heavy metals from wastewater. The polysaccharides are renewable resources which are currently being explored intensively for their applications in water treatment (Gupta and Ravikumar, 2000). Among the polysaccharide compounds such as chitin (Ravikumar, 2000), starch (WurzburgIn, 1986) and their derivatives, chitosan (Varma et al., 2004), deserve particular attention. These polysaccharides are abundant, renewable, and biodegradable, low-cost and are the best choice in water treatment and useful tool for protecting the environment (Bolto, 1995).

Advantages of using Natural Materials for Adsorption

Generally, a suitable adsorbent for adsorption process of pollutants should meet several requirements such as

  (a) Efficient for removal of a wide variety of target pollutants,
  (b) High capacity and rate of adsorption,
  (c) Important selectivity for different concentrations,
  (d) Granular type with good surface area,
  (e) High physical strength,
  (f) Able to be regenerated if required,
  (g) Tolerant for a wide range of wastewater parameters, and
  (h) Low cost.

Advantages of Biopolymers

The sorbents are low cost materials obtained from natural raw resources. The materials are versatile in properties with repetitive functional groups; biopolymers provide excellent chelating and complexing materials for a wide variety of pollutants including dyes, heavy metals, and aromatic compounds.

- The regeneration step is easy.
- Biosorption, in particular chitosan, is an emerging technology that attempts to overcome the selectivity disadvantages of adsorption processes (Crini, 2005).

CHITIN AND CHITOSAN

Chitin is the second most abundant natural biopolymer derived from exoskeletons of crustaceans and also from cell walls of fungi and insect (Peter in et al., 2005). The degree of acetylation in chitin can be as low as <10% and the molecular weight of this linear polysaccharide can be as high as $1-2.5*10^6$ Da, corresponding to a degree of polymerization of Ca 5,000–10,000. Chitosan is a product derived from N-deacetylation of chitin in the presence of hot alkali. In chitosan, the degree of deacetylation ranges from 40% to 98% and the molecular weight ranges between $5*10^4$ Da and $2*10^6$ Da. The degree of deacetylation and the degree of polymerization, which in turn decide molecular weight of polymer, are two important parameters dictating the use of chitosan for various applications (Hejazi and Amiji, 2003).

In spite of potential applications of chitin and chitosan, it is necessary to establish efficient and appropriate modifications to explore fully the high potential of these biomacromolecules. Chemical modifications of chitin are generally difficult owing to the lack of solubility, and the reactions under heterogeneous conditions are accompanied by various problems such as the poor extent of reaction, difficulty in selective substitutions, structural ambiguity of the products, and partial degradation due to severe reaction conditions. Therefore, with regard to developing advanced functions, much attention had been paid to chitosan rather than chitin (Jalal Zohuriaan Mehr, 2004).

Chitosan in Wastewater Treatment

Chitosan is a well known solid sorbent for transition metals because the amino groups on chitosan chain can serve as coordination sites (Guibal et al., 1994; Onsoyed and Skagrud, 1990). In addition to binding ability, it has a high content of functional groups and is produced very cheaply, since chitin is the second abundant biopolymer in nature next to cellulose (Chaufer and Deratani, 1988; Geckeler and Volchek, 1996).

As a functional material chitosan offers a unique set of characteristics like hydrophilic nature, biocompatibility, biodegradability, antibacterial properties, and remarkable affinity to proteins. It is biologically inert, safe for humans and the natural environment (Li et al., 1992; Kubota and Kikuchi, 1998). Amine groups present in chitosan are strongly reactive with metal ions. Indeed, nitrogen atoms hold free electron doublets that can react with metal cations. Amine groups are thus responsible for the uptake of metal cations by a chelation mechanism. However, the amine groups are easily protanated in acidic solutions. Hence, the protanation of these amine groups may cause electrostatic attraction of anionic compounds including metal anions (or) anionic dyes (Gibbs et al., 2003).

The binding mechanism of metal ions to chitosan is not yet fully understood. Various processes such as adsorption, ion exchange, and chelation are discussed as the mechanism responsible for complex formation between chitosan and metal ions. The type of interaction depends on the metal ion, its chemistry and the pH of the solution (Guibal et al., 2000; Inoue et al., 1993). Metal anions can be bound to chitosan by electrostatic attraction. It is likely that the chitosan-metal cation complex formation occurs primarily through the amine groups functioning as ligands (Roberts, 1992). It is well known that chitosan may complex with certain metal ions (Muzzarelli, 1977). Possible applications of the metal binding property are wastewater treatment for heavy metals and radio isotope removal with valuable metal recovery, and potable water purification for reduction of unwanted metals (Onsoyen and Skaugrud, 1990). Chitosan is a good scavenger for metal ions owing to the amine and hydroxyl functional groups in its structure (Alves and Mano, 2008; Sudha et al., 2008; Zhao et al., 2007).

Chitosan has a strong metal binding ability. It was found that their adsorption of uranium is much greater than of the other heavy metal ions (Sakaguchi et al., 1981). Chitosan, a polymer of biological origin has been reported to be an effective adsorbent for Cr (VI) removal from waste water (Bailey et al., 1999). Lasko and Hurst (1999) studied silver sorption on chitosan under different experimental conditions, changing the pH in the presence of several ligands (Dinesh karthick et al., 2009). Molybdate anions are selectively bound to chitosan in the presence of excess nitrate (or) chloride ions, with selectivity to chitosan in the presence of Ni^{2+}, Zn^{2-},Cd^{2+}ions, with selectivity coefficients in the range of 10–10,000 (Inger et al., 2003). Nair and Madhavan used chitosan for the removal of mercury from solutions and the adsorption kinetics of mercuric ions by chitosan was reported. The result indicates that the efficiency of adsorption of Hg^{2+} by chitosan depends upon the period of treatment, the particle size (Nair and Madhavan, 1984).

Jha studied the adsorption of Cd^{2+} on chitosan powder over the concentration range 1–10 ppm using various particle sizes by adopting similar procedures as for the removal of mercury (Jha et al., 1988). Mc Kay used chitosan for the removal of Cu^{2+}, Hg^{2+}, Ni^{2+}, and Zn^{2+}within the temperature range of 25°–60°C at neutral pH (Mc Kay et al., 1989). Further adsorption parameters for the removal of these metal ions were reported by Yang (Yang et al., 1984). Chitosan due to its high content of amine and hydroxyl functional groups has an extremely high affinity for many classes of dyes including disperse, direct, anionic, vat, sulphur, and naphthol (Crini and Badot, 2008; Martel et al., 2001).

The absorption spectrometry measurements proved the occurrence of interaction between the chitosan and acid dye in an aqueous solution. By assessment of chitosan/dye interaction it was possible to show that there is a 1:1 stoichiometry between protonated amino groups and sulfonate acid groups on the dye ions in low concentrated chitosan solutions. This interaction between chitosan and dye forms an insoluble product. With the excess of chitosan in the solution, the dye can be distributed between the different chitosan molecules and the chitosan/dye soluble products remain in the solution. Dye binding to chitosan involves mostly the adsorption on the active sites on chitosan macromolecules (Dragan Jocic et al., 2005).

MODIFICATIONS OF CHITOSAN

Modification of chitosan is to introduce special properties into these abundant bio-polymers and enlarge their fields of potential applications (Hong-Mei Kang et al., 2006). Physical and chemical modifications have been performed for improving metal sorption selectivity by template formation (or) the imprinting method (Baba et al., 1998; Cao et al., 2001; Tan et al., 2001).

Physical Modifications

One of the most interesting advantages of chitosan is its versatility. The material can readily be modified physically, preparing different conditioned polymeric forms such as powder, nanoparticles (Van der Lubben et al., 2000), gel beads (Guibal et al., 1999), membranes (Wang and Shen, 2001), sponge (Mi et al., 2001), honeycomb (Amiaike et al., 1998), fibers (Vincent and Guibal, 2000), hollow fibers (Agboh and Quin, 1997) for various fields of applications such as wastewater treatment, biomedical, textiles, and so on.

Chitosan Beads

Chitosan has a very low specific area ranging between 2 and 30 m^2g^{-1} (Dzul Erosa et al., 2001). Glutaraldehyde cross-linked chitosan gel beads have a higher specific surface area around 180–250 m^2g^{-1} (Milot, 1998). Gel bead conditioning significantly modifies the porous characteristics of the polymer, which may explain the differences in the sorption properties of the material (Guibal et al., 1995). The preparation of mag-netic chitosan gel beads (Rorrer et al., 1993) offers interesting perspective for the treat-ment of metal containing slurries. These kind of magnetic particles have been used for the recovery of cadmium and for dye sorption (Denkbas et al., 2002).

Sorption equilibria of Cu^{2+}, Ni^{2+}, and Zn^{2+} from single and binary-metal solutions on glutaraldehyde cross-linked beads were studied at 20°C. The amount of metal sorp-tion increased with increasing pH, confirmed the occurrence of competitive sorption of proton and metal ions (Ruey-Shin Juang et al., 2001). Liu prepared new hybrid materials that adsorb transition metal ions by immobilizing chitosan on the surface of non-porous glass beads. Column chromatography on the resulting glass beads re-vealed that they have strong affinities to Cu (II), Fe (II), and Cd (II) (Liu et al., 2002, 2003). Alginate-chitosan hybrid gel beads were prepared and shown to very rapidly adsorb heavy metal ions (Gotoh et al., 2004). Modified chitosan gel beads with phenol derivatives were found to be effective in adsorption of cationic dye, such as crystal violet and bismark brown (Chao et al., 2004).

The cross-linked chitosan beads had very high adsorption capacities to remove the anionic dyes whose maximum monolayer adsorption capacity ranges from 1,911 to 2,498 g/kg at 30°C. The adsorption capacities of the cross-linked chitosan beads are much higher than those of chitin for anionic dyes. It shows that the major adsorption site of chitosan is an amine group-NH_2, which is easily protonated to form-NH_3^+ in acidic solutions. The strong electrostatic attraction between the-NH_3^+ of chitosan and anionic dye can be used to explain the high adsorption capacity (Ming Shen Chiou et al., 2004). Chitosan bead is a good adsorbent for the removal of Congo red from its aqueous solution and 1 g of chitosan in the form of hydro gel beads can remove 93 mg of the dye at pH 6 (Sandi pan Chatterjee et al., 2007).

Membrane

Membranes and membrane processes were first introduced as an analytical tool in chemical and biomedical laboratories; they developed very rapidly in to industrial products and methods with significant technical and commercial impact. Today, membranes are used on a large scale to produce potable water from sea and brackish water, to clean industrial effluents and recover valuable constituents, purity and to separate gases, and vapor in petrochemical process (Baker, 2004; Bhattacharya and Misra, 2003).

Muzzarelli reported a decrease in the metal ion sorption efficiency of chitosan membranes compared to chitosan flakes and attributed this effect to a decrease in contact surface, despite the thickness of the membrane (Muzzarelli, 1974). Krajewska prepared chitosan gel membranes and extensively characterized their diffusion properties (Krajewska, 1996). The permeability of metal ions through these membranes was measured, Cu < Ni < Zn < Mn < Pb < Co < Cd <Ag (Krajewska, 2001).

Cross-linked chitosan membranes with epichlorohydrin have been proposed to improve pore size, distribution, mechanical resistance, chemical stability, and adsorption properties (Beppu et al., 2004; Vieira et al., 2006; Wan Ngah et al., 2002). The maximum Cr adsorption capacity occurred in epichlorohydrin-cross-linked chitosan at pH 6 (Baroni et al., 2007). The removal of divalent metal ions including Cu(II), Co(II), Ni(II), and Zn(II) from aqueous solutions by chitosan enhanced membrane filtration was studied. At neutral condition the removal of Cu(II) was more efficient compared to other metals (Ruey Shin Juang, 1999).

Flakes

Chitosan, a polymer of biological origin has been reported to be an effective adsorbent for Cr(VI) removal from wastewater (Ramnani, 2006; Rojas, 2005; Schmuhi, 2001). Cr(VI) removal ratio was optimized by surface response methodology. Accordingly a maximum of 92.9% Cr(VI) removal was attained at pH 3 with 13 g/l chitosan flakes from a solution initially concentrated as 30 mg/l (Yasar Andelib Aydin and Nuran Deveci Aksoy, 2009).

Maruca used chitosan flakes of 0.4–4mm for the removal of Cr(III) from wastewater. The sorption of arsenate on to chitosan flakes has been studied (Maruca, 1982). The maximum adsorption capacity occurs at an initial pH 3.5 (Katrina et al., 2009). Chitosan was chemically modified by introducing xanthate group onto its backbone using CS_2 under alkaline conditions. The chemically modified chitosan flakes was used as an adsorbent for the removal of Cd ions from electroplating waste effluent under laboratory conditions. The maximum uptake of Cd was found to be 357.14 mg/g at an optimum pH of 8 where as for plain chitosan flakes it was 85.47 mg/g (Divya Chauhan and Nalini Sankararamakrishnan, 2008).

Fibers

Scanty information is available onto the use of chitosan in fiber forms for wastewater treatment. Chitosan fibers were tested with aqueous solutions of copper sulfate and zinc sulfate for different periods of time to prepare samples containing different levels of metal ion contents. On chelation of metal ions the chitosan fibers gained substantial increase in the both dry and wet strength. The metal ions were readily removed from the chitosan fibers by treatment with an aqueous EDTA solution (Yimin Qin, 1993).

The recovery of direct dye by adsorption on cross-linked fiber was developed and appeared technically feasible. The concentration of amino group fixed in the adsorbent phase was 3.30 mol/kg dry fibers. A typical direct dye, brilliant yellow was used. The breakthrough curves for adsorption of the dye were measured for different flow rates, bed heights, influent concentration of the dye, and temperature (Hiroyuki et al., 1997). Chitosan fibers have been studied for the recovery of dyes and amino acids (Yoshido, 1993) but less attention has been paid to the use of this conditioning of the polymer for the recovery of metal ions.

Hollow Fibers

Hollow fibers have recently received attention with the objective of performing the simultaneous sorption and desorption of the target metal. Hollow chitosan fibers were prepared and the system was used for the recovery of chromate anions. The hollow fibers were immersed in the chromate solution while an extractant was flowed through the lumen of the fiber. Chromate anions adsorbed on the fiber were re-extracted by the solvent extractant. The hollow fiber acts simultaneously as a physical barrier that can make the extraction process more selective. (Vincent, 2000, 2001)

CHEMICAL MODIFICATIONS

Recently, there has been growing interest in the chemical modifications of chitosan in order to improve its solubility and widen its applications (Heras et al., 2002; Kurita et al., 1998; Sashiwa and Shigemasa, 1999). Chitin and chitosan have been modified via a variety of chemical modifications. Some authors have reviewed the methods (Kurita, 2001; Van Luyen and Huong, 1996). Robert has explained the modifications reactions in his source-book, *Chitin Chemistry*. Of the various possible modifications, a few to mention are nitration, phosphorylation, sulphonation, xanthation (De Smedt et al., 2000), acylation, hydroxyalkylation (Van Luyen and Huong, 1996), Schiff's base formation and alkylation (Avadi et al., 2003, 2004). These modification techniques, as foreseen by Kurita will likely find new applications in some fields including water treatment, metal cation adsorption, toiletries, medicine, agriculture, food processing, and separation (Kurita, 2001).

Derivatives of Chitosan

The high sorption capacities of modified chitosan for metal ions can be of great use for the recovery of valuable metals (or) the treatment of contaminated effluents. A great number of chitosan derivatives have been obtained with the aim of adsorbing metal ions by introducing new functional groups onto the chitosan backbone. The new functional groups are incorporated into chitosan to increase the density of sorption sites, to change the pH range for metal sorption and to change the sorption sites in order to increase sorption selectivity for the target metal (Alves and Mano, 2008).

Li reported the first synthesis of calixarene-modified chitosan. The adsorption properties of calixarene-modified chitosan were greatly varied compared with that of chitosan, especially with the adsorption capacity towards Ag^{2+} and Hg^{2+}, because of the presence of calixarene moiety (Li et al., 2003). Pyruvic acid modified chitosan had higher adsorption capacities for Cu^{2+}, Zn^{2+}, Co^{2+} than chitosan and salicylalde-hyde modified chitosan. Chitosan benzoyl thiourea derivative has been synthesized and used successfully for the removal of the hazardous ^{60}Co and ^{154}Eu radio nuclides from aqueous solutions (Metwally et al., 2009).

Multiple Modifications

There are some cases in which:

(i) The reactivity of chitin/chitosan itself is insufficient to participate in the desired reaction.

(ii) The modified chitin/chitosan does not possess the desired properties.

(iii) Some sites of chitin/ chitosan must be protected to sustain during the modification reactions. In such instants, there are two general approaches via chemical modification when the graft polymerization is a certain pathway to achieve desired characteristics:

(a) ----*In situ*-and/ or post-treatment of the graft copolymer,

(b) Graft copolymerization onto a previously modified chitin/chitosan. So, the products may generally be referred to as multiple modified materials (Jalal Zohuriaan Mehr, 2004). Chitosan can be modified as cross-linked chitosan, blends, composites and its derivatives.

Cross-linked Chitosan

Chemical modification of cross-linked chitosan through introduction of xanthate was investigated in order to achieve highly enhanced adsorption performance for lead ions under acidic solution conditions. The choice of xanthate group is due to the presence of sulfur atoms and it is well known that sulfur groups have a very strong affinity for most heavy metals, the metal-sulfur complex is very stable in basic condition (Divya Chauhan and Nalini Sankararamakrishnan, 2008).

Cross-linking of chitosan has been performed with many reagents like formalde-hyde, glutaraldehyde, ethylene glycol. Palladium sorption occured on glutaraldehyde cross-linked chitosan. The optimum pH for palladium sorption is close to pH 2 (Ruiz et al., 2000).Crown ether bound chitosan will have a strong complexing capacity and better selectivity for metal ions because of the synergistic effect of high molecular weight. Tang prepared the crown ether bound chitosan with Schiff's base type. It had not only good adsorption capacities for metal ions Pb^{2+}, Au^{3+}, Ag^+ but also high selectivity for the adsorption of Pd^{2+} in the presence of Cu^{2+} and Hg^{2+} (Tang et al., 2002). The spherical chitosan-tripolyphosphate chelating resins were used as sorbents for the removal of Cu(II) (Lee, 2001).

A new chitosan derivative has been synthesized by cross-linking a metal complexing agent, [6, 6′-piperazine-1, 4 diyl dimethylene bis (4-methyl-2-formyl) phenol] with chitosan. Adsorption towards various metal ions such as Mn(II), Fe(II), Co(II), Cu(II), Ni(II), Cd(II), and Pb(II) were carried out at 25°C. The maximum adsorption capacity was 1.21 mmol/g for Cu (II) and the order of adsorption capacities for the metal ions studied was found to be Cu(II) > Ni(II) > Cd(II) > Co(II) > Mn(II) > Fe (II) > Pb(II) (Krishnapriya and Kandaswamy, 2009).

Grafted Chitosan

Among the various methods of modifications, graft copolymerization has been mostly used method. Grafting of chitosan allows the formation of functional derivatives by covalent binding of a molecule, the graft, on to the chitosan backbone. Chitosan has two types of reactive groups that can be grafted. First, the free amine groups on deacetylated units and secondly, the hydroxyl groups on the C_3 and C_6 carbons on acetylated (or) deacetylated units. Recently researches have shown that after primary derivation followed by graft modification, chitosan would obtain much improved water solubility, antibacterial and antioxidant properties (Xie et al., 2001, 2002) and enhanced adsorption properties (Thanou et al., 2001).

Grafting is a method wherein monomers are covalently bonded onto the polymer chain. Two major types of grafting may be considered as

(i) Grafting with a single monomer.

(ii) Grafting with a mixture of two monomers. (Bhattacharya and Misra, 2004)

The grafting of carboxylic functions has frequently been regarded as an interesting process for increasing the sorption properties of chitosan (Holme and Hall, 1991). Carboxylic acids have also been grafted on chitosan through Schiff's base reactions (Guillen et al., 1992; Muzzarelli, 1985; Muzarelli et al., 1985; Saucedo et al., 1992). A Schiff's base reaction was used for the grafting of methyl pyridine on chitosan in order to prepare a sorbent for precious metal recovery (Baba and Hirakawa, 1992), and also for copper uptake (Rodrigues et al., 1998).

Becker prepared a sulfur derivative by a two step procedure consisting of pre-reaction of chitosan with glutaraldehyde followed by reaction with a mixture of formaldehyde and thioglycolic acid .These sulfur derivatives have been successfully tested for the recovery of mercury and the uptake of precious metals, owing to the chelating affinity of sulfur compounds for metal ions (Becker et al., 2000). Sulfonic groups have been also grafted on chitosan to improve sorption capacity for metal ions on acidic solutions (Weltrowski et al., 1996).

Chitosan grafted with poly (acrylonitrile) has been further modified to yield amidoximated chitosan, a derivative having a higher adsorption for Cu^{2+}, Pb^{2+} compared to cross-linked chitosan (Kang et al., 1996). Recently, a great deal of attention has been paid to the grafting of crown ether on chitosan for manufacturing new metal ion sorbents using a Schiff's base reaction (Peng et al., 1998; Tan et al., 2000). Azacrown ether grafted with chitosan and mesocyclic diamine grafted with chitosan crown ether showed high selectivity for Cu^{2+} in presence of Pb^{2+} (Yang et al., 2000).

Yang and Cheng (2003) have also reported the metal uptake abilities of macro cyclic diamine derivative of chitosan. The polymer has high metal uptake abilities, and the selectivity property for the metal ions was improved by the incorporation of azacrown ether groups in the chitosan. The selectivity for adsorption of metal ions on polymer was found to be $Ag^+ > Co^{2+} > Cr^{3+}$. These results reveal that the new type chitosan-crown ethers will have wide ranging applications for the separation and concentration of heavy metal ions in environmental analysis.

Chitosan grafted with poly(methylmethacrylate) is an efficient adsorbent for the anionic dyes (procion yellow MX, Remazol Brilliant violet and Reactive blue H5G) over a wide ph range of 4-10 being most at pH 7. The adsorbent was also found efficient in decolorizing the textile industry wastewater (Singh et al., 2009). Chitosan grafted with cyclodextrin, has ability to form complexes with a variety of other appropriate compounds, and are very promising materials for developing novel sorbent matrices (Sreenivasan, 1998). Martel showed that the adsorption of textile dyes from the effluent can be carried out with CD grafted with chitosan derivatives. Moreover these systems have superior rate of sorption and global efficiency than that of parent chitosan polymer and of the well-known cyclodextrin-epichlorohydrin gels (Martel et al., 2001).

Composites of Chitosan

Silicate-chitosan composite shows the greatest adsorption of Cd(II) and Cr(II) at pH 7. When Cr(II) was evaluated, pH 4 was optimal for its adsorption (Copello et al., 2007). Steen Kamp investigated the capacity of Cu(II) adsorption on alumina/chitosan composite, a new composite chitosan biosorbent prepared by coating chitosan, a glucosamine biopolymer, onto ceramic alumina (Steem Kamp et al., 2002). Chitosan coated on alumina exhibits greater adsorption capacity for Cr(VI). The ultimate capacity obtained from the Langmuir model is 153.85 mg/g chitosan (Boddu et al., 2003).

Chitosan/magnetite nano composite beads have the maximum adsorption capacities for Pb(II), and Ni(II) at pH 6 under room temperature which were as high as 63.33 and 52.55 mg/g respectively. Chitosan magnetite nano composite beads could serve a promising adsorbent not only for Pb(II) and Ni(II) but also for other heavy metal ions in wastewater treatment technology (Hoang Vinh Tran et al., 2010).

Chitosan/polyurethane porous composite-based on chitosan, polyether polyols, and tolylene diisocyanate were prepared. Adsorption of Cu^{2+} and Cd^{2+} on CS/PU was studied with atomic adsorption spectroscopy. The removing rate of Cu^{2+} reached 96.67% the removing rate of Cd^{2+} reached 95.67%; the adsorption capacity of CS/PU for Cu^{2+} and Cd^{2+} was 28.78 mg/g and 25.32 mg/g, respectively. The selective adsorption of CS/PU for Cu^{2+} is higher than that for Cd^{2+} when Cu^{2+} and Cd^{2+} co-exist (Liu et al., 2010).

Chitosan/kaolin/nanosized $\grave{Y}\text{-}Fe_2O_3$ composites were prepared by a micro emulsion process and characterized by TEM, SEM, and WAXRD. Many pores and pleats were visible on the surface of the composites and provided a good condition for dye adsorption. Methyl orange was selected as a model anionic azo dye to examine the adsorption behavior of the composites. About 71% of methyl orange was adsorbed within 180 min, from 20 mg/l at ph 6 by 1.0 g/l adsorbent dosage. The composites can be used as a low cost alternative for anionic dyes removal from industrial wastewater (Hua and Yue Zhu et al., 2010). A novel biocompatible composite (chitosan-zinc oxide nanoparticle) was used to adsorb the dye like AB26 and DB78. It concluded that the CS/n-ZnO being a biocompatible, eco-friendly and low-cost adsorbent might be a suitable alternative for elimination of dyes from colored aqueous solution (Raziyeh et al., 2010).

The effect of temperature on adsorption of reactive blue 19(RB19) by cross-linked chitosan/oil palm ash composite beads was investigated. It was observed that the uptake of this dye increased with increasing temperature (Masitah et al., 2009). A new type composite flocculant, polysilicate aluminum ferric-chitosan was prepared. The performance was analyzed by testing the removal of efficiency of Cu^2, Ni^2, Zn^2, Cd^2, and Cr^{6+} of heavy metals wastewater. For different heavy metal ions, the best removal efficiency of Cr^{6+} and Ni^{2+} were 100% and 82.2% respectively (Wu et al., 2010).

Blends of Chitosan

A N,O-Carboxy methyl chitosan/cellulose acetate blend nano filtration membrane was prepared in acetone solvent. It had been tested to separate chromium and copper from effluent treatment. The highest rejection was observed to be 83.40% and 72.60%, respectively (Alka et al., 2010). A chitosan/cellulose acetate/polyethylene glycol ultra filtration membrane was prepared with DMF as solvent. It was focused to be efficient in removing chromium from artificial and tannery effluent wastewater. The highest rejection rate was responding (Sudha et al., 2008).Cross-linked chitosan/polyvinyl alcohol blend beads were prepared and studied for the adsorption capacity of Cd^{2+} from wastewater. The maximum adsorption of Cd(II) ions was found to be 73.75% at pH 6 (Kumar et al., 2009).

The study of reuse of wastewater generated in the dyeing of nylon-6,6 fabrics and treated by adsorption process with nylon-6,6/chitosan (80/20) blend flakes were carried out. The efficiency in color removal of the adsorption process varied between 97 and 98% with exception of yellow erinoyl effluents which gave 65%. The reuse of treated wastewater from polyamide dyeing under the tested condition is feasible although with some restriction for yellow especially (Barcellos et al., 2008).

Adsorption of Remazol Violet on to the chitosan grafted with polyacrylamide was optimized, kinetic and thermodynamic studies were carried out taking chitosan as reference. The chitosan grafted with polyacrylamide was found to be very efficient in removing color from real industrial wastewater (Vandana Singh et al., 2009). Highly porous adsorption hollow fiber membranes were directly prepared from chitosan and cellulose acetate blend solutions and were examined for copper ion removal from aqueous solution in a batch adsorption mode. X-Ray photoelectron spectroscopic study confirmed that the adsorption of copper ions on the CS/CA blend hollow fiber membranes was mainly attributed to the formation of surface complexes with the nitrogen atoms of CS in the hollow fiber membranes (Chunxiu liu and Renbi Bai, 2006).

A new form of polymer blend, macroporous chitosan/poly vinyl alcohol foams made by a starch expansion process, exhibits the functionalities of chitosan while avoiding its poor mechanical properties and chemical instabilities. The chitosan/poly vinyl alcohol foams demonstrated interconnected and open cell structure with large pore size from tens to hundreds of micrometers and high porosities from 73.6% to 84.3%. Glutaraldehyde was employed to improve the retension of chitosan and copper adsorption of the chitosan/poly vinyl alcohol foams. While it increased the retension of chitosan and the adsorption capacities, glutaraldehyde decreased the pore size and porosity. The macro porous structure of the chitosan/poly vinyl alcohol foams indicate extensive application prospects in terms of the considerable adsorption of heavy metal ions (Xiao Wang et al., 2006).

SOME LIMITATIONS IN USING NATURAL MATERIALS AS ADSORBENTS

* The adsorption properties of an adsorbent depend on the source of raw materials. The sorption capacity of chitin and chitosan materials depend on the origin of the polysaccharide, the degree of N-acetylation, molecular weight and solution properties and varies with crystallinity, affinity for water, percent deacetylation and amino group content (Kurita, 2001). These parameters determined by the conditions selected during the preparation control the swelling and diffusion properties of the polysaccharide and influence its characteristics (Berger et al., 2004).

*Chitosan-based materials have high affinities for heavy metal ions. Hence chitosan chelation is a procedure of choice for extraction and concentration techniques in the removal of heavy metals. However, chitosan has low affinity for basic dyes.

*Pollutant molecules have many different and complicated structures. This is one of the most important factors influencing adsorption. There is yet little information in the literature on this topic. Further research is needed to establish the relationship between pollutant structure and adsorption in order to improve the sorption capacity.

The production of chitosan involves a chemical deacetylation process. Commercial production of chitosan by deacetylation of crustacean chitin with strong alkali appears to have limited potential for industrial acceptance because of difficulties in processing particularly with the large amount of waste of concentrated alkaline solution causing environmental pollution. However, several yeasts and filamentous fungi have been recently reported containing chitin and chitosan in their cell wall and septa. They can be readily cultured in simple nutrients and used as a source of chitosan. With advances in fermentation technology chitosan preparation from fungal cell walls could become an alternative route for the production of this biopolymer in an ecofriendly pathway (Crini, 2005).

CONCLUSION

Environmental requirements are becoming of great importance in today's society. Since, there is an increased interest in the industrial use of renewable resources such as starch and chitin, considerable efforts are now being made in the research and development of polysaccharide derivatives as the basic materials for new applications. In particular, the increasing cost of conventional adsorbents undoubtedly make chitosan-based materials one of the most attractive biosorbents for wastewater treatment. Recent and continuing interest in these macro molecules is evident from the number of papers that appear each year in the literature on this topic.

It is evident from our literature survey that chitosan and its derivatives have demonstrated outstanding removal capabilities of metal ions as compared to other low-cost sorbents and commercial activated carbons. Biopolymer adsorbents are efficient and can be used for the decontamination of effluents, for separation processes, and also for analytical purposes. The literature data show that the sorption capacity, specificity and adsorption kinetics are mainly influenced by chemical structure and composition of the bio polymer, and also by the accessibility of chelating or complexing groups.

Despite the number of papers published on natural adsorbents for pollutants uptake from contaminated water, there is yet little literature containing a wholesome study comparing various sorbents. Infact, the data obtained from the biopolymer derivatives have not been compared systematically with commercial activated carbon or synthetic ion exchange resins, which showed high removal efficiencies, except in recent publications. In addition, comparisons of different sorbents are difficult because of inconsistencies in the data presentation. Thus, much work is necessary to better understand adsorption and the possible technologies in the industrial scale.

KEYWORDS

- **Adsorption**
- **Azacrown ether**
- **Glutaraldehyde**
- **Polyacrylamide**
- **Stoichiometry**
- **Xanthate group**

Chapter 4

Maize–Natural Fiber as Reinforcement with Polymers for Structural Applications

Saravana D. Bavan and Mohan G. C. Kumar

INTRODUCTION

Nature has provided an immense source of fibers to the human kind, and among them plant based fibers are in great importance because these plant natural fibers are available in abundance which are able to replace the synthetic fibers in the present composite field due to their low density, non-abrasive, good-insulation properties, recyclability, biodegradable, and other mechanical properties (Jacob and Thomas, 2008; Mohanty et al., 2001, 2005). The above mentioned properties make them superior over the synthetic fibers or man-made fibers. Natural fibers are classified as plant fibers or vegetable fibers, animal fibers, and mineral fibers. Plant-based or vegetable natural fibers are lignocellulosic, consisting of cellulose micro fibrils in an amorphous matrix of lignin and hemicellulose (Sanadi et al., 1997). The fiber structure is hollow, laminated, with molecular layers and an integrated matrix. Fiber structure and properties of some agricultural residues is shown in Table1. Natural fiber reinforced polymer composite materials are an important class of engineering materials because they have better mechanical properties and ease of fabrication. Thermal properties of natural fiber reinforced polyester composites were carried out (Idicula et al., 2006) and indicated that the chemical treatment of the fibers reduces the composite thermal contact resistance.

Thermosetting resins are used today with plant fiber for panels suitable for inner door panels for the automotive industry and other applications in structures. The costs of these resins are low compared to thermoplastic resins and also the properties of the resin are well suited for structural application (Wool and Sun, 2005). Recently lot of research has been carried out on natural fibers using resin transfer molding (RTM) as a processing method of composites. Dynamic mechanical properties of sisal based natural fibers reinforced with polyester were investigated using RTM (Sreekumar, 2009, 2009a). Vacuum assisted resin transfer molding (VARTM) techniques have been developed for fabricating small and large components but mainly for complex shapes. It also gives good surface finish to the particular part. It is best suited for low cost components of complex part which should be pressurized by external source. It uses atmospheric pressure as a clamp to hold the laminates together.

Unsaturated polyester resins are the cheapest and best material in the composites industry and represent approximately 75% of the total resins used. Thermoset polyesters are produced by the condensation polymerization of dicarboxylic acids and difunctional alcohols (glycols). In addition, unsaturated polyesters contain an unsaturated material, such as maleic anhydride or fumaric acid, as part of the dicarboxylic acid component. The finished polymer is dissolved in a reactive monomer such as styrene to give a low viscosity liquid. When this resin is cured, the monomer reacts with the unsaturated sites on the polymer converting it to a solid thermoset structure.

Natural fiber materials have a long history of use in construction since Egyptian period. They were widely used in straw bale houses and straw mud houses. Biological renewable sources are novel materials for construction purposes (Herrmann et al., 1998). Polymers are widely used as reinforcement in structures of civil construction, in strengthening of bridge girders, bridge decks, cable stayed bridges, columns, and walls (Sheikh, 2002; Uomoto et al., 2002). Researchers (Dweib et al., 2004, 2006) developed structural members like panels from natural composite materials that can be used for roofs, floors, or in low-commercial building. A bio-based material reinforcement by natural fibers was carried out and these structural beams were successfully manufactured and mechanically tested giving good results.

The main aim of the work is to focus on maize fibers for reinforcement with thermosetting polymers in the process of vacuum assisted RTM and examine the prepared composite material. Natural fibers reinforced polymers have large advantages such as high specific properties, high mechanical properties, low density, low weight, and low cost. The study focused much on maize stalk based fiber its chemical and thermal properties and also its influence in thermosetting polymers.

Table 1. Fiber structure and properties of some agricultural residues (Reddy and Yang 2005).

Fiber	Fiber Structure				Fiber Properties	
	Cell dimensions		Crystallinity (%)	Elongation (%)	Moisture content (%)	
	Length (mm)	Width (μm)				
Corn husk	0.5–1.5	10–20	48–50	12–18	9	
Sorghum Stalk	0.8–1.2	30–80	-	-	8–12	
Rice straw	0.4–3.4	4–16	40	-	6.5	
Wheat straw	0.4–3.2	8–34	55-65	-	10	
Barley straw	0.7–3.1	7–24	-	-	8–12	

Maize is also known as corn in many English speaking countries. They are widely used for many purposes like starch products, food, and fodder uses. Maize stems almost resemble bamboo canes and the internodes can reach 20–30 cm, the stems are erect conventionally 2–3 m in height with many nodes casting off flag-leaves at every node. The top producing countries are United States of America, China, Brazil, Mexico, Indonesia, India, and other European countries.

Various researchers had found that stalk fibers have better properties than that of the other parts of plant fiber and indicated that they can be used for composite and other industrial applications (Reddy and Yang, 2009). Natural fiber composite laminates with distributed areca and maize stalk fibers using phenol formaldehyde were investigated (Kumar, 2008). Composite laminates were prepared with different proportions of phenol formaldehyde and fibers. Mechanical test such as tensile test, adhesion test, moisture absorption test and biodegradability test were carried out and found that these composite materials have good tensile strength and are promising materials for packaging and other general structural applications. The chemical constituents of maize stalk fiber are shown in Table 2. The ultimate strength of the maize stalk fiber is 152 MPa. The stress-strain diagram for maize fiber is shown in Figure1.

Table 2. Chemical constituents of maize stalk fiber (Reddy and Yang, 2005a).

Fibers	Chemical composition (%)			
	Cellulose	Hemi cellulose	Lignin	Ash
Maize stalk	38–40	28	7–21	3.6–7.0

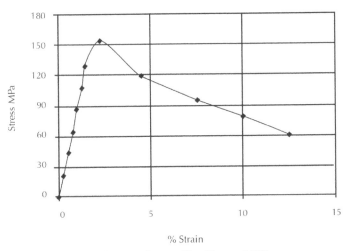

Figure 1. Stress–strain diagram for natural maize fiber (Kumar, 2008).

EXPERIMENTAL WORK

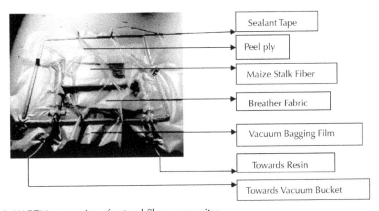

Figure 2. VARTM processing of natural fiber composites.

Maize stalk fibers are obtained from the farm field by local resources and it is cleaned manually and later thoroughly washed in running water and the fibers were sun dried. These fibers were decorticated properly and later chemically treated by 5% NaOH solution. The fibers were chopped into random pieces of uniform length. The obtained fibers were kept in an oven at around 60°C for 120 min and later used for the work. The natural fibers are now ready to use as a fiber material for processing. The VARTM method of processing composites is carried out efficiently with no voids and spaces around the mold. The laminate is sealed within an airtight envelope. When the bag is sealed, pressure on the outside and inside of this envelope is equal to the atmospheric pressure. The matrix material used is a thermosetting resin of unsaturated polyester resin mixed with methyl ethyl ketone peroxide as a catalyst and cobalt octoate as a promoter. The detailed work is depicted in Figure 2. The composites prepared were examined for integrity of fiber with the resin and confirmed that the maize fibers are suitable reinforcements for composites.

RESULTS AND DISCUSSIONS

Plant based natural fibers are lignocellulosic, consisting of cellulose micro fibrils in an amorphous matrix of lignin and hemicellulose. To improve the incorporation of natural fibers into polymers and to have higher fiber/matrix interfacial adhesion, natural fibers can be altered by different physical and chemical treatments.

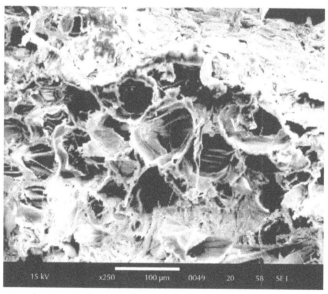

Figure 3. Scanning electron micrographs of raw maize fiber.

Fiber samples were sputtered and placed in for analyzing using scanning electron microscope. Figure (3, 4) reveals the cross section of raw maize fiber. It has a thick layer of protective material and cellular deposits and also presence of lumen in increasing the absorbency of the fibers. During, chemical treatment, the surface morphology of natural fiber changes and also the crystalline structure of cellulose is changed as shown in Figure 5.

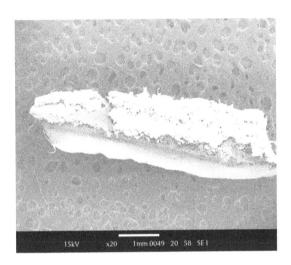

Figure 4. Scanning electron micrographs of cross-sectional view of maize fiber.

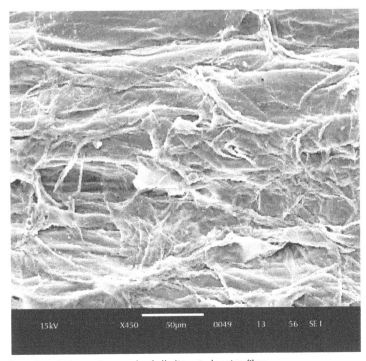

Figure 5. Scanning electron micrograph of alkali treated maize fiber.

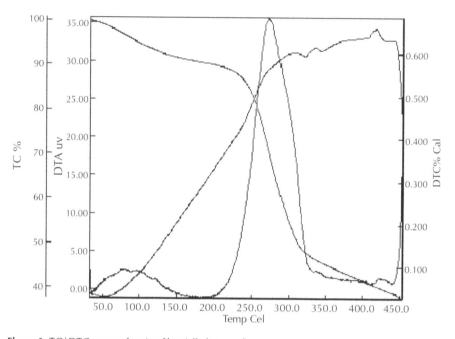

Figure 6. TG/ DTG curve of maize fiber (alkali treated).

Thermal analysis of the fiber was carried out using Thermo Gravimetric Analyzer (TGA). Thermo gravimetric (TG) curves and derivative thermogravimetric (DTG) curves of the fibers were obtained by heating the samples under a nitrogen atmosphere at a heating rate of 10°C/min. The TG/DTG curves of maize fiber (alkali treated) is shown in Figure 6. It confirms the increase of thermal stability of the fibers. The TGA curve shows two stages of decomposition, initial stage can be due to the decomposition of cellulose and hemicellulose segments and the later stage due to the degradation of lignin and other alkali segments on the fiber surface. The DTGA curves show a single peak at 272°C which may be due to decomposition of some flexible segments.

CONCLUSION

Vacuum bagging of maize natural fibers reinforced with polymers proved out to be a successful technique for initiating in small structural works. This kind of natural fiber composite reinforced polymers can be used in small existing low cost-structural applications. It can be potential substitutes for synthetic fibers in many applications where high strength and modulus are not preferred. Composite materials made from such natural fibers are in much demand in the present era since they are environment friendly green materials.

KEYWORDS

- Cellulose
- Hemicellulose
- Maize stalk fiber
- Synthetic fibers
- Thermoplastic resins

Chapter 5

Poly-β-hydroxybutyrate (PHB): A Biodegradable Polymer of Microbial Origin

N. S. Gangurde and R. Z. Sayyed

INTRODUCTION

Plastics have been an integral part of our life. World wide production of plastics is about 180 million tonnes/year (mt/year) with Asia, Australia, and India accounting for about 33% of the output, Western Europe 25% and North America 29%. By 2012, global consumption of plastics is expected to increase from the current 180 mt/year to 258 mt/year (Tuominen et al., 2002). Data released by the United States Environmental Protection Agency (2003) shows that somewhere between 500 billion and a trillion plastic bags are consumed worldwide each year. Less than 1% of bags are recycled. It costs more to recycle a bag than to produce a new one. The annual consumption of plastics in India is about to reach 4 kg/person/year, which is very small compared to a world average of 24.5 kg in 2000 and 12.5 kg for South and East Asia. But with a population of 1.0 billion people, the total demands for plastic in India are still around 4 mt and are growing fast. The current worldwide demand for plastics is in excess of 100 mt/year. The disposal of petrochemical derived plastics poses a threat to our environment. Liberal use of large amount of non-biodegradable synthetic polymers has created frightening scenario for environment. Further, more conventional petrochemical plastics are recalcitrant to microbial degradation. These non-degradable petrochemical plastics accumulate in environment at a rate of 25 mt/year. Therefore, replacement of non-biodegradable by biodegradable and eco-friendly polymers like Poly-β-hydroxybutyrate (PHB) will help to combat environmental problems created due to the use of synthetic polymers (Kumar et al., 2004). The PHB has been found as eco-friendly and best alternative biopolymer having variety of saturated or unsaturated and straight or branched chains containing aliphatic or aromatic side groups (Doi et al., 1992; Smet et al., 1983). The PHB has been found as eco-friendly substitutes and ideal candidate for making biodegradable plastics since its physical characteristics are similar to those of petrochemical polyesters such as polypropylene, in the natural environment.

The PHB as a biodegradable thermoplastic has captured the attention for more than 30 years, due to their similar properties to various thermoplastics and elastomers, which have been used in consumer products, and completely degraded to water and carbon dioxide upon disposal under various environments (Choi and Lee, 1999). However, the inefficient and expensive production and recovery methods, lack of detailed knowledge about the cultural conditions regulating PHB production, need of gene transfer from efficient PHB producer to easily cultivable organisms have hampered their use for wide range of applications and therefore, need to be explored fully for the commercialization of PHB.

It is easily degradable both aerobically and anaerobically by wide variety of bacteria and can also be produced from renewable resources. Due to these and many other useful properties, it has found technical, commercial and biomedical significance.

SYNTHETIC PLASTIC *VS.* NATURAL PLASTIC

The production and use of natural plastic is generally regarded as a more sustainable activity when compared with plastic production from petroleum (petroplastic), because it relies less on fossil fuel as a carbon source and also introduces fewer, net-new greenhouse emissions if it biodegrades. They significantly reduce hazardous wastes caused by oil-derived plastics, which remain solid for hundreds of years, and open a new era in packing technology and industry.

While production of most bioplastics results in reduced carbon dioxide emissions compared to traditional alternatives, there are some real concerns that the creation of a global bioeconomy could contribute to an accelerated rate of deforestation if not managed effectively. There are associated concerns over the impact on water supply and soil erosion and bioplastics represent a 42% reduction in carbon footprint. On the other hand, bioplastic can be made from agricultural byproducts and also from used plastic bottles and other containers using microorganisms.

SYNTHETIC PLASTIC

A synthetic plastic material is a wide range of synthetic or semi-synthetic organic solids used in the manufacture of industrial products and daily living purpose. They are typically polymers of high molecular mass and monomers of plastics are synthetic organic compounds.

Types of Synthetic Plastic

There are two types of plastics, thermoplastics and thermosetting polymers. Thermoplastics are the plastics that do not undergo chemical change in their composition when heated and can be molded again and again, examples are polyethylene, polystyrene, polyvinyl polystyrene, polyvinyl chloride, and polytetrafluroethylene (PTFE). The raw materials needed to make most plastics come from petroleum and natural gas.

Plastics can be classified by chemical structure, namely the molecular units that make up the polymer's backbone and side chains. Some important groups in these classifications are the acrylics, polyesters, silicones, polyurethanes, and halogenated plastics. Common thermoplastics range from 20,000 to 500,000 MW. These chains are made up of many repeating molecular units, known as repeat units, derived from monomers; each polymer chain will have several thousand repeating units. The vast majority of plastics are composed of polymers of carbon and hydrogen alone or with oxygen, nitrogen, chlorine or sulfur in the backbone. Some plastics are partially crystalline and partially amorphous in molecular structure, giving them both a melting point and one or more glass transitions. The so-called semi-crystalline plastics include polyethylene, polypropylene, poly (vinyl chloride), polyamides (nylons), polyesters, and some polyurethanes. Many plastics are completely amorphous, such as polystyrene and its copolymers, poly (methyl methacrylate), and all thermosets.

Table 1. Families of synthetic plastic.

Plastics families		
	Amorphous	Semi-crystalline
Ultra polymers	PI, SRP, TPI, PAI, HTS	PFSA, PEEK
High performance polymers	PPSU, PEI, PESU, PSU	Fluoropolymers: LCP, PARA, HPN, PPS, PPA
Mid range polymers	PC, PPC, COC, PMMA, ABS, PVC Alloys	PEX, PVDC, PBT, PET, POM, PA 6,6, UHMWPE
Commodity polymers	PS, PVC	PP, HDPE, LDPE

NATURAL PLASTIC

Natural plastics/bio-plastics or organic plastics are special type of biomaterials, made from renewable biomass sources such as vegetable oil, corn starch, pea starch, or microbiota, rather than fossil-fuel plastics which are derived from petroleum.

Starch Based Plastics

Starch based plastics constituting about 50% of the bioplastics market, thermoplastic starch, such as plastarch material, currently represents the most important and widely used bioplastic. Pure starch possesses the characteristic of being able to absorb humidity, and is thus being used for the production of drug capsules in the pharmaceutical sector.

Cellulose Based Plastics

Cellulose bioplastics are mainly the cellulose esters like cellulose acetate, nitrocellulose and their derivatives like celluloid.

Some Aliphatic Polyesters

The aliphatic biopolyesters are mainly polyhydroxyalkanoates (PHAs) like the poly-β-hydroxybutyrate (PHB), polyhydroxyvalerate (PHV), and polyhydroxyhexanoate (PHH). It not only resembles conventional petrochemical mass plastics (like PE or PP) in its characteristics, but it can also be processed easily on standard equipment that already exists for the production of conventional plastics.

Polylactic Acid (PLA) Plastics

Polylactic acid (PLA) is a transparent plastic produced from cane sugar or glucose. The PLA and PLA blends generally come in the form of granulates with various properties, and are used in the plastic processing industry for the production of foil, moulds, tins, cups, bottles, and so on.

Polyamide 11 (PA 11)

The PA 11 is a biopolymer derived from natural oil. It is also known under the trade name Rilsan B. The PA 11 belongs to the technical polymers family and is not biodegradable. Its properties are similar to those of PA 12, although, emissions of greenhouse gases and consumption of non renewable resources are reduced during its production. It is used in high-performance applications like automotive fuel lines, pneumatic airbrake tubing, electrical cable antitermite sheathing, flexible oil and gas pipes, control fluid umbilicals, sports shoes, electronic device components, and catheters.

Poly β-hydroxybutyrate and other Polyhydroxyalkanoates

Initially world came to know about PHB by discovery of Maurice Lemoigne of the Pasteur Institute, Paris in 1925 while studying *Bacillus megaterium* (Lemoigne, 1925). When PHB was extracted from the bacteria it crystallizes to form a polymer with similar properties to polypropylene as shown in Table 2 and is a biodegradable substitute for thermoplastics. The PHB is accumulated as a carbon reserve under nutrient limitation.

The PHB was found to be part of a larger family of poly(hydroxyalkanoates) or PHAs. The PHB is the first type of PHA to be identified (Lemoigne, 1925). The PHB is microbial polyester accumulated as lipoidic inclusions and that is osmotically inert of diameter 0.2 to 0.5 μm (Poirier et al., 1992). The P(3HB) inclusions contain approximately 97–98% P(3HB), 2% protein and 0.5% lipid. The lipid is a mixture of phosphatidic acid, triacetin, tributyrin, tripropionin, and other unidentified lipids (Kawaguchi and Doi, 1990). The P(3HB) granules *in vivo* are noncrystalline but once isolated are found to be 60–70 % crystalline.

In these microbial PHB, the ester bond is formed between carboxyl groups of one monomer with the hydroxyl group of the neighboring monomer. The PHB is active and isotactic due to the (R) stereochemical configuration (Ballistreri et al., 2001; Lee et al., 1999).

Extracted PHB are broadly divided into three classes, on the basis of Molecular weight (MW) after extraction, that are-

(a) Low MW (Reusch, 1995; Reusch et al., 1986; Reusch and Sadoff, 1983),

(b) High MW (Dawes, 1988; Dawes and Senior, 1973), and

(c) Ultra-high MW PHB (Kusaka et al., 1997).

The low MW PHB is widely accumulated by eukaryotes and archaebateria (Reusch, 1992; Reusch and Sadoff, 1983) and having MW < 12,000 Da. It is also called complexed PHB (cPHB). The MW of high MW PHB is in the range of 200,000–30,00,000 Da which is actually depending on microorganism and growth conditions (Byrom, 1994). In contrast to low MW PHB which accumulated as chloroform insoluble inclusion, high MW PHB is accumulated as water insoluble and chloroform soluble inclusion (Haung and Reusch, 1996). The Ultra high MW PHB is synthesized by recombinant *E. coli* when providing a specific growth condition (Kusaka et al., 1997). The MW of these polymers is >30,00,000 Da. It is particular use for blending and composite material.

According the carbon chain PHB is classified into two classes:

a) Short chain lengths (SCL) are comprised of monomers with 3 to 5 carbon atoms per repeat unit

b) Medium chain lengths (MCL) are comprised of monomers with 6 to 14 carbon atoms in the repeat unit. The conversion of SCL to MCL resulted in a change in physical properties from crystalline to low crystalline elastomeric thermoplastics.

The PHB is distinguished primarily by its physical characteristics. It produces transparent film at a melting point higher than 130°C and is biodegradable without residue. The PHB can be synthesized and accumulated by Gram negative and also by wide range of Gram positive bacteria, *cyanobacteria* and *archaea*. But the extensive cross linking and the thick peptidoglycan layer present in the gram positive species enables the bacterium to retain cell wall and it is more difficult to disrupt such cells to isolate the intracellular PHB. Wallen and Rohwedder (1974) reported the identification of polyhydroxyalkanoates other than P(3HB) especially poly-3-hydroxyvalerate P(3HV) and poly-3-hydroxyhexanoate P(3HHx). More than 80 different forms of PHAs can be synthesized by bacteria (Lee, 1996).

PROPERTIES OF PHB

Chemical structure and monomer composition are the most important factors in determining the physical and material properties of a polymer. Table 2 shows the properties of PHB in comparison of synthetic plastic.

(a) It is thermoplastic.

(b) It has biodegradability, easily degrade in natural environment

(c) Biocompatible and non-toxic

(d) Isotactic and optically pure

(e) Insoluble in water and high density

(f) Highly crystalline nature

(g) Piezoelectric

(h) Produced from renewable sources

Table 2. Properties of PHB.

Parameter	Polypropylene (pp)	PHB
Melting point Tm [°C]	171–186	171–182
Crystallinity [%]	65–70	65–80
Density [g cm⁻³]	0.905–0.94	1.23–1.25
Molecular weight Mw (x10-5)	2.2–7	1–8
Tensile strength [MPa]	39	40
UV resistance	poor	good
Solvent resistance	good	poor
Biodegradability	-	good

DETECTION OF PHB INCLUSIONS *IN VIVO*

Selection of efficient PHB producers among different microorganisms demands simple staining methods of PHB content estimation in living bacterial cells which can give preliminary indication for presence of PHB inside the cells. Different staining techniques are very useful for detection of physical state of PHB inclusion *in vivo*. The PHB inclusion is of 0.2–0.5 µm in diameter, in cytoplasm can be viewed microscopically due to their high refractivity (Dawes and Senior, 1973).

Sudan Black B Staining

A simple technique of PHB in bacterial cells is of Sudan black b staining in which bacterial cells were subjected to staining after growth in deficient minimal medium. Under bright field microscope, granules of PHB which revealed the blue color granules against pink cytoplasm. The number of granules varies according to microbial cells. By this staining, a single granule can be stained properly and observed under microbial field (Schlegel et al., 1970).

Phosphine R Staining

This is fluorescent technique of PHB staining in which bacterial colonies are grown on a solid medium with the presence of phosphine 3R (lipophylic due). According to this staining, the fluorescence of colonies in UV light shows the presence of the polymer. Colonies of the strains with the high content of the polymer produce bright-green fluorescence in UV light (Bonartseva, 2007). The method has been used for primary qualitative selection of the bacterial strains having high PHB content. Colonies of strains containing PHB in small quantities does not fluoresce.

Nile Blue Staining

In this method, oxazine dyes like Nile blue A is used which exhibits a strong fluorescence under the fluorescence microscope (Ostle and Holt, 1982). Nile red can also be used to detect PHB in growing cultures (Spiekermann et al., 1999). However, during purification, the membrane surrounding PHB granules is lost, therefore, cannot be stained by this technique (Dawes and Senior, 1973).

PHYSIOLOGICAL ROLE OF PHB

Physiologically, PHB is associated with the growth and cell division of bacteria. As PHB is stored as restored energy source, it directly relates with the physiological active state of cell. Author has observed that PHB accumulation in *Alcaligenes* sp. and *Pseudomonas* sp. increases at the exponential phase of growth. This observation led to the conclusion that bacteria make and store PHB when nutrients get exhausted and when cell is at its higher stage of growth. A deficiency of nitrogen can initiate PHB biosynthesis. Sayyed et al. (2009) Sayyed and Gangurde (2010) have reported that *Pseudomonas* sp. RZS1 accumulated higher amount of PHB after 30 hr of growth (Figure 1) and *A. faecalis* after 24 hr of growth (Figure 2).

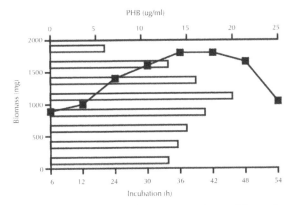

Figure 1. Growth and PHB production in *Pseudomonas* sp. RZS1 at different time interval.

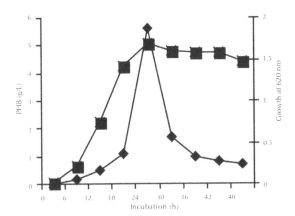

Figure 2. Growth curve and intracellular accumulation and mobilization of PHB in *A. feacalis*.

PHB PRODUCTION

Conditions and Nutrient Sources

The PHB accumulation in cytoplasm is due to the unbalance of essential nutrients, such as nitrogen, phosphate, potassium, iron, magnesium or oxygen. Feeding small quantity of nitrogen promotes accumulation of PHB than without nitrogen. The higher nitrogen leads to the intracellular degradation of PHB and also reduction of accumulation. Sayyed and Chincholkar (2004) have reported maximum accumulation of PHB in *A. faecalis* by providing nitrogen deficiency. Similar results were reported by Sayyed and Gangurde (2010) for *Pseudomonas* sp.

Growth Phase

The PHB accumulation is the growth associated property of bacterium, the exponential phase of culture shows higher state of PHB accumulation. In addition, it provides the idea about PHB productivity profile of bacterium (Figure 1 and 2) (Sayyed et al., 2009, Sayyed and Gangurde, 2010).

Carbon Source

Excess of carbon sources is required for efficient synthesis of PHA (Lee, 1996). Lee (1996) and Haywood et al. (1991) reported accumulation of PHB in *P. olovarans* from a carbon source containing 6–14 C atoms. *Fluorescent pseudomonads* have been reported to utilize the organic acids like acetic acid and lactic acid effectively for the accumulation of PHB (Bitar and Underhill, 1990). Although, PHB accumulation was evident in pH range, 5.5–9, the level increased towards neutral pH with an abrupt decline beyond 8. Maximum PHB levels have also been reported around neutral pH (Byrom, 1994). Rapske (1962) observed that optimum pH for growth and PHB production by *A. eutrophus* was 6.9.

The PHB accumulation is directly proportional to the Carbon: Nitrogen ratio (Macrae and Wilkinson, 1958). The hydroxyl acid monomer units depend on the carbon source utilized. Bacteria such as *Alcaligenes eutrophus* utilize various C4 and C5 sources to produce polymers with monomer compositions of 3HB, 4HB, 3-hydroxyvalerate (HV), and 5HV (Anderson and Dawes, 1990). The C1-C9 alcohols and C2-C10 monocarboxylic acid have also been tested as nutrient sources and proceed useful for PHB production and found that PHB could be obtained with the odd number of carbon sources.

On the other hand, PHB can be produced using renewable carbon sources such as sugars and plant oils. Various waste materials are also considered as carbon sources such as whey (Ahn et al., 2000; Wong and lee, 1998), molasses (Page, 1992; Page et al., 1997; Zhang et al., 1994) and starch (Hassan et al., 1998; Yu, 2001). Table 3 shows the microorganisms producing PHB from different substrates.

Table 3. Microorganisms producing polymers of Polyhydroxybutyrate from different substrates and biowastes.

Substrate	Polyhydroxybutyrate *(PHB)*			
	Gram-Positive	Reference	Gram-Negative	Reference
Glucose	*Bacillus*	Valappil et al., 2007; Porwal et al., 2008; Kumar et al., 2009;	*Azotobacter* *Comamonas* *Escherichia*	Page and Cornish, 1993; Lee et al., 2004; Nikel et al., 2006,
	Streptococcus *Streptomyces*	Yuksekdag and Beyatli, 2004;	*Pseudomonas* *Ralstonia*	Bertrand et al., 1990; Nurbas and Kutsal, 2004
		Valappil et al., 2007	*Vibrio*	Chien et al., 2007
Fructose	*Bacillus*	Kumar et al., 2009	*Comamonas* *Ralstonia*	Lee et al., 2004 Young et al., 1994
Sucrose	*Bacillus*	Kumar et al., 2009 Valappil et al., 2007 Anil et al., 2007	*Alcaligenes* *Comamonas* *Vibrio*	Shi et al., 2007 Lee et al., 2004 Chien et al., 2007
	Streptococcus	Yuksekdag and Beyatli, 2004		

Table 3. *(Continued)*

Substrate	Polyhydroxybutyrate *(PHB)*			
	Gram-Positive	**Reference**	**Gram-Negative**	**Reference**
Lactose	*Lactobacillus* *Streptococcus*	Yuksekdag and Beyatli, 2004	*Comamonas* *Hydrogenophaga* *Methylobacterium* *Paracoccus* *Pseudomonas* *Sinorhizobium*	Lee et al., 2004 Povolo and Casella, 2003 Yellore and Desai, 1998 Povolo and Casella, 2003 Young et al., 1994 Povolo and Casella, 2003
Fatty acids	*Bacillus*	Valappil et al., 2007, Valappil et al., 2007	*Brachymonas* *Comamonas* *Pseudomonas* *Spirulina* *Vibrio*	Shi et al., 2007 Lee et al., 2004; Zakaria et al., 2008 Ashby et al., 2003 Jau et al., 2005 Chien et al., 2007
Maltose			*Comamonas* *Protomonas*	Lee et al., 2004 Suzuki et al., 1986
Methanol			*Pseudomonas*	Young et al., 1994
Starch			*Azotobacter* *Haloferax*	Kim and Chang, 1998 Lillo and Rodriguez, 1998
Glycerol			*E. Coli* *Methylobacterium* *Ralstonia* *Vibrio*	Nikel et al., 2008 Bormann and Roth, 1999 Bormann and Roth, 1999 Chien et al., 2007
Xylose			*Burkholderia* *Methylobacterium*	Silva et al., 2004 Kim et al., 1996
Agricultural Waste	*Bacillus* *Staphylococcus*	Kumar et al., 2009 Valappil et al., 2007; Wu et al., 2001; Vijayendra et al., 2007 Wang et al., 2007	*Alcaligenes* *Azotobacter* *Burkholderia* *Escherichia* *Haloferax* *Klebsiella* *Ralstonia*	Wang et al., 2007, Page and Cornish, 1993 Silva et al., 2004 Liu et al., 1998 Huang et al., 2006 Zhang et al., 1994 Haas et al., 2008
Dairy Products			*Escherichia* *Hydrogenophaga* *Methylobacterium* *Pseudomonas* *Sinorhizobium*	Liu et al., 1998, Vijayendra et al., 2007 Povolo and Casella, 2003 Yellore and Desai, 1998 Jiang et al., 2008 Povolo and Casella, 2003
Oily Waste			*Ralstonia*	Kahar et al., 2004

Table 3. (Continued)

Substrate	Polyhydroxybutyrate *(PHB)*			
	Gram-Positive	**Reference**	**Gram-Negative**	**Reference**
Industrial Waste	*Actinobacillus* *Bacillus* *Rhodococcus*	Son et al., 2004 Vijayendra et al., 2007 Fuchtenbusch and Steinbuchel, 1999	*Azotobacter* *Burkholderia* *Pseudomonas*	Cho et al., 1997 Alias et al., 2005 Koller et al., 2008

Culture Systems

Mixed cultures or co-culture systems have been recognized to be important for several fermentation processes. Several studies have claimed the integrity and effectiveness of the system using mixed cultures. To date, considerable efforts have been carried out about PHB production using mixed culture by many researchers, in general, the PHB production of mixed cultures as summarized in following Table 4 by providing different growth conditions.

Table 4. PHB production under different growth conditions.

Conditions	PHB Production (%)	References
Aerobic	70.79	Sayyed and Gangurde, 2010
Semi aerobic	7.5	Sayyed and Gangurde, 2010
Micro aerophilic	62.0	Ueno et al., 1993
Feast-famine (aerobic)	62.0	Beccari et al., 1998
Feast-Famine Process	66.8	Dionisi et al., 2001
Fully aerobic	70.0	Punrattanasin, 2001
Feast-Famine Process	78.5	Serafim et al., 2004
Feast-Famine Process	37.9	M. Din et al., 2004

Under these growth conditions, the mixed culture subjected to consecutive periods of external substrate accessibility (*feast period*) and unavailability (*famine period*) generates unbalanced growth. During the excess of external carbon substrate, the growth of biomass and storage of polymer occur simultaneously.

Production of PHB from Waste Material

A major limiting factor in the development of biodegradable polyesters is the expense associated with the carbon substrate used in the fermentation, which can account for up to 50% of the overall production cost of PHB. The PHB production schemes based on relatively inexpensive agricultural sources may contribute significantly to lower manufacturing costs (Hass et al., 2008).

PHB BIOSYNTHESIS

In the case of normal metabolism and cell growth, organisms convert glucose into two pyruvate molecules via the glycolytic pathway. Acetyl coenzyme A is synthesized from this pyruvate, which then enters the citric acid cycle, releasing energy in the form of ATP, and GTP as well as NADH which then enters the electron transport chain and donates its electrons to O_2, where the energy released is trapped in the form of ATP.

The PHB synthesis is nothing but the conversion of acetyl-CoA to PHB as a mechanism for storing carbon. The P(3HB) biosynthesis, is the three step biosynthesis pathway, mainly consists of three enzymatic reactions catalyzed by three distinct enzymes. The enzymes are β-ketothiolase, acetoacetyl-CoA reductase and PHB synthase.

Enzymes Involved
β-ketothiolase

It catalyzes the first step in P(3HB) synthesis. It is a member of the family of enzymes of thiolytic cleavage of substrate into acyl-CoA and acetyl-CoA. The two acetyl CoA molecules are combined to form acetoacetyl-CoA catalyzed by 3-ketothiolase (acetyl-CoA acetyl transferase)

It is divided into two groups; first group of thiolases fall into the enzyme Class EC.2.3.1.16 and the second group, belonging to enzyme Class EC.2.3.1.9. The first group is involved mainly in the degradation of fatty acids and is located in the cytoplasm of prokaryotes and in the mitochondria. The second type is considered to take part in P(3HB) biosynthesis (Madison and Huisman, 1999).

In *Ralstonia eutropha*, two β-ketothiolases, enzyme A and enzyme B, have been discovered to take part in biosynthesis of PHB. Enzyme A is a homo-tetramer of 44-kDa subunits and enzyme B, a homo-tetramer of 46 kDa subunits. (Madison and Huisman, 1999). Slater et al. (1998) have shown that enzyme B is the primary catalyzer for the P(3HB-3HV) formation. The enzymatic mechanism of β-ketothiolase consists of biological condensation of two acetyl-CoA-moieties with formation of carbon-carbon bond.

Acetoacetyl-CoA Reductase

It catalyzes second step in the P(3HB) biosynthetic pathway by stereo-selective reduction of acetoacetyl-CoA formed by β-ketothiolase to 3-hydroxybutyryl-CoA. Two types of reductases are found in organisms (Haywood et al., 1988a, 1988b). First type is a NADH dependant reductase, EC.1.1.1.35, while the second type, EC.1.1.1.36, is a NADPH dependant. The former is a tetramer with identical subunits of 30 kDa. The latter is a homo-tetramer of 25 kDa subunits. (Schembri et al., 1994).

P(3HB) Synthase

The last reaction in the polymer formation is catalyzed by the enzyme PHB synthase which links D(-)-β-hydroxybutyryl moiety to an existing polyester molecule by an ester bond. This key enzyme determines the type of PHB synthesized. The PHB synthase is soluble only as long as no PHB synthesis and accumulation occurs in an organism. It, however, becomes granule associated under storage conditions (Haywood, 1989).

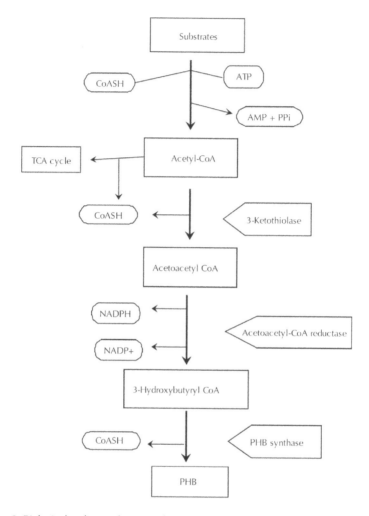

Figure 3. Biological pathway of PHB production.

MOLECULAR EXPRESSIONS OF PHB SYNTHESIS GENES

Expression in Bacteria

Many studies have demonstrated heterologous expression of PHB synthases into bacteria. *R. eutropha* PHB⁻ 4 (Schlegel et al., 1970) and *P. putida* GPp104 (Huisman et al., 1991) are defective in phaC gene used for physiological studies and cloning experiments which require only PHB synthase gene for accumulation and expression of PHB and such strains are easily identified by staining as mentioned earlier.

E. coli is the widely acceptable host for cloning experiments which offers a well defined physiological environment for the production of recombinant proteins and other bioproducts because this bacterium is widely studied in detail. Among the different strains of *E. coli*, XL1 blue and *E. coli* XL⁻ B have been the best for PHB production

(Lee et al., 1994). The PHB biosynthetic genes of *A. latus* have been used to clone and produce PHB in *E. coli* (Choi et al., 1998) and resulted in better PHB production in *E. coli* compared to the PHB biosynthetic genes of *R. eutropha.*

In recombinant microorganisms, plasmid stability is of crucial importance for continued PHB production which is a problem associated with the expression of PHB biosynthetic gene in bacteria

Expression in Eukaryotic Cells

Saccharomyces cerivisiae is the only eukaryotic organism which has been transformed to accumulate PHB so far with expression of the PHA synthase (phaC$_{Re}$) of *R. eutropha* in the *S. cerivisiae* cytoplasm. In contrast to *E. coli* and plants, yeast does not require the expression of β-ketothiolase and acetoacetyl-CoA reductase genes for the PHA accumulation and these steps are catalyzed by the native Erg10 and Fox2 proteins, respectively, which are involved in β-oxidation but functioning as β-ketothiolase and acetoacetyl-Co A reductase (Leaf et al., 1996).

Expression in Animal Cells

Expression of PHB biosynthesis genes in animal cells has been achieved in the cells of insect *Spodoptera frugiperda* and *Trichopulsiani*. In Spodoptera, an alternative pathway for the biosynthesis of PHB is created. The dehydratase domain mutant rat fatty acid synthase cDNA and phaC$_{Re}$ were expressed simultaneously which resulted into PHB synthesis (Williams et al., 1996). A baculovirus, *Autographa californica* nuclear polyhedrosis virus system has been also used to express PhaC$_{Re}$ in *Trichopulsiani* cells. These cells accumulated PHB synthase in 50% of the total cell protein.

Expression in Plants

The PHA biosynthesis genes from microorganisms have been successfully expressed in *Arabidopsis thaliana* (Nawrath et al., 1994; Poirier et al., 1992), *Brassica napus* (Valentin et al., 1999), *Gossypium hirsutum* (John and Keller, 1996), *Nicotiana tabacum* (Nakashita et al., 1999), *Solanum tuberosum* (Mittendorf et al., 1998) and *Zea mays* (Hahn et al., 1999). However, transgenic plants are impaired. For PHA biosynthesis in plants, phaC$_{Re}$ and phaB$_{Re}$ alone or together with phaA$_{Re}$ have been used. Sometimes, phaA expression is not required because it is present in plants. Replacement of phaA$_{Re}$ by bkt B$_{Re}$ in plants enabled them to accumulate co-polymers (Slater et al., 1998; Valentin et al., 1999). Transgenic plants harbour PHB synthesis gene and found to accumulate PHB in various compartments of plant cells and tissues such as nucleus, vacuoles and cytoplasm. However, in case of transgenic plant, *B. napus,* PHB accumulation occurred in leucoplasts (Valentin et al., 1999).

Transgenic plant may prove cost effective PHB producer because PHB are produced from carbon dioxide, water and sunlight. If efforts are concentrated on improving this process, it may be possible to see bioplastic agriculture in the future and agricultural bioplastic products may appear in the market.

ISOLATION AND PURIFICATION OF PHB

Solvent Extraction

The PHB is soluble in organic solvent; this property leads to the effective and higher PHB extraction from cells. Several solvents such as chloroform, dichloroethane, trichloroethane, ethylene, hexane propanol, and acetone: alcohol have been in use for isolation and purification of PHB to solubilize the PHB. The treatment of methanol or acetone to biomass, before the solvent extraction leads to denature low molecular weight proteins which enhance the purity of PHB.

Sodium hypochlorite was first used by Williamson and Wilkinson (1958) to isolate PHB granules from *Bacillus cereus*. However, sodium hypochlorite digestion of NPCM resulted in the lysis of cells without affecting the PHB. Sodium hypochlorite purification method was affecting the sudanophilic properties and molecular weight of the polymer granules.

The dispersion of chloroform and sodium hypochlorite is also used for the extraction of PHB. By this method, the degradation of PHB by sodium hypochlorite is slightly reduced but highly inconvenient for analytical purposes, which leads to improper data analysis. Extraction of PHB with a mixture of hexane and propanol resulted in poor recovery yield.

The solvent system consisting of 1:1 mixture of ethanol and acetone is known to be specific for efficient recovery method capable of specifically lysing the NPCM without affecting PHB. Rawte and Mavikurve (2002) have also reported the usefulness of acetone and ethanol in the extraction of PHA. However, author has reported that ethanol and acetone (1:1 v/v) is the best recovery method and gives high recovery yields. Table 5 gives the comparative account of recovery yield of PHB with various methods.

Table 5. Recovery of PHB with Different Recovery Methods.

Recovery Methods of PHB	PHB Recovery Yield (gL⁻¹)	Reference
Acetone: Alcohol	5.6	Sayyed et al., 2009
Hypochlorite Digestion	1.34	Reusch et al., 1987
Chloroform extraction method	0.63	Choi and Lee, 1997
Dispersion method	1.1	Lee, 1996

An inconvenience for solvent extraction methods is that large volume of solvents and non-solvents are needed to extract and to precipitate the polymer. However, PHB recovery by solvent extraction resulted in a pure white very crystalline powder of high molecular weight.

Enzymatic Extraction

In 1964, Merrick and Doudoroff isolated PHB granules from *B. cereus* using lysozymes and deoxyribonuclease to solubilize the peptidoglycans and the nucleic acids respectively. The enzymatic digestion of cell components usually released the nucleic acids in the suspension medium which made suspension highly viscous and impossible to treat any further. Heating prior to the enzymatic digestion step is used to denature and solubilize the DNA. Following this pretreatment, several enzymes (alcalase, phospholipidase, lecitase, and lysozymes) are used to solubilize the non-PHB cell components. The PHB from such an isolation procedure contain at least some cell debris. The purity can be achieved by an extra purification step by solvent extraction.

APPLICATION OF PHB

Packaging
For preparation of films, blow molded bottles and paper, disposal items such as razors, utensils, diapers and feminine, PHB are used.

Medical Field
Its slow and *in vivo* degradability proves it a potential candidate for use in reconstructive surgery. Development of cardiovascular products such as pericardial patches, vascular grafts, heart valve. It can also be used for controlled drug release system. The 4 HB units are pharmacologically active compounds, therefore are very suitable for treatment of alcohol withdrawal syndrome, narcolesy, chronic schizophrenia, catatonic schizophrenia, atypical psychoses, chronic brain syndrome, neurosis, drug addiction and withdrawal, Parkinsons's disease, circulatory collapse, radiation exposure, cancer, myocardial infarction, and other neuropharmacological illnesses.

Disposable Personal Hygiene
PHB can be used as sole structural materials or as parts of degradability plastics.

Tissue Engineering Scaffolds
The suitable material properties of PHB such as compatibility, support cell growth, guide and organize cells allow tissue in growth.

Agricultural
The PHB are biodegraded in soil. Therefore, the use of PHB in agriculture is very promising. They can be used as biodegradable carrier for long-term dosage of insecticides, herbicides, or fertilizers, seedling containers and plastic sheaths protecting saplings, biodegradable matrix for drug release in veterinary medicine, and tubing for crop irrigation. Here again, it is not necessary to remove biodegradable items at the end of the harvesting season.

KEYWORDS

- **Bioplastic**
- **Lysozymes**
- **Petrochemical polyesters**
- **Polyhydroxyalkanoate**
- **Pyruvate molecules**

Chapter 6

Barrier Properties of Biodegradable Bacterial Polyester Films

Nadège Follain, Frédéric Chivrac, Florent Girard, and Stéphane Marais

INTRODUCTION

With the increasing concern of human society to environmental and energy problems, the family of microbial synthesized polymers poly(hydroxyalkanoates) (PHA) have been attracting more and more attention in both academic and industrial fields due to their complete biodegradability and the renewable carbon resources used to produce them (Doi, 1990; Müller, 1993).

Undoubtedly, the use of long-lasting polymers for short-lived applications (packaging, catering, surgery, hygiene…) is facing the global growing conscience relating to the preservation of ecological systems. Most of the today's polymers are produced from petrochemicals and so are not biodegradable. Indeed, such alternative bio-based packaging materials have attracted considerable research and development interest for a significant length of time. In this context, tailoring new environment-friendly polymers based on PHA have numerous specific properties, which open new fields of applications. The PHA received increasing attention during the last decade due to its biodegradability, biocompatibility and physiologically benign behavior that makes them interesting for biomedical or food applications, for example. In the family of PHA, polyhydroxybutyrate (PHB) and its copolymers are the most produced and investigated products, and the molecular structure differences result in copolymers that are more ductile and processable than PHB.

The aim of this study is to evaluate the potential of bio-based packaging materials from microorganisms for film applications, and the most important properties can be narrowed down to four intrinsic properties of the material: mechanical, thermal, gas barrier and water vapor properties. This work will focus on the two latter properties which are few considered in the literature. To date, to our knowledge, very few researches, investigating barrier properties to diffusing water and/or gas molecules of PHA family, have been published despite some published studies devoted to biodegradable polyesters including PHA. Future bio-based materials must be able to mimic the gas and water vapor barriers of the conventional materials known today in the considered case of replacement materials. In addition, in storage or under operation conditions, polymer materials are often exposed to moisture and water molecules can penetrate a polymer matrix thus modifying its physical and transport properties. When compared to conventional polymers, PHBV copolymers showed a very good balance of barrier properties as regards to water and diffusing gas molecules. Besides, these copolymers have the great advantage of its moisture insensitivity and low hydrophilic character, common drawback of bio-based polymers.

These copolymers have not found extensive applications in the packaging industries to replace conventional plastic materials, although there could be an interesting way to overcome the limitation of the petrochemical resources in the future. Technically, the prospects for PHA are very promising. It is noteworthy that developments are still being made.

MATERIALS, PROCESSING, AND METHODS

Currently, some biodegradable polyesters are synthesized and/or formed in nature by organism growth. Different classifications of these biodegradable polymers have already been proposed in the literature (Averous, 2004; Bordes, 2009) depending on the synthesis process. For our purposes, the main families of polyesters are extracted and highlighted in Figure 1.

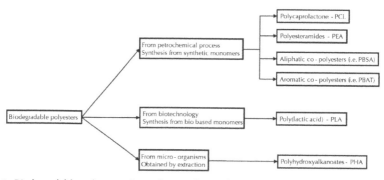

Figure 1. Biodegradable polyesters depending on the synthesis process.

Materials

In this section, the major features of biodegradable polyesters are reported and the chemical structures are reminded in Figure 2.

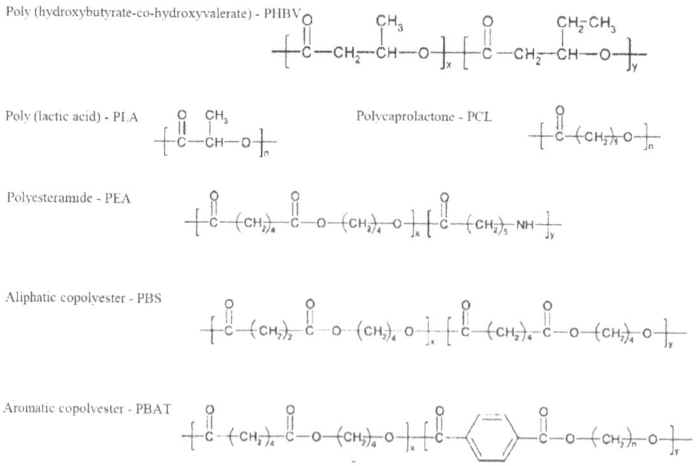

Figure 2. Chemical structures of available biodegradable polyesters.

Polyhydroxyalkanoates—PHA

As pointed out by Lenz and Marchessault in an excellent review (Lenz and Marches-sault, 2005), Lemoigne and co-workers published their observations and interpreta-tions on PHA family at the time when Herman Staudinger was proposing the existence of high molecular weight molecules or polymers, which he termed "macromolecules".

From the simple PHB homopolymers discovered by Maurice Lemoigne in the mid-twenties, a family of over 100 different aliphatic polyesters of the some general structure has been identified (Lenz and Marchessault, 2005; Steinbüchel, 2003). The main features are reported in Figure 3.

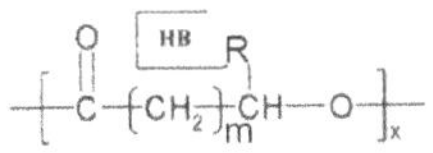

m	R groups	Name
1	hydrogen	Poly(3-hydroxypropionate)
1	methyl	Poly(3-hydroxybutyrate)
1	ethyl	Poly(3-hydroxyvalerate)
1	propyl	Poly(3-hydroxyhexanoate)
1	pentyl	Poly(3-hydroxyoctanoate)
1	nonyl	Poly(3-hydroxydodecanoate)
2	hydrogen	Poly(4-hydroxybutyrate)
3	hydrogen	Poly(5-hydroxyvalerate)

Figure 3. General structure of polyhydroxyalkanoates.

Depending on bacterial species and substrates used for feeding the bacteria and the conditions of growth, high molecular weight stereoregular polyesters have emerged as a new family of natural polymers, named polyhydroxyalkanoates. The way of sepa-ration and purification may also have a certain influence. In addition, the attainable molecular properties, the molecular weight, and molecular weight distribution are de-pendent on the bacterial fermentation conditions.

This possibility of preparing various PHA as a function of the substrate was first revealed by De Smet and co-workers (De Smet, 1983). For example, the polymer con-sisting principally of 3-hydroxyoctanoate units (PHO) can be formed from *Pseudomo-nas oleovorans* in n-octane. And, the polymer with 3-hydroxybutyrate units (PHB) can be obtained from *Alcaligenes eutrophus* or from *Azotobacter beijerinckii* as reported by Schlegel and co-workers, and Dawes and co-workers, respectively (Oeding, 1973, Senior, 1973). This family can thus be produced in many grades, differing in composi-tion, molecular weight, and other parameters.

Stanier, Wilkinson, and their co-workers (Macrae, 1958; Stanier, 1959; Williamson, 1958) have evenly reported that PHA acted as an intracellular food and energy reserve in bacteria. The polymer is produced by the cell in response to a nutrient limitation in the environment in order to prevent starvation; in the case where an essential element is unavailable. The nutrient limitation activates a metabolic pathway activating the PHA production: PHA represented as carbon-storage polymers.

The structure of PHA is based on polyester macromolecules bearing optically active carbon atoms. In fact, PHA family present the same three-carbon backbone structure (Figure 3) but differing in the type of alkyl group at the third position. All these polyesters have the same configuration for the chiral center which is very important for their ensuing physical properties. All PHA are thus completely isotactic with the [R] configuration (Figure 4), as expected by Dawes and co-workers in 1989 (Haywood, 1989).

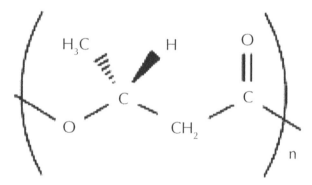

Figure 4. Poly(3-hydroxyburyrate) with *R* configuration.

Different structures, isotactic with random stereo sequences for these bacterial polyesters are then obtained. Although, PHB is the main polymer of PHA, different poly(hydroxybutyrate-*co*-hydroxyalkanoates) copolyesters exist such as PHBV poly(hydroxyl butyrate-*co*-hydroxyvalerate), or PHBHx poly(hydroxybutyrate-*co*-hydroxyhexanoate), PHBO poly(hydroxybutyrate-*co*-hydroxyoctanoate) and PHBOd poly(hydroxybutyrate-*co*-hydroxy octadecanoate).

Several serious drawbacks have hindered potential applications in numerous cases of PHB such as high glass transition temperature of PHB, its high crystallinity and large brittleness. Therefore, copolymers of PHB with hydroxyvalerate units have been developed showing an improved processing behavior, provides the possibility for technical use in many applications. Indeed, the first industrial production of PHBV took place (Holmes, 1981) although the patents on PHB as a potentially biodegradable polymer used in plastic industries had already been filed in 1962. In addition, to facilitate the film transformation, the PHBV is more adapted for the process.

In the context of barrier properties against small molecules (gas like liquid molecules) and to place some PHA polymers in an economic context, a comparison of their properties is essential to ascertain their potential with those of conventional polymers commonly used such as polyethylene, polypropylene and with those of polyesters available in industry such as poly(lactic acid), polycaprolactone, for instance, that at the present time present the highest potentials due to their availability and low price. For that, an introduction of these polyesters is thereafter proposed.

Other Bio-resources and Petroleum Based Polyesters

PLA – Lactic acid is an organic acid found in many products of natural origin such as animals, plants and microorganisms (Sodergard and Stolt, 2002). Industrially, lactic acid can be derived from intermediates with renewable origin or from chemical molecules based on coal or oil. Lactic acid (2-hydroxypropanoic acid) is one of the smallest optically active molecules which can be either of L(+) or D(−) enantiomer monomers. The properties of lactic acid based polymers depended on the ratio and distribution of the two enantiomer monomers and can be varied to a large extent: 100% L-PLA present a high crystallinity and copolymers of poly(L-lactic acid) and poly(D,L-lactic acid) are rather amorphous (Perego, 1996).

These polymers with high molecular weight (Albertsson and Varma, 2002) can be obtained by different routes such as the dimerization of polycondensated lactic acid into lactide (cyclic dimmer) or by ring-opening polymerization (ROP), as reported by Carothers and co-workers (Carothers, 1932). The polymers derived from lactic acid by polycondensation are generally referred to as poly(lactic acid) and the polymers resulted to lactide by ROP as poly(lactide). Both types are generally referred to as PLA. The PLA is presumed to be biodegradable, although the role of hydrolysis versus enzymatic depolymerization in the biodegradation process remains open to debate according to Bastioli (Bastioli 1998). But, according to Tuominen and co-workers (Tuominen, 2002), PLA does not exhibit any eco-toxicological effect during its biodegradation.

Indeed, a large number of biodegradable polyesters are based on petroleum resources, obtained chemically from synthetic monomers and polycaprolactone, polyesteramide, or aliphatic/aromatic copolyesters can be distinguished. Generally, these petroleum-based polyesters are soft at room temperature since their glass transition temperatures are lower.

PCL – The polycaprolactone is derived from ε–caprolactone by ring-opening polymerization catalyzed by transition metal compounds. Tokiwa and Suzuki (Tokiwa and Suzuki, 1977) have reported that PCL can be enzymatically degraded in presence of fungi and the biodegradation process has been discussed by Bastioli (Bastioli, 1998).

PEA – The highest polar polyester is represented by polyesteramide which is synthesized from statistical polycondensation between polyamide monomers with adipic acid. Different commercial grades, named BAK®, have been developed by Bayer but the production has been stopped in 2001. Contrary to PLA, PEA polymers have exhibited a negative eco-toxicological effect during composting (Averous, 2004).

Aliphatic and aromatic copolyesters – Large grades of copolyesters are obtained from the polycondensation of diols with dicarboxylic acid. Depending on the dicarboxylic acid used (terephtalic acid, adipic, or succinic acid), aromatic or aliphatic copolyesters are synthesized. Aromatic like aliphatic copolyesters are estimated totally degradable in microorganisms environment (Averous, 2004).

Processing

From a practical point of view, chloroform and dichloromethane are commonly used for the extraction of PHB or its copolymers from the bacteria. The solubility of PHA polymers differs depending on their composition and is also the case during the processing technology.

Solvent Casting

In this context, Lemoigne and co-workers firstly showed that PHB polymer could be cast into a transparent film like the well-known cellulose nitrate material referred as collodion. The solution casting technique from 1-3% chloroform solution is conventionally used to prepare polyester films. Kamaev and co-workers (Kamaev, 1999) reported the preparation of PHB film by using casting chloroform-soluble polymer. Tang and co-workers (Tang, 2008) studied the water transport properties of biomaterial incorporating PHBV (12 mol% HV) and nano-hydroxyapatite modified which is prepared by solution casting method from the dissolution in chloroform at 50°C to form 5 wt% polymer solution.

In blend context, since PHB can easily decompose at temperatures near its melting temperature (PHB melting temperature around 180°C with processing temperature to be at least 190°C), the thermal stability of PHB is largely approached by some authors, particularly through PHB-polyester blends. In this context, Erceg and co-workers (Erceg, 2005) added aliphatic/aromatic copolyester to PHB film by using chloroform-solvent casting method. They estimated that the thermal instability of PHB can be prevented by mixing it with biodegradable materials.

Melt Processing

The PHA polymers can be considered as thermoplastic polymers and can be processed by conventional plastic forming equipment. Considering the melting point of PHB, in particular, its processing temperature should be at least 190°C. At this temperature, thermal degradation can proceed rapidly and consequently a pronounced effect on the ultimate properties may arise. As expected from PLA behavior, PHA polymers are sensitive to the process conditions such as the increase of shear level, the processing temperature and/or the residence time (Ramkumar and Bhattacharia, 1998). The major problem in the manufacturing of PHA polymers concerns the limited thermal stability during the melt processing which is mainly extrusion process. It is worth to note that a rapid decrease of viscosity and molecular weight due to macromolecular linkage can be observed under extrusion. This thermal degradation is generally described either by a *cis*-elimination or a *trans*-esterification. The ester groups tend to degrade into smaller fractions when exposed to heat even for short times. However, the presence of a second monomer in PHA copolymers such as PHBV has a certain stabilizing effect in thermal degradation. Wang and co-workers (Wang, 2008) reported their work on the mechanical and degradation properties of ternary PHBV blends in order to improve the toughness of PHB polymer by using a co-rotating twin-screw microcompounder having screw length of 150 mm, L/D of 18, and barrel volume of 15 cm^3.

Precise conditions are required for industrial processing PHA polymers. For example, Biomer (Germany) has recommended to process PHA polymers in two-step melt-state transformation, for Biomer products in particular: first, melting the product when its introduction into the equipment and, second, to carry out a gradual decrease of temperature in the next zones until the die: 185°C in zone 1 down to 150°C in zone 4 and 145°C in the die for Biomer grade P226 (PHB polymer). As of hydroxyvalerate units incorporated into PHB polymer, the process temperatures were lower than that of PHB ones and was kept below 170°C. The temperature of the barrel at the bottom section and of the injection molder (155°C) was kept below the melting point of PHBV (at around 157°C). Bhardwaj and co-workers presented the fabrication of eco-friendly green composites from PHBV with 13 mol% valerate units (Zeneca Bio Products from Biomer) by co-rotating twin-screw microcompounder extrusion (Bhardwaj, 2006). The temperature profile used for a perpendicular barrel at top, center, and bottom sections are 155°C, 160°C, and 155°C, respectively. The lower melting point of PHBV improves the melt stability and broadens the processing window of this polymer.

Therefore, melt processing should be limited to a narrow temperature window and alternative processing technologies can be developed, such as solid-state processing, in order to minimize thermal degradation.

Solid-state Processing
Depending on the combination of pressures with temperatures used, the solid-state processing can resemble to powder sintering (metallurgy process) or to hot-press moulding. The solid-state processing consists in transform polymer powders at very high pressures and elevated temperatures but below the melting point.

A two-step solid-state transformation by sintering powders has been described by Lüpke and co-workers (Lüpke, 1998) and its principles are also applicable to other PHA polymers than PHB. The authors estimated that the ultimate mechanical properties depend substantially on the initial molecular weight of polymer as well as on the processing temperature. A slight decrease of crystallinity was also reported.

Chen and co-workers (Chen, 2007) reported the preparation of PHBV/hydroxyapatite nanocomposites by using hot-press molding for thermal and physical tests in order to consider potential medical applications.

Transport Properties Methods
From published reports and studies, we can note that some technical procedures depend on the background of scientific teams and available apparatus for sorption and permeation measurements (Alexandre, 2009; Erkske PEASC, 2006; Follain, 2010; JAPS, 1999a, 1999b; Marais, 1999; Miguel Yoon JAPS 77, 2000). In addition, the use of different equipments and dissimilar conditions for the measurements complicated the comparison of the different results obtained.

Water Vapor Sorption
Water sorption isotherm is useful for predicting water sorption properties of polymer and allowed to provide little insight into the interaction between water molecules and polymer tested.

The conventional approach to perform water vapor sorption kinetics is based on the measurement of the film sample mass until reaching constant value corresponding to a saturation level in a hydrated environment. For each hydrated environment, the mass of sample is measured allowing to obtain the mass gain, expressed in percent, or, in other word, the water concentration, expressed in mass of water sorbed per mass of polymer or in mmol of water sorbed per mass of polymer, inside sample. The combination of mass gain at each humidity level leads to build the water vapor sorption isotherm. The water vapor sorption and the isotherm shape are depending on the moisture resistance of polymer and its ability to interact with water.

By using Saturated Salt Solutions

After a drying time under vacuum, the polymer films were kept in saturated water vapors in separate containers at constant temperature during a period of time necessary for equilibrium saturation of the polymer film (Iordanskii, 1998, 1999).

Appropriate saturated salt solutions were used to provide constant water activities (a_w), or relative humidities, ranging from 0.05 to 0.98 according to standard UNE-EN ISO 483:1988. Ideal behavior is assumed, when activity is evaluated as the ratio of the water vapor pressure (p_w) to the saturated water vapor pressure (p^{sat}_w) at 25°C i.e.: $a_w = p_w / p^{sat}_w$.

Technically, samples were periodically weighed and water sorption equilibrium was considered to be reached when no mass change occurred. Then, the water concentration in the films was calculated from the sorption data in the stated conditions. This technique is less widely used for the benefit of automated dynamic gravimetric sorption system.

By using an Electronic Micro-balance

The water vapor sorption experiments were generally performed in a Cahn D-200 electronic microbalance (with a sensitivity of 10^{-5}g) enclosed in a thermostated reactor (Figure 5). The sample is placed in a pan and dried at 0% humidity. After reaching a plateau, the dry mass is achieved. Thereafter, the sample is exposed to vapor pressure and the mass gain is measured as a function of time until reaching the equilibrium state. The mass equilibrium is obtained at each humidity level tested.

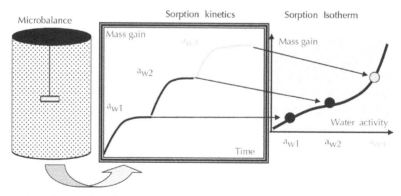

Figure 5. Contribution of an electronic microbalance for building sorption isotherm (Follain, 2010).

On the basis of water vapor isotherm, we can find three sorption modes. The first concave part of the curve indicates Langmuir-type adsorption behavior assuming a sorption on specific sites or micro-voids inside polymer (free volume). The following linear part of the isotherm corresponds to a Henry-type process involving random adsorption by dissolution and diffusion of the water molecules inside polymer. The combination of Langmuir-type and Henry-type sorptions is commonly attributed to the dual-mode sorption in glassy polymers. The convex part of the curve corresponds to sorption of water molecules leading to the water clustering formation. Usually the combination of dual mode sorption with aggregation sorption corresponding to Park's model was typical of water sorption in hydrophilic polymers. In general, the water sorption behavior in hydrophilic material is analyzed from BET (Brummer-Emmet-Teller) isotherm, and in particular GAB (Guggenheim-Anderson-de-Boer) model.

Water and Gas Permeation
The permeation properties of polymer film were studied with methods appropriate to the nature of the diffusing molecules. The water and gas diffusing molecules permeation measurements can also be performed from different techniques.

The gas molecules permeation measurements were carried out generally by the "time-lag" method, a variable pressure method, by using the permeation apparatus shown in Figure 6. The permeability coefficient, expressed generally in Barrer (10^{-10} cm^{3}STP cm/cm^2 s cmHg), was calculated from the slope of the steady state line imparting the saturation level by taking into account the exposed area of the film and the vapor pressure difference across the two sides of the film.

A preliminary high vacuum desorption was realized on both sides of the permeation cell. Then the upstream compartment was filled with gas at determined pressure. In the downstream compartment, the increase of pressure was measured as a function of time by using a datametric pressure sensor communicating with a data acquisition system.

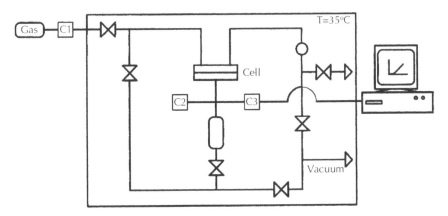

Figure 6. Apparatus based on a barometric permeation method (Joly, 1999).

For water permeation, the measurements were carried out by using a pervaporation method. This differential permeation method allow to obtain the diffusion and permeability coefficients by taking into account the exposed area of the film and the vapor pressure difference across the two sides of the film.

The permeation flux through the film is measured as a function of time. The permeation cell consisted of two compartments separated by the polymer film. Both compartments were flushed by dry nitrogen gas in order to dry the polymer film. Thereafter, a stream of water was introduced into the upstream compartment. The water concentration was measured in the downstream compartment by using a hygrometric sensor (a chilled mirror hygrometer) coupled to a data acquisition system.

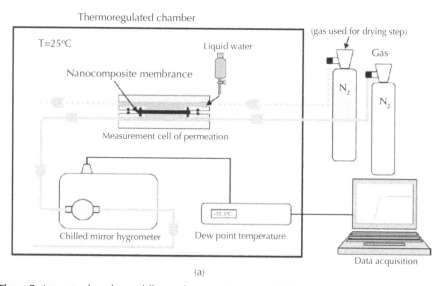

(a)

Figure 7. Apparatus based on a differential permeation method (Alexandre, 2009; Marais, 1999).

Both Water Sorption and Permeation

Different approaches were studied for obtaining both water sorption coefficient and water permeability coefficient. For that, some authors cut rectangular specimens of polymer films which are dried in vacuum oven at 50°C and their weight was measured until reaching no weight loss change. The films were immersed in a deionized water bath at 37°C. All specimens were weighed as a function of immersion time until the sorption process was complete. Then, the mass uptake at time t, and water solubility were calculated and thereafter water diffusivity, sorption and permeability coefficients.

Vacuum Quartz Microbalance

The water diffusion and permeability coefficients can be measured from a two-compartment cell by using a vacuum quartz spring microbalance techniques at controlled temperature. Iordanskii and co-workers have described the water resistance–PHB concentration dependence with this technique for PHB/LDPE blends made by melting in a single-screw extruder (Iordanskii, 1999). The effect of hydrophilic groups on water permeability was demonstrated. By virtue of limited water solubility in such polymers, water diffusion in moderately hydrophobic polymers is a structurally sensitive process. The authors estimated that the crystalline structure of PHB led to the decrease of water diffusivity.

Vapor Transmission Rate (Vpt) through Polymer Film

Some authors measured the vapor permeability by using a gravimetric cell (Miguel, 1997) which is basically a small container partially filled with diffusing liquid molecules, and with a polymer film sealing its top. The permeation process is reflected as a reduction in the overall weight of the cell. When the diffusing liquid molecules come in contact with the film (the gravimetric cell is placed downward) liquid permeation can be measured. This technique can be used when transport properties against vapors of diffusing molecules that are liquids at normal pressures and temperatures are tested.

For example, with PHB film obtained by 3% chloroform casting method, Miguel and co-workers (Miguel, 1999) have followed the weight change by using a computer-connected Sartorius analytical balance with a sensitivity of 10^{-5} g. Two different permeability coefficients have been calculated depending on the units in which the concentration gradient of the diffusing molecules is expressed. In both cases, the calculated values took into account the exposed area and the partial water vapor pressure difference across the two sides of the film. The first coefficient that is the permeability, expressed in Barrer, is calculated from the steady-state slope of the permeation measurement. The second coefficient corresponds to vapor transmission rate coefficient (VTR, expressed in g cm/cm^2 s) related the driving force of the permeation process to diffusing molecules activity. The VTR coefficient is proportional to the actual flux of the diffusing molecules passing through the film.

PHYSICAL PROPERTIES

The purpose of this section is to compare the properties of PHA polymers with those of polyesters commercially available. In fact, the thermal properties of materials are important for processing and also during the use of the products derived from these materials as well as the physical and mechanical properties of these polymers in order to design novel eco-friendly films. The properties of some biodegradable polymers are gathered in Table 1 and data available for two most common polymers (LDPE, PP) are added for comparison.

Due to its high degree of crystallinity of 60–80%, linear polyester PHB is a stiff material of high tensile strength and especially pronounced brittleness, although excellent mechanical strength and tensile modulus are present. It can be seen that the properties of PHB are rather close to those of polypropylene, outperforming polyethylene in some parameters. The primary negative importance of PHB is represented by its very low deformation at break related to its brittleness and low toughness. These serious drawbacks have hindered potential applications of PHB. The reason for the PHB brittleness comes from large volume-filled crystals in the form of spherulites from few nuclei due to their high purity. This stiffness is reduced by introducing hydroxyvalerate units. As observed in Table 1, the mechanical properties of PHBV were enhanced, whatever the hydroxyvalerate units used. In addition, these PHB copolymers showed an improved processing behavior (Table 1), provides the possibility for technical use in many applications. Copolymers then exhibited properties much closer to those of LDPE.

Interesting information can be obtained from the comparison with properties of various commercial biodegradable polyesters such as PLA (bio-resource based polyester), and PCL, PEA, PBSA, PBAT polymers (petroleum-based polyesters) given in Table 1. The density of all these polyesters is similar and above unity. Concerning the glass transition temperature (Tg), most polyesters, including PHB copolymers, exhibit values below the room temperature explaining the rubbery state of these polymers. An exception can be made for PLA which presents Tg value higher than the room temperature and thus leading to a glassy state to common temperature. That is why, polyesters can be considered as replacement of materials made from polyethylene and PLA for applications involving polystyrene. In addition, the melting temperatures of polyesters are within a reasonable range allowing a melt transformation of polyesters in order to tailor film materials. We can also note that the value of elongation at break for polyester materials is in relation with the glass transition temperature one: the highest value is obtained for the lowest glass transition temperature as exhibited by PBAT and PBSA polyesters in rubbery state. The same comment can be made for PCL polymer.

As observed in Table 1 and in the literature, the properties of PHB copolymers can be adjusted by varying the hydroxyvalerate unit content. An increase of this content resulted in a decrease of the melt and glass transition temperatures, of the tensile strength and of the crystallinity. The PHB copolymers are in general much more ductile and elastic than PHB. The crystallinity value for PHB copolymers is thus within a reasonable range common to thermoplastic polymers. Correlated with the crystallinity, the decrease of the glass transition temperature involved an increase of the elongation at break, even if the temperature value did not vary to a large extent. These results are in relation with the morphology based on a co-crystallization between hydroxyvalerate and hydroxybutyrate units which takes place inside PHB copolymers with a slow crystallization rate involving thinner crystal lamellae than in PHB.

Table 1. Physical and mechanical properties of some conventional polymers and some biodegradable polyesters.

	Density	Melting temperature (°C)	Glass transition temperature (°C)	Crystallinity (%)	Young's modulus (MPa)	Tensile strength at break (MPa)	Elongation at break (%)
PHB	1.25	175	4	60	3.5	0	5
PHBV 7 mol% HV Monsanto - Biopol D400G	1.25	153	5	51	900	–	15
PHBV 13 mol% HV	–	157.3	0.3	–	1186	25	10
PHBV 20 mol% HV	–	145	–1	–	0.8	20	50
LDPE	0.92	110	–30	50	0.2	10	600
PP	0.91	176	–10	50	1.5	38	400

Table 1. *(Continued)*

	Density	Melting temperature (°C)	Glass transition temperature (°C)	Crystallinity (%)	Young's modulus (MPa)	Tensile strength at break (MPa)	Elongation at break (%)
PLA Dow-Cargill - Natureworks	1.25	152	58	0-2	2000	–	9
PCL Solvay – CAPA680	1.11	65	–61	67	190	14	> 500
PCL Solvay – CAPA6500	–	59	–	41	383	36.1	668
PEA Bayer - BAK1095	1.07	112	–29	33	262	17	420
PBSA Showa - Bionolle3000	1.23	114	–45	41	249	19	> 500
PBAT Eastman - Eastar Bio 14766	1.21	110-115	–30	20-35	52	9	> 500

Source: Data extracted to Sudesh, 2000; Averous, 2004; Bardwaj, 2006; Duquesne, 2007.

TRANSPORT PROPERTIES

The objective of this section is to describe the barrier properties of PHA polymers compared to common polymers and some commercially available polyesters. To achieve this objective, the transport properties of the polymer films have been performed with diffusing molecules either at liquid state for sorption measurements or at gas and liquid state for permeation measurements.

Generally, water and carbon dioxide molecules, subjects of great importance in the field of packaging, are considered in the case of sorption measurements because polyesters contain ester groups on the backbone structure which can interact with water or carbon dioxide molecules.

For permeation measurements, diffusing molecules are classified as a function of the difference in their kinetic diameter and their interaction capacity: nitrogen for its chemical inertia, carbon dioxide, dioxygen for their molecule diameter, and water for its ability to interact with the polymer. The permeation mechanisms depend on the chemical nature of the polymer and the diffusing molecules characteristics, explaining why barrier properties are classified according to the type of diffusing molecules. Most important external parameters are temperature and humidity affecting the behavior and the structure of both the polymer and the diffusing molecule. Therefore, it is of prime importance to know in which conditions tests have been carried out.

Water Vapor Sorption

Even though results from different research teams are difficult to compare due to the differences in methods and measuring conditions, data shown in Figure 8 (data recalculated from original papers (Gouanve, 2007; Guptaa, 2007; Miguel, 1999a; Oliveira,

2006)) exhibit the following tendencies. There was no real significant difference in water sorption behavior between the PHBV copolymers. It can be stated that water sorption is practically independent of the hydroxyvalerate unit contents in the composition range studied in literature. The sorption results were close in magnitude of those of PLLA, PLA, and PCL polymers, for example. We can observe a slight reduction in water concentration compared with PET and PLA results. Of course, the water concentration of PLA, PCL, and PHBV polymers were larger than that of polypropylene since polyesters contain ester groups able to interact with water molecules. Indeed, these water sorption uptakes are the consequence of the affinity between the hydroxyl groups of the water molecules and the ester groups of the backbone even if the observed values are quite low. The small hydrophilic nature and the semi-crystalline structure are primarily the reason for this water like behavior. Compared to PET fiber or PLA polymers, the water sorption capacity of PHBV polymers is slightly reduced due to a higher crystallinity degree.

Although, PHB and PHBV polymers are generally considered as hydrophobic polymers according to Iordanskii and co-workers (Iordanskii, 1999), it seems that bacterial polymers slightly interacts with water molecules explaining the water sorbed concentration inside the film. This hydrophilic character is a function of the ratio between dispersive interactions (hydrophobic effect) and hydrophilic interactions (polar and electrostatic effects) and depends on the morphology and degree of orientation of the anisotropic units inside film. The presence of imperfect crystallites inside polymer may result in the formation of sites (essentially accessible carbonyl moieties) for the absorption of water molecules.

At higher water activities, the large water sorption uptakes are usually ascribed to the clustering of the water molecules inside the polymer to form aggregates because of the predominance of water-water hydrogen interactions over water-polymer hydrogen interactions.

The dependence of the ambient humidity on PHBV polymers structure, even if the impact is small, can dramatically affect the thermal and mechanical properties. This changing behavior is commonly observed for natural polymers such as cellulose and starch, and for natural polyesters such as PLA and PCL.

The CO_2 sorption was investigated gravimetrically and the behavior proved to be linear for all PHBV polymers studied according to Miguel and co-workers (Miguel, 1999b). The low sorption observed indicated that the PHBV polymers have good barrier properties mainly due to the high degree of crystallinity in this family of bio-resource based polymers. The authors mentioned a certain susceptibility of these polymers with respect to CO_2 reflecting some kind of structural rearrangement induced by diffusing molecules. In this case too, the values were comparable with those of common polymers.

The CO_2 and water sorptions induce similar trend for PHBV polymers due to the affinity of the backbone structure and the crystallinity degree exhibited. A major challenge for the material manufacturing concerns the resistance to moisture conditions. However, when comparing the water vapor resistance of various PHBV polymers to materials based on mineral oil, it becomes clear that it is possible to produce bio-based materials with water vapor resistance comparable to the ones provided by some conventional plastics. On the basis of these sorption results, PHBV polymers are promising materials for disposable packaging applications as it is also biodegradable.

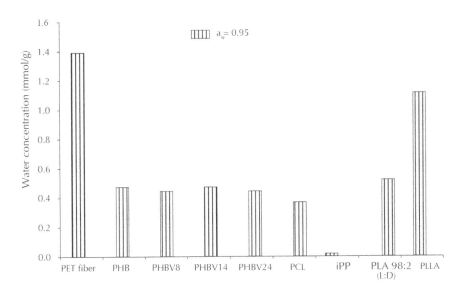

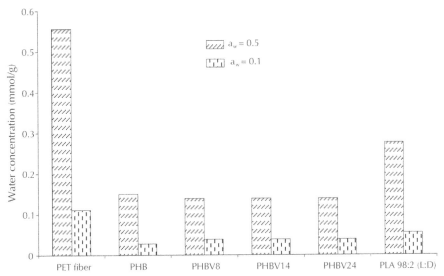

Figure 8. Water Vapor Sorption of PHBV Polymers Reported at Three Water Activities Compared to Common Polymers and Others Polyesters At 25°C.

Recalculated from the original papers.

Source: Miguel, 1999a; Oliveira, 2006; Gouanve, 2007; Guptaa, 2007.

Note: PHBVX correspond to PHBV with X mol % of hydroxyvalerate units.

Water and Gas Permeation

The permeation mechanisms depend mainly on the chemical nature of the polymer backbone and the characteristics of the diffusing molecules. It is well-known that diffusing gas molecules, except for CO_2, have no, or at least very few, interactions with polymer materials. It is important to know the conditions of permeation measurements carried out however this information is unfortunately often forgotten in the literature. Added to this fact, few data about permeation in literature are available as well as in great diversity of units which makes it difficult to establish comparisons.

Some selected data are presented in Table 2 with some common polymers.

Table 2. Gas Permeability of Common Polymers Available for Packaging Applications.

	CO_2 P (barrer)	O_2 P (barrer)	N_2 P (barrer)	H_2O P (barrer)
Linear PLA 96:4 (25°C) Cargill–Dow Polymers Crystallinity 1,5%	9.8	3.15	1.3	–
Linear PLA 98:2 (25°C) Cargill–Dow Polymers Crystallinity 2,5%	9.6	3.1	1.2	–
Linear PLA 96:4 (30°C) Cargill–Dow Polymers	10.2	3.3	1.3	6200
PET (30°C)	0.2	0.04	0.008	130
LDPE (30°C)	28.0	6.9	1.9	72
PS (30°C)	10.5	2.6	2.2	–
PP (30°C)	2.9	–	–	51
PHB (25°C) Biomer P209 Crystallinity 40%	9.3	1.4	0.74	4743
PHB (25°C) Biomer P226 Crystallinity 45%	2.3	0.5	0.13	2544
PHBV (25°C) Tianan Y1000P Crystallinity 57%	0.15	0.1	0.02	897

Data depended on temperature tested.

Source: Williams, 1969; Rogers, 1985; Miguel, 1997; Lehermeier, 2001; Follain, 2010.

Concerning gas permeation measurements, although PET, PLA, and PHBV copolymers are both polyester backbones, PET contains aromatic rings in the chain structure which reduces free volume and chain mobility. This results in the lower permeation for PET. The main difference between PLA and PHB polymers is based on the crystal organization inside material and the crystallinity degree which is found higher for PHB copolymers. This results in the lower permeation for PHB polymers. The measurements showed that crystallinity significantly impacts permeation process. The crystalline domains are thought to alter the diffusion pathway of diffusing molecules since the permeation is assumed not to occur in and through a crystal (Tsujita, 2003) involving the improvement of the barrier properties of the materials. The barrier properties are based on tortuosity concept which is one of the most important parameters that increases molecule migration pathways and consequently limits permeability. Compared to polymers with fossil origin, the barrier properties are increased due primarily to tortuous concept linked to presence of crystalline domains.

Concerning water permeation measurements, the presence of ester groups in backbone structure induced water affinity leading to an increase of permeability coefficient, as observed with PLA. The PHA are thermoplastics with high degree of crystallinity. The crystalline structure affected the transport properties of PHA films directly. It provides valuable informations for interpreting the water vapor permeation properties. The water vapor permeability is thus related to the chemical structure and also to the morphology of the PHA films. The higher hydrophilic nature of polymer is enhanced, the higher the water resistance is reduced. The water affinity of PHBV polymers helps the diffusion of water molecules within the film due to water plasticization effect. But compared to bio-based polyester such as PLA, the permeabilities are lower.

With moderate hydrophilic character such as PHBV copolymers, the water barrier properties are maintained close to values corresponding to medium barrier polymer for water. In particular, the permeation properties of PHBV copolymers reveal that they could serve as useful packaging material. The PHB copolymers showed the medium barrier properties with polar molecules and the highest with polar ones.

We can observe that the same behavior is observed with the three diffusing gas molecules and the PHB films are more permeable to CO_2 than O_2 and N_2 gas. This result is usually obtained whatever the polymer studied (Tsujita, 2003). The general behavior in permeation is maintained with bacterial based polymers.

In conclusion, the barrier effects towards gas and liquid molecules are mainly attributed to the tortuosity pathway through the polymer (due to the crystalline phase) independently to the hydrophilic character of polymer backbone structure. However, comparisons between different bio-based materials are complicated and sometimes not possible due to the use of different types of equipments and dissimilar conditions for the measurements.

CONCLUSION

In this work, the relationship between structure, physical and transport properties of bio-resources polymers obtained from bacteria was analyzed, showing the high potential presented by these eco-friendly polyesters. Currently, only three prominent polymers in PHA family [PHB, poly(3-hydroxybutyrate-co-3-hydroxyvalerate) and poly(3-hydroxyhexanoate-co-3-hydroxyoctanoate)] were produced to a relatively high concentration with high productivity. Because of its renewable character, its biodegradability, its low water affinity which is arguably the main advantages of PHA family, PHBV polymers can be considered in many applications by replacement of conventional polymers. In fact, PHBV copolymers exhibited properties much closer to those of polyethylene LDPE or polypropylene but their availability and price still can represent a hindrance for this family to be considered as serious competitors against common polyolefins. The resistances to water, gas, and high temperature of PHBV are considered as excellent. Bacterially synthesized PHBV became attractive for applications in the biomedical materials and environmental friendly materials. The PHBV materials are able to mimic the water vapor barriers of the conventional materials known today, necessary properties for this type of applications.

KEYWORDS

- **Crystallinity value**
- **Dual-mode sorption**
- **Hydroxyvalerate units**
- **Macromolecules**
- **Vacuum quartz**

ACKNOWLEDGMENTS

The authors are thankful to Dr. Corinne Chappey (UMR 6270 CNRS, Rouen, France) for the technical insight and support in gas permeation experiments.

Chapter 7

Silk Fibers and their Unidirectional Polymer Composites

Mansur Ahmed, Md Saiful Islam, Qumrul Ahsan, and Md Mainul Islam

---

INTRODUCTION

Natural fiber reinforced polymer composites have gained their popularity in the composite research because of versatility and diversified nature of their applications (Gay et al., 2003). This is due to a range of potential advantages of natural fibers, especially with regard to their environmental performance. Natural fibers are renewable resources and even when their composite wastes are incinerated, they do not cause net emission of carbon dioxide to the environment. They are inherently biodegradable, which may be beneficial (Aquino et al., 2007; Singleton et al., 2003; Verpoest et al., 2010). Among all the natural reinforcing fibrous materials, silk appears to be a promising fiber due to its high toughness and aspect ratio in comparison with other natural reinforcements. Moreover, the mechanical properties of silk fibers consist of a combination of high strength, extensibility, and compressibility. On the other hand, maleic anhydride grafted polypropylene (MAPP) and commercial grade polypropylene (PP) are commonly used matrices for fiber-reinforced composites due to their good interfacial bonding properties (Karmarkar et al., 2007; Wambua et al., 2003).

In this chapter, both raw and alkali treated silk fibers are taken into consideration to characterize their properties first. The specific objective is to determine the modulus and strain to failure of the single silk fiber. Other aims of this work are to fabricate the unidirectional silk fiber reinforced polymer (MAPP and commercial grade PP) matrix composites by varying fiber volume fraction and to determine their mechanical properties such as tensile strength, flexure strength, impact strength, and hardness.

MATERIALS AND METHODOLOGY

Collection of Silk Fiber and Polypropylene Resin

Silk fibers were collected from Sapura Silk Industries, Rajshahi, Bangladesh. Two types of silk—raw silk and alkali treated silk fibers were collected. Commercial grade PP resin was collected from a local chemical manufacturer and distributor. The MAPP was collected from Belgium.

Specimen Preparation for Tensile Testing of Fibers

At first, single silk fibers (both raw and alkali treated silk) were chosen randomly, which were cut down a particular length at prior. The diameters of single fibers were measured using Scanning Electron Microscope (SEM). The fibers were stacked between two-paper frame, as shown in Figure 1, to conform good gripping to the clamps of test machine and to provide straight direction during test. This paper frame was clamped in the machine jaws and cut the paper frame carefully before starting the test. The crosshead speed maintained for silk fiber was 3 mm/min and a 5 N load cell was used. A small tensile testing machine was used for testing with various span lengths of 5, 15, 25, and 35 mm.

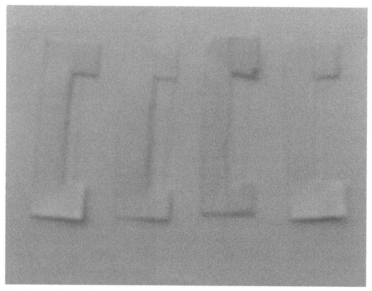

Figure 1. Specimens for tensile test.

Tensile strength is calculated by following formula:

$$\sigma = F_{max}/A \tag{1}$$

where,

σ = Tensile strength

F_{max} = Maximum force

A = Cross sectional area

Cross sectional area, A was measured by using this formula:

$$A = \pi(d/2)^2 \qquad \text{(For jute fibers (Mitra et al., 1998))} \tag{2}$$

where, d is diameter.

Young's modulus, E was measured from the stress-strain curve.

Fabrication of Composites

Four types of composites were manufactured. Unidirectional silk fiber composites were made using PP and MAPP matrix. By varying fiber volume fraction (30, 35, 40, and 45%), 2 mm thick composites were manufactured.

Steps for manufacture are given below:

(a) At first, alkali treated silk fiber was weighted according to the required volume fraction needed. The required amount of weight of silk fiber was weighted.

(b) Sufficient amount of commercial PP was taken in a beaker and weighted. The 20% MAPP was added with it. To prevent voids, water bubbles, poor fiber matrix adhesion, the PP was dried in an oven at about 100°C for 3–4 hr.

(c) Mold surface was cleaned very carefully and Teflon sheet was placed over the mold surface properly for the easy removal of the product.

(d) A layer of PP granule with MAPP was placed in a female mould and 2 mm thick die was placed surrounding the granule.

(e) Then one layer of continuous fibers was aligned parallel to each other onto PP granule using glue at the edges of fiber (see Figure 2) and another layer of fibers was placed over prior layer. And so on to produce as required fiber volume fraction.

(f) Finally, another layer of mixture of PP and MAPP was placed onto silk fiber layer.

(g) The female mould with aligned fiber with PP was covered by a male mould with Teflon sheet.

(h) Plates were placed in hot pressing machine, 185°C temperature and 85 kN pressure was applied simultaneously for about 15 min and then cooled slowly using water-cooling system.

(i) At last, the specimen was carefully discharged from the mould.

(a) PP Layer   (b) Fiber Alignment   (c) PP Layer on Fiber

Figure 2. Unidirectional silk fiber reinforced composite fabrication process.

DISCUSSION AND RESULTS

Tensile Testing of Fibers

Stress-strain curves of raw and alkali treated silk fibers (25 mm span) are shown in Figure 3. Also, uncorrected and corrected extrapolated curves (1/span vs. modulus, strain to failure and span vs. tensile strength) for the spans of 5, 15, 25, and 35 mm of raw and alkali treated silk fibers are shown in Figures 4–6. It seems that with an increase of span length, Young's modulus increased for uncorrected curves. Some slippage portion occurs during tensile testing (which is called machine constant and denoted by (α). Larger span length helps to minimize the slippage portion (α) compared to smaller one as shown in Figure 7. Therefore, modulus of longer span is higher and strain to failure is lower than smaller span. Total slippage portion of the machine can be calculated by using the following equations:

$$\Delta L_{total} = \Delta L_{fiber} + \Delta L_{grip} \tag{3}$$

$$\Delta L_{fiber} = \sigma \, Span/E_0 \Delta L_{grip} = \alpha.(A.\sigma) \tag{4}$$

$$\alpha = (\Delta L_{total} - \Delta L_{fiber})/(A \, \sigma) \tag{5}$$

Corrected modulus and strain to failure can be calculated by the following equations:

$$\alpha_i = \Delta L_{total}/F - L_0/E_0.A_i \tag{6}$$

$$\Delta L_{total}/F = \varepsilon \, L_0/\sigma \, A_i = 1/E \, L_0/A_i \tag{7}$$

Equations for strain correction:

$$\Delta L_{grip}/L_0 = \alpha_1 \, (A_i\sigma)/L_0 \tag{8}$$

$$\Delta L_{fiber}/L_0(Corrected) = (\Delta L_{total}/L_0 - \Delta L_{grip}/L_0) \tag{9}$$

where,

α_i = Machine constant for each fiber

L_0 = Original span length

E = Young's modulus (elastic modulus or E-modulus) for each fiber

E_0 = Extrapolated modulus

A_i = Cross-sectional area for each fiber

F = Force

ε = Strain

σ = Stress

On the other hand, tensile strength and strain to failure decreased with an increase of span length. As mentioned by Bledzki and Gassan (1999), the longer the stressed distance of the natural fiber, the more in homogeneity will be in the stressed fiber segment, weakening the structure. Thus, the strength decreased with fiber length. Tensile properties of silk fibers are summarized in Table 1.

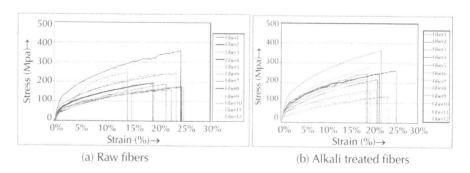

(a) Raw fibers (b) Alkali treated fibers

Figure 3. Stress-strain curves of silk fibers from tensile testing for 25 mm span length.

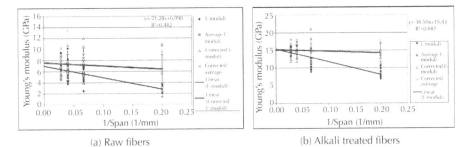

(a) Raw fibers   (b) Alkali treated fibers

Figure 4. Young's modulus vs. 1/Span for silk fibers.

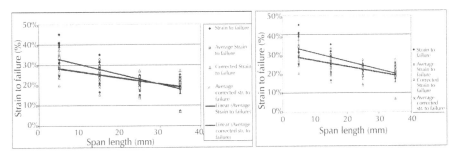

Figure 5. Strain to failure vs. span of silk fibers.

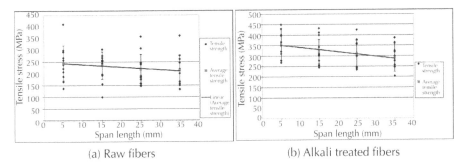

(a) Raw fibers   (b) Alkali treated fibers

Figure 6. Tensile stress vs. span for silk fibers.

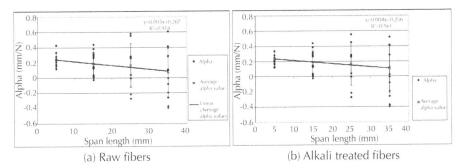

(a) Raw fibers   (b) Alkali treated fibers

Figure 7. Alpha (α) vs. Span curve for silk fibers.

Table 1. Tensile properties of silk fiber.

Fiber types	Average Tensile Stress (MPa)	Average E-modulus (GPa)	Average strain (%)
Raw silk	227	9.56	23.53
Alkali Treated silk	319	15.45	25.67

Surface Morphology of Fibers

Surface morphology of raw and alkali treated silk fibers was observed by using SEM. It is shown in Figure 8. Raw silk surface contains fibroin and siricin (Figure 8(a)). Siricin is a hard and gummy matter that binds fibroin together. Fibroin gives the fiber strength. Fiber defects can be seen in Figure 8(b), which reduce the fiber strength. Silk fibers after treated with alkali show rather smooth surface and reduction in diameter as compared to raw fibers. Alkali treated silk fiber contains small amount siricin than raw silk. For that reason, alkali treated silk fibers give better properties than raw silk fibers.

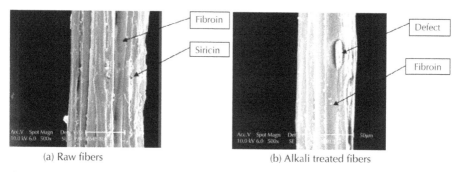

(a) Raw fibers   (b) Alkali treated fibers

Figure 8. Surface morphology of silk fibers (500X).

Thermo-gravimetric Analysis (TGA) of Silk Fibers

From TGA curves of raw and alkali treated silk fibers (Figure 9), it is seen that raw silk fiber is stable up to 281°C and alkali treated silk fiber is stable up to 265°C.

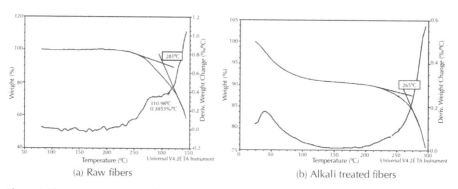

(a) Raw fibers   (b) Alkali treated fibers

Figure 9. Thermo-gravimetric analysis (TGA) curves of silk fibers.

Tensile Testing of Composites

Stress-strain curves of unidirectional fiber-reinforced composites under longitudinal loading are represented in Figure 10. It is seen from this Figure that ultimate tensile strength (UTS), modulus and percentage elongation are maximum for fiber volume fraction at highest end (45%), because the load applied to the fiber direction. Longitudinal strength is dominated by fiber strength. The stress-strain mode that operates for a specific composite will depend on fiber and matrix properties, and the nature and strength of the fiber-matrix interfacial bond. Figure 11 represents the comparison between theoretical and experimental strength and modulus; the experimental values are lower than theoretical values. This is due to the poor fiber-matrix interface and misalignment of continuous unidirectional fibers. Figure 12 shows that longitudinal strain increases with fiber volume fraction.

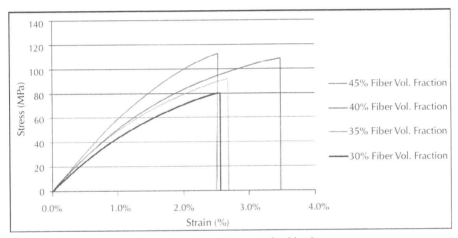

Figure 10. Stress-strain curve for composites under longitudinal loading.

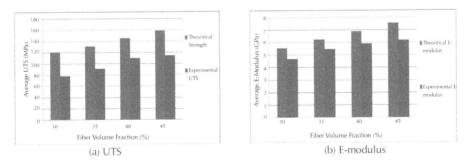

(a) UTS

(b) E-modulus

Figure 11. Comparison between theoretical and experimental longitudinal strength and modulus.

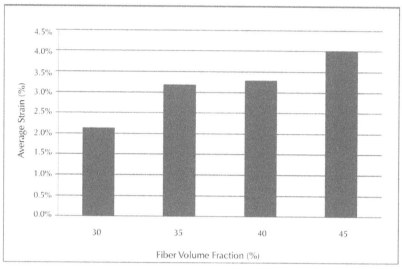

Figure 12. Comparison of longitudinal average strain.

Stress-strain curves of unidirectional fiber-reinforced composites under transverse loading are shown in Figure 13. With the increase of fiber volume fractions, UTS, modulus, and strain increases. A variety of factors will have a significant influence on the transverse strength; these factors include properties of the fiber and matrix, the fiber-matrix bond strength, and the presence of voids. Transverse UTS, modulus and strain is lower than longitudinal UTS, modulus and strain (Figure 14). Longitudinal strength depends upon the fiber strength and interfacial bonding between fiber and matrix. On the other hand, transverse strength depends upon the fiber-matrix interfacial strength.

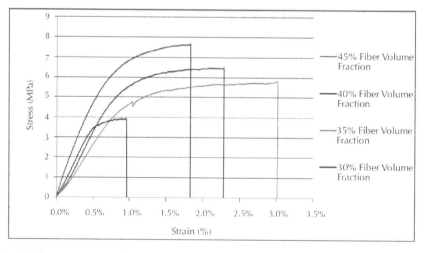

Figure 13. Stress-strain curve for composites under transverse loading.

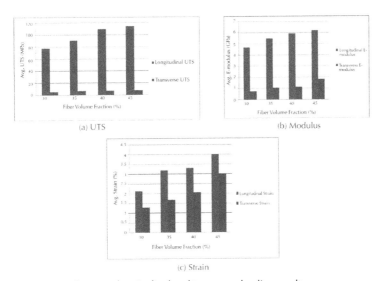

(a) UTS

(b) Modulus

(c) Strain

Figure 14. Comparison between longitudinal and transverse loading results.

Microscopic Observations of Tensile Specimens

The fracture surfaces of the tensile test specimens were examined by SEM. To study the surface morphology of the prepared composites, SEM photographs were taken for 30, 35, 40, and 45% fiber volume fraction, as shown in Figure 15. The SEM of fracture surface of composite indicates that fiber pullout occurs. This is due to lack of interfacial adhesion between silk and PP matrix. Due to intermolecular hydrogen bond formation between silk fibers and hydrophobic nature of PP matrix, hydrophilic silk fibers tend to agglomerate into bundles (Figure 15(d)), and become unevenly distributed throughout the matrix. It is seen that for 40% fiber volume fraction composite, less fiber pullout happens, better interfacial adhesion between silk and PP matrix occurs.

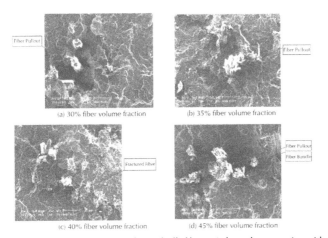

(a) 30% fiber volume fraction

(b) 35% fiber volume fraction

(c) 40% fiber volume fraction

(d) 45% fiber volume fraction

Figure 15. SEM micrographs of fracture surface of silk fiber-reinforced composite with various fiber volume fractions (200X).

Flexure Testing of Composites

Stress-strain curves achieved from flexure test for four volume fractions are shown in Figure 16. It is observed from the Figure that the values of flexural strength, modulus, and strain increase with the increase of fiber loading, as shown in Figure 17 by a comparison. Longitudinal strength depends upon the fiber strength and interfacial bonding between fiber and matrix. On the other hand, transverse strength depends upon only the fiber-matrix interfacial strength.

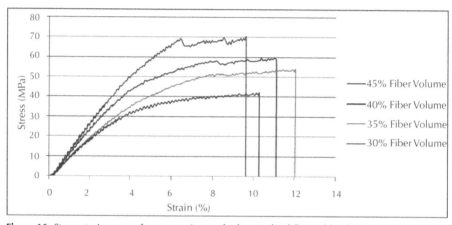

Figure 16. Stress-strain curves for composites under longitudinal flexural loading.

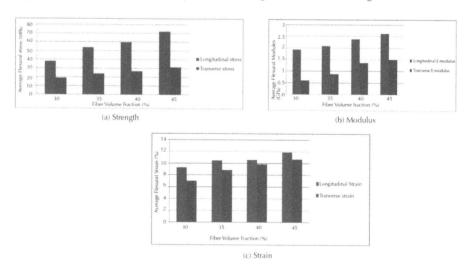

Figure 17. Comparison between longitudinal and transverse flexural test results.

Impact Testing of Composites

It is apparent from the test result presented in Figure 18 that the impact strength increases with the fiber loading. The impact strength is significantly more than that of PP polymer.

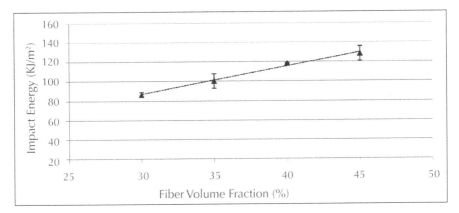

Figure 18. Variation of impact strength with fiber volume fractions of composites.

Hardness Testing of Composites

It can be noticed from the hardness test results presented in Figure 19 that the hardness increases with fiber loading. Hardness tests showed not very profound variation among all the composites.

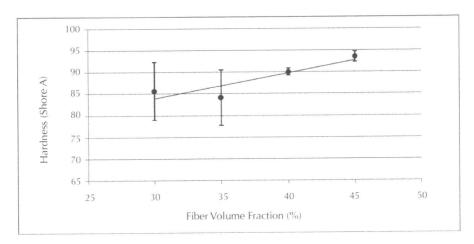

Figure 19. Variation of hardness with fiber volume fractions of composites.

CONCLUSION

The physical and mechanical properties of raw and alkali treated silk fibers were evaluated first. Based on the experimental results, it can be concluded that the Young's modulus of silk fibers increases when span length is increased. With an increase in the span length, tensile strength, strain to failure, and machine constant (α) decrease. In comparison of raw and alkali treated silk fibers, alkali treated ones give the higher Young's modulus and tensile strength.

Secondly, variation of properties by varying fiber volume fraction in the composites was observed. From the test results, it can also be concluded that composite containing 45% volume percentage shows higher strength than other lower fiber volume fraction composites. It was expected that with the increase of fiber volume fraction, strength of the composites will increase. As fiber contains inherent defects, strength of single fibers is not constant and poor adhesion exists between fiber and matrix; strength is less than theoretical strength. Hardness measurement shows not much variation in varying fiber volume fractions. Impact energy was greater in higher volume fraction composites as compared to lower volume fraction ones. In conclusion, it can be said that silk fibers should be incorporated in highly aligned manner and parallel to each other to get substantial mechanical properties. Fiber volume fraction may be increased keeping in mind that these fibers are not entangled, defect free and do not impair proper wetting with the matrix phase.

KEYWORDS

- **Longitudinal strength**
- **Natural fibers**
- **Polypropylene**
- **Scanning electron microscope**
- **Siricin**
- **Thermo-gravimetric analysis**
- **Ultimate tensile strength**

Chapter 8

Bionanocomposites for Multidimensional Brain Cell Signaling

Mark A. DeCoster, J. McNamara, K. Cotton, Dustin Green,
C. Jeyasankar, R. Idowu, K. Evans, Q. Xing, and Y. Lvov

INTRODUCTION

Cell communication in the brain is dynamic, including electrical and chemical components. These dynamics are utilized for maintenance of the highly complex communication pathways within and between cells of the brain, both in health and disease. Bionanocomposites allow us to study brain cell function conceptually from 0 to 3 dimensions (03D), with 0D representing the single cell, 1D representing a linear arrangement of two or more connected cells, 2D including cellular surface areas in both X- and Y-dimensions, and 3D taking into consideration tissue scaffolds that include X-, Y-, and Z-dimensions. These considerations are shown conceptually in Figure 1.

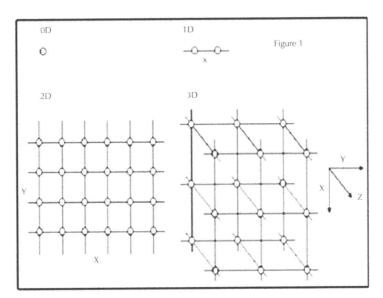

Figure 1. Visualization of cellular networks.

Using bionanocomposites, we are investigating how both normal brain cells and brain tumor cells are communicating in multiple dimensions, identifying biocompatible nanomaterials that can be used for experimental and modeling purposes.

Translational applications for this multidimensional approach include, for example, engineered environments for assessing drug delivery to cells and tissues. For the brain, we are also taking into consideration the added complexity of two physiological factors. A first consideration is the bloodbrain barrier (BBB), which prevents passive diffusion of most chemicals from the bloodstream into the brain (Weiss et al., 2009). For this, we are including rat brain microvascular endothelial cells (RBMVECs) in our engineered culture environments. Second, in the brain (once past the BBB), there is a major interplay between multiple cell types, for example, between neurons and glial cells.

We are considering the effect of the glial subtype, astrocytes on neuronal communication. Using a predator-prey concept (Baxendale and Greenwood, 2010), where astrocytes are the predator taking up glutamate, and neurons are the prey releasing glutamate, we are measuring the effect of astrocytes on intracellular calcium concentration($[Ca^{2+}]_i$)dynamics in these cultures. The ability to engineer the growth environments for both of these cell types will allow us to control both the ratio and position of these cell populations relative to each other, providing input control to the predator-prey model, backed up with experimental data. Using multiple nanotechnology approaches we have been able to define engineered environments that have established single glial cells (0D), connected networks of neurons, and networks of brain tumor cells (2D). In addition, we have recently used natural and bio-polymer composites of gelatin and pulp and paper fibers to form 3D micro- and macro-structures that guide and support the growth of brain tumor cells (Xing et al., 2010). We have identified at least 3 major benefits of these composites for tissue engineering in 3D: (1) the gelatin-cellulose composites are of sufficient porosity to allow for cell invasion and penetration of nutrients into the matrix; (2) these composites allow light transmission for phase- and fluorescence-microscopy; and (3) the composites are of sufficient strength to retain 3D structure for long-term *in vitro* studies, which in our case was up to 17 days *in vitro*.

From the standpoint of dynamic cell communication, we utilize fluorescent calcium indicators such as fluo-3 combined with digital imaging to monitor changes in both $[Ca^{2+}]_i$ in the individual cell and changes that are intercellular, such as the synchrony or asynchrony of $[Ca^{2+}]_i$ oscillations. We hypothesize that the ability to engineer the spatial arrangement of brain cells will aid in our analysis of these dynamics. We also anticipate that defining the spatial arrangement of normal and brain tumor cells will aid in image analysis of cell growth patterns with and without anti-cancer drug treatment.

METHODS

Nanofilm printing: Glass slides for nanofilm printing were prepared by sonication in deionized water using a Branson model 1510 sonicator for 30 min followed by sonication in 75% ethanol for 30 min. Slides were then removed, washed once in deionized water and blown dry using dry nitrogen. The newly cleaned slides were then silanized to increase covalent bonding of polymers using vapor deposition (Mandal et al., 2007). Silanization was accomplished by filling a micro centrifuge cap with 50 μl of (3-aminopropyl)-triethoxysilane (APTES; Sigma-Aldrich, USA) and then placing the filled cap in the center of a glass Pyrex dish. Cleaned glass slides were then positioned around the APTES filled cap and the Pyrex dish placed in an oven and baked at 60°C for 2 hr. After this step, APTES was removed from the dish, and the dish with slides was placed back in the oven and baked for 16 hr at 100°C. After baking, slides were allowed to cool to room temperature and then immersed in 6% gluteraldehyde in 1X phosphate buffered saline (PBS; Invitrogen) for 2 hr at room temperature. Slides were then washed with deionized water and blown dry using nitrogen.

For polymer printing using the Nanoenabler system (BioForceNanosciences, Inc. USA), fluorescein isothiocyanate (FITC)-labeled poly-L-lysine (PLL) was used (Sigma-Aldrich). To prevent premature drying, the FITC-PLL was mixed in a 1:1 ratio with a protein loading buffer (BioForceNanosciences). A stock solution of FITC-PLL was prepared consisting of 20 μl of FITC-PLL at 10 mg/ml with 20 μl of 1x PBS. This 40 μl of solution was then mixed with 40 μl of protein loading buffer, for a final FITC-PLL concentration of 2.5 mg/ml. The SPT-S-C30S cantilevers (BioForceNanosciences) were used for surface patterning. These cantilevers were equipped with micro-fluidic reservoirs for holding liquids. One microliter of FITC-PLL was used to fill the reservoir before printing.

[Ca²⁺]ᵢ Imaging: Cytoplasmic calcium was monitored using the InCytIm 1 fluorescentsystem(Intracellular Imaging, University of Cincinnati, OH) attached to a DVC 340 M12-bit camera after loading cells with fluo-3 (Invitrogen) as previously described (DeCoster et al., 2002).

Primary Neuronal Rat Cell Culture: The cortical rat neuronal cells were harvested by performing cervical disarticulation on 1-day-old rat pups. The cells were placed in 12- and 24-multiwell plates with a 1:1 coated surface of poly-L-lysine (Sigma Aldrich) and deionoized water (DI) to enhance cell affinity for attachment. The neuronal cells were cultured in Neuronal Culture Media (NCM) comprised of Fetal Bovine Serum (FBS), Horse Serum (HS), glucose solution, glutamine, antibiotics, Ham's F12-K (ATCC) and Basal Media Eagle's as previously described (Daniel and DeCoster, 2004).

Cell Culture of RBMVECs: The RBMVECs were purchased from Cell Applications Inc., San Diego, CA. The cells were grown on plates coated with Attachment Factor solution (Cell Applications Inc., San Diego, CA) which promoted their attachment and proliferation. The plates were coated with the Attachment Factor for 30 min at 37°C. The endothelial cells were cultured in Rat Brain Endothelial Growth Medium (Cell Applications Inc., San Diego, CA) fully supplemented with FBS, growth factors, trace elements, antibiotics and vascular endothelial growth factor according to the supplier's instructions. The cells were passaged when they reached 70% confluence, as they formed tight junctions at higher confluence and became difficult to trypsinate. Cells were used between passages 3 15.

Cell culture of glioma cells: Human glioma cells were obtained from ATCC (Manassas, VA, USA; ATCC number CRL-2020) and cultured in CRL-2020 complete media as indicated by the vendor.

Calcein staining and confocal imaging: Calcein staining for viable glioma cells in 3D biocomposites was carried out as previously described (Xing et al., 2010).

RESULTS AND DISCUSSION

Reflecting the conceptual framework of cellular networks as shown in Figure 1, we have also previously described nanotechnology methods combined with cell culture to demonstrate micropatterning of brain cells (Mohammed et al., 2004; Shaikh et al., 2006). Figure 2 shows the distinct morphologies of individual patterned brain astrocytes using new nanofilm printing methods (as described in Nanofilm printing methods section above).

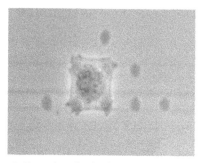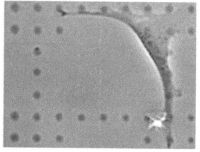

Figure 2. Examples of individual patterned brain astrocytes withdefined morphologies and positions engineered with the adhesive polymer poly-lysine (PLL) and the NanoEnabler molecular printer. The printed PLL spots are 10µm in diameter.

As shown in Figure 2, biocompatible nanofilms such as the charged polymer poly-lysine can be used to promote cell adhesion.

Using this system, we are studying the control of [Ca^{2+}]$_i$ oscillations in small networks of brain cell cultures with microscopy and intracellular fluorescent calcium indicators. As shown in Figure 3, these systems often have intrinsic oscillatory behavior which may be modulated by pharmacological intervention.

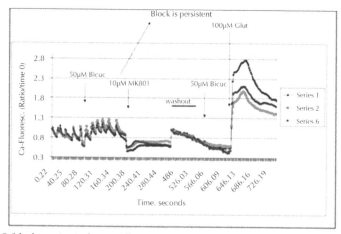

Figure 3. [Ca^{2+}]$_i$ dynamics in brain cell cultures treated in sequence with the inhibitory blocker bicuculline (50 µM), the glutamate receptor antagonist MK-801 (10 µM), bicuculline after washout (50 µM), and exogenous glutamate (100 µM), as indicated. Note break in time scale at washout. Series 1, 2, and 6 represent 3 individual neurons in the same dish.

Numerical analysis and mathematical techniques were utilized to identify $[Ca^{2+}]_i$ oscillations in mixed primary cultures of cortical neurons and astrocytes. We found that stimulation of cells with 50 mMKCl followed by glutamate induced synchronized and non-synchronized $[Ca^{2+}]_i$ oscillations. In contrast, addition of the gamma-aminobutyric acid (GABA) receptor antagonist bicuculline, converted spontaneously oscillating neurons into highly synchronized small networks (Figure 3). Since, GABA receptors are inhibitory, these results indicate that bicuculline is causing disinhibition, revealing more excitatory behavior, in this case, as revealed by more synchronized $[Ca^{2+}]_i$ oscillations.

Another level of control for the intact brain includes the blood–brain barrier (BBB). The BBB is a cellular structure that is composed of endothelial cells. The presence of tight junctions in the endothelial cells with support from astrocytic end feet makes the barrier impermeable to substances from the blood stream, thereby maintaining homeostasis (Ma et al., 2005). Many neurological diseases occur as a result of the dysfunction of the brain endothelium. The breakdown of the barrier may result in a vicious cycle of brain injury (Bowman et al., 1983). The endothelium lining the brain microvasculature is different from other vascular endothelia by the presence of tight junctions and limited permeability with a unique pattern of receptors, transporters, and ion pumps that protect the barrier from hydrophobic substances (Abbott, 2002). Adenine nucleotides like adenosine triphosphate (ATP), adenosine diphosphate (ADP), adenosine monophosphate (AMP) and adenosine mediate important signaling events like inflammatory pathways, neurotransmission, and energy-driven processes, causing calcium influx into the cell (Gordon et al., 1986). This calcium increase occurs in the form of a stream of calcium ions coming into the cell, called calcium waves. These waves provide long-range signaling and act as a control mechanism providing feedback to events like vasodilation. Hence, the study of these calcium waves on the microvasculature of the brain provides insight into the role it has on normal brain function, neurological diseases, and the effect it has on other brain cells (Janigro et al., 1994). As shown in Figure 4, we are utilizing rat brain microvascularendothelial cells (RBMVECs) *in vitro* which show intrinsic morphologies such as curved and circular areas which may reflect their function *in vivo* of making up blood vessels.

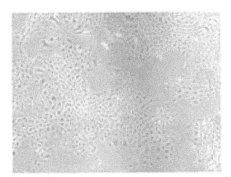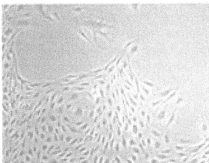

Figure 4. Rat brain microvascular endothelial cells demonstrating differentiation *in vitro*. Left panel shows cells at original magnification of 40x. Right panel shows a zoomed area of the left panel, original magnification= 100x.

We anticipate that engineered environments for cell patterning to combine RBM-VECs, astrocytes, and neurons will provide a better model and experimental testbed for multidimensional analysis of brain cell signaling dynamics.

Finally, as shown in Figure 5, and as we have recently described (Xing et al., 2010), biocomposites may be used in three dimensional matrices to grow brain cells over extended periods of time.

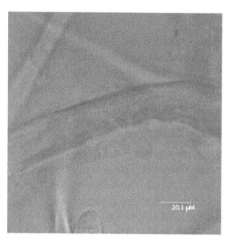 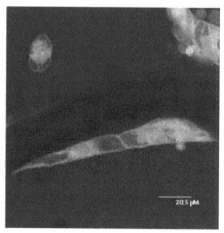

Figure 5. Confocal microscopy images of brain tumor cells grown in 3D matrices of gelatin-cellulose nanocomposites from wood fibers. Left panel shows the white light phase image of the 3D matrix with nanocomposite fibers and cells visible. Right panel shows fluorescent image of the same field with cells stained with the vital dye calcein. Scale bar indicates 20.55 microns.

Thus far we have utilized CRL-2020 brain tumor cells to test these 3D matrices, since they have intrinsic invasive behavior and high growth potential (Sehgal et al., 1999). In their own right, these 3D matrices may provide important growth environments for modeling brain tumor growth and for testing the efficacy of drug delivery and diffusion in all dimensions. We chose cellulose-derived fibers from wood as an important component of our developed 3D matrices due to their intrinsic aspect ratio that includes diameters in the 1020 micron range, which may be ideal for binding and guiding brain cell growth. Furthermore, although we have not tested the idea, these cellulose fibers do have intrinsic lumen structures which might help retain nutrients permissive for developing cells in three dimensions.

CONCLUSION

Putting together the different dimensions of engineered growth environments with $[Ca^{2+}]_i$ dynamics experiments and modeling has benefitted from bionanocomposites such as gelatin-cellulose matrices shown in Figure 5, and from printed nanofilms as shown in Figure 2. By controlling the growth environments of cells of the brain and by varying which cells are present, we may better understand and model brain cell communication in both health and disease. For example, the release of glutamate from glioma cells in brain tumors may alter neuronal network communication and lead to more glutamate release, exacerbating neuronal injury and favoring glioma proliferation as has been previously described (Takano et al., 2001). In our model systems, a 3D matrix such as shown in Figure 5 may be useful to elucidate the effect(s) of brain tumor cells on normal brain cell function. Furthermore, in a predator-prey concept, we suggest that excess glutamate release from glioma cells may "swamp" the normal balance of astrocyte glutamate uptake and removal following normal neuronal glutamate release. This is exacerbated by the impaired uptake of glutamate by glioma cells (Ye and Sontheimer, 1999) and by indications that glutamate may serve as an autocrine factor for glioma cell growth (Lyons et al., 2007). Thus, in the three dimensional environment of the brain during tumor growth, the normal predator-prey balance between astrocytes and neurons is altered by growing glioma cells which take up space and alter the glutamate load on the system. A better understanding of these dynamics may lead to better anti-cancer approaches.

While, much work is being done on cellular calcium modeling (see, for example, (Falcke, 2004); (Guisoni and de Oliveira, 2005); (Means et al., 2006); and (Patterson et al., 2007)), little effort is focused on single compartment calcium models of brain cells in 0 (single cell), 1, 2, or 3 dimensions as we suggest in Figure 1, validated against fluorescence data. The work in Means et al. (2006) has a similar flavor to that presented here in that a 2D network, or lattice, is constructed to examine calcium dynamics. However, it must be noted that this lattice, which is made up of two sub-lattices, one consisting of calcium channels and the other of calcium ions, is used to represent the calcium exchange of the endoplasmic reticulum membrane of a single cell, not a cellular network.

Single compartment models are desirable in this work, because any interaction between the mathematical models and the corresponding experimental systems should be based on the same known quantities. It would not be reasonable to include variables in the models that either cannot be determined from separate equations or cannot be provided as an input from the available experimental data. To check and validate models, we are using microscopy and imaging techniques to capture dynamic calcium events taking place within a cell as a whole, as opposed to calcium diffusion across a cell or intracellular calcium waves, for example (Patterson et al., 2007).

The work in Ventura et al. (2006), which does focus on single compartmental modeling of a single cell, is of particular relevance to that described here. To address the inherent difficulty of numerous unknown parameters arising in traditional intracellular calcium modeling, Ventura et al. (2006), present a data-driven approach where models are simplified based on reasonable assumptions. The mathematical equations are s implied but not abandoned in their entirety, which is desirable due to our other research interests related to control and estimation techniques. Results indicate that the model does reveal quantitative characteristics like amplitude and kinetics of a calcium signal whenever calcium enters through "one or several distinguishable spatially localized regions (channels or clusters of channels)." This requirement of the calcium passing through particular channels is problematic from the perspective that available fluorescence data does not provide this level of detail for contribution to the mathematical model. Still, the positive results from Ventura et al. (2006), indicate that a simplified approach should not be overlooked for model consideration. Having multidimensional tools (including biocomposites), for placing brain cells into defined networks, may help in our analysis and modeling of intracellular calcium dynamics.

KEYWORDS

- **Bionanocomposites**
- **Deionized**
- **Microvasculature**
- **Neurological diseases**
- **Predator-prey concept**

ACKNOWLEDGMENTS

This work was supported by the Louisiana Board of Regents PKSFI contract no. LE-QSF (2007-12)-ENH-PKSFI-PRS-04 and NSF Awards #0701491 and 1032176.

Chapter 9

Salix Viminalis as a Source of Nanomaterials and Bioactive Natural Substances

A. Cyganiuk, A. Olejniczak, A. Kucinska, R. Klimkiewicz,
J. P. Lukaszewicz

INTRODUCTION

Plant *Salix viminalis* (Figure 1) belongs to the group of so called short-rotation coppices agriculturally cultivated as a renewable source of energy (Borjesson et al., 1997).

Figure 1. Cultivating of *Salix viminalis* in Poland.

Salix viminalis is also applied for purification of soil and water as an effective phytoremediator, that is a plant which easily accumulates high quantities of heavy metal ions through capillary action involving roots and transporting tissues in stem. The absorbed metal ions are accumulated in the plant cells which were proved in several studies analysing metal ion content in different parts of the plant (Lukaszewicz et al, 2009; Mleczek et al, 2009a, 2009b). The use of *Salix viminalis* for other purposes, beside energetic use and/or phytoremediation of waters and soil, is a unique case. Besides the expected and proven high tolerance to heavy metal ions, *Salix viminalis* grows very fast in the mild climate zone yielding hard wood in high quantities from 1 ha (Labrecque et al., 1997). In light of the above statements, *Salix viminalis* seems to be a valuable but underestimated raw material for some fabrication processes. The high agricultural yield of *Salix viminalis* (Labrecque et al., 1997) will reduce considerably the cost of products obtained from this source if one would like to expand the currently described methods to industrial scale.

Contemporary materials science puts in picture the on numerous materials of unique properties potentially applicable in many areas of industry and science. However, most of them are materials which are hardly accessible in larger amount which results either from complexity of processing and/or cost of material fabrication. For example, template synthesis of porous carbons and silica (Ryoo et al., 2001; Schlottig et al., 1999) yields adsorbents of steerable pore dimension (pore size distribution function PSD) and volume but transfer of such a technology to mass scale seems to be rather questionable. Similar limitations may be recalled in the case of carbon nanotubes (CNTs) which for nearly 20 years are named as "very promising" material regarding an impressive number of described possible applications (Bakshi et al., 2010; Upadhyayula et al., 2009). However, a wide practical application has not materialized yet since the cost of 1 g CNTs of high purity (essential demand in most applications) and properly functionalized (tailoring of chemical and physical) is ranging from ca. hundred to above thousand of dollars (http//www.cheaptubes.com, 2011; http//www.sigmaaldrich.com, 2011). Therefore, despite discrete pore structure (very narrow PSD) CNTs cannot be regarded as a practical adsorbent for large scale use despite of numerous papers and patents (Hu and Ruckenstein, 2004; Pilatos et al., 2010) claiming such a possibility.

Thus, there is an obvious need for functional materials fabricated in a low-cost process from widely accessible raw materials. Renewable resources are preferred due to environment protection. The current report should be seen in light of the above statement that is it describes a multidirectional chemical exploitation of *Salix viminalis* wood grown on agriculture plantations (Figure 1) with high economical efficiency. This suits regional policy in the area of Torun (Poland) but can be easily expandable to other regions of Poland and adopted by other countries, as well. The proposed elaboration is practically waste-less since all main products (solid and liquid phase) of the process are in fact valuable and functional materials. In summary, the proposed concept of *Salix viminalis* wood elaboration is environmentally friendly being waste-free and based on renewable resources.

Thermal treatment in oxygen-free conditions is the essence of *Salix viminalis* wood processing described in the current paper. The treatment of dried wood is apparently an ordinary technological measure. However in some cases, this process may trigger several interesting phenomena like those which may be accounted to widely understood nanotechnology because of the properties of materials obtained during pyrolysis. Two basic products may be obtained after pyrolysis of the wood: porous carbonaceous solid and vapors that evolve during the treatment. The first product is an active carbon. In general, such carbons are capable to adsorb selectively various species from gas and liquid phase. Adsorption properties of carbons and their application range depend (among other) on pore size distribution (PSD) and pore volume. Narrowed PSD of solid adsorbents let them consider as potential molecular sieves. Numerous experiments proved that sieving properties of active carbons can be exploited to solve problems like (beside many other); air separation by PSA method (Hassan et al., 1986; Rege and Yang, 2000), noble gas separation (Gorska, 2009), hydrogen and carbon monoxide separation (Gorska, 2009), methane and hydrogen storage (Cheng et al., 1997; Dillon and Heben, 2001), and so on. Released vapors partly liquefy after cooling. The distillate of *Salix viminalis* wood has not been investigated, yet. Therefore, there is a need to know whether the distillate is a useless waste, a hazardous byproduct or a valuable product containing precious organic and inorganic substances.

EXPERIMENTAL

Harvested *Salix viminalis* stems were dried and ground into shavings ca. 1 cm long, then pyrolysed in inert gas atmosphere. Pyrolysis (carbonization) was carried out in two stages: (i) the preliminary stage 1 hr at 600°C for expelling some volatile species, (ii) the secondary stage 1 hr at arbitrarily chosen temperature (600, 700, 800, 900°C) for expelling residual volatile fractions and for the formation of micropore-rich polycrystalline carbon matrix.

Micropore analyzer ASAP 2010 was applied to the collection of nitrogen adsorption data (N_2 adsorption at −196°C versus relative pressure of nitrogen). Nitrogen adsorption data were regressed according to Horvath–Kawazoe theory (Horvath and Kawazoe, 1983) using commercial software provided by equipment manufacturer (Micromerotics, Inc.).

Volatile products were collected during the first stage of carbonization in the temperature range 140–600°C. The volatiles were cooled to room temperature yielding a dark brown viscous liquid. The product was subjected to water and organic solvent extraction aiming at separation of its components (mainly polyphenols). Finally in this study, two phenolic fractions were isolated through solvent extraction: extract (A) ether-soluble and extract (B) methylene chloride-soluble. The content of extracts was characterized by chromatographic analysis. The GC-MS analyses were performed on a Autosystem XL gas chromatograph with MS TURBOMASS mass selective detector (Perkin Elmer) and a 30 m long, 0.25 mm ID, 0.25 mm thick SLB-5MS column (Supelco). Helium was the carrier gas. 700 MHz NMR and FT-IR spectrometers were used to characterize extract composition and in the experiment testing antioxidant activity of the extracts. The ^{1}H NMR experiments were recorded on a Bruker Avance spectrometer operating at 700 MHz.

Chromatographic separation of binary gas mixtures was performed using a gas chromatograph Schimadzu GC-14B supplied with a TCD detector kept at constant temperature of 110°C. Volume flow of carrier gas (helium) was set at 10, 15, 20, 25, 30, 40, and 50 cm^3/min. The flow rate value was set regarding the results of van Demter optimization procedure preformed prior to the separation tests. A. glass chromatographic loop was 2.5 m long and its inner diameter was 2.6 mm. The separation tests were performed at several temperatures (70, 60, 50, 40, 30°C) and consisted in recording of chromatograms for injected samples (pure gases, two component mixtures, and three component mixtures). For better separation of gases two measures were undertaken:

- Instead of bar carbon molecular sieves whose average surface area is placed within the range 300–400 m^2/g, activated carbon were obtained via conventional zinc chloride method (final surface above 1,400 m^2/g).
- Active carbon was grained, sieved and 0.02–0.25 mm fraction was collected for filling of chromatographic loop.

The antioxidative effect of the obtained phenolics fractions on a selected diester DBS (dibutyl sebacate) was carried out following the ASTM D 4871 standard. This universal test allows the examination of oxidation stability of a liquid test sample. The samples of uninhibited DBS and after addition of ca. 1,000 ppm of a selected antioxidant agent (BHT, extract A, extract B) were prepared and examined. Commercial BHT (2,6-ditertbutyl-4-methylphenol) was used as a standard antioxidant for comparison. Investigated samples with a mass of 100 g were subsequently placed in the reaction vessel and oxidized at 150°C by flowing air stream (100 cm^3/min).

RESULTS

The *Salix viminalis* originated active carbons are in fact solids containing a network of very tinny pores. Average calculated linear dimension (Horvath and Kawazoe, 1983) was below 1 nm. In practice, these pores nearly exclusively contribute to the total pore volume of the obtained carbons. Moreover PSD is very narrow suggesting that such materials are potential molecular sieves providing molecular sieving abilities (Figure 2). This property of the obtained carbons needed an experimental verification.

We assumed that molecular sieving effect over a porous solid should be more efficient if the gas molecules (to be separated) are of possibly different dimensions and when the size of pores becomes comparable to dimensions of the molecules. In our study, average pore size approaches sub-nanometer range which in general is only few times bigger than average dimensions of some simple molecules. Dimensions of simple molecules and atoms in gas phase (Table 1) are given in literature (Shirley and Lemcoff, 2002). The dimensions were estimated basing on experimental results and theoretical considerations.

Some studies (Corma, 1997; Ivanova et al., 2004) point out the importance of geometric factor as the main reason of gas separation leading to differentiated adsorption rate of molecules from gas phase in static condition (determination of adsorption isotherms) (Emmett, 1948). Gas separation may proceed in dynamic conditions when a gas mixture is passing through a membrane or along the surface of a porous solid. For the latter phenomenon, we have selected chromatographic tests.

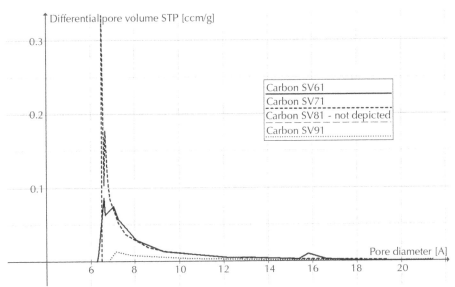

Figure 2. Typical pore size distribution (PSD function) for carbon molecular sieves obtained by pyrolysis of *Salix viminalis* wood. Symbols 6, 7, 8, 9 denote carbonization temperature of 600, 700, 800, 900°C, respectively. The PSD functions calculated basing from N_2 adsorption data at −196°C and Hortvath-Kawazoe model.

Table 1. Atomic and molecular dimensions of frequently used gas adsorbates.

Molecule	Kinetic diameter [nm]	MIN-1 Smallest dimension of the molecule [nm]	MIN-2 Intermediate dimension perpendicular to MIN-1 [nm]
H_2O	0.27	0.29	0.32
CO_2	0.33	0.32	0.33
Ar	0.34	0.35	0.36
Ne	0.28*	–	–
Kr	0.36*	–	–
O_2	0.35	0.29	0.30
N_2	0.36	0.30	0.31
CH_4	0.38	0.38	0.39
C_3H_8	0.43	0.40	0.45
$n\text{-}C_4H_{10}$	0.43	0.40	0.45
$iso\text{-}C_4H_{10}$	0.50	0.46	0.60

Table 1. *(Continued)*

Molecule	Kinetic diameter [nm]	MIN-1 Smallest dimension of the molecule [nm]	MIN-2 Intermediate dimension perpendicular to MIN-1 [nm]
SF$_6$	0.55	0.49	0.53
CCl$_4$	0.59	0.57	0.58
C$_6$H$_6$	0.59	0.33	0.66
Cyclohexane	0.60	0.50	0.66
CO	–	0.28	–

We performed several separation tests targeting on separation of gases (Table 2) which should be incapable to specific molecular interactions like dipole orientation. For better illustration of separation capability of the fabricated carbons, we have chosen chemically inert gases of similar symmetry that is neon and krypton. Nitrogen was picked-up regarding its inertia but also its specific geometry (longitudinal two-atom molecule) differing from a spherical shape ascribed to Ne and Kr atoms. Figure 3(a) and 3(b) present chromatographic peaks attributed to particular gases resulting from gas mixture separation (Kr/N$_2$ and Ne/N$_2$ respectively).

It is visible that Kr/N$_2$ mixture undergoes separation more effectively at both temperatures (30°C and 70°C) while Ne/N$_2$ cannot be separated at 70°C (peaks overlapping). The effect may be explained basing on atomic and molecular dimensions given in the Table 1. Minimal dimensions (MIN-1) of Kr and N$_2$ differ considerably (0.36 nm and 0.30 nm respectively) while minimal dimensions of Ne and N$_2$ are very similar (0.30 nm and 0.28 nm respectively). Thus, the unique pore structure of *Salix viminalis* originated carbons helped to separate effectively atoms/molecules whose shape and size differed only by 0.06 nm size difference. The results also suggest that N$_2$ molecules penetrate into pores of investigated carbons based on their minimal dimension.

Table 2. Effectiveness of binary gas mixture separation over *Salix viminalis* originated carbon molecular sieve.

Temp [°C]	CH$_4$/N$_2$	CO/N$_2$	Ne/N$_2$	Kr/N$_2$	CH$_4$/CO	Ne/CO	Kr/CO	CH$_4$/Kr	Ne/Kr	CH$_4$/Ne
70	+	–	+	+	+	+	+	–	+	+
60	+	–	+	+	+	+	+	–	+	+
50	+	–	+	+	+	+	+	–	+	+
40	+	–	+	+	+	+	+	–	+	+
30	+	–	+	+	+	+	+	–	+	+

"+"—separation coefficient > 1
"–"—separation coefficient < 1

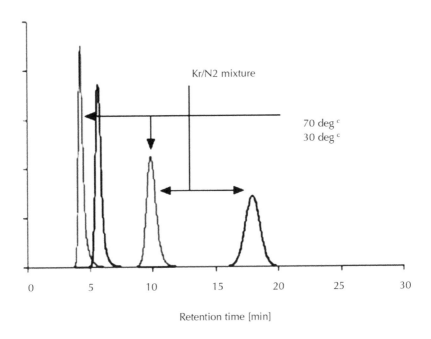

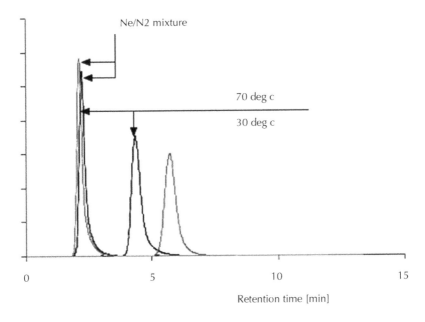

Figure 3. Chromatographic separation of Kr/N$_2$ (a) and Ne/N$_2$ (b) mixture over a *Salix viminalis* originated carbon DSV.

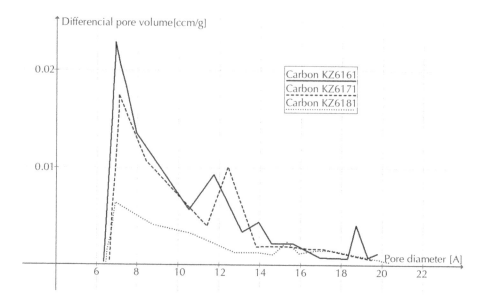

Figure 4. PSD function determined for a series of carbons obtained from *Acer platanoides*. Symbols 6, 7, 8 denote carbonization temperature of 600, 700, 800°C, respectively. The PSD functions calculated basing from N_2 adsorption data at –196°C and Hortvath-Kawazoe model.

The narrowed PSD in the case of *Salix viminalis* wood-originated carbon occurs in relatively limited range of carbonization temperatures (600–700°C) and its limited duration (typically 1 hr + 1 hr). More severe carbonization conditions leads to a collapse of this specific system of pores (Gorska, 2009; Ohata et al., 2008) due to graphitization of carbon matrix.

The formation of mentioned pore structure of narrowed PSD is not a common phenomenon that is similar heat-treatment of other sorts of wood often leads to qualitatively and quantitatively different results (Lukaszewicz et al., 2009). Figure 4 depicts the run of a PSD function determined for carbon obtained from Norway maple wood (*Acer platanoides*) in identical carbonization process. It is visible that the maximum of PSD is shifted towards pores of bigger diameter and is bimodal that is two local maxima exist: the first in sub-nanometer range and the second at ca. 1.2–1.4 nm.

Pyrolysis of biomass, as well as pyrolysis of *Salix viminalis* wood, yields volatile products besides the mentioned solid product—active carbon (Figure 5). Figure 5 documents that the proportions between basic products of biomass heat-treatment depend on process dynamics. High speed pyrolysis prefers the formation of non-condensable gases while slow carbonizations increase the share of solid product (char or active carbon).

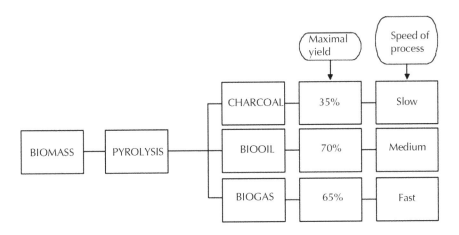

Figure 5. Typical composition of products obtained from pyrolysis biomass. Maximal achievable content of different products depends on the speed of pyrolysis.

In fact, volatile products which start to evolve intensively at 140°C during heat treatment of *Salix viminalis* wood, partially condensate after cooling in the form of a dark brown viscous liquid. Volatile products in condensed form seem to be useless but besides water, alcohols, ketones, aldehydes, acids, the condensate contains a mixture of polyphenols called "bio oil" (Figure 6). Volatiles evolution intensity depends on temperature reaching maximum in the range 260–340°C (Figure 7). Antioxidant properties are routinely ascribed to many polyphenols (Rice-Evans et al., 1995; Rosicka-Kaczmarek, 2004) of biological origin. Within the current project authors performed separation of polyphenols in condensate and identified them by means of NMR, GC-MS, and FTIR (Table 3). Finally, chemical antioxidants test were done using purified condensate that is extracts from "bio oil" and some test samples that is BHT and "biodiesel" for engines. The BHT served as a reference commercial antioxidant.

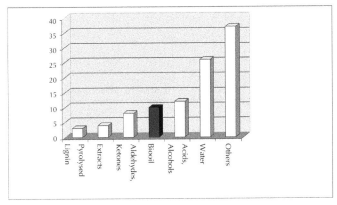

Figure 6. Average composition of liquid condensate obtained by carbonization of *Salix viminalis* wood.

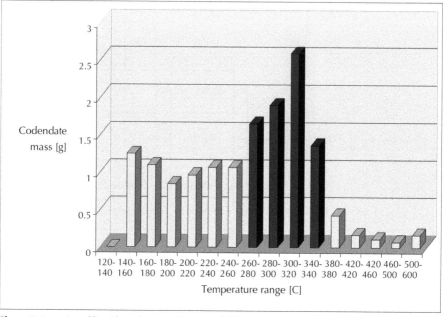

Figure 7. Intensity of liquid products evolution at different temperature ranges.

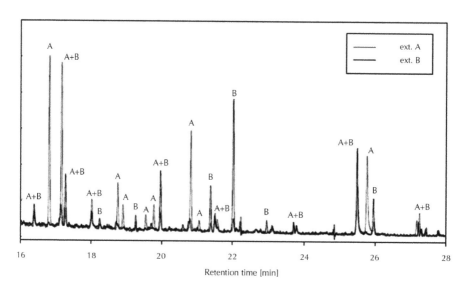

Figure 8. Visualization of differences in extract A and extract B composition: overlay of two chromatograms.

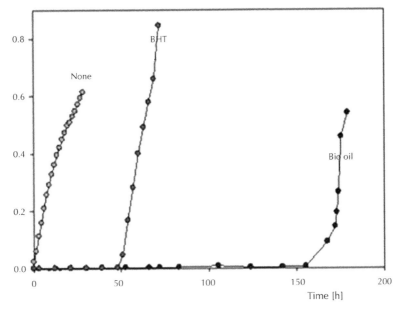

Figure 9. Antioxidant activity of "bio oil" (extract B) and a reference commercial antioxidant (BHT 6-ditertbutyl-4-methylphenol)-) against oxidation of DBS (2,6-di-tertbuthylo-4-methylophenol). Concentration of DBS oxidation products versus time of oxidation in liquid phase at 150°C in the presence of various antioxidants ("bio oil" and BHT) under investigation.

Table 3. Polyphenols identified in A and B extracts fraction obtained during pyrolysis of *Salix viminalis* wood.

Mass	Forumula	Structure	M S ions	Mass	Forumula	Structure	M S ions
124	$C_7H_8O_2$		124 109 81 63 53	194	$C_{11}H_{14}O_3$		194 147 131 119 91 77
122	$C_8H_{10}O$		122 107 91 77	196	$C_{11}H_{16}O_3$		196 167 123 77
138	$C_8H_{10}O_2$		138 123 95 77 67 55	194	$C_{11}H_{14}O_3$		194 179 131 119 91 77 65
126	$C_6H_6O_3$		126 109 97 81 69 53	182	$C_9H_{10}O_4$		182 167 139 111 96 65
110	$C_6H_6O_2$		110 81 63 53	194	$C_{12}H_{10}O_2$		194 179 131 119 91 77
140	$C_7H_8O_3$		140 125 97	186	$C_{11}H_{14}O_3$		186 130 103 93 65
152	$C_9H_{12}O_2$		152 137 122 91 77				

Figure 8 presents a selected part of a gas chromatogram recorded for the extracts A and B. The extract B consisting of furan derivative contains more oxygen heterocyclic compounds compared to extract A. The excess of O-heterocycles is responsible for the observed (better) antioxidant activity of extract B (Figure 9). Careful analysis of GC, FT-IR, and NMR results let to identify more than 50 compounds which could be accounted to the compounds potentially exhibiting antioxidant activity. Some of them are listed in Table 3 which contains names, formulas and information on most typical MS ions observed during the GC-MS identification. In general, most of the identified compounds may be seen as derivatives of the three basic structures (Figure 10): (A) cumarol alcohol, (B) coniferyl alcohol , and (C) synapine alcohol.

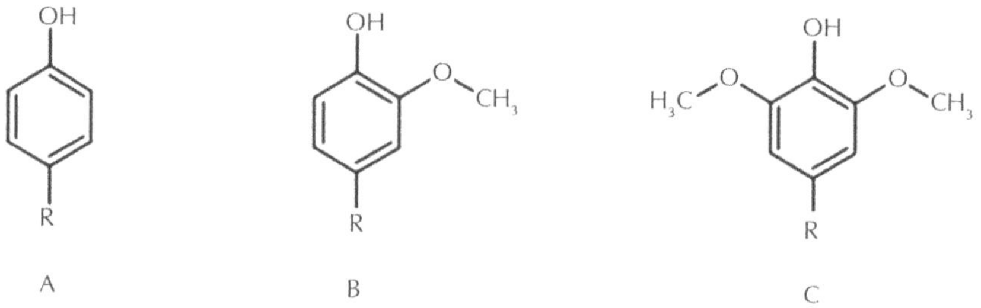

A   B   C

R= -H, -CH$_3$, -C$_2$H$_3$, -C$_2$H$_5$, -C$_3$H$_5$, -C$_3$H$_7$, -COCH$_3$, -C$_3$H$_4$OH, -CHO

Figure 10. Three basic phenolic structures occurring in "bio oil."

Analysis on Figure 9 proves the superiority of *Salix viminalis*-originated "bio oil" over commercially available antioxidant BHT. The advantage is very spectacular since DBS protection time is more than 3-times longer in the case of "bio oil". It has to be stated that "bio oil" fabrication is relatively simple and does require complex procedures and expensive reagents. It is nearly completely environment friendly because of the following reasons:

- Basic raw material comes from renewable resources that is may be grown on low-quality soils with minor use of agriculture chemicals,
- All products of the proposed thermal treatment may be successfully applied in different domains of everyday life and industry,
- Solid product may serve as a sort of active carbon and/or burn as fossil fuel,
- Liquefied fraction of gas products is a source of precious chemicals including very efficient antioxidants,
- A part of non-liquefied gas products may be burned as a fossil fuel,
- A part of non-liquefied gas products consists of species which are not toxic (at level of limited concentration) for people and other forms of *flora* and *fauna*.

In literature, it is hard to find other examples of processes which produce less waste.

CONCLUSION

Heat-treatment of widely accessible and inexpensive raw material that is *Salix viminalis* yields precious products of potentially wide application in practice. Solid product is a strictly nanoporous active carbon which resembles a perfect molecular sieve. Its sieving properties were successfully proved in the case of Ne/N_2 and Kr/N_2 mixtures separation. Liquefied distillate is a source of "bio oil" containing a mixture of numerous polyphenols. "Bio oil" exhibits excellent antioxidation activity even if compared to commercial antioxidants like BHT.

KEYWORDS

- **Carbon nanotubes**
- **Minimal dimensions**
- **Pore size distribution**
- **Purification**
- **Pyrolysis**

Chapter 10

Properties of Differently Compatibilized Bamboo Fiber Reinforced PE Composites

Sandeep Kumar

INTRODUCTION

Process of "grafting" is a very useful method for making branches or chemical modification in polymeric materials. Natural fiber reinforced thermoplastic composites grafting, in context of compatabilizer, can be done by two ways: one is pre-grafting (the polymer is first taken and grafted as compatabilizer) and another is ----*in situ* grafting. In ----*in situ* grafting process, polymer is grafted during the fabrication of composites/ blends. The later kind of compatibilization is also called reactive compatibilization. Polyolefins are commonly functionalized with maleic anhydride (MA) to enhance adhesion to polar materials and are used as a compatibilizer. The bamboo-fiber is often brittle compared to other natural fibers, because the fibers are covered with lignin and it undergoes degradation above 200°C (Seema and Kumar, 1994). Polyethylene (PE) and polypropylene (PP) are among the commonly used thermoplastics for the preparation of natural fiber reinforced composites due to their relatively low prices and good processability (Deshpande et al., 2000; Kim et al., 2006).

The synthesis of polyolefin graft copolymers by melt processing was reviewed and interdependent factors (include, mixing efficiency, temperature, pressure, residence time, venting, polyolefins, monomer(s), initiator(s), and screw/extruder design) are to be optimized so as to maximize the degree of grafting (DG) and minimize the side reactions (Moad, 1999). Various coupling agents or compatibilizers have been reported in the literatures to improve the interface between the thermoplastic matrix and natural fibers. Among them, maleic anhydride-grafted PE or PP is the most commonly available compatabilized matrix because of their ability to effectively enhance the mechanical properties of the composites (Liu et al., 2008). The solution surface grafting of MA onto PP in toluene solvent by using benzoyl peroxide as an initiator in nitrogen atmosphere to remove dissolved oxygen has been done to prepare bamboo-fiber PP composites. These composites revealed improved mechanical properties when compared with commercially available wood pulp composite (Chen et al., 1998). Literature reported the free radical melt grafting of glycidyl methacrylate onto a range of PP grade in a Haake Rheomix 600 batch mixer and then grafted product was purified with toluene and acetone (Chen et al., 1996). The grafted structures of MA onto PE have been investigated (Heinen et al., 1996). Reduction in melting temperature and crystallization percentage or increase in crystallization peak temperature was observed to be an effect of MA grafted onto PE (Martínez et al., 2004). The mechanism of cross-linking or grafting by melt blending technique has been discussed by researchers (Gaylor et al., 1989; Heinen et al., 1996; Shi et al., 2001).

Most of the grafting reaction studies have been carried out using different grafting recipes (type and amount of peroxide and MA content) and different processing conditions (type of reactor, screw speed and temperature) (Machado et al., 2001; Qi et al., 2004; Sclavons et al., 2000; Shi et al., 2001) for the polymer blend system. Considerable efforts have been made in producing newly polymeric composite materials with an improved performance/costs balance. This can be achieved by using a direct grafting of MA/GMA instead of pre-grafting of MA/GMA onto existing polymers in the process of fabrication of natural fiber composites. From a research and development point of view, the direct grafting routes of monomer (MA/GMA) are usually more efficient and less expensive.

The objective of this research is to compare the thermal and tensile properties of ----*in situ (i-)* grafted polymer compatibilized composites with pre (p-) grafted polymer compatibilized composites. In turn, this research gives an idea of advantages and disadvantages of both the process in terms of the manufacturing methods employed to fabricate the composites.

MATERIALS AND METHODS

Linear low density polyethylene (LLDPE) (G-Lene) (MFI 0.9, ρ 0.92g/cm³) was obtained from GAIL (India) Limited. The LLDPE-g-maleic anhydride (TP-568/E) (MFI 2-3@190°C-2.16Kg, MAH content (%) 0.4–0.6 and ρ 0.923 g/ml) was obtained from Pluss Polymers (P) Limited (India). Glycidyl methacrylate (GMA) (ρ 1.450 g/ml, bp142°C, Aldrich) and MA (mp 52–55°C, Aldrich) was used for ----*in situ* grafting. The MA or GMA was chosen for ----*in situ* melt grafting because it has been reported that the anhydride or methacrylate groups strongly associates with the hydroxyl groups (Chen et al., 1996; Chen et al., 1998). The present research was carried out by using bamboo species-*Ochlandra travancorica* (one year old)-grown in India. The process of delignification of bamboo to produce deglinified bamboo fiber and their physical properties has been reported by earlier research work (Kumar et al., 2010).

Composite Preparation with i-/p-grafted Polymer

Short delignified bamboo-fiber (SDBF) having an average length of 1-2 mm were pre-dried for processing in an oven at 110°C for 8 hr to expel moisture. In order to distinguish the properties of differently compatibilized composites, the ratio of matrix to fiber was kept constant in all the formulations that is, 30% w/w fiber content depending on the optimized properties of fiber reinforced composites at this fiber weight fraction (Kumar et al., 2010). To compare the properties with ----*in situ* grafted polymer as a compatibilizer, MA-g-LLDPE content have therefore converted into MA content. The LLDPE was replaced by pre-grafted polymer (LLDPE-g-MA having MA content 0.75–5%) and the effect of MA content in graft copolymer on the properties was investigated. ----*in situ* grafted composites were prepared by melt grafting of 1%, 3%, and 5% (w/w) GMA or MA onto LLDPE with 1% dicumyl peroxide (DCP) as an initiator and SDBF. The low level of DCP was chosen as it gives best result for grafting reaction of MA/GMA (Ho et al., 1993; Jang et al., 2001). It was considered that the excess of initiator results in a greater extent of degradation/side reactions. In a typical melt-grafting experiment, LLDPE granular polymer was first melted by dosing into the Haake chamber and then bamboo-fibers were added to this molten polymer. The powdered MA and DCP as a radical initiator were dosed after 5 min via a hopper into a Haake Rheomix 600 (equipped with a pair of high shear rollers). The melt grafting reaction was done for i-GMA (30 min) and for i-MA (15 min) at melting temperature of 170°C and 50 rpm speed. The grafting reaction mechanism of GMA onto PE in melt phase process when DCP was used as an initiator was discussed earlier (Jang et al., 2001). No special precautions were taken to exclude air from the rheomixing chamber. The mixing torque and melt temperatures were monitored on-line and recorded for all the samples. The ejected products before making composite sheets were heated for 30 min to remove the unreacted MA/GMA in an oven at 190°C. The composite sheets (having average thickness of 3 mm) were prepared by controlled press molding at 200°C and 6 ton pressure for 15 min. The sheets were cooled around for 15 min under pressure with water circulatory system and after removing from mold, cut for further testing.

Determination of Grafting Degree

Maleic anhydride (MA)

The grafting degree (GD) of MA was determined by the back-titration method including the dissolution of a small amount of composite sample (1.0 g) in 100 ml of boiling xylene in a conical flask and a few drops of water was added to hydrolyze all anhydride functionability. Then, 10 ml ethanol solution of NaOH (0.1 mol/l) was added and mixed gently by stirrer. After refluxing for 30 min under stirring, this solution was back-titrated with HCl (0.1 mol/l) using methyl red as the indicator. The GD_{MA} was defined (Nakason et al., 2004; Qi et al., 2004; Shi et al., 2001) as the amount of grafted MA as a percentage of LLDPE-g-MAH and is calculated by equation given below:

$$GD_{MA}(\%)=\left(\frac{(V_0-V_g)\times C\times M}{2W_s\times 1000} \right)\times 100$$

where, V_0 is the amount of HCl consumed by using LLDPE as reference (ml),
V_g is the amount of HCl consumed by the grafted LLDPE (ml),
C is the molar concentration of HCl (mol/l),
M (98) is the molecular weight of maleic anhydride,
W_s is the weight of sample (g).

Glycidyl methacrylate (GMA)

After melt grafting of GMA onto LLDPE the weight was recorded, the grafted LLDPE were extracted with acetone by means of a Soxhlet extractor for 48 hr to remove residual GMA and possible homopolymer that is poly-GMA. To evaluate the degree of grafting (GD) of GMA, the left over products, pure graft copolymers, were dried in vacuum at 60°C and weighed. The GD_{GMA} was calculated by the following equations (Chen et al., 1996; Jang et al., 2001; Kim et al., 2001):

$$GD_{GMA}(\%)=\left(\frac{W_{Grafted\ LLDPE}-W_{LLDPE}}{W_{LLDPE}}\right)\times100$$

where $W_{Grafted\ LLDPE}$ is the dry weight of grafted LLDPE,
W_{LLDPE} is the weight of LLDPE.

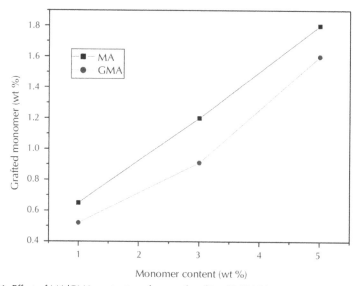

Figure 1. Effect of MA/GMA content on degree of grafting (DCP1%).

Figure 1 shows the effect of monomer (MA/GMA) content on the DG using 1% DCP as an initiator. It was observed that the percentage of GMA grafted onto LLDPE was less compared to percentage of MA grafted onto LLDPE under the similar conditions. This may be due to the formation of poly-GMA in the composite product.

Characterization of i-/p-grafted Compatibilized Composites

The degradation/decomposition behavior of i-/p-grafted LLDPE/SDBF composites was investigated by using Perkin Elmer Pyris 6 having TG module and TG/DTG traces in nitrogen atmosphere were recorded by taking the average weight (10±2 mg) of composite in each experiment. Dynamic TG/DTG scans were run from 50°C to 700°C at a heating rate of 20°C/min. Perkin Elmer Pyris 6 was used for recording DSC scans for composite samples (weight of 5±2 mg). The samples were heated to 150°C at a rate of 20°C/min, and maintained for 3 min to remove the thermal history. The samples were then cooled to 40°C at a cooling rate of 10°C/min, and heated again to 150°C at the rate of 10°C/min. The crystallinity index (CI_{DSC}) of LLDPE was calculated as per discussion in literature (Kumar et al., 2010). Tensile tests were performed with a universal testing machine (UTM), Zwick (Z010) model. Tensile specimens of differently compatibilized composites were machined in dumb-bell shape according to ASTM D638-03 type 1V. Five specimens for each composite composition were tested.

The cross-head speed of 5 mm/min was used for tensile specimens. Tensile properties were measured from the stress-strain curve. The tensile strength and modulus of composites was normalized with the neat LLDPE tensile strength and modulus. The interfacial morphology of the varying compatibilized composites was investigated by doing cryogenic fracture of tensile specimen in liquid nitrogen. Before fracture, all specimens were dried at a temperature of 80°C in a vacuum oven. The fractured surface was analyzed using scanning electron microscopy (SEM) (Cambridge stereoscan 360) after coating with gold to avoid the electrostatic charging and poor image resolution. The extent of interfacial adhesion between bamboo-fiber and the matrix can be investigated by examining the SEM microphotographs.

RESULTS AND DISCUSSIONS

Degradation Behavior of Compatibilized Composites

Figure 2 and 3 shows the degradation behavior of i-grafted and p-grafted polymer composites. All the composites showed two steps mass loss. First step mass loss was related to the degradation of cellulose and degradation of remaining lignin (at approximately 400°C) and second step mass loss was related to matrix (at approximately 500°C). It has been reported that LLDPE showed a single step decomposition in the temperature range of 400–500°C with almost no char yield. It was found that in uncompatibilized composites, the first weight loss was increased from 11.4 to 23.9% and matrix loss was decreased from 87.5 to 71.6% as a function of fiber content (from 20 to 50 weight percent). The first weight loss for 30 wt% fiber in uncompatibilized composite, as TGA traces reported in literature (Kumar et al., 2010), was 13.5 while the matrix loss was 85.0%. From the Figure 2 and 3, it is evident that the initial decomposition temperature (IDT) of p-grafted/----in situ grafted compatibilized composites decrease with increase of the grafted content/compatibilizer.

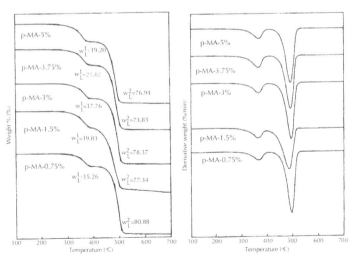

Figure 2. TGA traces of short delignified bamboo fiber reinforced composites in presence of pre-grafted polymer as a compatibilizer (w_L^1=f irst weight loss, w_L^2=second weight loss).

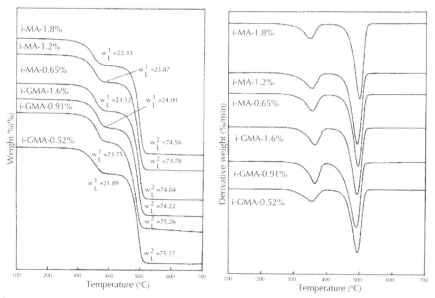

Figure 3. TGA traces of short delignified bamboo fiber reinforced composites in presence of ----*in situ* melt grafting of maleic anhydride (MA)/ glycidyl methacrylate (GMA) (w_L^1 = first weight loss, w_L^2 = second weight loss).

The composites, reported in this chapter, are multi-component system after using compatibilizer and the degradation patterns depend on the extent of hydrogen bonding network of interface. There is no definite trend in the weight loss of composite component as a function of p-grafted polymer while increasing first weight loss as a function of i-grafted polymer is observed (Figure 2 and 3). The thermal stability is good at concentration of 0.91 for i-grafted GMA as well as 0.75 for p-grafted GMA. ----*in situ* grafted MA polymer composites have slightly lower thermal stability in comparison to p-grafted MA. This might be due to the presence of free MA group/unreacted MA in the final product.

Melting Behavior and Crystallinty Index of Differently Compatibilized Composites

The melting behavior curve of p-grafted MA polymer composites is shown in Figure 4. The endothermic transition is characterized by noting the peak of melting endotherm (T_m) and the heat of fusion (calculated from the area under the melting peak). The shoulder before the melting peak in the composite increased as a function of p-grafted polymer. As the MA content increased, the composites started to melt at a lower temperature as shown in Figure 4. The melting peak increased around 4°C as a function of increasing MA content. Figure 5 shows the melting behavior of i-GMA/i-MA grafted polymer composites. The melting and crystallization trend are not similar in the composites due to the different extent of hydrogen bonding network formation at the interface. The fusion peak of 30 wt% filled bamboo fiber LLDPE composite was at 103°C (Kumar et al., 2010). The melting peak increased around 10°C in both kind of ----*in situ* MA/GMA compatibilized composites with the increase in the content of monomer (MA/GMA). The second heating scans were used to determine the crystallinity index of the matrix in the composites. The crystallinity index of uncompatibilized LLDPE/30 wt%-SDBF composite was 62.99% (Kumar et al., 2010). The crystallinity index of matrix decreased with increasing amounts of p-MA and i-GMA grafted polymer while it increased for i-MA grafted polymer with increasing of MA monomer content. With the former one, the similar result of the decrease in crystallinity with the GMA graft content was observed in PCL-g-GMA polymer system (Kim et al., 2001). This may be due to the hindrance of LLDPE crystallization by the increase of chain structural irregularity caused by grafting reaction of GMA. In the latter case (i-MA grafted polymer composite system), the structural regularity might not be disturbed due to the appendage of single MA unit to the PE chain. The effects of p-MA and i-GMA/i-MA on the crystallization behavior of polymer composites are delineated in Figure 6 (left) and 6 (right). The crystallization peak temperature (T_c) values of composites are reported in Figure 6. The crystallization peak also increased around 10°C for i-MA composites and 3°C for i-GMA grafted composites. The positive change in the onset of melt crystallization, T_{omc}, in presence of compatibilizer indicate a faster crystallization process, which may be basically due to the formation of hard phase near the surface of fibers. There was not a substantial change in the crystallization process with the addition of compatibilizer excepting i-MA compatibilized composites. The crystallization peak of 30 wt% filled SDBF/LLDPE composite was at 107°C or T_{omc} at 110.55°C as reported in literature (Kumar et al., 2010). The p-grafted composites showed an increment in their crystallization temperature, as seen in Figure 6 (left). This effect is basically due to the presence of the MA in the polymer matrix and preventing the chain folding and producing changes in the form and size of crystals (Martínez et al., 2004). The lower crystallinity index of compatibilized composites compared to uncompatibilized composites might be giving a support to the above mentioned fact (Figure 5).

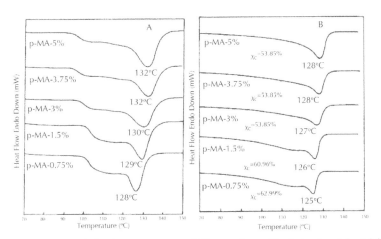

Figure 4. Effect of pre-grafted polymer as a compatibilizer on the melting behavior of short delignified bamboo fiber reinforced composites: (A) first and (B) second heating scans (χ_c represents crystallinity index).

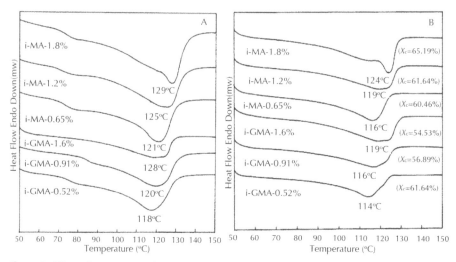

Figure 5. Effect of ----*in situ* melt grafting of maleic anhydride (MA)/glycidyl methacrylate (GMA) on the melting behavior of SDBF reinforced composites: (A) first and (B) second heating scans (χ_c represents crystallinity index).

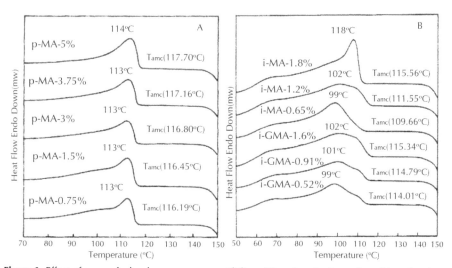

Figure 6. Effect of pre-grafted polymer as a compatibilizer (A) and ----*in situ* melt grafting of maleic anhydride (MA)/ glycidyl methacrylate (GMA)(B) on the crystallization behaviour of SDBF reinforced composites.

Tensile Properties of ----*In situ* Grafted and Pre-grafted Polymer Composites

The results of normalized strength and modulus as a function of monomer content (%) in compatibilized composites are shown in Figure 7. The strength first increased in pre-grafted LLDPE composite due to the improvement at the interface of the matrix and bamboo-fiber and then showed no trend with the addition of compatibilizer. It was clear that the p-grafted composites held good strength with the addition of 1.5%

(w/w) of compatibilizer. The strength was increased for ----*in situ* grafted composites with grafting content but the lower strength was observed in case of i-grafted GMA at similar compositional level. The modified law of mixtures equation for discontinuous short fiber composites is given below (Chen et al., 1998):

$$\sigma_c = KV_f\sigma_f\left(1-\frac{L_c}{2L}\right)+(1-V_f)\sigma_m^*$$ (1)

$$L_c = \frac{D\sigma_f}{2\tau_i}$$ (2)

where σ_c is the tensile strength of the composite, which is the empirical fiber efficiency parameter, V_f is the volume fraction of fiber, σ_f is the tensile strength of fiber, σ_m^* is the tensile strength of the matrix at the breaking strain of the fiber, L is the fiber length, L_c is the critical fiber length, D is the diameter of the fiber, and τ_i is the interfacial shear strength. From equations (1) and (2), it is seemed that the σ_c value rise up with increasing of τ_i value which in turn depends on the improvement of interfacial adhesion between the fiber and matrix, when K, L, and D are assumed constant in all composite systems. The fiber fraction (V_f) was also constant in all composites composition. It indicates that better adhesion gives higher τ_i value resulting in the increase in the tensile strength of the composites.

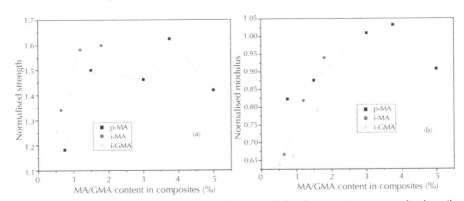

Figure 7. Effect of monomer content (MA/GMA) in compatibilized composites on normalised tensile properties.

The tensile modulus increased almost linearly but at higher compatibilizer content got decreased. The mixing time was higher in i-GMA grafted LLDPE composites and thus, the chances of the breakage of fibers increased which in turn reduced the efficiency of fibers. This may be the possible cause for the reduction of the modulus of these composites.

Effect of ----*in situ* and Pre-grafted Polymer on Interfacial Adhesion of Composites

The fracture morphology of composites modified with ----*in situ* and pre-grafted polymer as a compatibilizer was shown in the Figure 8. The morphology of 30 wt%

filled SDBF uncompatibilized composite was shown in literature (Kumar et al., 2010). Figure 8 delineated that there is a little sign of LLDPE matrix and delignified bamboo fiber decohesion for the ----*in situ* grafting or p-grafting even after increasing the percentage of grafted content. In ----*in situ* grafting, the GMA/MA molecules at the interface on one hand react with the hydroxyl groups on the bamboo fiber dispersed phase and on the other hand, they are grafted on the LLDPE chain. The reactive GMA/MA molecules act as bridges between the matrix and reinforcement phase. Likewise MA-g-LLDPE on one side entangled with other LLDPE chains in the matrix phase, due to structural similarity while other side react with the bamboo fibers at the interface. Therefore, adhesion will depend on the number of bond formed at the interface and the effectiveness of these entanglements which ultimately depends on the DG or chain length. The extent of adhesion was like in that order adhesion ----*in situ* GMA<adhesion ----*in situ* MA~adhesion p-MA as per tensile properties evaluation.

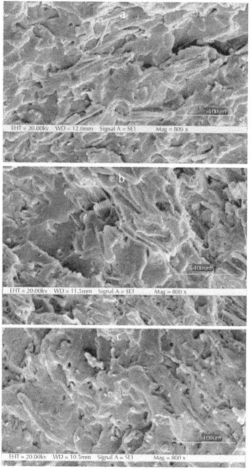

Figure 8. SEM photographs of ----*in situ* compatibilized composites [(0.91% GMA, a), (1.2% MA, b) and pre (p)-grafted compatibilized composites [5% MA, c].

Comparison of ----*in situ* Grafting Polymer with Literature Pre-grafting Polymer Technique

The melt grafting (Chen et al., 1996; Jang et al., 2001; Kim et al., 2001)/ radiation-induced grafting (Choi et al., 1998) of glycidyl methacrylate (GMA) or solution surface grafting (Chen et al., 1998; Heinen et al., 1996; Machado et al., 2001) / UV pre-irradiation step grafting (Martínez et al., 2004)/ melt grafting (Heinen et al., 1996; Machado et al., 2001; Shi et al., 2001) /Ball- milling grafting technique (Qiu et al., 2005) of MA onto PP/PE (various form like powder, fiber and granules) are discussed in literature. These pre-grafting techniques are very tedious, time taking process and wastage of money or energy. Therefore, it is good to fabricate composites with ----*in situ* grafting techniques. Figure 9 shows the flow chart of making bamboo-fiber composites with p-grafted or ----*in situ* grafted polymer as a compatibilizer. The ----*in situ* grafting revealed a considerable grafted weight of MA or GMA. Finally, ----*in situ* MA grafted polymer (≥1.2%) composites showed a good tensile strength comparable to pre-grafted polymer composites but the thermal stability was little bit poor because free MA group might be present in the system while in p-grafted system free MA group was removed by purification technique. Hence, if better optimal control on the ----*in situ* grafting is maintained in a way so that no free MA/ GMA group are present, the composites compatibilized with ----*in situ* grafting can give better properties.

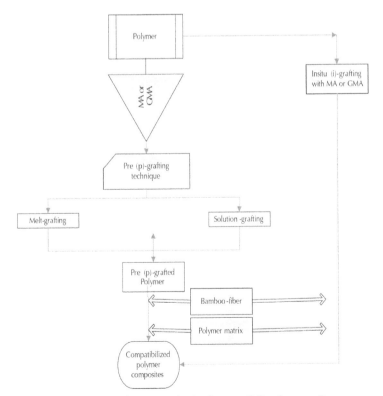

Figure 9. Flow chart of grafting technique to obtained compatibilized composites.

CONCLUSION

In general, it can be concluded that the requirement for proper matrix adhesion is optimal balance between fibers, matrix and compatibilizer. Here, for 30 weight % of SDBF composites, it was recommended that ≥1.2% i-MA content was adequate for good mechanical properties and robustness of composite when compared to p-MA grafted polymer composites while ----*in situ* grafted GMA gave poor mechanical properties. The cost of i-grafted polymer composites should be lower due to the reduction in number of process while preparing the composites via p-grafting of polymer.

KEYWORDS

- **Bamboo-fiber**
- **Compatibilization**
- **Crystallization**
- **Homopolymer**
- **Polyolefins**

Chapter 11

Jute/Polyester Composites

S. K. Malhotra

INTRODUCTION

Glass, carbon, Kevlar, and boron fibers are used as reinforcing materials for fiber reinforced plastics (FRP), which are widely accepted as materials for both structural and non-structural applications. However, these materials are resistant to biodegradation and pose environmental problems. Natural fibers from plants such as jute, bamboo, sisal, coir, and pineapple have high strength and can be effectively used for many load-bearing applications. These fibers have special advantage over synthetic fibers in that they are abundantly available from a renewable resource and are biodegradable (de Albuquerque et al., 2000). Also, they have low density, high toughness and acceptable specific strength. Among all natural fibers, jute is more promising because it is relatively inexpensive and commercially available in many forms (Gowda et al., 1999). One of the most easily available forms of jute fabric is the Hessian cloth (Uddin et al., 1997).

All natural fibers are hydrophilic and their moisture content ranges from 3 to 13%. This leads to a very poor interface between natural fiber and the hydrophobic matrix and very poor moisture resistance. Several fiber surface treatments are used to improve the interface viz-thermal treatment, chemical treatment, and use of coupling agents (Bledzki and Gassan, 1999). Fillers are added to composites for technical and chemical reasons (Milewski and Katz, 1980). Calcium carbonate has maximum usage as filler in plastics and FRP due to its low cost, non-toxicity and lack of odor. Present work gives effectiveness of jute fabric treatment and addition of filler on properties of jute reinforced polyester composites (JFRP).

Several treatments have been suggested to improve fiber/matrix interface in natural fiber composites. Also, fillers are added in composites to reduce costs with acceptable performance. By effective treatment of readily available natural fiber fabrics and suitable filler addition, utilization of natural fiber composites can be increased. Present work examines the effectiveness of surface treatment methods when applied to whole fabrics. The effect of filler on performance of jute fiber composites is also investigated. Evaluation of mechanical properties and moisture absorption of jute fabric reinforced thermosetting polyester is carried out. Calcium carbonate filler up to 40% of resin weight was used. It is observed that mechanical properties of composite are significantly improved by fabric treatment. Alkali treatment of jute fabric is found to reduce moisture absorption. It is observed that addition of calcium carbonate filler reduces tensile strength significantly but flexural strength is not much affected. Whole fabric treatment and filler addition can be effectively used to produce low cost natural fiber composites.

Jute fabric when used with general purpose polyester resin and large percentage of cheap filler like calcium carbonate can find many low cost applications. Some of the applications are grain storage silos, fishing trawlers and temporary/transportable shelters. For protection against environment and water one outer layer of glass fabric/polyester can be used. Also, jute with thermoplastics matrices finds many applications in automotives, rail coaches, etc.

EXPERIMENTAL

Materials

Two types of commercially available jute fabrics were used: Hessian cloth with starch and Hessian cloth without starch. Isophthalic polyester resin was used as matrix. Methyl ethyl ketone peroxide (MEKP) and cobalt naphthenate were used as catalyst and accelerator, respectively. Calcium carbonate (200 mesh) was used as filler.

Fabric Treatment

The above two fabrics were given two types of treatments:

Alkali treatment: Jute fabrics were soaked in 2% (w/w) NaOH solution for 2 hr. Then, they were washed thoroughly with distilled water and dried at 60°C for 24 hr in hot air oven.

Thermal treatment: Jute fabrics were dried at 100°C for 2 hr in hot air oven.

Nomenclature for composite specimens is as given below:

Hessian starch untreated/polyester-HSUNTR

Hessian starch alkali treated/polyester-HSAL

Hessian starch alkali treated with mechanical constraint/polyester-HSALMC

Hessian starch thermal treated/polyester-HSTHER

Hessian untreated/polyester-HUNTR

Hessian untreated/polyester-HAL

Polyester–POLYESTER

Processing

Hand lay-up (contact molding) method was used for preparation of jute/polyester composite laminates. 1.5% catalyst (MEKP) and 1.5% accelerator (cobalt naphthenate) was used for room temperature cure of isophthalic polyester resin. Filler (calcium carbonate) was added up to 40% by weight of resin. Fiber content in composite was maintained at around 30%.

Mechanical Testing

Mechanical properties were evaluated using an INSTRON universal testing machine (Model 4301). Tensile, flexural and interlaminar shear properties were determined according to relevant ASTM standards. Izod impact strength was determined using impact tester (Zwick, Germany).

Water Absorption

The test specimens (25 mm x 25 mm) were cut from the laminates and tested for water absorption as per ASTM-D570.

DISCUSSION AND RESULTS

Tensile properties of various jute/polyester composites are given in Table 1.

Table 1.

S. No.	Composite type	Tensile strength, MPa	Tensile modulus, GPa
1	HSUNTR	48.5	9.6
2	HSAL	35.4	–
3	HSALMC	50.3	5.5
4	HUNTR	33.6	–
5	HAL	11.3	–
6	HSTHER	51.2	10.2
7	Polyester	15.2	2.1

As observed from Table 1, both tensile strength and tensile modulus are maximum for HSTHER.

Flexural properties of various jute/polyester composites are given in Table 2.

Table 2.

S. No.	Composite type	Flexural strength, MPa	Flexural modulus, GPa
1.	HSUNTR	54.2	3.1
2.	HSALMC	55.3	2.8
3.	HSTHER	68.2	4.4
4.	POLYESTER	30.6	1.3

As observed from Table 2, flexural properties (both strength and modulus) are maximum for HSTHER.

ILSS and izod impact strength of various jute/polyester composites is given in Table 3.

Table 3.

S. No.	Composite type	ILSS, MPa	Izod impact strength, J/m
1.	HSUNTR	4.5	103.2
2.	HSALMC	6.2	123.3
3.	HSTHER	5.3	85.6
4.	POLYESTER	–	88.4

ILSS and impact strength are maximum for HSALMC.

Effect of filler content on tensile and flexural strength of HSALMC jute/polyester composite is given in Table-4.

Table 4.

S. No.	Filler content, %	Tensile strength, MPa	Flexural strength, MPa
1.	0	47.2	65.6
2.	10	42.8	62.8
3.	20	37.6	59.5
4.	30	33.2	57.6
5.	40	28.3	54.2

It is observed from Table 4 that the effect of filler addition on tensile strength is far more than that on flexural strength. At 40% filler content, tensile strength is reduced by 40% while reduction in flexural strength is only 17.4%.

Moisture absorption by various jute/polyester composites is given table-5.

Table 5.

S. No.	Composite type	Moisture absorption (24 hrs), %
1.	HSUNTR	1.4
2.	HSALMC	0.4
3.	HSTHER	1.6
4.	POLYESTER	0.2

Moisture absorption is the least for HSALMC.

CONCLUSION

In this chapter, the effect of fabric treatment and filler addition on mechanical performance and moisture absorption of jute/polyester composites is investigated. Alkali and thermal treatments were carried out on commercially available whole fabrics. Jute fabric/polyester composites with calcium carbonate filler find many low cost applications.

KEYWORDS

- **Biodegradation**
- **Filler**
- **Hydrophilic**
- **Isophthalic polyester**
- **Thermosetting polyester**

Chapter 12

Control of the Hyaluronidase Activity Towards Hyaluronan by Formation of Electrostatic Complexes: Fundamental and Application Relevance

Brigitte Deschrevel

INTRODUCTION

Hyaluronan (HA) is a glycosaminoglycan (GAG) widely distributed in vertebrate tissues. Due to its unique biophysicochemical properties, it is involved in many biological processes under both normal and pathological conditions. In addition, HA and its derivatives are used in various medical and aesthetic applications. However, the biophysicochemical properties and biological functions of HA strongly depend on its chain size. By catalyzing HA hydrolysis hyaluronidases (HAases) play an important role in the control of the HA chain size. Our study of the kinetics of the HA hydrolysis catalyzed by HAase led us to the conclusion that two phenomena should be taken into account to properly describe the behavior of the hyaluronan/hyaluronidase system: (i) formation of catalytic complexes which leads to HA hydrolysis and (ii) formation of electrostatic hyaluronan hyaluronidase complexes in which hyaluronidase is catalytically inactive. As a consequence, the hyaluronidase activity can be strongly modulated by formation of electrostatic complexes involving either HA and/or hyaluronidase. The present report shows that this knowledge of the behavior of the hyaluronan/hyaluronidase system is important with respect to the detection, quantification, characterization, and use of hyaluronidase. It also strongly suggests that this behavior could be of importance under *in vivo* conditions, since, for example, it was shown that, according to its concentration, a hyaladherin (protein able to specifically bind HA) is able to either enhance or suppress the activity of a tumoral hyaluronidase.

The HA, also known as hyaluronic acid is a polysaccharide which belongs to the GAG family. It is widely distributed in vertebrate tissues and fluids and is the most abundant GAG in mammalian tissues. In human body, HA concentration varies over a wide range; for example, it ranges from 1,420 to 3,600 mg l^{-1} in synovial fluids, it is around 200 mg l^{-1} in dermis and between 0.02 and 0.04 mg l^{-1} in blood serum and plasma (Chichibu et al., 1989; Engstrom-Laurent et al., 1985; Laurent and Fraser, 1986). In connective tissues, HA is one of the main components of the extra-cellular matrix (ECM). The HA is composed by the repetition of D-glucuronic acid-$\beta(1,3)$-N-acetyl-D-glucosamine disaccharide units linked through $\beta(1,4)$ glycosidic linkages (Figure 1). It is unbranched, not sulfated nor modified in any other way throughout its length, all of which makes it the simplest of the GAGs. Under normal physiological conditions, this linear polymer has high molar mass, usually exceeding 10^6 g mol^{-1} (2,500 disaccharide units) with values reaching 10×10^6 g mol^{-1} (25,000 disaccharide units) as, for example, in synovial fluid (Noble, 2002). However, HA molar mass varies according to the species and the tissue and, even in a given tissue, it can have a great polydipersity (Laurent, 1987).

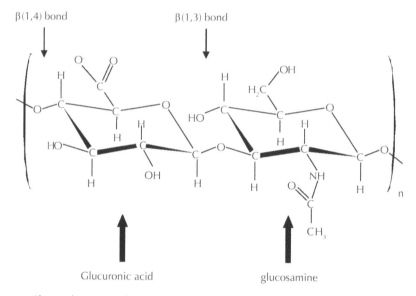

Figure 1. Chemical structure of HA: D-glucuronic acid-β(1,3)-N-acetyl-D-glucosamine disaccharide units linked through β(1,4) glycosidic bonds (Deschrevel, in press).

While HA has long been considered as an inert component acting only as a space filler, it is now established that HA has many biological functions related to its unique biophysicochemical properties, among which, its viscoelastic behavior (Cowman and Matsuoka, 2005; Fouissac et al., 1993) and its ability to retain large amount of water (Laurent, 1970) and to interact with various molecular species (Hardingham and Muir, 1972; Hascall and Heinegard, 1974; Pasquali-Ronchetti et al., 1997; Underhill and Toole, 1979). The HA is actually an excellent space-filling molecule able to adjust to surfaces and to act as a lubricant. It can undergo deformation, as required during growth and tissue remodelling, and has energy-absorbing properties, as required for energy dissipation under dynamic shock imposed to articulations (Cowman and Matsuoka, 2005; Lapčik Jr. et al., 1998; Laurent and Fraser, 1992). The HA plays a significant role in the water homeostasis of the tissues and contributes to the regulation of the distribution and transport of smalls ions and proteins (Gribbon et al., 1999; Laurent, 1987; Laurent and Ogston, 1963). The HA provides a scaffold on which various molecular species can reversibly assemble. In particular, HA forms aggregates with proteoglycans containing sulfated polysaccharides, such as chondroitin sulfate and keratan sulfate (Hardingham and Muir, 1972; Hascall and Heinegard, 1974). The formation of such aggregates involves link proteins (Day and Prestwich, 2002; Yang et al., 1994). These laters belong to a family of proteins termed hyaladherins because of their ability to specifically bind HA. Hyaladherins also include membrane receptors such as CD44 (cluster of differentiation 44) and RHAMM (receptor for hyaluronan-mediated motility) (Day and Prestwich, 2002; Yang et al., 1994). Binding of HA to such receptors can activate intracellular signaling pathways. Thus, hyaladherins play an important role in the involvement of HA in cell proliferation, differentiation, adhesion, and migration (Evanko et al, 1999; Knudson and Knudson, 1993).

As one may have expected, the viscosity of HA solutions strongly increases as the HA chain length is increased (Cowman and Matsuoka, 2005). Interestingly, it was shown that the ability of HA to bind hyaladherins as well as the stability of the resulting complexes also depends on the HA chain length (Courel et al., 2002; Deschrevel et al., 2008b). In other words, the biological functions of HA strongly vary according to its chain length (Stern et al., 2006). From the general point of view, HA of high molar mass is believed to play a homeostatic role, whereas, HA chains of low molar masses act as endogenous danger signals (Scheibner et al., 2006; Stern et al., 2006). For example, it is established that high molar mass HA is anti-angiogenic, anti-inflammatory, and immunosuppressive, while HA chains of low molar masses (below 2.5×10^5 g mol^{-1}) are pro-inflammatory and activate the innate immune system, and HA oligosaccharides (4–25 disaccharides units) are angiogenic (Delmage et al., 1986; Gately et al., 1984; Hogde-Dufour et al., 1997; Liu et al., 1996; McKee et al., 1996, 1997; Takahashi et al., 2005; Taylor et al., 2004; Termeer et al., 2002; Trochon et al., 1997; West and Kumar, 1989).

The many biological functions of HA make it plays important roles under normal conditions but also under pathological conditions since HA is involved, for example, in immune response (Delmage et al., 1986; Gately et al., 1984; Scheibner et al., 2006), wound healing (David-Raoudi et al., 2008) and tumor proliferation and invasion (Stern, 2008). Moreover, the absence of immunogenicity of HA, its biocompatibility, biophysicochemical properties, and biological functions have led to its use in an increasing number of medical and aesthetic applications. However, for several of these applications, HA have two major drawbacks: a high rate of *in vivo* degradation and poor mechanical properties. In order to overcome these drawbacks and to extend the application range of HA, the latter has been the subject of various chemical modifications (Deschrevel, in press). In the medical field, HA or its derivates are used in ophtalmology, osteoarthritis, laryngology, urology, embryo implantation, adhesion prevention, wound healing, tissue engineering, and drug delivery engineering (Deschrevel, in press). Molar mass of HA has always been taken into account with respect to the various applications. However, as the knowledge about the influence of the chain length of HA on its biological functions has increased, molar mass of HA has become a crucial parameter and the demand of HA fragments of well-defined size has increased (Deschrevel, in press).

The knowledge of the pathways of HA synthesis and catabolism (Deschrevel, in press) strongly suggests that HA molar mass distribution is most probably regulated through the catabolic pathway. In addition, according to Mio and Stern (2002), such regulation is kinetically and energetically more efficient through the catabolic, rather than the anabolic, pathway. The catabolic pathway involves HAases, enzymes that catalyze HA cleavage. On the basis of biochemical data, Meyer classified HAases into three differents categories: (i) testicular-type HAases (EC 3.2.1.35), which hydrolyze the $\beta(1,4)$ glycosidic linkages (Figure 2), (ii) leech-type HAases (EC 3.2.1.36), which hydrolyze the $\beta(1,3)$ glycosidic bonds, and (iii) bacterial-type HAases (EC 4.2.99.1), that cleave the $\beta(1,4)$ glycosidic linkages by a β elimination reaction (Deschrevel, in press; Meyer, 1971). In the human, as in all mammals, all the HAases found until now are of the testicular-type. The human genome includes six HAase genes: *Hyal1*, *Hyal2*, *Hyal3*, *Hyal4*, *PH-20/Spam1*, and *HyalP1* (Girish and Kemparaju, 2007; Stern et al., 2007). Among the products of these genes, *Hyal1*, *Hyal2* and *PH-20*, or *SPAM1* (Sperm Adhesion Molecule 1), show HAase activity and are the best characterized. *Hyal3* is widely expressed, but no enzymatic activity has yet been detected for this protein using the available HAase assays. *Hyal4* seems to be a chondroitinase with no activity towards HA. *HyalP1* is a pseudogene, transcribed but not translated in the human (Deschrevel, in press; Girish and Kemparaju, 2007; Stern et al., 2007). However, although the knowledge about HAases did not ceassed to increase over the years, the way by which HAases are involved in the control of the HA molar mass distribution is still to be elucidated. Indeed, rather little is known about the species, the parameters and the phenomena that control HAase activity. Mio and Stern (2002) suggest that HAase could be present in tissues together with inhibitors which may allow HAase to be rapidly activated or inactivated. However, even though some HAase inhibition activities have been detected *in vivo*, very little is known about their molecular basis and mechanism of action (Mio and Stern, 2002).

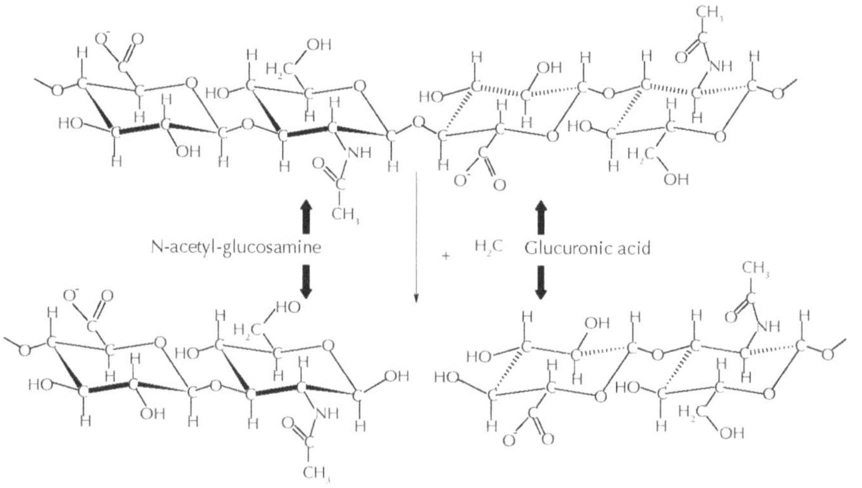

Figure 2. Hydrolysis of the $\beta(1,4)$ glycosidic bonds of HA as catalyzed by testicular-type HAases (EC 3.2.1.35) (Deschrevel, in press).

In the course of the last 10 years, we have studied the kinetics of the HA hydrolysis catalyzed by HAase. An atypical kinetic behavior was observed whose extent varied according to the experimental conditions used. We then showed that, in fact, to properly describe the behavior of the HA/HAase system, the ability of HA to form electrostatic complexes with proteins, including HAase, should be taken into account in addition to the formation of the HA-HAase catalytic complexes. Our studies also showed that, as a consequence, the HAase activity towards HA can be strongly modulated by the formation of electrostatic complexes involving either HA or HAase or both. In addition to review these results and their main conclusions, we discuss here their relevance under both *in vitro* and *in vivo* situations.

HA AND HAase: CATALYTIC COMPLEXES VERSUS ELECTROSTATIC COMPLEXES

For our study of the kinetics of the HA hydrolysis catalyzed by HAase, we used bovine testicular HAase (BT-HAase), which is commercially available, as a model enzyme. Indeed, this enzyme is the bovine counterpart of the human PH-20 and, like all the human HAases, it is of the testicular type. In addition, it was shown that the human HAases share 33.1–41.2% aminoacid sequence identities and a higher degree of structural similarity (Chao et al., 2007; Jedrzejas and Stern, 2005). Among the methods used to study the HA hydrolysis catalyzed by HAase, we chose the colorimetric assay for N-acetyl-D-glucosamine reducing ends, also known as the Reissig method (Reissig et al., 1955) or the Morgan-Elson method (Elson and Morgan, 1933), because it is the only one for which the measurement is directly related to the number of $\beta(1,4)$ bonds cleaved. Indeed, each hydrolysis of a $\beta(1,4)$ bond leads to a one unit increase in the number of HA chains and so, the same for the number of N-acetyl-D-glucosamine reducing ends (Figure 2). We used our improved version of the method which allows to deduce the turbidimetric contribution from the total absorbance measured at 585 nm and thus, to obtain the actual colorimetric contribution. This colorimetric component of the absorbance at 585 nm is proportional to the concentration of N-acetyl-D-glucosamine reducing ends in the sample (Asteriou et al., 2001). Thus, for each kinetic experiment, the concentration of N-acetyl-D-glucosamine reducing ends was plotted against the reaction time. The experimental points were fitted by the bi-exponential equation we developed (Vincent et al., 2003) and the initial hydrolysis rate was calculated as being equal to the value of the first derivative of that function at time zero. Initial hydrolysis rates are thus expressed in μmol of N-acetyl-D-glucosamine reducing ends released per liter of reaction medium and per minute.

Our study of the influence of the HA concentration on the initial rate of the HA hydrolysis catalyzed by BT-HAase under low ionic strength conditions (5 mmol l⁻¹) revealed an atypical behavior (Figure 3) for increasing HA concentrations, the initial hydrolysis rate successively increased, reached a maximum and then decreased to a very low level, close to zero, at high HA concentration, instead of reaching a plateau, as it does for a Michaelis-Menten type enzyme (Asteriou et al., 2002, 2006). The same atypical behavior was observed whatever the BT-HAase concentration used (Figure 3).

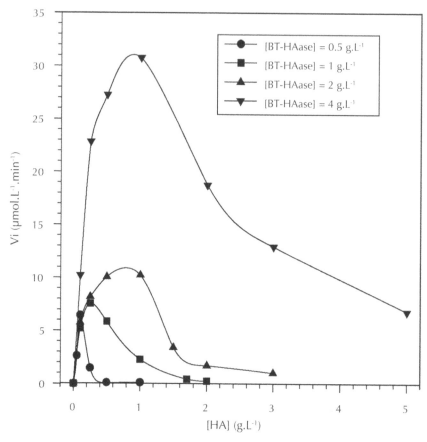

Figure 3. Substrate-dependence of the HA hydrolysis catalyzed by BT-HAase in 5 mmol l⁻¹ ammonium acetate, at pH 5 and at 37°C, for different BT-HAase concentrations ranging from 0.5 to 4 g l⁻¹. The number average molar mass of HA was 1.45×10^6 g mol⁻¹. Data from Asteriou et al. (2006).

However, for any HA concentration, the initial hydrolysis rate increased when the BT-HAase concentration was increased. In fact, analysis of the substrate-dependence curves obtained for various BT-HAase concentrations indicated that the atypical behavior depended on the ratio of the BT-HAase concentration over the HA concentration since there was a roughly constant ratio between (i) the HA concentration for which the initial hydrolysis rate was maximum and the BT-HAase concentration and, (ii) the HA concentration above which the initial rate became close to zero and the BT-HAase concentration (Asteriou et al., 2006). The substrate-dependence curves obtained under various ionic strength conditions showed that the existence and the extent of the atypical behavior of the HA/BT-HAase system strongly depended also on the ionic strength level (Figure 4). The higher the ionic strength level, the higher the HA concentration above which the initial hydrolysis rate decreased and, for the highest ionic strength, no decrease in the initial rate was observed. Moreover, for any

HA concentration, the initial hydrolysis rate increased when the ionic strength was increased. In fact, Bollet et al. (1963) already reported such a low hydrolysis rate at high HA concentration with rat kidney HAase in 0.10 mol l⁻¹ acetate buffer at pH 3.8. Similarly, Aronson and Davidson (1967) mentioned that rat liver HAase in 0.10 mol l⁻¹ acetate buffer at pH 3.9 exhibited inhibition at high HA concentration and that this inhibition was prevented by 0.15 mol l⁻¹ sodium chloride. Nevertheless, these authors did not provide any explanation for their results.

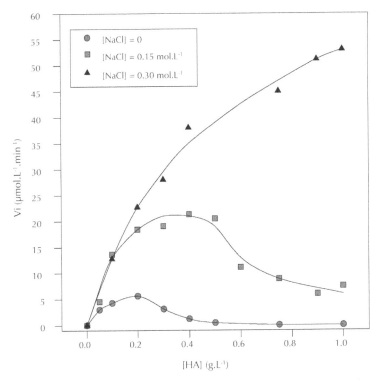

Figure 4. Substrate-dependence of the HA hydrolysis catalyzed by BT-HAase at pH 4 and at 37°C, for different sodium chloride concentrations ranging from 0 to 0.3 mol l⁻¹. The BT-HAase concentration was 0.2 g l⁻¹. The number average molar mass of HA was 0.97 × 10⁶ g mol⁻¹. (unpublished personal data).

In order to explain the atypical behavior of the HA/BT-HAase system, we investigated several hypotheses including a classical inhibition by an excess of substrate, a steric exclusion effect so that the enzyme would not be able to get close enough to any cleavable bonds of the HA molecule and a viscosity effect which would reduce the HAase diffusion rate. However, none of these hypotheses could explain all the experimental results and so, they could not be responsible by themselves for the atypical behavior of the HA/BT-HAase system (Asteriou et al., 2006). In fact, the only one receivable explanation for the atypical behavior was based on the idea according which HA and BT-HAase are able to form electrostatic complexes different from the catalytic complexes.

The fact that each glucuronic acid residue bears a carboxyl group makes HA chains behave as polyelectrolytes whose pK_0 was estimated to be equal to 2.9 (Berriaud et al., 1998; Cleland et al., 1982). This means that, except under very acidic pH conditions, HA is a polyanion. Thus, according to the ionic strength level, HA may be able to form electrostatic complexes with polycations. In fact, the ability of HA to form electrostatic complexes with proteins was shown more than 60 years ago. Indeed, the very first methods developed to assay HAase were based on the measurement of the turbidity resulting from the formation of complexes between long HA chains and serum proteins under acidic conditions: the lower the turbidity, the higher the HAase activity (Deschrevel et al., 2008a). The turbidimetric method was further developed by Tolksdorf et al. (1949), introducing the definition of the turbidity reducing unit based on the turbidity obtained when HA is mixed at pH 3.8 with the bovine serum albumin (BSA). This method still constitutes the current United States Pharmacopeia XXII assay for HAase. More recently, investigations (Gold, 1980; Grymonpré et al., 2001; Malay et al., 2007; Moss et al., 1997; Van Damme et al., 1994; Xu et al., 2000) were devoted to the characterization of the complexes formed between HA and some proteins. For example, Xu et al. (2000) examined the influence of pH on the solubility of the electrostatic HA-BSA complexes and Moss et al. (1997) studied the electrostatic complexes formed between HA and lysozyme (LYS).

For our kinetic experiments, there was neither protein nor any other polycation added into the reaction media. The only one species which could form electrostatic complexes with HA was thus BT-HAase. By using an electrophoretic method, we estimated the isoelectric point (pI) of BT-HAase to be close to 7, which mean that, under acidic pH conditons, the net charge of BT-HAase is positive. Thus, HA and BT-HAase should be able to form electrostatic complexes at least between pH 2.9 and 7. In fact, the first time we experimentally observed the ability of HA and BT-HAase to form electrostatic complexes under acidic pH conditions was when we investigated the origin of the turbidity component of the total absorbance measured at 585 nm in the N-acetyl-D-glucosamine reducing end assay (Asteriou et al., 2001). Then, by using HA and BSA, as a model system for the formation of electrostatic HA protein complexes, and by performing turbidimetric measurements, we confirmed the existence of electrostatic HA-BT-HAase complexes under the pH conditions we used in our kinetic studies (Deschrevel et al., 2008a; Lenormand et al., 2008). In fact, in the case of the HA/BT-HAase system, study of the electrostatic complexes is complicated by the fact that the enzymatic hydrolysis of HA leads to a decrease in the HA chain length and, we showed, using the HA/BSA system, that the formation of electrostatic complexes depends on the size of HA chains (Lenormand et al., 2010b).

The investigations we carried out in order to explain the kinetic atypical behavior of the HA/BT-HAase system thus led us to the conclusion that, in fact, two phenomena must be considered to properly describe the behavior of that system: on one hand, the formation of catalytic complexes between HA and BT-HAase which leads to HA hydrolysis, and, on the other hand, the formation of electrostatic complexes between

HA and BT-HAase in which BT-HAase is catalytically inactive (Deschrevel et al., 2008a). Let us see how this conclusion allows us to explain the atypical shape of the substrate-dependence curves. When we studied the influence of the HA concentration on the initial rate of HA enzymatic hydrolysis, we used a fixed BT-HAase concentration. However, when using experimental conditions under which BT-HAase formed electrostatic complexes with HA, the higher the HA concentration, the higher the part of BT-HAase involved in electrostatic complexes. But, since BT-HAase involved in electrostatic complexes was catalytically inactive, increasing the HA concentration also meant a decrease in the concentration of active BT-HAase. Thus, in the first part of the substrate-dependence curve (Figure 5, part 1), the increase in the HA concentration enabled the initial hydrolysis rate to increase despite of the decrease in the concentration of active BT-HAase. On the contrary, in the second part of the curve (Figure 5, part 2), the increase in the HA concentration was no-longer sufficient to compensate for the decrease in the concentration of active BT-HAase and thus, the initial hydrolysis rate decreased. Then, when all the BT-HAase molecules had formed electrostatic complexes with HA, the initial hydrolysis rate became equal to zero, which meant that the concentration of active BT-HAase was nil. It should be noted that if BT-HAase involved in electrostatic HA-BT-HAase complexes was catalytically active, no decrease in the initial hydrolysis rate would have been observed at high HA concentrations (Deschrevel et al., 2008a).

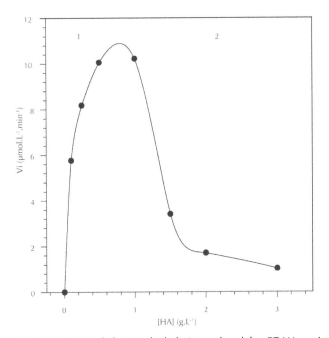

Figure 5. Substrate-dependence of the HA hydrolysis catalyzed by BT-HAase in 5 mmol l^{-1} ammonium acetate, at pH 5 and at 37°C, for a BT-HAase concentration of 2 g l^{-1}. The number average molar mass of HA was 1.45×10^6 g mol^{-1}. The dashed line allows to distinguish between the two characteristic parts of the atypical curve (see text). Data from Asteriou et al. (2006).

As we have observed on Figure 3, the initial rate of HA hydrolysis catalyzed by BT-HAase depended on the ratio of the BT-HAase concentration over the HA concentration. Indeed, the higher the ratio of the BT-HAase concentration over the HA concentration, the higher the concentration of active BT-HAase and the higher the initial hydrolysis rate. Nevertheless, the ability of HA and BT-HAase to form electrostatic complexes foremost depends on both pH and ionic strength conditions. As mentioned above, HA and BT-HAase are able to form electrostatic complexes at least between pH 2.9 and 7, pH domain in which simultaneously HA behaves as a polyanion and BT-HAase as a polycation. Moreover, the higher the ionic strength level, the lower the stability of the electrostatic HA-BT-HAase complexes. Indeed, an increase in the ionic strength level leads to an increase in the screening of the charges borne by HA and BT-HAase by small ions and thus, to a decrease in the ability of these two polyelectrolytes to form electrostatic complexes. This is in very good agreement with the fact that the atypical behavior tended to decrease, up to disappear, when the ionic strength of the reaction medium was increased (Figure 4). However, our experiments showed that the formation of electrostatic complexes between HA and BT-HAase can occur under physiological-type ionic strength since the atypical behavior was observed with an ionic strength equal to 0.15 mol l^{-1} (Lenormand et al., 2008).

ENHANCEMENT/SUPPRESSION OF HAase ACTIVITY TOWARDS HA BY POLYCATIONS AND POLYANIONS

As mentioned above, various proteins are able to form electrostatic complexes with HA. We thus decided to investigate the effect of the presence of a protein with no catalytic activity towards HA on the kinetics of the HA hydrolysis catalyzed by BT-HAase. For that purpose, we chose to first use BSA. Indeed, BSA is known as to be able to form electrostatic complexes with HA. In addition, BSA is a major protein of synovial fluid (Scott et al., 2000), a fluid which is also rich in HA (see Introduction section). Figure 6 shows BSA-dependence curves (that is to say, initial hydrolysis rate plotted as a function of the BSA concentration) obtained for various HA concentrations, at pH 4 and at low ionic strength (5 mmol l^{-1}). We can observe (Figure 6) that (i) the initial rate of HA hydrolysis catalyzed by BT-HAase strongly depended on the BSA concentration and (ii) the BSA-dependence curves had all the same shape whatever the HA concentration (Lenormand et al., 2009).

In fact, the BSA-dependence curves can be divided into four domains (Figure 7) (Deschrevel et al. 2008a; Lenormand et al., 2009):

- The first domain corresponds to BSA concentrations ranging from zero to A. When the BSA concentration is nil, nearly all the BT-HAase molecules form electrostatic complexes with HA. The concentration of catalytically active BT-HAase is thus close to zero, which makes the initial hydrolysis rate extremely low. For increasing BSA concentrations up to A, the added BSA molecules use the space remaining free on HA molecules to form electrostatic complexes. This has no effect on the initial hydrolysis rate which remains close to zero. It should be noted however that domain 1 exists only for low values of the ratio of the BT-HAase concentration over the HA concentration.

- In domain 2, the initial hydrolysis rate strongly increases when the BSA concentration is increased. In fact, as expected, a competition between BT-HAase and BSA for the formation of electrostatic complexes with HA occurs and leads to the release of BT-HAase molecules. So, for increasing BSA concentrations, the concentration of active BT-HAase increases and this allows the initial rate to increase. Under these conditions, the added BSA clearly acts as an enhancer of the BT-HAase activity. Then, when the BSA concentration is equal to B, all the BT-HAase molecules are free, and thus catalytically active and the initial hydrolysis rate reach its maximum value.

- In domain 3, all the BT-HAase molecules are free. The concentration of active BT-HAase is equal to its maximum value whatever the BSA concentration. Nevertheless, added BSA molecules continue to form electrostatic complexes with HA and this leads to a progressive decrease in the β(1,4) bonds accessible to the BT-HAase molecules. In other words, in that BSA concentration range, when the BSA concentration is increased, the substrate concentration progressively decreases and, as a consequence, the initial rate of HA hydrolysis decreases too. In these cases, BSA acts as a suppressor of the BT-HAase activity.

- In domain 4, the HA molecules being saturated by the BSA molecules, the concentration of β(1,4) bonds accessible to BT-HAase is close to zero. Thus, although all the BT-HAase molecules are catalytically active, they cannot catalyze the HA hydrolysis and the initial hydrolysis rate is nil. In this domain, in the same way as in domain 3, BSA acts as a suppressor of the BT-HAase activity.

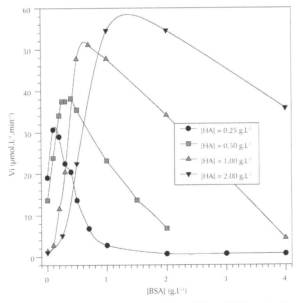

Figure 6. BSA-dependence of the HA hydrolysis catalyzed by BT-HAase in 5 mmol l⁻¹ sodium chloride, at pH 4 and at 37°C, for different HA concentrations ranging from 0.25 to 2 g l⁻¹. The BT-HAase concentration was 0.5 g l⁻¹. The number average molar mass of HA was 0.97×10^6 g mol⁻¹. Data from Lenormand et al. (2009).

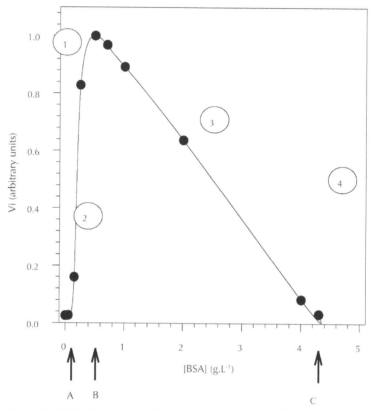

Figure 7. Generalized BSA-dependence of the HA hydrolysis catalyzed by BT-HAase. The dashed lines allow to distinguish between the four characteristic domains of the curve, and A, B and C are characteristic BSA concentrations corresponding to the junctions between the domains (see text). Model curve from Lenormand et al. (2009).

The effects of serum proteins, in particular BSA, on HAase activity were already reported by some authors. According to the first studies, reviewed by Mathews and Dorfman (1955), serum, whatever its origin, inhibited HAase. Such inhibitions may be explained by the fact that the total protein concentration in serum was high as compared to the HA and HAase concentrations used, so that the formation of electrostatic complexes between HA and many serum proteins reduced the concentration of HA β(1,4) bonds accessible to HAase. On the contrary, Gacesa et al. (1981) observed a marked enhancement of BT-HAase activity upon addition of serum to incubation mixtures at pH below 5 and in the presence of 0.1 mol l⁻¹ sodium chloride. These authors noted that the extent of activation was largely dependent upon the ratio of enzyme over serum quantities. They further showed that the greatest activation effect was obtained by using BSA or human serum albumin. Gold (1982) showed that both BT-HAase and human liver HAase exhibited increased activity in the presence of BSA at pH 4. All these observations are in total agreement with our above-mentioned results.

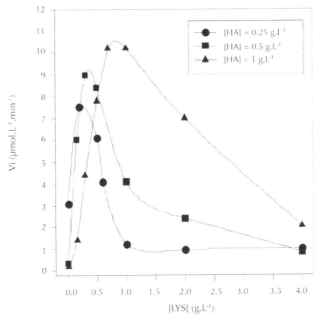

Figure 8. LYS-dependence of the HA hydrolysis catalyzed by BT-HAase in 5 mmol l⁻¹ sodium chloride, at pH 5.25 and at 37°C, for different HA concentrations ranging from 0.25 to 1 g l⁻¹. The BT-HAase concentration was 0.5 g l⁻¹. The number average molar mass of HA was 0.97 × 10⁶ g mol⁻¹. Data from Lenormand et al. (2010a).

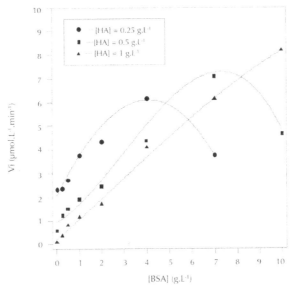

Figure 9. BSA-dependence of the HA hydrolysis catalyzed by BT-HAase in 5 mmol l⁻¹ sodium chloride, at pH 5.25 and at 37°C, for different HA concentrations ranging from 0.25 to 1 g l⁻¹. The BT-HAase concentration was 0.5 g l⁻¹. The number average molar mass of HA was 0.97 × 10⁶ g mol⁻¹. Data from Lenormand et al. (2010a).

In order to verify that such a modulation of the BT-HAase activity was not specific to albumin, we performed experiments with LYS, another non-catalytic protein, present together with HA in cartilage. The LYS-dependence curves obtained for various HA concentrations, at pH 5.25 and at low ionic strength (5 mmol l⁻¹) are shown on Figure 8. We can observe that these LYS-dependence curves had exactly the same shape as the BSA-dependence curves shown on Figure 6. This means that, in the same way as BSA, LYS is able to either enhance or suppress BT-HAase activity according to its concentration. We also performed experiments with BSA by using exactly the same experimental conditions as those used to study the effect of the LYS concentration on the initial rate of HA hydrolysis catalyzed by BT-HAase. The BSA-dependence curves thus obtained are shown on Figure 9. Comparison between the curves on Figures 8 and 9 clearly shows that under the experimental conditions used, LYS had a higher ability to form electrostatic complexes with HA than BSA since, for any given HA concentration, the concentration of non-catalytic protein giving the maximum value of the initial hydrolysis rate was higher for BSA than for LYS. This difference in behavior between LYS and BSA with respect to their ability to form electrostatic complexes with HA comes from the difference in their pI values: pI of LYS was estimated to 10.6 (Hoon Han and Lee, 1997) and that of BSA is close to 5.2 (Wang et al., 1996; Xu et al., 2000). Thus, at pH 5.25, the net charge of LYS is positive whereas that of BSA is nearly nil. In fact, the ability of BSA molecules to form electrostatic complexes with HA at pH 5.25 was due to the existence of positive patches on the protein surface (Grymonpré et al., 2001). Moreover, we performed experiments by using poly-L-lysine, a synthetic polycation, instead of a non-catalytic protein. According to its concentration, poly-L-lysine was able to either increase or decrease the initial rate of HA hydrolysis (Figure 10). In other words, the presence of poly-L-lysine in the reaction medium had exactly the same effect on the BT-HAase activity as the addition of the non-catalytic proteins.

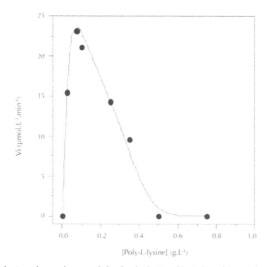

Figure 10. Poly-L-lysine-dependence of the hydrolysis of HA (1 g l⁻¹) catalyzed by BT-HAase in 5 mmol l⁻¹ sodium chloride, at pH 4 and at 37°C, for a BT-HAase concentration of 0.5 g l⁻¹. The number average molar mass of HA was 0.97 × 10⁶ g mol⁻¹. (Unpublished data).

All these results clearly indicate that the BT-HAase activity towards HA can be either enhance or suppress by formation of electrostatic complexes. The BT-HAase activity can be enhance by polycations, which *in vivo* are mainly proteins, to the condition that (i) they are able, by forming electrostatic complexes with HA, to avoid, or at least to limit, the formation of electrostatic complexes between HA and BT-HAase and, (ii) the ratio of the polycation over HA quantities in the HA-polycation complexes allows enough HA $\beta(1,4)$ bonds to remain accessible to BT-HAase. Suppression of the BT-HAase activity may result from the formation of two types of electrostatic complexes: (i) HA-polycation complexes in which the accessibility of BT-HAase to the HA $\beta(1,4)$ bonds is hindered because of a too high value of the ratio of the polycation over HA quantities in the complexes and, (ii) polyanion-BT-HAase complexes in which BT-HAase is catalytically inactive. In our study, the only one polyanion was HA and we showed that it is able to form electrostatic complexes with BT-HAase in which BT-HAase is catalytically inactive. Nevertheless, many polyanions, including GAG other than HA (heparin, heparan sulfate, dermatan sulfate), HA derivatives (O-sulfonated HA) and synthetic polyanions (poly(styrene-4-sulfonate)) are known to inhibit HAase (Aronson and Davidson, 1967; Girish and Kemparaju, 2005; Isoyama et al., 2006; Mathews and Dorfman, 1955; Toida et al., 1999). In the case of heparin, Maksimenko et al. (2001) demonstrated that the inhibition results from the formation of electrostatic heparin-HAase complexes.

RELEVANCE OF THE PROCESS OF ENHANCEMENT/SUPPRESSION OF THE HAASE ACTIVITY BY FORMATION OF ELECTROSTATIC COMPLEXES

The results of our studies of the BT-HAase activity towards HA clearly show that the enzymatic activity of BT-HAase can be strongly modulated by formation of electrostatic complexes involving either HA or BT-HAase or both. The question then is what could be the role of the process of enhancement/suppression of the HAase activity by formation of electrostatic complexes under *in vitro* and *in vivo* conditions? Concerning *in vitro* conditions, we show below that it could play an important role in the detection, quantification, characterization and use of HAase. In the case of *in vivo* conditions, even though it is much more difficult to unambiguously demonstrate the role of the process of enhancement/suppression of the HAase activity by formation of electrostatic complexes, this question is discussed below on the basis of experimental data related to the involvement of HA and HAase in cancer.

Detection, Quantification, Characterization and use of HAase

The ability of HA and HAase to form electrostatic complexes either with each other or with other polyelectrolytes very probably accounts for some of the difficulties encountered in attempts to detect, to quantify and to characterize HAase activity. Indeed, in addition to being present at exceedingly low concentration (Stern, 2005), it was observed by some authors that for highly purified HAase preparations no enzymatic activity was detectable in the absence of added proteins (Maingonnat et al., 1999; Mathews and Dorfman, 1955). According to our results, we may suggest that when these purified HAase preparations were mixed with HA, HAase underwent electrostatic complex formation with HA and was thus unable to catalyze HA hydrolysis. We can observe on Figure 4 that, even in the presence of 0.15 mol l^{-1} ionic strength, for low ratios of the BT-HAase concentration over the HA concentration the initial hydrolysis rates became close to zero. Under the same ionic strength condition and with the same BT-HAase concentration, the initial hydrolysis rate could be much higher to the condition that the HA concentration used was lower, so that the ratio of the BT-HAase concentration over the HA concentration was higher (Figure 4). When HAase was assayed in purified preparations, the presence of added proteins such as BSA allowed the formation of electrostatic complexes between HA and the added protein and thus, prevented electrostatic HA-HAase complex formation. On the contrary, when HAase is assayed in biological samples, addition of proteins to the reaction mixture may be not appropriated. In such cases, the total protein content may be high enough for HA to be saturated by proteins, so that HA $\beta(1,4)$ bonds are not accessible to HAase. Thus, although HAase is present, HAase activity may be not detected in reaction mixtures containing too high levels of proteins.

Concerning the characterization of the HAase activity, a matter of interest is that of the influence of pH. Indeed, a great diversity of pH-dependence curves of HAase activity were published. Considering a given HAase, as for example, BT-HAase, there are two main reasons for this diversity. The first one concerns the experimental methods used to assay HAase activity. Indeed various methods were used to study the effect of pH on the HAase activity and, using exactly the same experimental conditions, different pH-dependence curves can be obtained. For example, Hofinger et al. (2008) found a maximum BT-HAase activity at pH 3.5 when using the N-acetyl-D-glucosamine reducing end assay whereas it was around 5.5–6 with a turbid metric method. The second reason is that substrate-dependence curves were obtained by using a broad range of experimental conditions in terms of ratio of the HAase concentration over the HA concentration, purity of the HAase preparation, ionic strength and added species (buffer, BSA, …). According to our above-mentioned results, all these experimental conditions are precisely important with respect to the HAase activity and also to the HA and HAase abilities to form electrostatic complexes. During our study of the pH-dependence of the BT-HAase activity, we performed HA enzymatic hydrolysis experiments in the presence of a low ionic strength (5 mmol l^{-1}) by adding either BSA or LYS at various concentrations into the reaction medium. The initial hydrolysis rates we determined from the kinetic curves are reported on Figures 11 and 12 (Lenormand et al., 2010a). We can observe that in the absence of non-catalytic protein, the BT-HAase activity was detectable only between pH 3.75 and 4.75, whereas, in the presence of BSA (Figure 11), it was detectable at least between pH 3.5 and 5.25 and, in the presence of LYS (Figure 12), BT-HAase activity could be detected in a pH domain ranging at least from 3 to 9. These results allow to explain those obtained by Maingonnat et al. (1999) when they studied the activity of a tumoral HAase produced by a cell line derived from a brain metastasis. They observed that, whatever the pH between 3.3 and 4.4, the tumoral HAase activity was strongly increased in the presence of either BSA or an HA binding protein. Similarly, our results on Figure 11 are in total agreement with those of Gacesa et al. (1981) according which, in a pH domain ranging from 3 to 5, BT-HAase activity was strongly increased by addition of human serum or albumin to the reaction mixtures. Moreover, results on Figure 4 make it possible to explain the increase in BT-HAase activity between pH 3.5 and 5 reported by Gacesa et al. (1981) when they increased the sodium chloride concentration from 0.176 to 0.44 mol l⁻¹.

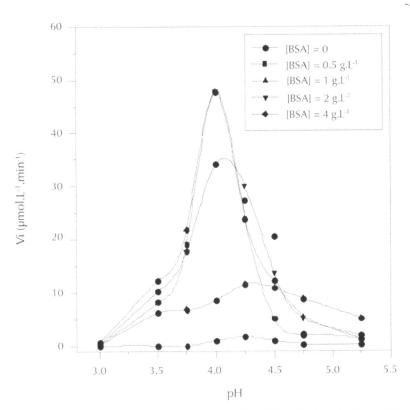

Figure 11. pH-dependence of the HA hydrolysis catalyzed by BT-HAase in 5 mmol l⁻¹ sodium chloride and at 37°C, for different BSA concentrations ranging from 0 to 4 g l⁻¹. The HA concentration was 1 g l⁻¹ and the BT-HAase concentration was 0.5 g l⁻¹. The number average molar mass of HA was 0.97×10^6 g mol⁻¹. Data from Lenormand et al. (2010a).

In fact, to be able to characterize the intrinsic HAase activity, the formation of electrostatic complexes involving either HA and/or HAase should be avoided. From the experimental point of view, the only way to achieve this is the use of a high level of ionic strength. According to our results (Figure 4), ionic strength should be higher than 0.15 mol l⁻¹, an ionic strength level usually considered as to be the physiological one. This probably constitutes one of the key reasons which allowed some authors in the recent years to detect BT-HAase and human testicular HAase (pH-20) activities at pH values equal to or higher than 7 (Franzmann et al., 2003; Hofinger et al., 2008). For example, by using an ionic strength of at least 0.25 mol l⁻¹ and 0.2 g l⁻¹ BSA, Franzmann et al. (2003) found an enzymatic activity for the human testicular HAase only about 2.5 lower at pH 8 than at pH 4.5.

For many years, HAases have been used as a spreading factor. The HAase preparations are used as adjuvant to increase the absorption and dispersion of injected drugs into the tissue ECM. On the basis of our results, we may suggest that efficiency of such HAase preparations would strongly depend on the composition of the ECM of the tissues where they are injected. Moreover, HAases are also used to prepare HA fragments with well-defined sizes (Deschrevel, in press). We showed that with a HA concentration of 5 g l^{-1}, at pH 4 and at 37°C, HA fragments of different ranges of molar masses can be produce by properly selecting the HAase concentration, the ionic strength and the end reation time (Tranchepain et al., 2006). The HA fragments were then purified by using size exclusion chromatography followed by dialysis for salt elimination. More recently, we showed that BSA could be efficiently used instead of salt to increase the rate of HA hydrolysis catalyzed by BT-HAase. Thereby, we were able to enzymatically produce high quantities of HA tetrasaccharide within a rather short reaction time and, above all, because the reaction mixture did not contain any salt, purification could be rapidly and efficiently performed by using an ultra filtration method (unpublished results).

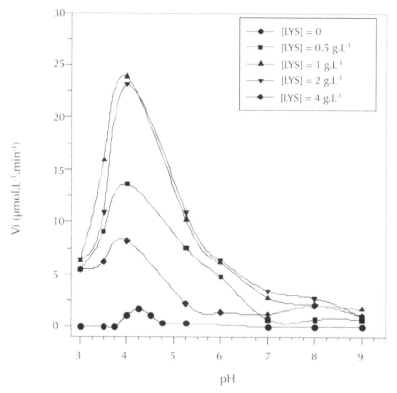

Figure 12. pH-dependence of the HA hydrolysis catalyzed by BT-HAase in 5 mmol l^{-1} sodium chloride and at 37°C, for different LYS concentrations ranging from 0 to 4 g l^{-1}. The HA concentration was 1 g l^{-1} and the BT-HAase concentration was 0.5 g l^{-1}. The number average molar mass of HA was 0.97 × 10^6 g mol^{-1}. Data from Lenormand et al. (2010a).

Involvement of HA and HAase in Cancer: The Role of HA Binding Proteins

For several years, the involvement of the HA/HAase system in cancer have been the subject of many research works. Indeed, many reports indicate elevated levels of both HA and HAase in several tumors including carcinomas of the breast, colon, prostate, bladder, lung, and ovary (Auvinen et al., 2000; Bertrand et al., 1997; Hautmann et al., 2001; Lokeshwar and Selzer, 2008; Lokeshwar et al., 1996; Paiva et al., 2005; Simpson, 2006; Stern, 2008). According to various results, HA and HAase are involved in cancer progression and invasion. Elevated levels of HAase in tumor are often correlated to tumor aggressiveness (Bertrand et al., 1997; Kovar et al., 2006; Lokeshwar and Selzer, 2008; Lokeshwar et al., 1996; Stern, 2008). The activity of HAase is higher in the most invasive forms of cancer and metastases than in primary tumors (Bertrand et al., 1997; Delpech et al., 2002). For example, HAase activity is more elevated in brain metastases than in primary tumors of the brain (Delpech et al., 2002). Identical differences are found when primary breast tumors are compared to node or muscle metastases of breast tumors (Bertrand et al., 1997). Moreover, cancer cells, which regularly produce metastases when grafted to nude mice, are cells that produce high HAase activity in culture (Delpech et al., 2001). Conversely, cell lines that do not express HAase activity do not produce metastases when grafted to nude mice. Quantification of HA in human breast carcinomas showed that it is significantly predominant in the peripheral invasive area of the tumor (Bertrand et al., 1992). In addition, angiogenic HA fragments were found in the saliva of patients with head and neck squamous cell carcinoma, in the urine of patients with high-grade bladder cancer and in the tissue extracts of high-grade prostate tumors (Franzmann et al., 2003; Lokeshwar and Selzer, 2008). All these results strongly suggest that the HA/HAase system plays an active role in tissue invasion by cancer cells. The role of HAase in tumor invasion could be related to the angiogenic switch (Folkman, 2002): high molar mass HA could form a hydrated HA matrix appropriate for tumor cell attachment and enabling the flow of nutriments at the primary site. When simple diffusion becomes no longer sufficient, HAase activity generates angiogenic HA fragments that induce the neovascularization required for tumor progression.

In order to investigate the behavior of a tumoral HAase as compared to that we observed for BT-HAase, we performed experiments with a human cancer cell HAase (H460M-HAase) purified from cultures of a lung carcinoma cell line (H460M) by Delpech and his collaborators (Maingonnat et al., 1999). The substrate-dependence curve obtained with the H460M-HAase (Figure 13) had exactly the same atypical shape as that obtained with BT-HAase in the presence of 0.15 mol l^{-1} ionic strength (Figure 4). This strongly suggests that, like BT-HAase, H460M-HAase has the ability to form electrostatic complexes with HA in which it is catalytically inactive. As a consequence, the H460M-HAase activity decreases as the ratio of the H460M-HAase concentration over the HA concentration is decreased. According to these results, with a high HA concentration, as found in tumor tissues (Auvinen et al., 2000; Hautmann et al., 2001; Paiva et al., 2005; Simpson, 2006), a high HAase level is required to obtain a significant HAase activity.

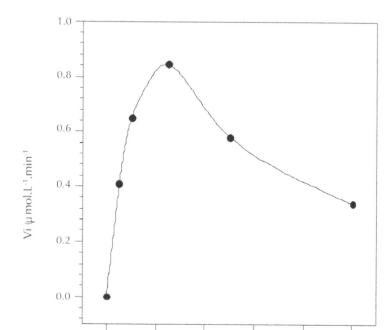

Figure 13. Substrate-dependence of the HA hydrolysis catalyzed by H460M-HAase in 0.15 mol l⁻¹ ammonium nitrate, at pH 3.5 and at 37°C, for a H460M-HAase concentration of 51 U l⁻¹. The H460M-HAase concentration was determined according to the procedure described in Delpech et al. (1995). The number average molar mass of HA was 1.45×10^6 g mol⁻¹. (unpublished personal data).

Moreover, Delpech and his collaborators showed that the H460M-HAase activity, in the presence of a 0.2 mol l⁻¹ ionic strength and at pH 3.8, was dependent on the BSA concentration in the reaction medium: when the BSA concentration was increased, H460M-HAase activity increased reached a maximum and then decreased (Maingonnat et al., 1999). In good agreement with the results we obtained with LYS and poly-L-lysine, they found that foetal calf serum immunoglobulins, human serum albumin, hemoglobin and transferrin were also able to increase the H460M-HAase activity (Maingonnat et al., 1999). All these results clearly indicate that, in the same way as for BT-HAase, the enzymatic activity of the human tumoral HAase (H460M-HAase) can be either enhanced or suppressed by non-catalytic proteins according to their concentration.

The group of Delpech extensively studied hyaluronectin (HN), a hayladherin which is the HA binding moiety of the proteoglycan versican. They showed that HN has a great and specific affinity for HA (Delpech et al., 1995). Like for many other hyaladherins, electrostatic interactions play an important role in the HN binding to HA (Day, 2001). Very interestingly, Delpech and his collaborators showed that, according to its concentration, HN was able to either enhance or suppress the H460M-HAase activity (Maingonnat et al., 1999). In other words, it means that a hyaladherin behave exactly in the same way as a polycationic non-catalytic protein, such as BSA, with respect to the enzymatic activity of a human tumoral HAase towards HA.

When studying HN, Delpech and his collaborators also obtained interesting results about the involvement of HN in tumors. They found that HN concentrations were lower (i) in invasive tumors and metastases as compared to poorly-invasive tumors (Delpech et al., 1993) and (ii) in invasive peripheral areas than in the central areas of breast cancer (Bertrand et al., 1992). Similarly, using H460M cells transfected by human HN cDNA, it was shown that the HN-rich clones developed much fewer metastases than HN-poor clones (Paris et al., 2006). In addition, HN was shown to be able to reduce the tumor growth rate of grafted human glioblastoma, while it was unable to do so when previously heat-inactivated so that it was unable to bind HA (Girard et al., 2000).

Altogether these results strongly suggest that, at rather low concentration, by forming complexes with HA, HN could prevent the formation of electrostatic complexes between HA and tumoral HAase and thus, enable tumoral HAase to be active. HA hydrolysis catalyzed by tumoral HAase would thus generate angiogenic HA fragments which in turn would enhance tumor progression. On the contrary, high levels of HN, leading to the formation of dense HA-HN complexes, would hinder tumoral HAase accessibility to HA $\beta(1,4)$ bonds. Under such conditions, even though tumoral HAase would be present, it could not catalyze HA hydrolysis and, as a consequence, tumor progression would be reduced. So, by modulating tumoral HAase activity through the formation of HA-HN complexes, HN could act either as a promoter of tumor growth and metastasis or as a metastatic spread suppressor.

The HN is not the only protein that can play a role in the HAase hydrolytic activity associated with tumor progression. Indeed, like HN, metastatin, and HA-binding complex from cartilage, inhibited angiogenesis, tumor growth, and metastasis (Liu et al., 2001). These inhibitions were not observed with heat-inactivated metastatin and with metastatin preincubated with HA, which suggests that metastatin must bind to HA to exhibit inhibition. In addition, it was shown that the binding of the link protein on HA molecules of the proteoglycan aggregates from cartilage protected HA from degradation by HAase (Rodriguez and Roughley, 2006). Inhibition of tumor growth and metastasis was also reported with the soluble form of CD44 and RHAMM. Overexpression of soluble CD44 in mouse mammary carcinoma cells or in human malignant melanoma cells led *in vivo* to inhibition of growth, local invasion and metastasis, while no significant effects were obtained if the soluble CD44 was mutated such that it could not bind to HA (Toole, 2002). All these results strongly support the idea that binding of various proteins to HA can control its hydrolysis catalyzed by HAase and thus are involved in the aggressiveness of tumors. Indeed, tumor progression and metastasis is dependent on the HA size distribution, which is related to HAase activity, and, according to our results, the HAase activity is itself dependent on the relative levels of HA, HAase and proteins.

CONCLUSION

The main conclusion of our study of the HA/HAase system is that the HA hydrolysis catalyzed by HAase is strongly modulated by the formation of electrostatic complexes. Indeed, on one hand, polyanions, such as HA, other GAGs and synthetic polyanions, by forming electrostatic complexes with HAase, can suppress its enzymatic activity. In such a case, HAase is present but is catalytically inactive. On the other hand, polycations, such as proteins or synthetic polycations, able to compete with HAase for the formation of electrostatic complexes with HA, can either enhance or suppress HAase activity according to their concentration. At rather low concentrations, by preventing HAase to undergo electrostatic complex formation with HA, polycations allow HAase to be free and thus active. At high concentrations, the electrostatic HA-polycation complexes are too much dense to allow the formation of catalytic complexes between HA and HAase. In that case, while HAase is present and catalytically active, no HAase activity can be detected. Moreover, it should be noted that, as shown in the present study, the modulation of the HAase activity by formation of electrostatic complexes can occur in a large pH domain, including pH 7, and at an ionic strength at least equal to 0.15 mol l^{-1}, ionic strength value usually considered as the physiological one.

The role played by the formation of electrostatic complexes involving either HA or HAase or both in the behavior of the HA/HAase system allows to explain some of the difficulties encountered in attempts to detect HAase activity and some of the discrepancies in the results obtained when quantifying and characterizing the HAase activity. This knowledge about the behavior of the HA/HAase system should be useful for improving the selection of the experimental conditions used to detect, quantify and characterize the HAase activity. Concerning the use of HAase, we showed that this knowledge was really useful for developing efficient enzymatic methods allowing us to produce HA fragments of well-defined size.

The results described in the present report also allow to suggest that the process of enhancement/suppression of the HAase activity by formation of electrostatic complexes could be of importance *in vivo*, as for example, in cancer invasion. Indeed, it was shown that the activity of a human tumoral HAase can be either enhance or suppress by a hyaladherin according to its concentration. Through the data related to the levels of HAase, HA, and the considered hyaladherin in tumors, this result was well correlated to tumor aggressiveness. Similar correlations, between tumor invasion and hyaladherin levels, were also established for several other hyaladherins. According to these results, rather low levels of hyaladherins would thus allow HAase to be able to produce angiogenic HA fragments required for tumor progression, while high levels of hyaladherins would suppress HAase activity and prevent tumor progession. If hyaladherins play such a role in the control of tumoral HAase activity and, as a consequence, in tumor progression, they could constitute the basis of new therapeutic possibilities.

KEYWORDS

- **Electrostatic complexes**
- **Hyaladherin**
- **Hyaluronan**
- **Hyaluronidase activity**
- **pH-dependence**
- **Tumor invasion**

Chapter 13

Spider Silk-production and Biomedical Applications

Anna Rising, Mona Widhe, and My Hedhammar

INTRODUCTION

Spider silk is Nature's high performance fiber with an unprecedented ability to absorb energy, due to a combination of strength and extensibility. Also, spider silk has been ascribed abilities to stop bleedings and promote wound healing (Bon, 1710-1712; Newman and Newman, 1995). These traits have made spider silk an attractive material for biomedical applications. While the outstanding mechanical properties of spider silk have been well documented (see e.g., Gosline et al., 1999; Hu et al., 2006), the suggested utility of spider silk has been hampered by lack of large scale production. Spiders are territorial and produce low amounts of silk and can therefore not be employed as such for industrial silk production. An alternative then is recombinant expression of spider silk proteins (spidroins), or designed proteins with sequences inspired by the overall properties of spidroins. The nature of spidroins (i.e. they are large, repetitive, and aggregation prone) pose significant challenges to their production. Several quite different production strategies have been described, the most common being prokaryote expression systems (e.g., *Echerichia coli*) and processes that involve use of precipitation and resolubilization procedures, *cf* below and (Rising et al.) potential applications of recombinant spider silk include implants to restore the function of damaged tissue (e.g., tendon repair) as well as matrices for cell culture and tissue engineering. In line with this, some recombinant spider silks have been used in cell culture studies and also there are a few reports on the *in vivo* tolerance of this material.

Spider Silk

Some spiders have up to seven different types of silk glands (Candelas and Cintron, 1981), each producing a silk with a specific purpose and unique mechanical properties. The dragline silk is the most studied silk, it is a strong and extendible fiber used to make the framework of the webs and as a lifeline (Gosline et al., 1999). The dragline silk is composed mainly of two similar proteins, major ampullate spidroin (MaSp) 1 and 2 (Hinman and Lewis, 1992). These proteins are large, approximately 3500 amino acid residues long and can be encoded by several gene loci (Ayoub et al., 2007; Rising et al., 2007). The MaSps have a tripartite composition of a non repetitive N-terminal domain (~130 amino acid residues) (Rising et al., 2006), an extensive repetitive region (Ayoub et al., 2007), and a non repetitive folded C-terminal domain of ~110 residues (Hagn et al., 2010). In the repetitive region, tandem repeat units of a glycine rich repeat followed by an alanine block appear consecutively about 100 times (Ayoub et al., 2007).

The MaSps are produced by a single layered epithelium in paired major ampullate glands that are connected via secretory ducts to the spinnerets on the ventral surface of the abdomen (Kovoor, 1987). The spidroins can be stored in the glandular lumen until they are converted into a solid fiber in the end of the secretory duct (Chen et al., 2002; Work, 1977). Even though the dragline silk proteins are prone to assemble, the spider manages to keep them in soluble form at very high concentrations (30-50%, w/v)(Chen et al., 2002; Hijirida et al., 1996). The fiber forming process is well controlled, enabling an almost instantaneous formation of the fiber in a defined segment of the duct, thus avoiding a fatal spread of the assembly process to the dope in the gland (Work, 1977). This process probably involves shear forces, dehydration, a drop in pH and changes in the ion composition along the duct (Chen et al., 2002; Dicko et al., 2004; Knight and Vollrath, 2001).

Production of Recombinant Spidroins in Different Hosts

Two strategies have been most commonly used for recombinant production of spidroins; to express parts of the native sequences, example (Arcidiacono et al., 1998; Stark et al., 2007), or iterated consensus repeats, example (Bini et al., 2006; Huemmerich et al., 2004a). Both these strategies have lead to problems such as truncations, translation pauses and rearrangements (Arcidiacono et al., 1998; Fahnestock and Irwin, 1997). Moreover, impoverishment of certain tRNA, proteolysis and low solubility of the protein has been observed (Fahnestock and Irwin, 1997; Prince et al., 1995; Winkler et al., 1999). There are no known essential post translational modifications of spidroins, although phosphorylation of tyrosine and serine residues in spidroins have been reported (Michal et al., 1996). Therefore, several expression hosts, including prokaryotes, have been attempted (reviewed in (Rising et al.)). From the Gram-negative enterobacterium *Escherichia coli* (*E. coli*), several successful fermentation processes have been reported (Arcidiacono et al., 1998, 2002; Bini et al., 2006; Brooks et al., 2008; Fukushima, 1998; Huang et al., 2007; Huemmerich et al., 2006; Slotta, 2006; Slotta et al., 2008; Stephens et al., 2005; Winkler et al., 2000; Zhou et al., 2001). However, low yields (Arcidiacono et al., 1998; Prince et al., 1995), accumulation of inclusion bodies (Liebmann, 2008) and low protein solubility (Bini et al., 2006; Fukushima, 1998; Mello et al., 2004; Szela et al., 2000; Winkler et al., 2000; Wong Po Foo et al., 2006) often occur. Moreover, the outer membrane of Gram-negative bacteria such as *E. coli* contains lipopolysaccharides (LPS) that are endotoxic and thus often have to be removed before the product is used in biomedical applications. Gram-positive bacteria, example *Bacillus subtilis* and *Staphylococcus aureus*, lack an outer membrane and therefore LPS (Sandkvist and Bagdasarian, 1996). This trait also simplifies secretion of recombinant proteins, although genetic manipulations are not as easy as with *E. coli* (Wong, 1995). Since expression in yeast gives rise to less translation stops, this host system might be more suitable than bacteria for long and repetitive sequences, although the production rate is slower. The methylotrophic yeast *Pichia pastoris* has been used in order to produce spidroins as well as amphiphilic silk-like protein (Fahnestock and Bedzyk, 1997; Werten et al., 2008). Recombinant expression of spidroins in various plants (tobacco, potato, *Arabidopsis*) has also been attempted (Scheller et al., 2001; Scheller et al., 2004) (Menassa et al., 2004; Patel et al., 2007; Yang et al., 2005). However, rather low expression levels are often reported (Scheller et al., 2001; Scheller et al., 2004; Patel et al., 2007; Yang et al., 2005) and large scale production of spidroins has so far only resulted in fairly low yields (Menassa et al., 2004). For biomedical applications, plant antigens, mycotoxins, pesticides, and herbicides can be problematic (Doran, 2000). Various mammalian cell systems have been used for expression of spidroins, the most promising being a 60 kDa ADF-3 protein secreted from cell lines grown in a hollow fiber reactor (Arcidiacono et al., 2002). Since most mammalian cells require complex media with biological supplement, this production system may cause contamination by virus, prions, and oncogenic DNA (Ma et al., 2003). Spidroin production in mammary glands and secretion into milk has been tried in mice (Xu et al., 2007) and goats (Williams, 2003), although potential contaminations are also a major drawback with production in transgenic animals. So far, cytocompatibility studies have been performed on recombinant spidroins produced in bacteria, yeast, and plants (see Table 1).

Table 1. Recombinant spider silk proteins in cytocompatibility studies.

Construct name	Size (kDa)	Production host	Type of silk	Species	Added motif	Sterilization technique	Tested cell type	References
SO1	51	Tobacoo	Dragline silk (MaSp1)	*Nephila clavipes*	ELP	filtration	human primary chondrocytes, CHO-K1	Scheller et al 2004
4RepCT	21	*E.coli*	Dragline silk (MaSp1)	*Euprosthenops australis*	none	filtration, autoclaving	HEK 293, human primary fibroblasts	Stark et al 2007, Hedhammar et al 2008, Widhe et al 2010
IF9	94	*E.coli*	Dragline silk (MaSp1)	*Nephila clavipes*	none	96% ethanol	3T3 fibroblasts	Agapov et al 2009
15mer	51	*E.coli*	Dragline silk (MaSp1)	*Nephila clavipes*	RGD, RGD-R5 (from silaffin)	ethylene oxide, 70% ethanol, UV	hMSC, MC3T3-E1 osteoblastic precursors, mesenchymal stem cells (hMSCs)	Bini et al 2006, Morgan et al 2008, Mieszawska et al 2010
pNSR16 pNSR32	not specified	*E.coli*	Dragline silk	not specified	RGD	not specified	3T3 fibroblasts	Wang et al 2009

Control of Assembly of Recombinant Spidroins

As mentioned above, expression of recombinant spidroins has often resulted in water insoluble products. Often the purification processes in these cases are obligated to include solubilisation steps using urea, guanidine hydrochloride, lithium bromide, hexafluoroisopropanol (HFIP) or formic acid (Agapov et al., 2009; Arcidiacono et al., 1998; Arcidiacono et al., 2002; Bini et al., 2006; Bogush et al., 2008; Exler et al., 2007; Fahnestock and Bedzyk, 1997; Fahnestock and Irwin, 1997; Fukushima, 1998; Geisler et al., 2008; Huemmerich et al., 2004a; Huemmerich et al., 2006; Junghans, 2006; Lammel, 2008; Lazaris et al., 2002; Lewis et al., 1996; Liebmann, 2008; Lin et al., 2009; Mello et al., 2004; Rammensee et al., 2008; Scheller et al., 2001; Slotta, 2006; Slotta et al., 2008; Teulé et al., 2009; Wong Po Foo et al., 2006; Zhou et al., 2001). After removal of the solvents used for solubilisation, it has generally been difficult to control assembly of the denatured spidroins into solid structures. The spider can store the spidroins in an aqueous solution and precisely regulate the conversion into fibers during spinning. Several strategies have been tried in order to mimic this process; to first prevent and then promote assembly and thereby avoid resolubilisation steps. For example introduction of methionine residues for controlled oxidation/reduction (Szela et al., 2000; Valluzzi et al., 1999; Winkler et al., 1999) have been utilised. A kinase recognition motif has also been used for controlled assembly *via* enzymatic phosphorylation (Winkler et al., 2000). When fused to the solubility tag thioredoxin a miniature spidroin composed of four poly-Ala/Gly-rich co-segments and the C-terminal domain, 4RepCT, has been successfully produced in soluble form (Stark et al., 2007). Upon proteolytic release from the solubilising thioredoxin partner the miniature spidroin 4RepCT spontaneously forms macroscopic fibers that resemble native spider silk (Stark et al., 2007). Together with minispidroins recombinantly produced in the cytosol of insect cells (Huemmerich et al., 2004b) this is so far the only report of spontaneous self assembly into silk-like fibers, indicating the importance of non-denaturing conditions during production.

Processing into Solid Spidroin Structures

Several different methods to convert denatured recombinant spidroins into fibers have been used, example wet spinning (Arcidiacono et al., 2002; Bogush et al., 2008; Brooks et al., 2008; Lazaris et al., 2002; Lewis et al., 1996; Teulé et al., 2007; Teulé et al., 2009; Yang et al., 2005), hand drawing (Exler et al., 2007; Teulé et al., 2007), spinning though microfluidic devices (Rammensee et al., 2008) and electrospinning (Bini et al., 2006; Bogush et al., 2008; Stephens et al., 2005; Wong Po Foo et al., 2006; Zhou et al., 2008). Other formats than fibers are preferred for some applications, example matrices for cell culture and tissue engineering. Therefore, materials have been processed into films (Arcidiacono et al., 2002; Bini et al., 2006; Fukushima, 1998; Huang et al., 2007; Huemmerich et al., 2006; Junghans, 2006; Metwalli, 2007 Scheller et al., 2004; Slotta, 2006; Szela et al., 2000; Valluzzi et al., 1999; Winkler et al., 1999; Wong Po Foo et al., 2006), microbeads (Liebmann, 2008), microspheres (Lammel, 2008; Slotta et al., 2008), and microcapsules (Hermanson et al., 2007). Three-dimensional porous scaffolds have also been produced by salt leaching methods (Agapov et al., 2009). Most of these techniques require post treatment in dehydrating or salting-out solvents, to increase the β-sheet contents and stability. In contrast, the recombinant miniature spidroin 4RepCT can be used to produce fibers directly in physiological buffers (Figure 1). The self-assembled fibers form bundles of varying thickness with a surface with lengthwise pattern of irregular grooves and ridges. The 4RepCT can also be casted into a coherent and flexible film with a mostly smooth surface. A stable 3D structure with heterogeneous mounds and craters can also be produced by foaming the 4RepCT protein (Figure 1) (Widhe et al., 2010).

If the intended use of the recombinant spidroin involves biomedical applications, the scaffolds should be sterilized before usage. This has been achieved by using filtration, ethanol, ethylene oxide or by autoclaving (Table 1). In some cases, the scaffolds have been pre-incubated in cell culture medium before cell culture. Whether traces of the agents used during production still remain in the scaffolds has not been investigated. However, the processes seem to result in scaffolds suitable for cell culture (see further below).

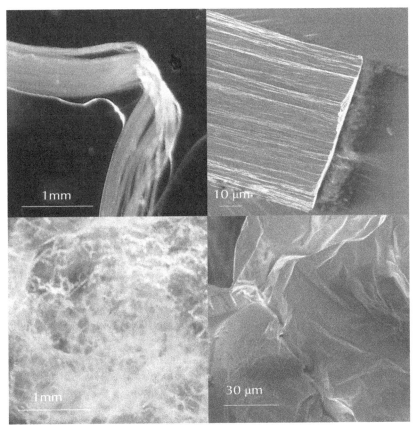

Figure 1. Scaffolds of recombinant spider silk. Upper row: photograph of a wet fiber (left) and scanning electron micrograph of a dried fiber (right). Lower row: photograph of a wet foam (left) and scanning electron micrograph of a dried foam (right). All scaffolds were made from the miniature spidroin 4RepCT (see Table 1).

Spider Silk for *in vitro* Applications

Native spider silk has been shown to support the growth of cells *in vitro*, for example chondrocytes (Gellynck et al., 2008b), Schwann cells (Allmeling et al., 2006) and mouse fibroblasts (Kuhbier et al., 2010). Recombinant spider silk has several benefits over the native silk, not only the better supply, but also the greater possibility to process it into scaffolds of different formats, for example film, foam and porous scaffolds. Such scaffolds offer a three dimensional (3D) structure and mechanical properties that differ substantially from the plastic surfaces traditionally used for cell culture. Since, substrate stiffness has been suggested to play a major role in cell responses (Nemir and West, 2010), softer substrates than cell culture plates are probably beneficial, making recombinant spider silk an interesting material. Recombinant spider silk scaffolds prepared from 4RepCT have been shown to support the growth of human primary fibroblasts (Widhe et al., 2010). These scaffolds, in formats of film, foam, fiber and fiber mesh, differ substantially in 3D structure and surface topography, but all provide a suitable *in vitro* environment for these cells. Another recombinant spider silk protein, IF9, can be used to form porous 3D scaffolds with good interconnectivity by salt leaching, allowing for proliferation of mouse fibroblast (Agapov et al., 2009). Cells attach to the scaffolds, proliferate and migrate into the deeper layers of the scaffolds within one week. Both these types of recombinant spider silk originate from dragline silk, and scaffolds prepared from them are apparently stable, cytocompatible and can offer a 3D microenvironment. To further improve the cell supporting capacity, recombinant spider silk has been genetically engineered by the introduction of cell binding domains or motifs originating from proteins of the extracellular matrix (see Table 1). Most studied is the introduction of the cell binding motif RGD (Arg-Gly-Asp) into recombinant spider silk derived sequences. The effects of this motif have been evaluated regarding cellular growth or differentiation using mesenchymal stem cells (Bini et al., 2006), murine osteoblastic cell line (Morgan et al., 2008) and mouse embryonic fibroblasts (Wang et al., 2009). Also, two larger peptides have been tried, that is an elastin like peptide (ELP) (Scheller et al., 2004) and a silaffin-derived peptide from a silica producing diatom (Mieszawska et al., 2010). The ELP was combined with the dragline silk derived from recombinant spider silk protein SO1, and was shown to support growth and prevent dedifferentiation of human chondrocytes (Scheller et al., 2004). The silaffin-derived peptide R5 was introduced in order to increase osteogenesis of mesenchymal stem cells growing on recombinant spider silk films (Mieszawska et al., 2010). Generally, it seems possible to prepare scaffolds from functionalized recombinant spidroins, and they are apparently cytocompatible.

Though, so far relatively sparsely explored, functionalized recombinant spider silk scaffolds for cell culture might provide a broad range of solutions for *in vitro* cell culture and possibly also for tissue engineering applications.

Spider Silk for *in vivo* Applications

There are only a few studies available on native spider silk implanted in living tissue. In one of these, spider dragline silk from *Nephila clavipes* was used as a component in artificial nerve constructs. The grafts were used to replace a 2 cm deficit of the sciatic nerve in rats, and shown to promote regeneration of peripheral nerves with high functionality, while the controls that received similar grafts without spider silk gained nearly no myelinated nerve fibers and showed distinctive muscle degeneration (Allmeling et al., 2008). Furthermore, no signs of inflammatory response or foreign body reaction were found, indicating *in vivo* tolerance of the silk (Allmeling et al., 2008). Another type of spider silk, egg sac silk, has been implanted subcutaneously in rats (Gellynck et al., 2008a). This study showed that during the first four weeks of implantation, egg sack silk treated with Proteinase K and/or trypsin evoked the same degree of inflammation as the widely used suture material Vicryl® (polyglactin). Egg sack silk that had not been subjected to enzymatical treatment evoked a severe inflammatory reaction, suggesting that the egg sack silk contains an immunogenic coating and/or was contaminated with agents that was removed by proteolysis. Also, Vollrath et al. (2002) have shown that both native dragline silk and swathing silk are degraded with time when implanted subcutaneously in pigs (Vollrath et al., 2002).

For the recombinantly produced spider silks, there are two studies where the *in vitro* cytocompatibility has been complemented with *in vivo* studies (Baoyong et al., 2010; Fredriksson et al., 2009). In the first one, recombinant spider silk fibers were implanted subcutaneously in rats (Fredriksson et al., 2009). The fibers did not give rise to any macroscopic signs of inflammation, and the histological examination revealed an inflammatory reaction at the same level as observed for the control (silk worm silk suture, Mersilk®). Surprisingly, newly formed capillaries could be detected in the centre of the recombinant spider silk fiber bundle already after one week of implantation. This suggests that the recombinant spider silk may have angiogenic properties, which would be beneficial in example tissue engineering and wound dressings. In line with this, recombinantly produced spider silk (pNSR-32 and pNSR-16, both containing the tripeptide RGD) used as dressings of deep burns in rats, have been suggested to be as efficient in promoting healing as clinically used collagen sponges (Baoyong et al., 2010).

CONCLUSION

Recombinant spider silk represents an interesting material for cell culture matrices, implants and other medical devices. Both *in vitro* and *in vivo* studies show that the material is cytocompatible and tolerated when implanted. However, the nature of spider silk proteins still makes them difficult to produce–a problem that must be solved before the material can be used in any commercial application.

KEYWORDS

- **Elastin like peptide**
- **Hexafluoroisopropanol**
- **Lipopolysaccharides**
- **Major ampullate spidroin**
- **Spiders**

ACKNOWLEDGMENTS

We are grateful to prof. Jan Johansson for constructive comments on the manuscript. This work was financially supported by the Swedish Research Council.

Chapter 14

New Nanocomposite Materials Based on Cellulose Fibers and Other Biopolymers

Carmen S. R. Freire, Susana C. M. Fernandes, Armando J. D. Silvestre, Alessandro Gandini, and Carlos Pascoal Neto

INTRODUCTION

The use of natural (cellulose) fibers as reinforcing components in polymeric composite materials has been extensively explored during the last few years (Belgacem, 2008), mainly in response to the economic and environmental concerns associated with the extensive exploitation of petroleum-derived products. The main advantages of natural fibers, when compared with their synthetic or inorganic counterparts, are their biodegradability, high availability, diversity, abundance, renewability, low cost, low energy consumption, low density, high specific strength and modulus (with fibers possessing an adequate aspect ratio), high sound attenuation and comparatively easy processing ability, due to their flexibility and non-abrasive nature (Bledzki, 1999; Pommet, 2008). Additionally, cellulose-based composites are also very attractive materials because of their good mechanical properties sustainability and environmental-friendly connotation, and have been used in a wide range of applications, such as in building, engineering, and automobile industries, as well as for the processing of furniture, packaging materials, recreation boats and toys, among others (Bledzki, 1998).

In addition to "conventional" vegetal cellulose fibers (Figure 1A), other forms of cellulose have been assessed in the last few years. The use of micro and nano-cellulose fibers, namely whiskers, obtained from a marine species (Samir, 2005), bacterial cellulose produced by some bacterial strains (Pecoraro, 2008) as well as micro- or nano-fibrillated cellulose prepared by mechanical, enzymatic or chemical treatments of the vegetal fibers (Nakagaito, 2004), for the development of high performance composite materials is attracting researchers from diverse fields (Dufresne, 2008; Lee, 2009), as the addition of very modest amounts of nano fibers leads to new composite materials with superior mechanical properties and new functionalities (Klemm, 2009) when compared with their conventional cellulose fibers counterparts.

Nanofibrillated cellulose (NFC) (Figure 1B) can be obtained from conventional cellulose fibers by different methods as mentioned before, in the form of aqueous suspensions bear the appearance of highly viscous, shear-thinning transparent gels. The fibrils have high aspect ratios and specific surface areas combined with remarkable strength and flexibility. Depending on the disintegration process, their dimensions vary, and fully delaminated NFC consists of long (in the micrometer range) nanofibrills (diameter =10–20 nm) (Dufresne, 2008; Nakagaito, 2004).

Bacterial cellulose (BC) (Figure 1C), also known as microbial cellulose, is produced by the biosynthesis of different genus of bacteria, for example of the *Glucona-cetobacter* genus. The BC is generated as a three-dimensional network of cellulose nano- and microfibrils with 10–100 nm diameter, that is about 100 times thinner than typical vegetal cellulose fibers. The microfibrillar structure of bacterial cellulose is responsible for most of its properties. Bacterial cellulose possesses very high purity (free of lignin, hemicelluloses, and the other natural components usually associated with plant cellulose) a high degree of polymerization and crystallinity, extremely high water binding capacity, high tensile strength and of course a higher surface area, as compared to the widespread plant-based counterparts (Pecoraro, 2008).

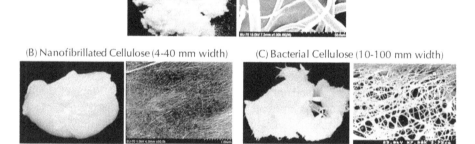

Figure 1. (A) Visual and SEM micrographs of vegetal cellulose, (B) nanofibrillated, and (C) bacterial cellulose.

The use of cellulose in different forms as reinforcing agent for matrices of other polysaccharides that bear promising functional properties but, poor mechanical features opened wide perspectives on the preparation of a wide range of promising functional materials through simple and green processes: due to the structural similarity of both cellulose fibers and polysaccharide matrices, the former can be used as reinforcing elements without any chemical modification. Two of the most interesting examples of polysaccharides matrices in which various cellulose forms can play an interesting role as reinforcing agents are starch and chitosan.

Starch (Figure 2), a mixture of amylose and amylopectin, is also one of the most abundant natural polymers and is considered as a promising raw material for the development of novel materials, including biocomposites. It can be converted into a thermoplastic material, known as thermoplastic starch (TPS), through the disruption of the molecular chain interactions under specific conditions, in the presence of a plasticizer (Carvalho, 2008). Water and glycerol are the most widely used plasticizers in the production of TPS.

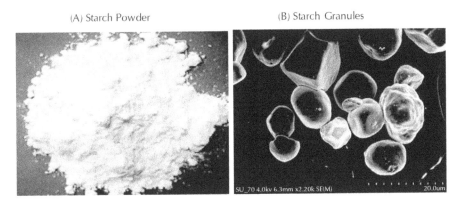

Figure 2. Visual (A) and SEM micrographs (B) of starch.

Chitosan (CH) (Figure 3), the major, simplest, and least expensive chitin deriva-tive, is a high molecular weight linear polymer obtained by deacetylation of chitin and is therefore composed of 2-amino-2-deoxy-*D*-glucose units linked through β (1⊕4) bonds. Chitin, on its hand is the main component of the exoskeleton of crustaceans and considered as the second most abundant natural polymer on earth (Peniche, 2008). Chitosan exhibits unique physicochemical properties like biocompatibility, antimicro-bial activity, biodegradability and excellent film-forming ability, which have attracted scientific and industrial interest in fields such as biotechnology, pharmaceutics, bio-medicine, packaging, cosmetics, among others (Peniche, 2008).

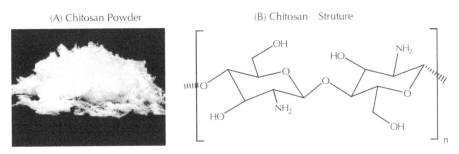

Figure 3. Chitosan sample (A) and its chemical structure (B).

A brief overview of composite materials based on cellulose fibers (vegetal cellu-lose, bacterial cellulose and nanofibrillated cellulose) with other natural polymers such as chitosan and starch will be presented.

Cellulose-Chitosan Nanocomposites Films
A significant number of studies dealing with the combination of chitosan with cel-lulose (Hasegawa, 1992; Lima, 2000; Shih, 2009; Twu, 2003), in solution or using the cellulose fibers in solid state, and its derivatives (Mucha, 2003, 2005), have been

published. One practical application of chitosan-cellulose mixtures is their processing into films, having high strength parameters and also good biocompatibility, biodegradability and hydrophilicity. More recently, the incorporation of micro and nanofibers cellulose into several polymeric matrices, gave materials with superior mechanical, thermal and barrier properties and transparency when compared with the conventional fibers. Some studies have been published dealing with the preparation and characterization of NFC or BC based-nanocomposites with chitosan (Ciechanska, 2004; Dubey, 2005; Fernandes, 2009, 2010; Hosokawa, 1990, 1991; Nordqvist, 2007).

Transparent films have been obtained from mixtures of chitosan and both BC and NFC nanofibers (Fernandes, 2009, 2010). The chitosan solutions were shown to be an efficient media to prepare stable suspensions of NFC or BC, and to produce transparent films with a very homogeneous distribution of BC and NFC (Figure 4).

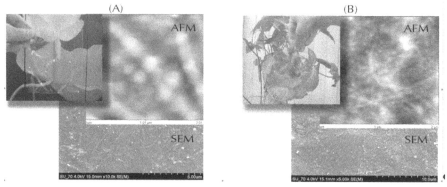

Figure 4. Images and AFM/SEM micrographs of transparent nanocomposite films of chitosan with BC (A) and NFC (B).

Chitosan- and water-soluble chitosan-cellulose nanofiber nanocomposite films were prepared by a simple and green procedure based on casting water (or 1% acetic solutions) suspensions of chitosan with different contents of NFC (up to 60%), and BC (up to 40%). The transparency of the films obtained indicated that the dispersion of the NFC and BC into the chitosan matrices was quite good. The nanocomposite films prepared with BC showed higher transmittance than the corresponding films prepared with NFC, because of the higher purity of BC.

The ensuing materials were in general very homogenous and presented better thermo-mechanical and mechanical properties than the corresponding unfilled chitosan films. With the NFC and BC addition to the chitosan's matrices, tensile strength and Youngs modulus were completely dominated by the NFC and BC network. The superior mechanical properties of all nanocomposite films, compared with those of the unfilled CH films, confirmed the good interfacial adhesion and the strong interactions between the two components. These results can be explained by the inherent morphology of BC with its nanofibrillar network, the high aspect ratio of NFC and the similar structures of the two polysaccharides. The nanocomposite films presented better thermal stability than the corresponding unfilled chitosan films. The nanocomposites prepared with the water soluble chitosan derivative are particularly interesting for future studies, since they have an attractive combination of properties, including a high optical transparency.

Globally, the properties of CH/NFC nanocomposite films were better than those displayed by similar chitosan films reinforced with BC Nano fibrils. This behavior could be due to the better dispersion of NFC into the chitosan matrices, related to the individual fiber morphology, contrasting with the tridimensional network fibers structure of BC, as well as to the higher aspect ratio of the NFC compared with BC.

The prominent properties of these nanocomposite films could be exploited for several applications, such as in transparent functional, biodegradable and anti-bacterial packaging, electronic devices and biomedical applications.

Cellulose-Starch Composites

The preparation and characterization of TPS-based composites with different cellulose substrates have been strongly explored, namely commercial regenerated cellulose fibers (Funke, 1998), vegetable fibers (Alvarez, 2005; Curvelo, 2001; Funke, 1998;), microcrystalline cellulose (Ma, 2008), microfibrillated cellulose (Dufresne, 2000) and cellulose nanocrystallites (Weng, 2006). Apart from the enhanced mechanical properties of these reinforced TPS materials, a significant improvement in water resistance is also obtained by adding cellulose crystallites (Lu, 2005) or microfibrillated cellulose (Dufresne, 2000). In general, these TPS/fiber composites also displayed improved thermal stability due to the higher thermal resistance of cellulose fibers (Curvelo, 2001). The preparation and characterization of bacterial cellulose-starch nanocomposites has been also developed (Grande, 2008; Martins, 2009; Orts, 2005).

Plasticized starch/bacterial cellulose nanocomposite sheets have been obtained by hot-pressing (Grande, 2008). The ensuing films were characterized in terms of their morphology, whereas other important parameters, such as their mechanical and thermal properties, were not investigated. The incorporation of bacterial cellulose microfibrils, obtained by the acid hydrolysis of the cellulose network, into extruded TPS and starch-pectin blends was also studied (Orts, 2005). However, in this work, the peculiarity of this cellulose substrate was not fully exploited, since the nano- and microfibril three-dimensional network morphology was partially destroyed during the hydrolysis step. Indeed, the microfibrils derived from bacterial cellulose did not improve the modulus to the same extent as cotton or softwood counterparts.

The work developed by Martins et al. (2009) described the preparation and characterization of biocomposite materials obtained by the incorporation of bacterial cellulose into a TPS matrix during the gelatinization process. Similar composite materials reinforced with vegetable cellulose fibers were also prepared for comparison purposes.

The starch-BC nanocomposites were prepared in a single step with cornstarch by adding glycerol/water as the plasticizer and bacterial cellulose (1 and 5% w/w) as the reinforcing in a melting mixer. All nanocomposites showed good dispersion of the nanofibers and a strong adhesion between the fibers and the matrix (Figure 5), as well as improved thermal stability and mechanical properties. For example, the Young modulus increased by 30% (with 5% BC), while the elongation at break was reduced from 144 to 24%. These results can be explained by the inherent morphology of BC with its nano- and micro-fibrillar network.

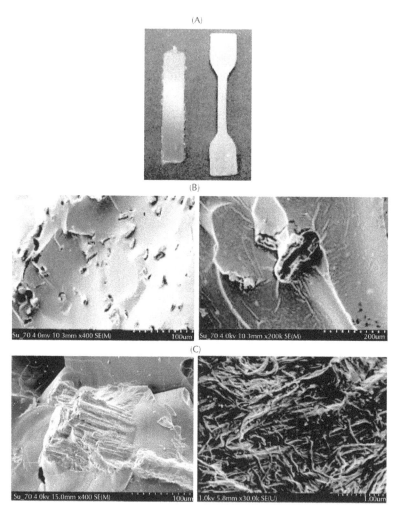

Figure 5. Images of composites (A) and SEM micrographs of TPS nanocomposites with 5 wt% of VC (B) and BC (C).

This study provides an initial insight into the use and characteristics of bacterial cellulose in starch-based composites. Bacterial cellulose acts efficiently as reinforcement, even in relatively low quantities, since 5% produced a significant increase in both modulus and tensile strength. These materials are promising candidates in applications like food packaging and biodegradable artifacts. However, the ensuing composites displayed a strong sensitivity to high relative humidity.

CONCLUSION

The development of new polymeric materials based on cellulose nanofibers and other polysaccharides is a promising strategy within the context of new sustainable and environmental-friendly materials. As exemplified in this brief review, the properties of these materials suggest that they could be successfully applied in such areas as packaging materials, particularly because bacterial cellulose is becoming a progressively more available raw material at reasonable price.

KEYWORDS

- **Biosynthesis**
- **Nanofibrills**
- **Plasticizers**
- **Polysaccharides**

Chapter 15

Cross-linking Chemistry of Type I and Type III Collagen with Anionic Polysaccharides

A. Gnanamani, Tapas Mitra, G. Sailakshmi, and A. B. Mandal

INTRODUCTION

The present study describes cross-linking chemistry of Type I and Type III collagen with anionic polysaccharides (alginic acid), followed by characterization of the resultant cross-linked material for, degree of cross-linking, thermal, and mechanical stability. Required concentration of alginic acid (AA) and collagen (Type I and Type III) was optimized. Results reveal, irrespective of the nature of collagen, 75% degree of cross-linking was observed with AA at 1.5% concentration (w/v). Cross-linking with AA increases the mechanical strength and thermal stability of both Type I and Type III collagen. About 5–6-fold increase in tensile strength compared to plain collagen was observed upon cross-linking with AA. The nature of bonding pattern and the reason for thermal stability of AA cross-linked collagen biopolymer was discussed in detail with the help of bioinformatics.

Collagen, the most abundant body protein and a biopolymer, till date, identified as 28 different types and grouped under collagen super family. Fibrillar collagen, one of the subfamily comprises, Type I, II, III, V, and XI as members. About 40% of skin, 10–20% of bone and teeth and 7–8% of blood vessel shared by fibrillar collagen (Kadler et al., 2007). Fibril-forming collagen, Type I, II, and III (Van der Rest and Garrone, 1991) constitute the majority of the internal structure of the body. Type I collagen comprises two α_1 chains and one α_2 chain whereas, Type III collagen comprises of three α_1 chains. Collagen is an important biomaterial finding several applications as prosthesis, artificial tissue, drug carrier, cosmetics, and in wound healing and in all these applications stable collagen is required.

However, the low denaturation temperature of collagen restricts the application as suitable biopolymers for clinical applications (Ru et al., 2005). As a result, most of the research is focused on to increases the thermal stability of the biopolymer materials. Recently, use of cross-linking agent to increase and also to improve the mechanical properties and thermal stability is reported by number of researchers (Sung et al., 1997).Collagen cross-linked with glutaraldehyde introduced by Carpentier in the year 1960 showed high stability and after him, almost all the biopolymer materials prepared from collagen is cross-linked with glutaraldehyde. Nevertheless, realization on various demerits such as; calcification, induction of local cytotoxicity due to depolymerization; release of unreacted glutaraldehyde monomer; incomplete suppression of immunological recognition and their negative effect on osteoblast attachment and proliferation, necessaciates the requirement of suitable cross-linking agent for the preparation of biopolymer with required thermal stability. As an alternative, attempt was made on to exploit the cationic and anionic polymers of both synthetic and natural polymers for cross-linking with collagen. Though, reports on interaction of cationic polysaccharides with collagen are available, however, the available reports on anionic polysaccharides involve additional chemical agents to enhance the cross-linking ability of AA with collagen. Furthermore, in the said study, sodium alginate was used as a source material. Nevertheless, in order to avoid the use of external agent for induction, in the present study, we made the attempt on to use plain AA for cross-linking with collagen. The results of our study demonstrate, cross-linked biopolymer has high thermal and mechanical properties. Thus, the detailed description of the study has been summarized below;

The AA, a linear anionic copolymer of 1, 4 linked β- D- mannuronic acid (M) 1,4 linked α- guluronic acid (G) arranged as homopolymeric or heteropolymeric block (GG, MM and GM), constitutes major structural polysaccharide of brown seaweeds (*Phaeopyta*), found non-toxic, non-carcinogenic, biocompatible, sterilizable, and offers cheap processing technique. The AA and its sodium/calcium salts have long been used in food, cosmetics, drugs, drug delivery, tissue engineering, and so on. Alginate wound dressing has recently been introduced for heavy exudation wounds as occlusive dressing by Pharmacy industries. It is reported that exchange of ions by calcium salts of AA and wound exudates accelerates the healing process (Albarghouthi et al., 2000). Further, retaining the moist condition during wound healing and increases the reepithelization process and healing rate by the use of salts of AA is also in reports (Gombotz and We, 1998). Salts of AA also found application as dewatering agent for the collagen films used for food casings. The mixture of collagen, sodium alginate and konjac glucommnnan increases the thermal stability of the mixture appreciably according to Wang et al. (2007).

EXTRACTION AND RECONSTITUTION OF TYPE I COLLAGEN

Type I collagen from bovine skin was extracted as per the method of Mitra et al (in press) and reconstituted according to Nomura et al. (1997). Since collagen was extracted using acetic acid, the resultant collagen was designated as acid soluble collagen (ASC) and all the cross-linking studies were carried out only with reconstituted ASC (RASC).

EXTRACTION OF RECONSTITUTION OF TYPE III COLLAGEN

Type III Collagen from avian intestine was extracted as per the steps followed in the flow chart given below;

Flow chart for extraction of Collagen from Avian intestine

Avian intestine obtained from slaughter house
↓
Washed and cleaned the waste materials inside the intestine
↓
Washed with chloroform for complete removal of fat materials
↓
Chopped in to pieces
↓
Soaked in 0.5(M) Sodium acetate buffer with continuous stirring at 4°C for 12-24h
↓
Homogenized with 0.5 (M) Acetic acid
↓
Centrifuged at 10,000rpm for 20 min at 4°C
↓
Supernatant precipitated with 5% NaCl
↓
Centrifuged at 10,000rpm for 20 min at 4°C
↓
Residue collected & Re-dissolved in 0.05(M) acetic acid
↓
Dialyzed against 0.005(M)acetic acid
↓
Lyophilized
↓
Acid soluble collagen

The molecular profile of Type I and Type III collagen were shown in Figure 1(a) and 1(b), and it demonstrate, presence of two types of α chains (α$_1$ and α$_2$) (100Kda) and β dimer (200 Kda) in Type-I collagen and one type of α chain (α$_1$) (100 KDa) and β dimer (200 KDa) for Type-III collagen (Kuga et al., 1998; Lin and Liu, 2006).

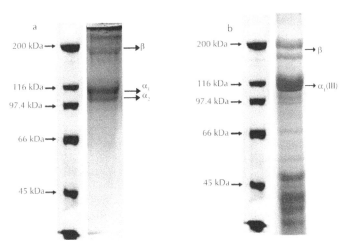

Figure 1. The molecular profile of (a) Type I and (b) Type III collagen.

PREPARATION OF CROSS-LINKED BIOPOLYMER MATERIAL

Different concentrations (0.5, 1.0, 1.5, 2.0, 3.0, 4.0, and 5.0%) of homogenized solution of AA were prepared by dissolving the required quantities in 70 (mM) sodium phosphate buffer (pH 6.5) under stirring for overnight at room temperature. About 0.5% (w/v) RASC (dissolved in 0.005M acetic acid) was mixed with different concentration of AA at 3:1 (RASC:AA) ratio respectively and the homogenized solution obtained upon stirring for 30 min at room temperature, incubated for overnight at 4°C. Followed by incubation, the reaction mixture was then transferred to a polypropylene plate (Tarson, India) and air-dried at 37°C for 12 hr. The biopolymer material obtained in the form of sheets from the above process was designated as AA cross-linked Type I (AACC 1) and Type III collagen (AACC 3). In addition, a separate collagen sheet material without cross-linker was also made accordingly and used for comparative analyses. The dried polymer sheets were further subjected to percentage of cross-linking degree, FT-IR, TGA, and Tensile strength analyses.

Figure 2 demonstrated, degree of cross-linking measurements (Bubnis and Ofner, 1992) of AA cross-linked biopolymer material with increasing concentration of AA. Maximum degree of cross-linking of 75% was observed with 1.5% concentration of AA with 0.5% of either Type I or Type III collagen. No further increase in degree of cross-linking with increasing concentration of AA was observed.

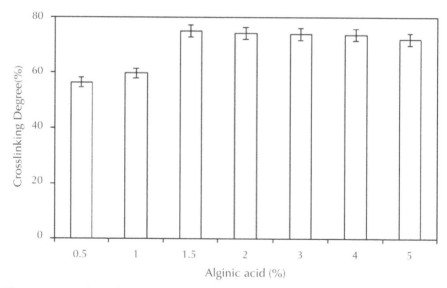

Figure 2. Degree of cross-linking measurements.

The FT-IR studies were conducted to monitor the chemical modifications in collagen structure due to cross-linking with AA. Figure 3 illustrate, FT-IR spectral details of AA, Type I collagen, Type III collagen, AACC 1, and AACC 3. When comparing the FT-IR spectral details we observed a complete different spectrum for AACC 1 and

AACC 3. In general, the amide I, II and, III peaks of Type I collagen was located at 1,658, 1,556, and 1,240 cm^{-1}. However, after cross-linking with AA, a sharp intense peak at 1,658 disappears with the overlapping peak of –C=O and –N-H bands in the range of 1,600–1,650 cm^{-1} in AACC 1 were observed. This might be due to the formation of covalent linkage between –COOH group of AA and ε-NH$_2$ group of collagen (Pavia et al., 2001). In addition, deep and sharp intense peak observed at 1,556 cm^{-1} in Type I collagen and the related overtone of 1,553 at 3,081 cm^{-1} was reduced without any overtone peaks in AACC 1. Simultaneously, the intensity of amide III was also reduced in AACC 1 compare to Type I collagen. However, in Type III collagen the amide I, II and III peaks were located at 1,655, 1,546, and 1238 cm^{-1} and after cross-linking with AA these three sharp amide signals have been reduced due to the above said reason. In addition, one significant change was observed for Type III collagen, that is, the overtone peak at 2,924 cm^{-1} which, was remarkably reduced in AACC 3 spectrum, suggests, the interaction of both types (Type I and Type III) of collagen with AA.

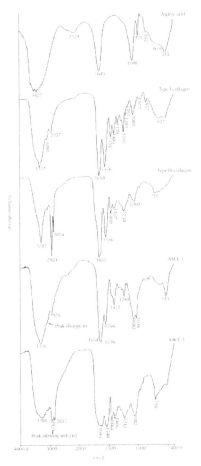

Figure 3. Illustrate, FT-IR spectral details of alginic acid, Type I collagen, Type III collagen, AACC 1 and AACC 3.

The derivative peaks obtained from TGA for all the experimental samples at optimized concentration (1.5% of AA) were illustrated in Figure 4(a) and 4(b). In Figure (a) and (b) maximum weight loss of AA was observed at 252°C and for Type I and Type III collagen it was 327°C whereas in AACC 1 and AACC 3 it was observed at 345 and 370°C respectively. When AA were cross-linked with Type I and Type III collagen at different concentrations, thermal stability increases as the percentage of AA increased up to 1.5%. Further increase in AA reduces the thermal stability of the resulting material.

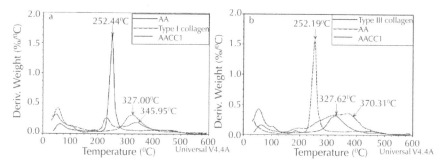

Figure 4. (a) and 4(b) in maximum weight loss of AA was observed at 252°C and for Type I and Type III collagen

Table 1. Tensile strength measurements of Type I, Type III collagen, AACC 1and AACC 3 by Universal Testing Machine (INSTRON model 1405) at a crosshead speed of 5mm/min.

Material Type	Tensile Strength (MPa)	Young's modulus (Gpa)	Elongation at break (%)
Type I collagen	2.110	7.8	31.99
Type III collagen	2.10	7.3	30.89
AACC 1	11.56	47.08	5.31
AACC 3	10.73	45.12	5.21

With regard to mechanical property, about 5 to 6-fold increase in tensile strength was observed after cross-linking with AA. The tensile strength measurements of native, AACC 1 and AACC 3 were taken and the ultimate tensile strength (MPa) and Young's modulus (Gpa) values were represented in Table 1.

INTERMOLECULAR H-BOND INTERACTION BETWEEN TYPE I, TYPE III COLLAGEN AND AA BY BINDING ENERGY CALCULATION USING BIOINFORMATICS TOOL SOFTWARE

For binding energy and bonding pattern assessment, docking study was followed. For the docking study, chemical structures were generated using ACD/ChemSketch (ACD/ChemSketch Version 12 Advanced Chemistry Development, Inc., Toronto, ON, Canada. 2009.). The 3D structure of collagen was generated using gencollagen program (http://www.cgl.ucsf.edu./cgi-bin/gencollagen.py). To find out the interaction between collagen with AA, AUTODOCK software has been used and AutoDock 4.2 used to calculate (Morris et al., 2009) the free energy of binding between the AA and collagen. Results on binding energy calculations based on bioinformatics tools for the cross-linking of Type-I, Type III collagen with AA using AutoDock software (Figure 5(a), 5(b) and 6(a), 6(b)) and, Table 2, 3 depicts the values for the binding energy, interaction sites, hydrogen bond sites, and bond distance of Type I and Type III collagen respectively.

In AACC 1, the binding energy of –7.28 was observed when Ala (11) residue of A chain of collagen interacting with AA through Nitrogen of alanine and forming three hydrogen bond with bond distance of about 3.08, 2.97, and 2.92. Glycine (7) of C chain of collagen interacted with AA through oxygen (of –OH group) and forming three hydrogen bonds with the bond distance of 3.98, 2.67, and 2.55. Similarly Serine (9) of C chain of collagen also forms one hydrogen bond through Nitrogen with bond distance of 3.15.

Table 2. Interaction sites, binding energy, hydrogen bond sites, and bond distance, calculated based on the bioinformatics tool for the cross-linking between collagen (Type I) and alginic acid.

Interaction site	Binding energy (kcal/mol)	H-bond	Bond distance (Å)
Thirteenth residue Gly of A-chain (α_1)	–6.0	Gly A 13-(C_a)-O ...H-O(AA)	3.13
Fifteenth residue Ala of A-chain (α_1)		Ala A 15-(C_a)-N ...H-O(AA)	2.88 & 2.89
Thirteenth residue Gly of A-chain (α_1)	–6.17	Gly A 13-(C_a)-O ...H-O(AA)	2.91 & 2.62
Eleventh residue Ala of B-chain (α_1)	–6.31	Ala B 11-(C_a)-N ...H-O(AA)	2.83 & 3.11
Tenth residue Gly of A-chain (α_1)		Gly A 10-(C_a)-O ...H-O(AA)	2.61 & 3.39
Seventh residue Gly of C-chain (α_2)	–6.51	Gly C 7-(C_a)-O ...H-O(AA)	2.44
Eleventh residue Ala of A-chain (α_1)		Ala A 11-(C_a)-N ...H-O(AA)	3.17
Eleventh residue Ala of A-chain (α_1)	–7.28	Ala A 11-(C_a)-N ...H-O(AA)	3.08,2.97 & 2.92
Seventh residue Gly of C-chain (α_2)		Gly C 7-(C_a)-O ...H-O(AA)	3.98,2.67 & 2.55
Nineth residue Ser of C-chain (α_2)		Ser C 9-(C_a)-N ...H-O(AA)	3.15

In the case of AACC 3, binding energy of –7.14 observed when Alanine (11) residue of A chain of collagen interacting with AA through Nitrogen of alanine and forming three hydrogen bond with bond distance of about 3.29, 3.04, and 2.85. Alanine (9) of C chain of collagen interacted with AA through oxygen (of –OH group) and forming two hydrogen bonds with the bond distance of 3.54 and 2.78. Similarly glycine

(7) of C chain of collagen also forms one hydrogen bond through Nitrogen with bond distance of 2.88 and 2.68.

Table 3. Interaction sites, binding energy, hydrogen bond sites and bond distance, calculated based on the bioinformatics tools for the cross-linking between collagen (Type III) and alginic acid.

Interaction site	Binding energy (kcal/mol)	H-bond	Bond distance (Å)
Eleventh residue Ala of A-chain (α_1)	−7.14		3.29,3.04,2.85
		Ala A 11-(C_a)-N ...H-O(AA)	
Ninth residue Ala of C-chain (α_1)		Ala C 9-(C_a)-N ...H-O(AA)	2.78,3.54
Seventh residue Gly of C-chain (α_1)		Gly C 7-(C_a)-O ...H-O(AA)	2.88,2.68
Seventh residue Gly of C-chain (α_1)	−6.89	Gly C 7-(C_a)-O ...H-O(AA)	3.27
Eleventh residue Ala of A-chain (α_1)		Ala A 11-(C_a)-N ...H-O(AA)	3.02,2.97,3.03
Eleventh residue Ala of A-chain (α_1)	−6.5	Ala A 11-(C_a)-N ...H-O(AA)	3.08,2.98
Ninth residue Ala of C-chain (α_1)		Ala C 9-(C_a)-N ...H-O(AA)	2.78
Seventh residue Gly of C-chain (α_1)		Gly C 7-(C_a)-O ...H-O(AA)	2.57
Twelfth residue Lys of A-chain (α_1)	−6.07	Lys A 12-(C_a)-N ...H-O(AA)	2.88,3.10
Thirteenth residue Gly of A-chain (α_1)		Gly A 13-(C_a)-O ...H-O(AA)	2.89,3.12
Thirteenth residue Gly of A-chain (α_1)	−5.92	Gly A 13-(C_a)-O ...H-O(AA)	3.22,2.79,2.59
Fourteen residue Pro of A-chain (α_1)		Pro A 14-(C_a)-N ...H-O(AA)	3.53
Twelfth residue Lys of A-chain (α_1)		Lys A 12-(C_a)-N ...H-O(AA)	3.25,3.32

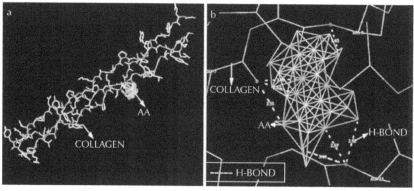

Figure 5. (a) and (b).

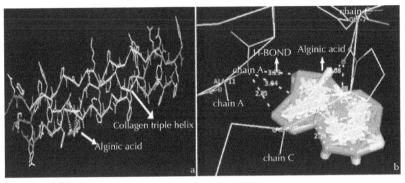

Figure 6. (a) and (b).

CHEMISTRY BEHIND THE CROSS-LINKING MECHANISM OF COLLAGEN WITH AA

From the above schematic representation, we found collagen and AA cross-linked through covalent linkage (chemical cross-linking). When the required concentration of collagen and AA (I) was mixed, the reaction starts from the ester formation in AA (II) due to loss of water according to Anson et al. (2009). Since the resultant lactones are biologically active, ready to interact with free –NH$_2$ group of lysine in collagen chain (III) and (IV) as shown above.

With regard to the thermal analyses of cross-linked biopolymer material (AACC 1 and AACC 3) and the plain (native) material, we observed an increase in thermal stability upon cross-linking with AA. Salome Machado et al. (2002) observed similar results for the cross-linking of collagen with chitosan. Derivative information's shown in Figure 4(a) and 4(b) emphasizes, sharp peak at 252°C corresponds to AA alone and a broad peak at 327°C corresponds to collagen alone. When collagen was mixed with AA at optimized concentration, the peak shifted towards right side for the cross-linked polymer material. This implies the cross-linking between Type I, Type III collagen, and AA. Nevertheless, when the concentration of AA increased to more than 1.5%, there was no appreciable increase in stability due to the non-availability of free molecules of collagen to interact with higher concentration of AA. Wu et al. (2007) observed the covalent linkages between glutaraldehyde and collagen. Further, with reference to chitosan (a natural polycationic polymer) cross-linking with collagen, bond formation was between -NH$_2$ groups of chitosan with –COOH group of amino acids of collagen. However, in the case of AA (a polyol) (Yang et al., 2007), having more number of –OH groups, ready to interact with both types of collagen (Type I, Type III) through intermoleculer multiple hydrogen bonds as shown in Figure 5(a), 5(b) and Figure 6(a), 6(b). The inter molecular multiple hydrogen bonding observed between AA and Type I, Type III collagen in the present study was similar to the observations made by Madhan et al. (2005) for the reaction between collagen and catechin. Further, the binding energy calculations (Table 2 and Table 3) received from the bioinformatics tools also confirm the intermolecular multiple hydrogen bonding. Thus, reverse to the bonding pattern of chitosan with collagen, both covalent and hydrogen bonding interactions occurs between collagen and AA, which in turn increases the stability and mechanical property (tensile strength) of the resulting biopolymer material appreciably.

To further understand the mechanism of stabilization of cross-linked collagen, we carried out experiments on measurements of percentage of cross-linking degree using TNBS. The procedure adopted in the present study reveals; TNBS interact only with the free amino groups as shown below;

As we discussed earlier, due to covalent bonding (between –COOH group of AA and –NH$_2$ group of amino acid residue of collagen), the amino groups get interacted with AA which is reflected in the percentage of cross-linking degree (75%) at 1.5% of AA. When the concentration of AA increases, there was no further increase in percentage degree of cross-linking. As shown schematically below, the –NH$_2$ groups in lysine residue of collagen was out of site when the concentration of AA increases, which results with no further increase in percentage degree of cross-linking (Figure 2).

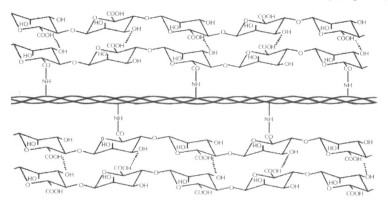

Followed by the preparation and characterization of the cross-linked biopolymer material using collagen and AA.

CONCLUSION

In the present study the choice for the preparation of thermally stable and mechanically stable biopolymer material was Type I and Type III collagen of bovine skin and avian intestine with a natural polymer AA.The reason for the selection of Type I and Type III collagen was due to its interaction with number of cells and its involvement in most of the human and animal diseases. Though numbers of studies were available for the different use of AA, its cross-linking potential with collagen has not yet in reports.The present study describes, the cross-linking ability and chemistry of collagen with AA was assessed using FT-IR, TGA, and degree of cross-linking measurements using standard procedures. The results obtained and the schematic representation of the reaction mechanism summarized proves, AA was cross-linked with collagen in two pathways one through multiple intermolecular hydrogen bonding (proved by bioinformatics study) and second is by covalent linkage (proved by TNBS/percentage of cross-linking degree assay).The biopolymer material (sheet) prepared upon cross-linking of AA was the green method of preparation. No toxic compounds were involved in this preparation and the resultant material found application as wound dressing sheet in clinical applications.

KEYWORDS

- **Acid soluble collagen**
- **Alginic acid**
- **Anionic polysaccharides**
- **Collagen**
- **Cross-linking**

Chapter 16

Cellulose Fiber Reinforced Poly(lactic acid) (PLA) Composites

Nina Graupner and Jörg Müssig

INTRODUCTION

The demand for environmentally friendly products based on renewable raw materials has increased in the last years. Examples of biodegradable materials include natural fiber reinforced biopolymers (Anonymous, 2007; Avella et al., 2009; Bhardwaj and Mohanty, 2007; Bledzki et al., 2009; Bodros et al., 2007; Cheung et al., 2009; Nampoothiri et al., 2010). As an alternative to natural fibers, man-made fibers based on renewable resources can be used as reinforcement. The advantage of the man-made over the natural fibers is their reproducible quality; a disadvantage is the higher price in comparison to for example bast fibers. An example of a man-made cellulose fiber is lyocell which is regenerated from 100% wood cellulose in the NMMO process using an organic solvent N-methyl-morpholine-N-oxyde. The NMMO process is based on the ability of amino oxide to dissolve cellulose under specific conditions. The cellulose can be regenerated out of these solutions and the fibers are produced by a wet fiber formation process (Albrecht et al., 1997). Lyocell fibers display special force elongation characteristics. They combine a high tensile strength and a high elongation at break in one fiber which is an advantage for the production of composites which must display high tensile and impact strength at the same time. A typical force-elongation curve of lyocell fibers in comparison to kenaf (bast fiber) and cotton (seed hair fiber) is shown in Figure 1.

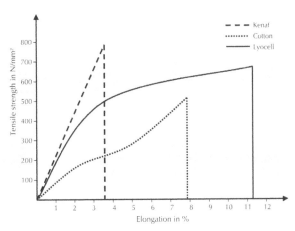

Figure 1. Force-elongation characteristics of kenaf, cotton and lyocell fibers.

One biopolymer that is already used as a matrix for composite production is poly(lactic acid) (PLA). The PLA is a thermoplastic biopolymer with lactic acid which is derived from starch by a fermentation process as its basic monomer. High molecular weight PLA is polymerized by the lactide ring opening polymerization to PLA (Garlotta, 2001; Gupta et al., 2007; Lim et al., 2008).

While cellulose fiber reinforced polypropylene (PP) is already used by default for example in the automobile industry for interior parts (Karus and Kaup, 2005), the conventional use of cellulose fiber reinforced PLA is still at the beginning. But there are also some products such as biodegradable urns, mobile phone shells or prototypes of spare tyre covers made from natural fiber reinforced PLA at the market (Anonymous, 2007; Iji, 2008; Grashorn, 2007). Many studies deal with the use of natural fibers as reinforcements in PLA composites. An overview about the mechanical characteristics and application areas of natural fiber-reinforced PLA can be found for example in Bhardwaj and Mohanty (2007), Avella et al. (2009), Ganster and Fink (2006), Jonoobi et al. (2010), and Graupner et al. (2009). For the improvement of the composite characteristics it is still necessary to carry out optimization processes for fibers, PLA matrix and the interactions of both. Moreover the processing parameters, force elongation characteristics of fibers and matrix as well as the use of additives like plasticizers or adhesion promoters have decisive influences on the mechanical characteristics of the composites.

For improving the mechanical characteristics of cellulose fiber reinforced PLA some studies have been carried out which deal with an improvement of fiber/matrix interactions. An important factor is the optimization of the fiber surface by modification methods. Examples for surface treatments on natural fibers and their effects on the mechanical PLA composite characteristics are given for example in the studies of Tokoro et al. (2008), Cho et al. (2007), Hu and Lim (2007), Lee et al. (2009), Masirek et al. (2007), Yu et al. (2010) and Huda et al. (2008). Beside fiber surface treatments, the bonding between fiber and matrix can be optimized by additives. A well known additive for improved fiber reinforced PP composite characteristics is anhydride maleic acid (MA). Some studies tested the use of MA in cellulose fiber reinforced PLA composites. But the most studies reported no improved tensile strength by using MA (Bourmaud and Pimbert, 2008; Plackett, 2004). Lee and Ohkita (2004) used bamboo fiber esterifies malefic anhydride reinforced PLA in the presence of dicumyl peroxide as a radical initiator. These results proved improved tensile characteristics of the composites. Lee and Wong (2006) studied the effects of lysine-based diisocyanate (LDI) as a coupling agent for compression molded bamboo fiber PLA composites. The results showed improved tensile characteristics of the composites. And a study of cotton fiber reinforced PLA with a fiber load of 40 mass% has shown that lignin as a natural additive increased the tensile strength by 9% and the Young´s modulus by 19%, whereas the impact strength was decreased by 17% in comparison to the pure PLA sample (Graupner, 2008). A previous study has also shown that a mixture of brittle and stiff hemp fibers and ductile lyocell fibers lead to improved composite characteristics (Graupner, 2009).

The admixture of lyocell fibers as a force elongation modifier for kenaf fiber reinforced PLA composites as well as the optimization of the processing parameters and the use of lignin as a kind of natural additive for improved fiber/matrix interactions will be described in the present chapter. The study is focused on compression molded lyocell and kenaf fiber reinforced PLA composites.

MATERIALS AND METHODS

Fibers and Matrix

For the composite production regenerated cellulose fibers (lyocell) provided by the Lenzing AG (Lenzing, Austria) with different fineness values (1.3, 3.3, 6.7, and 15.0 dtex) and yield retted kenaf fibers (Holstein Flachs, Mielsdorf, Germany) were used as reinforcements. As matrix PLA fibers (Ingeo fibers, type SLN 2660D (Far Eastern Textile Ltd., Taipai, Taiwan)) with a fiber fineness of 6.7 dtex and a density of 1.24 g/cm^3 (produced from a Nature Works PLA type 6202D; Cargill Dow LLC, Minnetonka, Minnesota, USA) were used. The PLA and cellulose fibers were mixed during the carding process with 20 and 40 mass% reinforcing fibers.

Fiber Characteristics

Fibers and fiber bundles were tested for their tensile characteristics and width values. Prior to the investigations the fibers were conditioned for at least 24 hr in a standard climate at 20°C and 65% relative humidity according to DIN EN ISO 139.

The tensile characteristics of single elements (single lyocell fibers and single kenaf fiber bundles) were tested by a testing instrument Fafegraph M (Textechno, Mönchengladbach, Germany) working with a load cell of 100 cN (lyocell fibers) and 10,000 cN (kenaf fiber bundles), respectively. Fibers and bundles were clamped into clamps covered with PVC under a pressure of approximate 3–4 bar under a preload of 200 mg. They were tested at a gauge length of 3.2 mm with a testing speed of 2 mm/min. At least 80 single elements were tested per sample.

For the investigation of the fiber width the Fiber shape image analyzing system was choosen. The elements were distributed on a slide frame (40 x 40 mm^2, glass width of 2 mm; company Gepe, Zug, Switzerland). Four slide frames were prepared per sample and scanned with a Canoscan scanner FS 2710 (Canon, New York, USA) at a resolution of 4,000 dpi. The images were analyzed with the image analysis software Fiber shape 4.1 (IST AG, Vilters, Switzerland). The evaluation was done with an IWTO-wool calibrated measuring mask based on a standard long-fiber measuring mask.

Composite Production

To investigate the influence of the kind of semi finished product composites were produced from multilayer webs or needle felts. Multilayer webs were produced by using a lab roller card (company Anton Guillot, Aachen, Germany) with a working width of 30 cm. The fibers were oriented predominantly in length direction to the machine direction (MD). The area weight per unit of the multilayer webs used for composite production was approximate 2,000 g/m². For the needle felt production one part of the multilayer webs was processed on an industrial machine to needle felts with an approximate area weight per unit of 2,000 g/m². The multilayer web was pre-needled with the needling density at the rate of 10 punches/cm² by a Heuer & Sohn (Tespe, Germany) pre-needle machine (type ROV 12S/250/11/900; one needle board, 800 needles per m working width, 7 needle lines, 320 pinpricks per minute) with a working width of 2,600 mm. For the pre-needling process needles of the type 15 x 20 x 4 R333 G 1012 (Groz & Beckert KG, Gummersbach, Germany) were used. The main needle felt production was carried out by a Fehrer (Linz, Austria) needle felt machine type NL1226 with a working width of 2,600 mm (two needle boards, 312 needles per line, 14 lines per needle board, feeding rate of 1.6, and 672 pinpricks per minute). Needle-punching density in the main needle process was per throughput 74 punches/cm². The roller carded webs were punched twice (from both sides, 148 punches/cm²). For the needle-punching process needles from the type 15 x 16 x 36 x 3 R222 G 53437 (Groz & Beckert KG, Gummersbach, Germany) were used.

The influence of the compression moulding step was analysed using two different compression moulding techniques:

CP-1: Ten minute pre-heating of the semi-finished product in the open press at 180°C (instead of a pre-drying step), 10 min compression moulding at 3.2 MPa (hydraulic press type KV 214.01 (Rucks Maschinenbau GmbH, Glauchau, Germany) and cooling at room temperature (20–25°C) for at least 30 min (Graupner and Müssig, 2009).

CP-2: The semi-finished products were pre-dried for 2 hr at 105°C in a forced air oven (Thermicon® (Heraeus GmbH, Hanau, Germany), compression molded at 180°C for 5 min at 5.8 MPa in a Joos hot press (hydraulic press type HP-S10, Joos, Pfalzengrafenweiler, Germany) and cold pressed for 3 min at 25°C at 5.8 MPa in a Joos cold press (hydraulic press type HP-S60, Joos, Pfalzengrafenweiler, Germany).

The treatment of one part of the multilayer webs with lignin was carried out by using lignin of the plant *Eukalyptus globulus* (type Hardwood kraft eucalyptus; BFH, Hamburg, Germany). One gram lignin was solved in 100 ml ethanol (96%) and the lignin solution was sprayed on and between the single layers of the multilayer webs by an aerosol can.

Composite Characteristics

From the resulting composite boards different test specimens were manufactured. The investigation of the tensile characteristics of the composites was carried out with a Zwick/Roell testing machine type Z 020 (Ulm, Germany) working with a load cell of 20 kN and pneumatic clamps according to DIN EN ISO 61. The testing speed was set to 2 mm/min. Unnotched Charpy impact characteristics were analyzed using a Zwick impact tester type 5,102 with a bearing distance of 40 mm, a pendulum length of 225 mm and a pendulum size ranging between 0.5 and 4 J according to DIN EN ISO 179.

Morphological examinations were carried out using a Zeiss DSM 940A scanning electron microscope (SEM) manufactured by Zeiss (Jena, Germany). All specimens were sputtered for 90 sec. with 56 mA with a layer of gold prior to SEM observations

(Bal-Tec sputter coater SCD 005, Bal-Tec, Liechtenstein) and mounted on aluminium holders using double-sided electrically conducting carbon adhesive tabs. A high voltage of 10 kV was used for making the photographs. The SEM investigation was used to study the fracture surface of the tensile specimens.

DISCUSSION AND RESULTS

Fiber Characteristics

The characteristics of lyocell and kenaf were investigated by determining fineness and tensile properties. The results show clear differences. Figure 1 gives an overview of the tensile characteristics and the width.

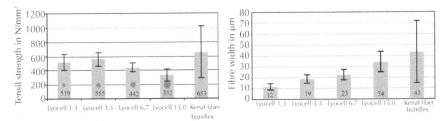

Figure 2. Tensile strength (left) and fineness measured as fiber and fiber bundle width (right) of lyocell and kenaf (mean values, standard deviations are shown as error bars; dots show the dimension of the fiber diameter).

The best tensile characteristics were achieved by lyocell fibers with 0.33 tex with a tensile strength of 555 N/mm², followed by lyocell 0.13 tex with 519 N/mm², lyocell 0.67 tex with 442 N/mm² and lyocell 1.5 tex with 332 N/mm². The highest tensile strength was measured for the kenaf fiber bundles with a tensile strength of 653 N/mm². Due to the bigger fiber diameter and the probability of more defects in the cross-section the tensile strength decreases in trend with an increasing fiber diameter for the lyocell fibers (Vetrotex, 1995).

Composite Characteristics

Influence of the Semi-finished Product

Composites reinforced with multilayer webs and needle felts were produced by the compression molding technique CP-1 with fiber reinforcements of 20 and 40 mass%.

The tensile characteristics have shown clear influences of the composites whether multilayer webs or needle felts were used for the composite production, while the Charpy impact strength has shown lower differences. Considering the tensile strength a better reinforcement effect to the pure PLA sample was achieved by using the needle felts as reinforcements in comparison to multilayer web reinforced composites (Figure 3). As shown in Figure 3 a higher tensile strength was proved with increasing fiber fineness for a fiber reinforcement of 40 mass%. For the multilayer webs a reinforcement effect could only be measured for the webs produced from the lyocell 1.3 fibers with 42 N/mm². The tensile strength of the pure PLA matrix was determined to be

40 N/mm². Composites reinforced with lyocell 6.7 and lyocell 15.0 multilayer webs reached values of 36 N/mm² and 28 N/mm² respectively. For the needle felt reinforced composites clear reinforcement effects were determined (lyocell 1.3-PLA = 63 N/mm², lyocell 6.7-PLA = 61 N/mm² and lyocell 15.0-PLA = 57 N/mm²). These results lead to the assumption that the kind of the semi-finished product has a clear influence on the mechanical characteristics of the composites produced by the compression molding technique CP-1. This effect could be attributed to a more homogeneous fiber distribution in the PLA matrix due to the additional needling process and the lower thickness of the needle felts compared to the multilayer webs. It is assumed that the lower thickness lead already in the pre-heating phase to a better heat conductivity and drying in the press.

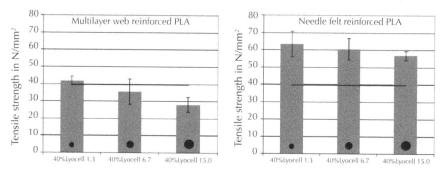

Figure 3. Tensile strength of 40 mass% lyocell fiber reinforced PLA in dependence of the fiber fineness. Composites reinforced with needle felts (left); Composites reinforced with multilayer webs (right) (mean values, standard deviations are shown as error bars; dots show the dimension of the fiber diameter).

By a reduction of the reinforcing fibers from 40 to 20 mass% significant better reinforcement effects were measured (compare Table 1). This phenomenon is in contrast to the composite theory because the higher value should be measured for the composites with the higher fiber load (Bunsell and Renard, 2005; Vetrotex, 1995). For the lyocell 1.3-PLA composites the tensile strength could be increased from 42 to 51 N/mm² and for lyocell 15.0-PLA from 28 to 54 N/mm². This effect can be explained by the worse coating and the bad interfacial interactions between the lyocell fibers and the PLA matrix due to the lower degree of compaction of the multilayer web reinforced composites with higher fiber loads. Because of the low degree of compaction the reinforcing fibers can act as weak spots and the reinforcement effect can be decreased by higher fiber loads. The SEM investigations approve the assumptions that the 40 mass% multilayer web reinforced composites show a lower degree of compaction which lead to a high number of fiber pull-outs in comparison to the 20 mass% reinforced composites (Figure 4).

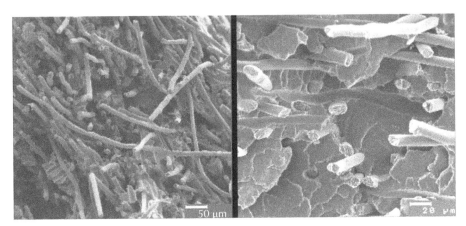

Figure 4. The SEM micrographs of tensile fractured surfaces of 40 (left) and 20 mass% (right) reinforced multilayer web reinforced composites.

Ibrahim et al. (2010) investigated similar trends. They investigated compression molded kenaf fiber reinforced PLA with fiber loads ranging between 10 and 50 mass%. The tensile strength decreased from 27 N/mm² measured for the 20% kenaf PLA composites to 16 N/mm² measured for the 40% kenaf PLA composites. And Oksman et al. (2003) described similar effects for compression molded flax-PLA. The tensile strength of 40% flax-PLA decreased compared to 20% flax-PLA. They partially attribute this behavior to poor adhesion between fiber and matrix.

A more detailed presentation of the influence of the semi-finished product on the mechanical composite characteristics is given in Graupner and Müssig (2009).

Lignin Treatment

An improvement of the tensile characteristics of the multilayer web reinforced composites produced by compression molding technique CP-1 was reached by the treatment of the multilayer webs with lignin. While the pure PLA sample and the lignin treated PLA sample showed no clear differences between tensile strength, Young´s modulus and Charpy impact strength (Table 1) clear influences of the lignin treatment were determined for the composites. This supports that the lignin interacts just between fiber and matrix or improves the compatibility of the molded composites and has no influence on the mechanical characteristics of the pure PLA matrix.

Mechanical investigations of 40 mass% lignin treated multilayer web reinforced PLA composites have shown improved tensile strength and Young's modulus values in comparison to the untreated composites. Considering the tensile strength the values increased from 28 to 57 N/mm² for the lyocell 15.0-PLA composites, from 36 to 53 N/mm² for the lyocell 6.7-PLA composites and from 42 to 52 N/mm² for the lyocell 1.3-PLA composites (compares Figure 5, Table 1).

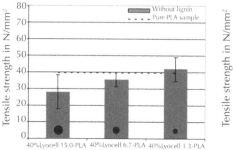
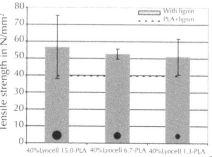

Figure 5. Multilayer web reinforced PLA composites (40 mass%) reinforced with untreated (left) and lignin treated fibers (right) (mean values, standard deviations are shown as error bars; dots show the dimension of the fiber diameter).

In contrast to the untreated composites the tensile strength decreased with reduced fiber diameters. It is assumed that this phenomenon is based on the low lignin concentration. The concentration (0.01 g lignin/ml ethanol) used could be too low to coat the big specific surface of the finer fibers while a better and more homogenous coating took place on the smaller specific surfaces of the coarser fibers. It is supposed that a raising lignin concentration could lead to a better coating of the finer fibers and thus a similar trend as described for the untreated fibers should be determined. The fractured surfaces of lignin treated composites show less and shorter fiber pull-outs. This effect is demonstrated in Figure 6. Which shows the fractured surface of a 40% Lyocell 6.7-PLA composite next to a fractured surface of a lignin treated 40% Lyocell 6.7-PLA composite.

Figure 6. Tensile fractured surface of lyocell 6.7-PLA composites (40 mass%). Left: untreated, right: lignin treated.

In contrast to the untreated composites higher tensile strength values were proved for the lignin treated composites with a higher fiber load. Due to a better compaction of the composites and better fiber/matrix interactions a higher reinforcement effect was measured for 40 mass% fiber reinforced multilayer web PLA composites compared to the 20 mass% reinforced composites. According to statistical investigations with U- and t-tests ($\alpha = 0.05$) the tensile strength values were increased significantly by the lignin treatment (Table 1). The tensile strength of 40% lyocell 1.3-PLA was increased from 41.9 to 51.0 N/mm², 40% lyocell 6.7-PLA from 35.6 to 52.7 N/mm² and 40% lyocell 15.0-PLA was increased from 28.2 to 56.7 N/mm². A similar trend was observed in a previous study for cotton-PLA composites reinforced with 4 mass% fibers. In contrast to the present study, the lignin used was in the form of powder with a higher concentration (2.5 g/100 g sample). The tensile strength of the cotton-PLA composites was increased by adding lignin. A disadvantage of these composites was the high odor intensity caused by the lignin (Graupner, 2008). Due to this in the present study the lignin concentration was reduced (1 g/100 g sample) and the powder was dissolved in ethanol. No atypical odor development was observed but similar improvements of tensile strength values were also measured with a lower lignin concentration.

Table 1. Mechanical characteristics of PLA composites and lignin treated PLA composites produced by compression moulding technique 1 CP-1 (mean values, standard deviations are given in brackets).

Kind of Composite	Lignin treatment	n	Tensile strength in N/mm²	Young's modulus in N/mm²	Elongation at break in %	n	Charpy impact strength in kJ/m²
PLA reference sample	No	11	39.6 (±1.8)	2243.9 (±62.2)	2.2 (±0.2)	9	15.8 (±3.8)
PLA reference sample	Yes	13	39.5 (± 1.8)	2299.0 (± 92.3)	2.2 (± 0.2)	7	15.6 (± 4.0)
20% Lyocell 1.3-PLA	No	7	50.9 (±3.2)	4199.4 (±260.9)	2.0 (±0.5)	8	18.1 (±4.4)
20% Lyocell 1.3-PLA	Yes	7	60.3 (± 2.4)	4271.8 (± 149.1)	2.2 (± 0.2)	5	22.6 (± 5.7)
40% Lyocell 1.3-PLA	No	7	41.9 (±2.6)	4142.1 (±292.8)	3.3 (±0.4)	7	33.1 (±5.5)
40% Lyocell 1.3-PLA	Yes	6	51.0 (± 3.6)	5457.3 (± 504.9)	2.6 (± 0.5)	7	31.6 (± 3.4)
20% Lyocell 6.7-PLA	No	6	35.3 (±4.3)	3648.5 (±328.5)	3.3 (±0.6)	8	23,7 (±5.2)
20% Lyocell 6.7-PLA	Yes	6	38.5 (± 6.5)	4388.5 (± 452.3)	2.1 (± 0.5)	10	18.7 (± 4.1)
40% Lyocell 6.7-PLA	No	7	35.6 (±7.3)	5164.4 (±485.6)	3.5 (±0.4)	7	38.8 (±5.6)
40% Lyocell 6.7-PLA	Yes	6	52.7 (± 10.8)	6221.8 (± 200.3)	3.9 (± 0.5)	9	28.0 (± 6.5)
20% Lyocell 15.0-PLA	No	7	53.9 (±3.4)	4611.8 (±248.5)	2.2 (±0.1)	7	26.9 (±7.1)
20% Lyocell 15.0-PLA	Yes	7	54.5 (± 3.7)	4877.1 (± 268.8)	2.0 (± 0.3)	5	20.3 (± 4.4)
40% Lyocell 15.0-PLA	No	6	28.2 (±4.4)	4293.5 (±239.8)	4.6 (±0.9)	5	30.5 (±3.4)
40% Lyocell 15.0-PLA	Yes	7	56.7 (± 3.1)	6144.9 (± 292.3)	2.7 (± 0.4)	5	23.7 (± 1.2)

In contrast to the tensile strength a clear reinforcement effect was achieved for the Young's modulus in comparison to the neat PLA sample. As observed for the tensile strength, higher Young's moduli were measured with the lignin treated composites

(Table 1). According to statistical investigations with the U-test, t-test and Lord-test (α = 0.05) there was a significant increase of the Young´s modulus for all composites in comparison to the pure lignin treated PLA sample. The highest values were achieved for the untreated and lignin treated 40% lyocell 6.7/PLA composites. The Young´s moduli of the Lyocell-PLA composites were increased significantly apart from 20% Lyocell 1.3-PLA due to the lignin treatments. This result was also similar to that presented in a previous work (Graupner, 2008) where lignin was added to cotton-PLA multilayer webs with a fiber load of 40 mass%.

The progress of the Young´s modulus of composites without lignin treatment shows no uniform trend. For lyocell 1.3 and lyocell 15.0 a lower value was measured for the composites with a higher fiber load. As discussed previously, for the tensile strength weak interactions between fibers and matrix and a low degree of compaction caused this result. As shown in Figure 7, with the lignin treated composites a higher reinforcement effect was achieved for the composites with the higher fiber load. Similar to the tensile strength, the highest Young´s modulus values were achieved in the composites reinforced with the coarsest lyocell fibers. The Young´s modulus of the lignin treated composites with fiber loads of 20 and 40 mass% was slightly decreased with increasing fiber fineness due to the better lignin accretion and the better coating of the fibers with the matrix on the coarser fibers.

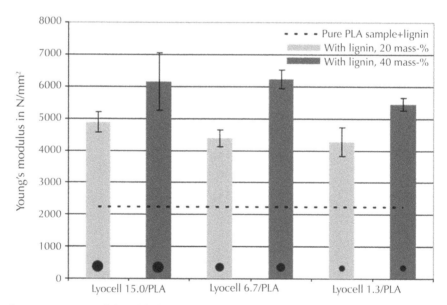

Figure 7. Young´s modulus of the lyocell-PLA composites with lignin treatment reinforced with 20 and 40 mass% (mean values, standard deviations are shown as error bars; dots show the dimension of the fiber diameter).

In contrast to the tensile characteristics which are strongly influenced by good interfacial interactions between fibers and matrix good impact characteristics are mainly based on the energy absorption by using fibers with high elongation, like lyocell fibers, and fiber pull-outs from the matrix (Gassan and Bledzki, 1999; Mieck et al., 2000). The lignin leads to better interfacial interactions between fibers and matrix and stronger bonding, so less energy can be absorbed by fiber pull-outs and debonding, and impact strength values were decreased. Due to this, the lignin treatment leads to a decline of the impact strength. However, the lignin treated composites showed still a clear reinforcement effect compared to the pure PLA sample with an impact value 16 kJ/m². The impact strength of the 40 mass% reinforced lyocell 1.3-PLA composites was reduced from 33 to 32 kJ/m², of lyocell 6.7-PLA from 39 to 28 kJ/m² and of lyocell 15.0-PLA from 31 to 24 kJ/m² (Table 1). Similar to the untreated PLA composites, the better impact strength values were measured for the composites with higher fiber loads (Figure 8).

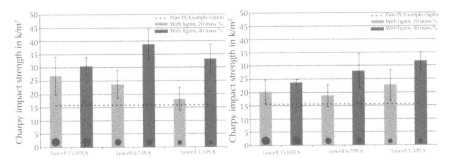

Figure 8. Unnotched Charpy impact strength of the untreated (left) and lignin treated (right) lyocell/PLA composites with fiber loads of 20 and 40 mass% (mean values, standard deviations are shown as error bars; dots show the dimension of the fiber diameter).

Due to the low elongation at break of the PLA matrix (2.2%) and the high elongation of the Lyocell fibers (10–15%), a direct investigation of the fiber matrix interactions by using the single fiber fragmentation test could not been carried out. But a better adhesion between cotton fibers and a PP matrix by the addition of lignin could be proved through a single fiber fragmentation test according to Huber and Müssig (2008). For the lignin treatment the cotton fibers were stored for 30 min in a lignin solution (1 g lignin in 100 ml ethanol). Figure 8 shows the results of the fragment lengths of raw cotton in the PP matrix and lignin treated cotton in the PP matrix. It can be seen that the fragment length of raw cotton of up to 10 mm is significantly longer than the fragment length of lignin treated cotton fibers which is of approximate 7 mm (proved by the t-test with α = 0.05). This proves a better adhesion of the lignin treated cotton fibers in the PP matrix compared to the raw cotton fibers. Due to the great structural and chemical differences of PLA and PP it cannot be proved that the same effect exists for PLA but the higher mechanical values of the lignin treated composites and lower fiber pull-outs allow us to assume that there is an influence of the lignin on the fiber/matrix interactions.

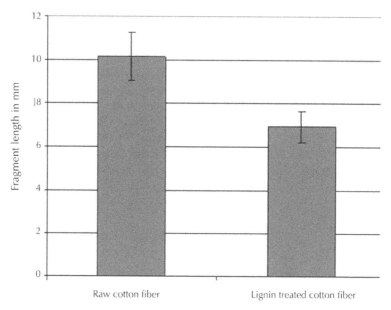

Figure 9. Fragment lengths of raw and lignin treated cotton fibers in a polypropylene matrix determined by the single fiber fragmentation test (standard deviations are given as error bars).

Modified Process Parameters

In the general outline it is obvious that the mechanical composite characteristics of the composites produced by compression moulding technique CP-1 were not improved as much as possible. This was mainly caused by the suboptimal process parameters used in the study. A previous study (Graupner et al., 2009) has shown tensile strength values more than twice as high for 40 mass% Lyocell 1.3 fiber reinforced PLA produced using other process parameters (3 hr pre-drying at 105°C in a forced air oven, 20 min. compression molding at 180°C). According to a model given by Moser for stereo-scopic, randomly oriented fiber layers (Moser, 1992) a theoretical Young´s modulus of 7,105 N/mm² could be calculated for 40 mass% Lyocell 1.3 fiber reinforced PLA. In the previous study a value of 6,784 N/mm² could be achieved while in the present study a value of only 4,142 N/mm² was measured. The aim of the changed processing parameters was a reduction of cycle time and process complexity. So, the pre-drying step which was used in the previous work was changed into a 10 min pre-heating step in the open press at 180°C and the press time was reduced from 20 to 10 min (compression moulding technique CP-1). Our results prove that these processing parameters affect the mechanical composite characteristics negatively (compare chapter Influence of the semi-finished product and Lignin treatment).

A clear improvement of the mechanical characteristics with regard to tensile and impact characteristics of multilayer web reinforced PLA composites was achieved by using modified process parameters (compression moulding technique CP-2). In comparison to the processing technique CP-1 a pre-drying step was inserted (2 hr at 105°C) and the compression molding time in the hot press was reduced to 5 min. The cooling of the composites occurred in a cold press for 5 min at 25°C.

The tensile strength of the composites reinforced with fiber loads of 40 mass% reached 99 N/mm² for the lyocell 1.3-PLA composites, 100 N/mm² for the lyocell 6.7-PLA composites and 85 N/mm² for the lyocell 15.0-PLA composites (Figure 10). The Charpy impact strength increased from 33 to 66 kJ/m² for lyocell 1.3-PLA composites, from 39 to 43 N/mm² for the lyocell 6.7-PLA composites and from 31 to 54 kJ/mm² for the lyocell 15.0-PLA composites.

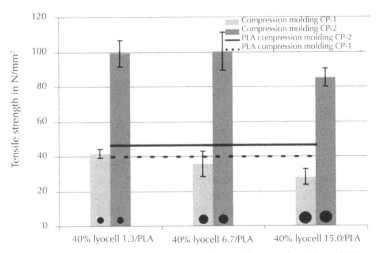

Figure 10. Influence of the processing parameters on the multilayer web reinforced PLA composites and the neat PLA matrix (mean values, standard deviations are shown as error bars; dots show the dimension of the fiber diameter).

A comparison of the mechanical characteristics of 40 mass% lyocell 1.3-PLA composites produced by different compression molding techniques is shown in Figure 11. The highest mechanical characteristics were achieved by using compression molding technique CP-2 followed by the compression molding technique used in the study Graupner et al. (2009). By far the lowest composite characteristics were measured for the composites produced by compression molding technique CP-1 due to a low degree of compaction and weak interfacial interactions in the composites.

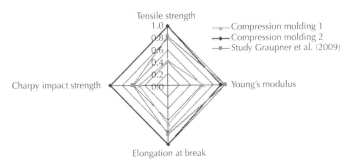

Figure 11. Comparison of the mechanical characteristics of 40 mass% multilayer web reinforced lyocell 1.3-PLA composites (highest values = 1).

Lyocell as an Admixture for Kenaf Fiber Reinforced PLA Composites

Due to the brittle character of the most bast fibers and the PLA matrix the resulting impact strength of bast fiber reinforced PLA composites is mostly worse than that of the pure PLA matrix (Bax and Müssig, 2008). Based on the lower price and the wish to produce stiff components of bast fibers like hemp, flax or kenaf they are used for many applications in the automobile industry like indoor panels (Anonymous, 2007; Karus et al., 2006). A limiting factor for some applications is the bad impact behavior of composites produced from bast fibers and a polymer matrix. Hence, an admixture of lyocell fibers to bast fibers in a PLA matrix with regard to the impact characteristics was investigated. The composites were produced by the carding technique with fiber loads of 40 mass% with a following compression molding step (compression mould-ing technique CP-2). Different fiber mixtures in a PLA matrix were investigated: 36% kenaf + 4% lyocell 3.3, 30% kenaf + 10% lyocell 3.3, 20% kenaf + 20% lyocell 3.3, 40% kenaf, and 40% lyocell 3.3.

The results of the impact characteristics are shown in Figure 12. Composites pro-duced from 40 mass% kenaf fibers show impact strength values of 14 kJ/m² which are significant (Welch test, $\alpha = 0.05$) lower than that of the pure PLA matrix (16 kJ/m²) while composites produced from 40 mass% lyocell 3.3 fibers show significantly (Welch test, $\alpha = 0.05$) the highest Charpy impact strength with a value of 85 kJ/m². Mixing 36% kenaf with 4% lyocell 3.3 lead to values on the level of the pure PLA matrix (Welch test, $\alpha = 0.05$). A mixture of 30 mass% kenaf and 10 mass% lyocell 3.3 lead to a significant (U-test, $\alpha = 0.05$) increase up to a value of 25 kJ/m². The 20 mass% kenaf mixed with 20 mass% lyocell 3.3 resulted in an impact strength more than 2.5 times higher (37 kJ/m²) than the values measured for 40 mass% kenaf-PLA.

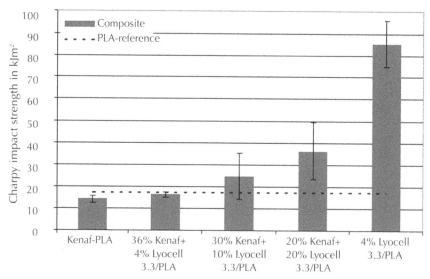

Figure 12. Unnotched Charpy impact strength of kenaf, lyocell and mixed kenaf/lyocell reinforced PLA composites (mean value as column, standard deviation as error bars).

A similar trend was described by Mieck et al. (2000). They investigated the influence of the admixture of lyocell fibers with different fiber loads on the mechanical characteristics of flax fiber reinforced PP composites. The impact strength values increased with a rising fiber load ranging between 0 and 40 mass%. By the admixture of 40 mass% lyocell the impact characteristics could be nearly tripled.

Figure 13 shows the tensile characteristics of the composites. A significant (U-test, $\alpha = 0.05$) reinforcement by adding the fibers could be proved with regard to the pure matrix shown as black line. A smal increase of the tensile strength with increasing lyocell fiber load was measured. The differences between the 40% kenaf/PLA composites and 20% kenaf + 20% lyocell 3.3/PLA composites were significant according to a Weir test with $\alpha = 0.05$. At the first view this looks a bit curious because the kenaf fiber bundles have the higher tensile strength. But this effect could be based on the higher specific surface of the finer lyocell fibers in comparison to the coarser kenaf fiber bundles (Vetrotex, 1995; comparison for glass fibers).

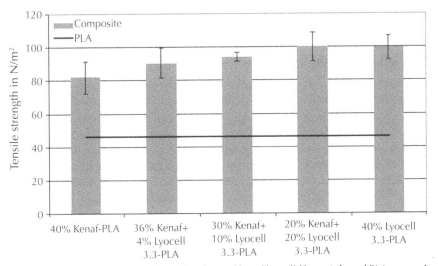

Figure 13. Tensile strength of kenaf, lyocell and mixed kenaf/lyocell fiber reinforced PLA composites (mean value as column, standard deviation as error bars).

CONCLUSION

The investigations have shown a great reinforcement potential for cellulose fibers in PLA matrices. The kind of semi-finished products as well as the composite production parameters has a big influence on the mechanical composite characteristics. Results proved a clear increase of the tensile strength by a modification of the process parameters up to values more than double as high (A pre-drying step was inserted for 2 hr at 105°C, the compression moulding time was reduced from 10 to 5 min and the cooling was carried out in a cold press at 25°C for 3 min instead of 30 min at room temperature).

Lignin as a natural additive has shown a potential for improving tensile characteristics in cotton or man-made cellulose fiber reinforced PLA composites and lead to composites with a higher degree of compaction. The tensile strength of lignin treated composites could be improved by a factor of up to 2.0 and Young´s modulus values by a factor of up to 1.4 compared to untreated PLA composites. The unnotched Charpy impact strength was decreased but the values are still higher than those of many bast fiber reinforced PLA composites. In comparison to a previous study dealing with lignin treated cotton fiber reinforced PLA, the odor behavior could be improved by reducing the lignin concentration and using the lignin in an ethanol solvent rather than in powder form. Despite the lower lignin concentration similar improvement effects for the tensile characteristics could be measured.

Brittle bast fiber reinforced composites were improved with regard to the impact characteristics by using lyocell fibers as additives. The combination of 20% kenaf and 20% lyocell in a PLA composite leads to stiff and strong composites reaching impact properties double as high as the pure matrix. In the general outline the presented PLA composites with their good mechanical characteristics are well suitable for higher stressed automotive applications if problems with regard to the temperature resistance are solved in the future.

ACKNOWLEDGMENT

We acknowledge Dipl.-Ing. André Decker (BIK, University of Bremen, Germany) for supplying the compression moulding technique as well as Georg Goeddecke (NAFGO GmbH, Neerstedt, Germany) for the support in the needle felt production. For supplying lyocell and PLA fibers we thank Dr. Gunther Sames, Markus Gobl and Dr. Joseph Innerlohinger (Lenzing AG, Lenzing Austria) as well as Norbert Schultz (RMB fibers AG, Gütersloh, Germany). For supplying lignin powder Dr. Britta Lohmeyer (Faserinstitut Bremen e.V., Bremen, Germany) and cand. Ph.D. B.Sc. Tim Huber (University of Canterbury, New Zealand) for carrying out the fragmentation tests should also be acknowledged.

KEYWORDS

- **Fiber bundles**
- **Fiber fragmentation**
- **Lyocell fibers**
- **Machine direction**
- **Poly lactic acid**
- **Polypropylene**
- **Scanning electron microscope**

Chapter 17

Isolation of Cellulose Nanowhiskers from Kenaf Fiber

Hanieh Kargarzadeh, SitiYasmine, Z. Z., Ishak Ahmad,
and Ibrahim Abdullah

INTRODUCTION

Recently, there has been wide interest in producing nanocellulose from natural fiber to replace the synthetic fiber. This phenomenon is caused by the urge and awareness from the society towards the green environment. Furthermore, the excitement is due to the extraordinary properties of the materials because of the nanosize effect of the reinforcement.

Natural fibers usually consist of cellulose, hemicelluloses, lignin, pectin, and waxy materials. It has good mechanical properties and can compete the strength and modulus of the synthetic fiber (Mallick, 1988; Mohd. Sapuan, 1999). Unlike synthetic fiber, natural fiber has more advantages such as low density, lightless, non-toxic, non-abrasive, biodegradable, high flexibility level, good dynamic mechanical, electrical and thermal properties to compared with other commercial fibers and available as abundant nature materials in large quantity which makes the interest for the researcher to replace the synthetic fiber (Bondeson and Oksman, 2007; Fahmy and Mobarak, 2008; Franco and Gonzalez et al., 2005; Kentaro et al., 2007; Roohani et al., 2008; Seydibeyoğlu and Oksman, 2008; Sun Young Lee et al., 2009).

Kenaf or *Hibiscus cannabinus* is a plant that belongs to the Malvaceae family. In Malaysia, kenaf is cultivated for its fiber due to the good weather, and the support and encouragement from the government to replace the tobacco plantation. This plant has a single, straight, and branchless stalk. Kenaf stalk is made up of a core and an outer fibrous bark surrounding the core. The fiber derived from the outer fibrous bark is also known as bast fiber. Kenaf bast fiber has superior flexural strength combined with its excellent tensile strength that makes it the material of choice for a wide range of extruded, molded and non-woven products.

Cellulose is the main component in natural fiber which this linear semi crystallite polysaccharide is built up of repeating unit of D-glucopyranose that consist three hydroxyl groups. The diameter of this fibril is ranged from 5 till 10 nm, meanwhile, the length depends on source from which the cellulose is obtained (Xue Li et al., 2007; Yousef Habibi et al., 2008). Extraction of cellulose from the natural fiber and applied as filler in nano range sized has find its place not just among the researcher but also to the industry sector.

One way to obtain cellulose whiskers is by acid hydrolysis where the cellulose is exposed to sulfuric acid for a controlled period of time and temperature. This process removes the amorphous parts of the cellulose, leaving single and well-defined crystals in a stable colloidal suspension. The rod-like cellulose particles from different sources have a diameter and length ranging from 5 to 20 nm and 100 nm to several micrometers, respectively, after the acid hydrolysis (Candanedo, Roman, and Gray, 2005). The introduction of sulfate groups along the surface of the crystallites will result in a negative charge on the surface as the PH is increased. This anionic stabilization via the attraction/repulsion forces of electrical double layers at the crystallites is probably the reason for the stability of the colloidal suspension of crystallites (Figure 1) (de Souze Lima and Brosali, 2004).

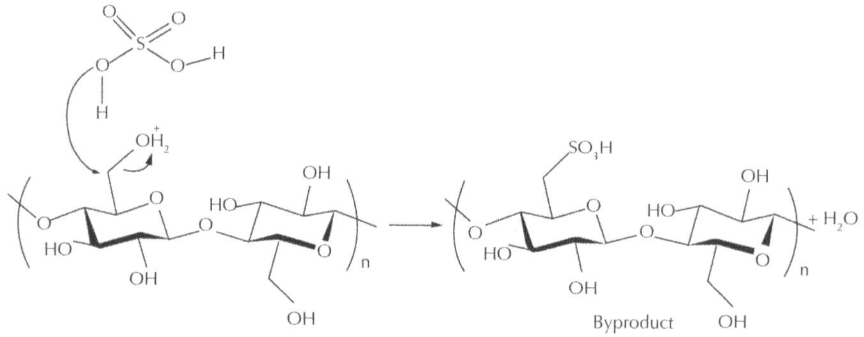

Figure 1. Esterification of cellulose chain in hydrolysis process.

The whiskers with high stiffness, surface area, and crystallinity are suitable for application in polymeric matrices, acting as reinforcing element (Gardner et al., 2008). Also, it can be used as a rheology modifier in foods, paints, cosmetics, and pharmaceutical products (Turbak et al., 1983).

The aim of this study was extraction of cellulose from kenaf fiber which is available in Malaysia and preparation of nanocellulose whiskers from extracted cellulose in controlled hydrolysis condition. Chemical composition, size of particles, morphology, crystallinity, and thermal analysis of cellulose whiskers were investigated.

MATERIALS AND METHODS

Materials

The materials used for the study includes raw kenaf bast (*Hibiscus cannabinus*) fibers were supplied by KFI Sdn. Bnd. (M). Sodium chlorites, sodium hydroxide, sulfuric acid, glacial acetic acid used were purchased from SYSTERM (M) Bhd.

Preparation of Cellulose Fibers

Kenaf fibers were cut into small pieces before treated with a 4 wt% NaOH solution at 130°C, under mechanical stirring, followed by washing with adequate distilled water to remove all alkali soluble components. A subsequent bleaching treatment was carried out to bleach the fibers. The solution used in this treatment consisted of equal parts of acetate buffer (a mix solution of 2.7g NaOH and 7.5 ml glacial acetic acid in 100 ml distilled water), aqueous chlorite (1.7% w/v) and distilled water. The bleaching treatment was performed at 130°C for 4 hr and 3 times, under mechanical stirring. The bleached fiber were washed respectively by distilled water and subsequently air-dried.

Preparation of Cellulose Nanowhiskers

Colloidal suspension of cellulose whiskers in water prepared as described elsewhere (Azizi Samir et al., 2004). Acid hydrolysis was done at 45°C with 65 wt% H_2SO_4 (pre-heated), for 40 min, under mechanical stirring. The suspension was diluted with ice cubes to stop the reaction and washed with water until neutrality before successive centrifugation at 10,000 rpm at 10°C for 10 min each step and dialyzed against distilled water in the sequence. The dispersion of whiskers was completed by an ultrasonic treatment. The resulting suspension was subsequently stored in a refrigerator after adding several drops of chloroform in order to avoid bacterial growth.

Characterization

Chemical Composition

Chemical compositions of fibers were estimated after various stages of chemical treatment. α-cellulose content was calculated by further treating the cellulose fibers with NaOH after the holocellulose content determination (ASTM D1103-55T and ASTM D1104-56). The difference between the values of holocellulose and α-cellulose gives the hemicelluloses content of the fiber. The lignin content was measured according to ASTM D1106-56.

Fourier Transform Infrared (FTIR) Spectroscopy

The Fourier transform infrared (FT-IR) spectroscopy (Perkin-Elmer, GX Model) was used to examine the changes in functional groups that may have been caused by the treatments. The samples were ground and mixed with KBr. The resultant powder was pressed into transparent pellets and analyzed in transmittance mode within the range of 4000600 cm^{-1}.

Field Emission Scanning Electron Microscope (FESEM)

The morphology of fibers after each treatment was investigated by using field emission scanning electron microscopy (FESEM) Zeiss Supra 55VP model with the magnification of 400x and all samples had been sputter-coated with gold to avoid changing.

X-Ray Diffraction Analysis (XRD)

The X-ray diffraction (XRD) measurements were performed on a Bruker AXS-D8 Advance model system. The diffracted intensity of Cu Kα radiation (0.154 nm, 40 kV and 40 mA) was measured in a 2θ range between 10° and 70°. Ground kenaf fibers and hydrolyzed cellulose sample were subjected to crystallinity analysis.

Termogravimetric Analysis (TGA)

The thermal stability of kenaf fibers after each treatment was characteraized using Mettler Toledo, SDTA 851e model TGA under linear temperature conditions. The samples were heated from room temperature up to 500°C at a heating rate of 10°C/min under a nitrogen atmosphere.

Transmission Electron Microscopy (TEM)

A Transmission Electron Microscope (TEM) observation was performed using a Philips CM30 to study the morphology of cellulose nanowhiskers. One droplet of 1% suspension was put on a Cu-Grid covered with a thin carbon film. To enhance contrast in TEM, the nano-saized whiskers were negatively stained in a 2 wt% solution of uranyl acetate (a heavy metal salt) in de-ionized water for one min.

RESULTS AND DISCUSSION

Chemical Composition

The chemical compositions of kenaf fiber at different stages of alkali and bleaching treatments are presented in Table 1. It is clear that raw kenaf fiber has the highest percentage of hemicelluloses and lignin and the lowest percentage of α-cellulose. When, the fiber is subjected to alkali and bleaching treatments, the percentage of hemicellulose and lignin decrease, whereas, the percentage of α-cellulose increased. The NaOH was found to be efficient in removing hemicellulose from kenaf fibers with the reduction of less than half of raw kenaf from 34.7 to 13.3 wt% after alkali treatment while the amount of lignin was reduced from 11.5 to 9.3 wt%.

Table 1. Chemical composition of kenaf fibre after each stage of treatment.

Materials	Cellulose (%)	Hemicelluloses (%)	Lignin (%)
Raw kenaf fiber	43.7±1.2	34.7±1.2	11.5±0.5
Alkali treated kenaf fiber	65.7±0.3	13.3±0.3	9.3±0.5
Bleached kenaf fiber	91±0.2	6±0.2	0

The bleaching process helps to remove the majority of the lignin component. As shown in Table 1, a further reduction in the percentage of hemicellulose and lignin contents is observed. Nevertheless, the percentage of α-cellulose content increased from 43 to 91%. The final fibers obtained after bleaching process is found to have the highest percentage of cellulose content.

Therefore, the reinforcement ability of these fibers is expected to be much higher than other fibers, which have less percentage of cellulose content. Figure 2, shows the optical micrographs of kenaf fibers after each stage. It is clear that the color of raw kenaf fiber turns to be lighter after alkali treatment and completely white after bleaching treatments. This indicates that there is still a significant amount of lignin after alkali treatment and removal of most of the hemicelluloses and lignin is by bleaching.

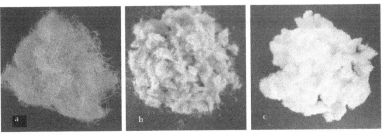

Figure 2. (a) Photo of raw fiber, (b) alkali treated fiber, and (c) bleached fiber.

FTIR Spectroscopy

The efficiency of the chemical treatments on the fibers was studied by FTIR (Figure 3). The absorbance peak at 3,400–3,300 cm^{-1} region and 1,634–1,640 cm^{-1} were attributed to the stretching and bending vibrations, respectively, of the hydrogen bonding – OH groups of cellulose, while those around 2,900–2,800 cm^{-1} were due to the stretching of C-H (Sain and Panthapulakkal, 2006). The prominent peak at 1,731 cm^{-1} that is seen in the spectrum of raw fiber was assigned to the C=O stretching of the acetyl group and uranic ester groups of the hemicelluloses or the ester linkage of carboxylic group in the ferulic and p-coumeric acid of lignin and/or hemicelluloses (Rongji et al., 2009). This peak disappeared completely in the spectrum of alkali treated fiber and bleached fiber. Therefore, it showed that most of hemicelluloses and lignin were removed with chemical treatment.

The peak at 1,250 cm^{-1} (Figure 3(a)) corresponds to the C-O stretching of the aryl group in lignin which disappeared in other spectra due to the removal of lignin after chemical treatment (Troedec et al., 2008). The peaks observed in the range of 1,420–1,430 cm^{-1} and 1,330–1,380 cm^{-1} in all spectra are attributed to the CH$_2$ symmetric bending and the bending vibration of C-H and C-O groups of aromatic ring in polysaccharides, respectively.

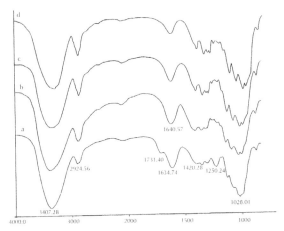

Figure 3. (a) FTIR spectra of raw kenaf fibre, (b) Alkali treated fiber, Bleached fiber (c), and cellulose nano whiskers (d).

Finally, the absorbance peaks in range of 1,0281,161 cm⁻¹ were due to C-O stretching and C-H rocking vibration of pyranose ring skeletal (Alemdar and Sain, 2008). The differences in the FTIR spectrum between raw kenaf fiber and CNW illustrates that CNW has been extracted successfully from kenaf fiber by hydrolysis treatment. No obvious changes were observed in the FTIR spectrum of the CNW at different time of hydrolysis.

Morphological Analysis

Figure 4 shows the FESEM micrograph and surface morphology of kenaf fibers after chemical treatments. It is clearly seen in Figure 4(a) that raw kenaf fiber bundles are composed of individual fibers linked together by massive cement material. Figure 4(b) reveals the morphology of alkali treated fiber. At high temperature alkali treatment, the hemicelluloses were hydrolyzed and became water soluble and lignin was depolymerized. It can be seen that all impurities around the fibers was removed and causes the fiber bundles to separate into individual fibers. Figure 4(c) illustrates the micrograph of bleached kenaf fiber. Bleaching process helps to remove most of the lignin present in the kenaf fiber, which helps in further defibrillation. Sodium chlorite and sodium acetate buffer allow the removal of lignin. Lignin is rapidly oxidized by chlorine. Lignin oxidation leads to lignin degradation and to the formation of hydroxylcarbonyl and carboxylic groups, which facilitate the lignin solubilization in an alkaline medium (Bibin Mathew Cerian et al., 2010).

Figure 4. FESEM images from raw kenaf fiber (a), Alkali treated fiber (b), Bleached fiber (c).

Furthermore, the image analysis demonstrated a significant reduction in the diameter of fiber from between 40170 to 2060 μm and 512 μm in raw fiber, alkali treated and bleached fiber, respectively.

Acid hydrolysis of kenaf fibers after bleaching, not only helps to disintegrate the fibers, but also helps in defibrillating the fibers diameter to nano range. Figure 5 shows the TEM micrograph of obtained cellulose whiskers. The cellulose whiskers displayed diameter between 6 and 22 nm. Moreover, the length of nanofibers was in 90150 nm range whereas for the raw fiber, the length was around 12.5 mm. The aspect ratio is one of the most important parameters in determining reinforcing capability of the nanofibers. Aspect ratio for obtained cellulose whiskers based on the average diameter and length was calculated to be around 13.

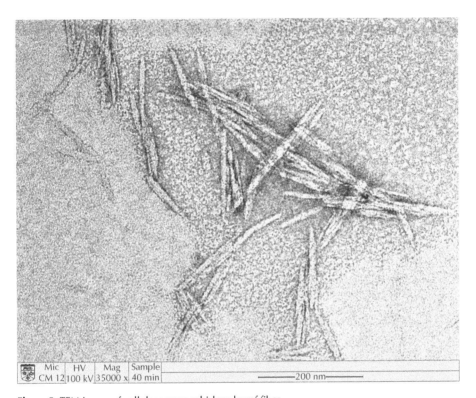

| Mic | HV | Mag | Sample |
| CM 12 | 100 kV | 35000 x | 40 min |

200 nm

Figure 5. TEM image of cellulose nano whiskers kenaf fiber.

XRD Analysis

In order to analyze the crystallinity of the cellulose obtained in this work x-ray diffractometry was carried out (Figure 6). The crystallinity index was determined based on the empirical method by using the following equation (Mwaikambo and Ansell, 2002; Segal et al., 1959).

$$C_{Ir(\%)} = (I_{200} - I_{am}) / I_{200} \times 100$$

where I_{200} is the peak intensity corresponding to the crystalline material and I_{am} to the amorphous material in cellulose fibers. Results show an increase in the crystallinity from the raw kenaf fiber to the cellulose fiber which indicates the removal of hemicelluloses and lignin that have random amorphous structure (Rongji et al., 2009). The pattern indicated that cellulose obtained from kenaf fiber presented a typical form of cellulose *I* since there was no doublet in the main peak at $2\theta = 22.5°$ (Klemm et al., 2005). The crystallinity index for raw kenaf fiber after alkali treatment, after bleaching treatment and cellulose nanowhiskers were found to be 60.8%, 68.2%, 72.8%, and 81.8%, respectively.

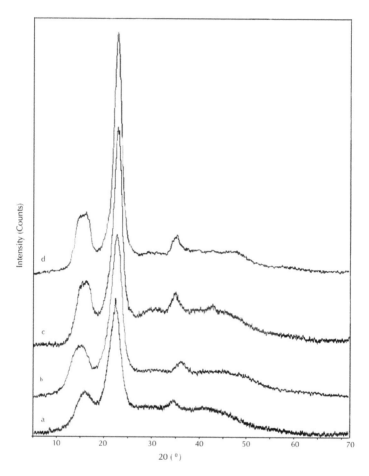

Figure 6. X-ray diffraction pattern of (a) Rae kenaf fibre, (b) Alkali treated kenaf fibre, (c) Bleached kenaf fibre, (d) cellulose nanowhiskers.

Moreover, after hydrolysis, the hydronium ions could penetrate into the amorphous regions of cellulose promoting the hydrolytic cleavage of glycosidic bonds and finally releasing the individual crystallines (de Souze Lima and Brosali, 2004) and increase the degree of crystallinity. It is believed that the higher crystallinity leads to higher tensile strength of the fiber and thereby improved mechanical properties of the corresponding nanocomposites (Alemdar and Sain, 2008).

TGA Analysis

The thermal degradation behavior of raw fiber, alkali treated fiber, bleached fiber and CNW investigated from TGA and DTA measurements (Figure 7). The TGA curve illustrates an initial weight loss about 4% during heating to 100°C which corresponds to vaporization and removal of water in fibers. Due to the differences in the chemical structure between hemicelluloses, cellulose and lignin, they usually decompose at different temperatures (Yang et al., 2007).

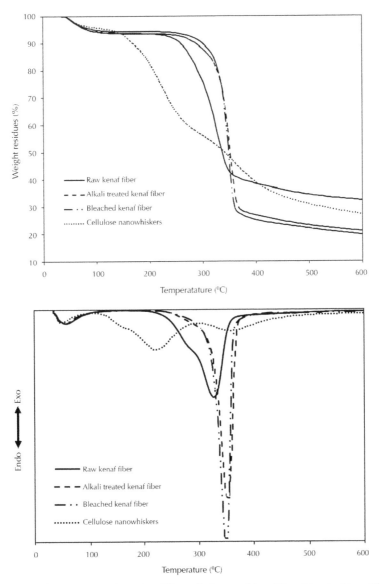

Figure 7. TG and DTG curves of raw kenaffiber, alkali treated kenaffibre, bleached kenaffiber and cellulose nanowhiskers.

There is an earlier weight loss started at around 230°C in raw fiber curve due to low decomposition temperature of hemicelluloses (Morán et al., 2008). A dominant peak at 328°C on the DTA curve accounts for the pyrolysis of cellulose. It is known that the hemicelluloses decompose before lignin and cellulose. The low thermal stability of hemicelluloses is due to the presence of acetyl groups (Shebani et al., 2008). On the other hand, by removing most amounts of hemicellulose and ligning in alkali treated and bleached fiber, the decomposition temperature increases to 350°C in the DTG curve of bleached fiber which corresponds to the thermal decomposition of cellulose. The higher amount of residue in the raw kenaf fiber as compared to fiber after alkali and bleaching treatment was due to the presence of ash as well as lignin (Ashori et al., 2006).

The cellulose whiskers show significantly different degradation behavior. There are two degradation stages in TGA curve starting from 198°C before the main peak at 335°C. The lower temperature stage may corresponds to the degradation of more accessible, and therefore more highly sulfated amorphous regions, whereas the higher temperature stage is related to the breakdown of unsulfated crystal. In fact, the introduction of sulfated groups into the crystals in the sulfuric acid hydrolysis process could reduce the thermal stability of whiskers (Kim et al., 2001). On the other hand, the increased weight residue of cellulose whiskers is because of the sulfate groups acting as the flame retardants (Maren and William, 2004).

CONCLUSION

The study has been done to extract cellulose and to prepare cellulose nanowhiskers from kenaf bast fiber. Results from morphological studies confirmed that the strong effects of extraction method and a significant decrease in the size of fibers to nanoscale. The FTIR results not only supported the results from FESEM analysis but also revealed that the chemical structure of obtained cellulose nanowhiskers did not change during hydrolysis process. Crystallinity was also increased by removal of the amorphous part of fiber during cellulose extraction as well in cellulose nanowhiskers by releasing individual cellulose crystals during hydrolysis process. Finally, thermogravimetry analysis showed that thermal stability of fiber is also increased due to the removal of lignin and hemicelluloses from raw fiber. However, the thermal stability after hydrolysis is reduced due to the formation of sulfated group at surface of cellulose which has low thermal degradation.

ACKNOWLEDGMENT

Authors would like to acknowledge the financial support prepared by Ministry of Higher Education (MOHE) under FRGS grant UKM-ST-07-FRGS0041-2009. We are also like to thank Prof. Dr. Alain Dufresne and Gilberto Siqueiraat Grenoble Institute of Technology, France for providing us with necessary equipment and for great discussions in the research.

KEYWORDS

- **Holocellulose**
- **Kenaffiber**
- **Nanowhiskers**
- **Non-toxic**
- **Thermogravimetry**

Chapter 18

Bio-hybrid Nanocomposite Roll-to-roll Coatings for Fiber-based Materials and Plastics

Jari Vartiainen, Vesa Kunnari, and Annaleena Kokko

INTRODUCTION

Development of innovative products starts from the raw materials. Packaging industry relies heavily on oil-based materials in certain applications. Replacing the oil-based material with bio-based products might give a competitive advantage due to more sustainable and greener image. In addition, nanotechnology in food packaging is expected to grow strongly over the next 5 years as the increased globalization sets demands for shelf life enhancing packaging. Recently, a lot of effort has been aimed at developing new bio-based polymer containing films and nanocomposites which can act as for example barriers in packaging materials (Arora and Padua, 2010; Lagarón and Fendler, 2009; Vartiainen et al., 2010a, 2010b). Unlike synthetic plastics, in dry conditions, the films and coatings from natural polymers exhibit good barrier properties against oxygen and grease due to the high amount of hydrogen bonds in their structure. However, natural polymers are hydrophilic in nature, thus films and coatings produced from these materials are often hygroscopic, resulting in partial loss of their barrier properties at high humidity (Hansen and Plackett, 2008). A major challenge for the packaging developers is to overcome the inherent hydrophilic behavior of biomaterials. Among the potential fillers for nanocomposites, clay platelets have attracted a particular interest due to their high performance at low filler loadings, rich intercalation chemistry, high surface area, high strength and stiffness, high aspect ratio of individual platelets, abundance in nature, and low cost (Blumstein, 1965). Clays are naturally occurring materials composed primarily of fine-grained minerals. Nanoclays (or nanolayered silicates) such as hectorite, saponite, and montmorillonite are promising materials with high aspect ratio and surface area (Lan et al., 1994; Messersmith and Giannelis, 1995; Yano et al., 1997). Because of their unique platelet-like structure nanoclays have been widely studied regarding the barrier properties. Such nanoclays can be very effective at increasing the tortuosity of the diffusion path of the diffusing molecules, thus significant improvement in barrier properties can be achieved with the addition of relatively small amounts of clays (Pavlidou and Papaspyrides, 2008). When the nanoclay layers are completely and uniformly dispersed in a continuous polymer matrix, an exfoliated or delaminated structure is obtained. Full exfoliation (single platelet dispersion) of nanoclay by using existing/traditional compounding techniques is very difficult due to the large lateral dimensions of the layers, high intrinsic viscosity of the polymer and a strong tendency of clay platelets to agglomerate (Hussain et al., 2006). Most of the clays are hydrophilic, thus mixing in water with water-soluble polymers results good dispersion, especially when the sufficient amount of mixing energy is used. The degree of exfoliation can be improved by the use of conventional shear devices such as extruders, mixers, ultrasonicators, ball milling, fluidizators, and so on.

MATERIALS AND METHODS

Natural polysaccharides (chitosan and pectin) were used as continuous polymeric matrixes in which an inorganic nanosized material, >98% montmorillonite, was dispersed. In order to ensure the sufficiently defoliated and nanosized structure of the nanoclay platelets, the ultrasonication and high pressure fluidization were used for homogenization polysaccharide-nanoclay dispersions. Coatings and films were prepared by wet coating of chitosan onto plasma-activated LDPE coated paper (coatings) and solvent casting of pectin on Petri dishes (films). Oxygen transmission rates were measured with Oxygen Permeation Analyser Model 8001 (Systech Instruments Ltd.). The oxygen transmission tests were carried out at 23°C and 0%, 50%, and 80% relative humidity.

Chitosan Coatings

Nanoclay (0.2, 1, and 2 wt%) was swelled in 30 ml of distilled water and dispersed using ultrasonification tip (Branson Digital Sonifier) for 10 min. The dispersion was added into 30 ml of 1% chitosan in 1% acetic acid, followed by sonication for 10 min. For reducing the surface tension and increasing the wettability, 60 ml of ethanol was added and mixed under rigorous mixing. For coated multilayer structures, the solutions were applied onto plasma-activated LDPE coated paper using the standard coating bar no. 6 (wet film deposit of 60 µm). Chitosan solutions with initial nanoclay concentrations of 0, 17, 50, and 67 wt%, and total coating dry weight of 0.2–0.6 g/m² were used.

Pectin Films

Aqueous solutions of pectin were prepared into distilled water by mixing pectin and glycerol at final concentrations of 5 and 1.75 wt%, respectively. The pH of the mixture was adjusted to pH 4.5 and the mixture was heated at 60°C for 2 hr to increase fluidity. 0.5, 1 or 2 wt% nanoclay was added to pectin solutions and immersed for 2 days under constant mixing. Fluidizer (Microfluidics M110Y) was used for homogenization of nanoclay/pectin dispersions. Feed solutions were pumped from inlet reservoir and pressurized by an intensifier pump to high pressure (900–1350 bars) and fed through fixed geometry chambers with inside micro channel diameter varying between 100 and 400 µm. Films of pectin and nanoclay were prepared by casting 15 ml of each solution in polystyrene Petri dish (Ø 8.5 cm) and dried for 2 days at room temperature. Final nanoclay concentrations in the solvent cast hybrid films were 10, 20, and 30 wt%.

DISCUSSION AND RESULTS

Nanoclay was delivered as dry powder with particle size of 2–15 µm. Nanoclays typically tend to be agglomerated when mixed into water. The agglomerates are held together by attraction forces of various physical and chemical natures, including van der Waals forces and water surface tension. These attraction forces must be overcome in order to deagglomerate and disperse the clays into water. Ultrasonication and high pressure fluidization were used to create alternating pressure cycles, which overcome the bonding forces and break the agglomerates. As can be seen in Figure 1(a), dry nanoclay powder consisted of round particles with coarse and platelety surface. By ultrasonic dispersing (Figure 1(b)), and high pressure fluidization (Figure 1(c)) the nanoclay platelets were effectively ripped off and distributed on the surface. The diameter of the intercalated nanoplatelets varied between 100 and 500 nm.

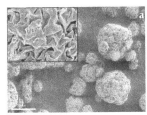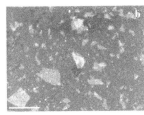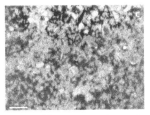

Figure 1. The SEM images of (a) large undispersed nanoclay aggregates. The excerpt shows the surface structural features: laminar fine structure can be seen on the aggregate surfaces, (b) spincoated nanoclay platelets after dispersing with the ultrasonic microtip, and (c) spincoated nanoclay platelets after high pressure fluidizer treatment.

Nanocomposite chitosan coatings effectively decreased the oxygen transmission of LDPE coated paper under all humidity conditions (Figure 2). In dry conditions, over 99% reductions and, at 80% relative humidity, almost 75% reductions in oxygen transmission rates were obtained. Highest concentration of nanoclay (67 wt%) offered the best barrier against oxygen, whereas the 17 wt% concentration of nanoclay performed almost as good as 50 wt% of nanoclay. Barrier effects of nanoclay became less evident in dry conditions. Presumably higher nanoclay concentrations were partly agglomerated, which hindered the crystallization and hydrogen bonding formation between chitosan chains, especially in dry conditions.

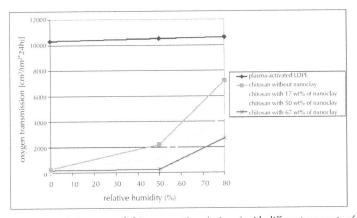

Figure 2. Oxygen transmission rates of chitosan coatings (< 1 μm) with different amounts of nanoclay.

Nanoclay addition also clearly improved the oxygen barrier properties of pectin films in high humidity conditions (Figure 3). Oxygen transmission rate was reduced by 80% with pectin films containing 30 wt% of nanoclay as compared with the pectin film without nanoclay. Water vapor transmission results indicated the improved barrier properties as well. However, the water-soluble pectin was lacking the capability of fully preventing the transmission of water vapor, and thus, total barrier effect of films with 30 wt% nanoclay was not more than 23%. Pectin itself formed an excellent barrier property against grease and nanoclay addition did not improve this barrier property anyhow. All films were totally impermeable to grease under the conditions tested. Barrier improvements are explained using tortuous path theory which relates to alignment of the nanoclay platelets. As a result of the sufficient defoliation, the effective path length for molecular diffusion increases and the path becomes highly tortuous to reduce the effect of gas transmission.

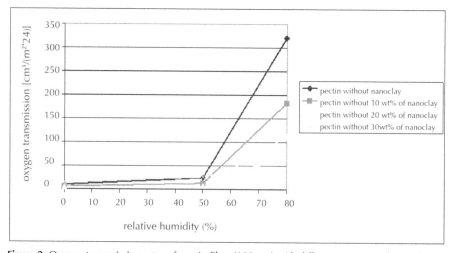

Figure 3. Oxygen transmission rates of pectin films (100 μm) with different amounts of nanoclay.

Up-scaling the Production with Roll-to-roll Pilot Line

The VTT has a roll-to-roll surface treatment concept that can be used for coating and surface treatment of fiber-based webs and plastics. It consists of separate one meter wide operational units build from aluminum frames. Each frame has four wheels placed under the frame. All operational units are placed on rails to eliminate need for leveling of units after changing the order of the units. Rails also eliminate need for guidance and other roll alignment after re-arranging the units. The units are attached together by a system called Click and Coat developed by manufacturer of equipment, Coatema and Germany. The concept includes several coating methods, pre-treatments and curing options. Coating methods include flexo-type roll coating, soft bar coating, kiss type coating and spray coating. All three first mentioned methods can be followed by spray coating to make wet on wet coatings. The sealable coating unit followed by exhaust air removal during drying makes it possible to test also coatings containing harmful volatile substances. Surface treatments can be cured using cold or hot air drying, IR-drying or UV-curing. The main benefits are low cost, low coating material demand and versatility of used substrates including plastics, paper and carton board. Basic layout including all units of the SUTCO surface treatment concept is shown in Figure 4.

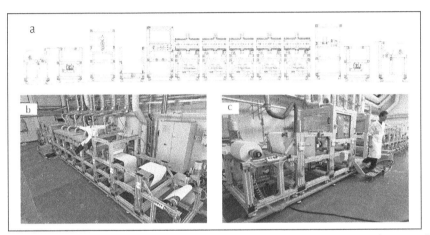

Figure 4. (a) Basic layout of the SUTCO surface treatment concept with configuration of 13 units, (b) drying section and rewinding, and (c) unwinding.

Pectin solutions were roll-to-roll flexo-coated onto argon/nitrogen plasma-activated oriented polypropylene (OPP) film (thickness 25 μm, roll width 38 cm) obtained from Amcor Flexibles Oy, Lieksa, and Finland. Plasma-activation was carried out in-line to increase the surface energy of OPP film prior the coating with water-based pectin solution. Effects of plasma-activation on water contact angle and surface energy can be found from Figure 5. Pectin was easily spread onto activated OPP surface forming an even and smooth dry coating (Figure 6).

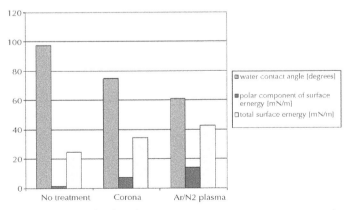

Figure 5. Effects of corona and Ar/N$_2$ plasma pre-treatments on water contact angle and surface energy of OPP film.

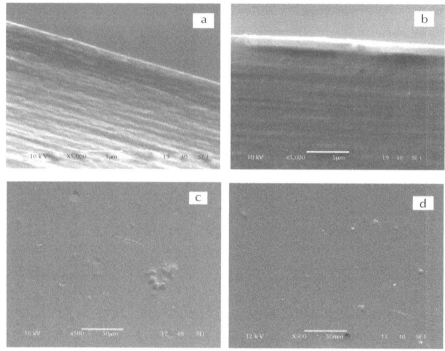

Figure 6. The SEM images of cross-section and surface topography of roll-to-roll coatings. Uncoated OPP film (a and b) and pectin coated OPP film; pectin thickness ~1 μm (c and d).

CONCLUSION

As a conclusion, montmorillonite nanoclay was successfully dispersed in aqueous polysaccharide solutions using ultrasonication and high pressure fluidization. Nanocomposite coatings and films showed improved barrier properties against oxygen, water vapor and UV-light transmission. Materials were also totally impermeable to grease. The developed bio-hybrid nanocomposite materials can be potentially exploited as safe and environmentally sound alternative for synthetic barrier packaging materials. The SUTCO surface treatment concept was used for up-scaling the production of pectin coatings. Pectin coating improved the smoothness as compared to uncoated OPP film surface.

KEYWORDS

- **Food packaging**
- **Montmorillonite**
- **Nanoclays**
- **Plasma-activated**
- **Ultrasonication**

Chapter 19

Advances in Collagen-based Tissue Engineering

Krystyna Pietrucha

INTRODUCTION AND BACKGROUND

Transplantation of organs or tissues is a widely accepted therapy to treat patients with damaged organs or tissues as a result of an accident, trauma, cancer or disease. However, autologous transplantation is limited because of donor site morbidity and infection or pain to patients because of secondary surgery. Alternative tissue sources that have origins from other humans remain problematic mainly due to immunogenic responses by the patients upon implantation and a shortage of donor organs. Each year, millions of people die still due to shortage of organs to transplantation. Tissue engineering, which applies methods from engineering and life sciences to create artificial constructs to direct tissue regeneration, has attracted many scientists, surgeons with hope to treat patients in a minimally invasive and less painful way. The aim of tissue engineering, with an interdisciplinary approach, is to break through the barriers which have limited the performance and application of artificial biomaterials over past decades. The research examinations lead to develop, design and synthesize novel multi-component structures, using biocompatible and biodegradable natural/synthetic polymers with/without bioactive molecules that may be used as efficient scaffolds for engineering tissue. For this ambitious purpose mainly collagen, poly(lactic-co-glycolic acid) (PLGA), cellulose and glycosaminoglycans (GAG) such as hyaluronic acid (HA) and chondroitin sulfate (CS), which have different physicochemical properties and degrade at different rates in human body have been used. The components of scaffolds are cross-linked/grafted by chemical, physical, and radiation methods. These scaffolds mimic closely the molecular and structural properties of native extracellular matrix (ECM). This review focuses on the progress in using some of the scaffolds for the specific regeneration of tissues. Among it, an advanced tissue engineering of collagen based materials for therapeutic treatment of injuries to central and peripheral nervous systems is presented

SOME ASPECTS OF TISSUE ENGINEERING

Tissue engineering applies scientific principles to the development, design, and construction of biological substitutes that restore, maintain, and improve natural tissue or organ function (Morrison, 2009). Its scientific input is derived from materials science, cell biology, physics, chemistry, and importantly, from clinical research. Restoration of the normal tissue structure and function occurs through the production of new tissue with replicates exactly that which has been lost as a result of degenerative disease, cancer, an accident or trauma. The restorative process depends on a balanced combination of cell culture growth with biomaterial/scaffold to support it and with bioactive agents to enhance and direct it (Jagur-Grodzinski, 2006). There are several approaches to creating tissue constructs. A well known one is to isolate specific cells through a small biopsy from the patient, to grow them on a three dimensional (3D) biomimetic scaffold, under precisely controlled culture conditions and to deliver the construct to the appropriate site in the patient's body, and to direct new tissue formation into the scaffold that would itself disappear, in due course, through biodegradation.

STRATEGY FOR CONSTRUCTION OF BIOMATERIALS/SCAFFOLDS

In this way, biomaterials or scaffolds play a pivotal role in tissue engineering (Ma, 2008). They serve matrices for cell migration, spreading, in growth, proliferation, and stereo formation of new tissue. An ideal artificial scaffold should resemble the structural and functional profile of the natural ECM (Wen et al., 2005; Williams, 2006).

Selection of Polymer for Scaffold

The strategy for the tissue engineering is to select a polymer with the best biocompatibility, extend its time-scale of biodegradation, increase its mechanical strength and convert it into nano-fiber structure resembling the structure and properties of natural tissue.

Collagen

Many natural and synthetic polymers have been extensively studied for preparation of biomaterials/scaffolds for tissue engineering applications but collagen occupies a unique place in the way it interactions with body tissues. Collagen is an essential component of ECM in many tissues such as skin, bone, cartilage tendon, and so on. General reviews of this natural polymer for biomedical applications, with references to numerous review papers dealing with various specific aspects of this subject have been published (Kolacna et al., 2007; Wess, 2008). Collagen is the primary structural material of vertebrates and is the most abundant mammalian protein accounting for ~ 30% of total body proteins. Reconstituted collagen from xenogeneic sources due to it is the excellent biocompatibility and due to its biological characteristic, is regarded as one of the most useful biomaterials. Collagen has a unique structure and amino acid sequence. Among the 29 isotypes of collagen, type I is composed of two α_1 chains and one α_2 chain. The underlying α chains that form these natural polymers are arranged into repeating motif exhibits a 67 nm interval that forms a coiled structure. The collagen molecule consists of three polypeptide chains twined around one another as in a three-stranded rope. The specific complement of α subunit present within the fibril defines the material properties of the polymer. Besides, in native ECM, collagen exists in a 3D network structure composed of multi-fibrils in the nano-fiber scale of 50–500 nm. The fibrillar structure of Type I collagen has been known for a long time to play an important part in cell attachment, proliferation, and differentiated function in tissue culture.

Collagens from different species and living environments have various properties. An alternative to mammalian collagen is fish collagen. Production of fish collagen is actually not new it has been produced for many years (Zhang et al., 2009). Nowadays, the importance of fish collagen is growing and there is growing number of possibilities to use it in medical application. One major advantage of collagen from aquatic organism sources in comparison to land animals is that they are not associated with the risk of outbreak of bovine spongiform encephalopathy. Unfortunately, the great differences exist between fish collagen and bovine collagen. The relatively fast biodegradation rate and low thermo-stability of fish collagen in comparison to bovine collagen has been one of the crucial factors that limit the usage of fish collagen in tissue engineering. There have been few reports on the modification of fish collagen by chemical and physical methods (Nalinanon et al. 2011; Pietrucha and Banas, 2010; Yunoki et al., 2007). The attention has been particularly paid to the cross-linking efficiency, hydrothermal stability and porosity of the new class of matrix.

Glycosaminoglycans

In creation of functional scaffolds for tissue engineering the role of both collagen and GAG such as HA and CS are important (O'Brien et al., 2005; Pieper et al., 2000; Zhong et al., 2007) HA and CS belong to the connective-tissue mucopolysaccharide group of substances that are present mainly in articular cartilage, vitreous humour and synovial fluid but they are widely distributed in other tissues and most body liquids. They have unique physicochemical properties and distinctive biological functions. Composites collagen-GAG has been used for a wide variety of *in vitro* studies of cellular migration, contraction, and tissue growth.

Other Components for Scaffolds

Cellulose is a linear homopolymer of glucose and is the most widespread polymeric material in nature. It is degradable by enzymes and its solubility in water depends on its chain length. The biocompatibility of cellulose and good match of their mechanical properties with those of hard and soft tissue accounts for its medical applications (Czaja et al., 2007; Fatimi et al., 2008). It has been used in research in both bone and cartilage tissue engineering. Cellulose derivatives such as carboxymethylcellulose (CMC), hydroxypropylmethylcellulose (HPMC) and oxidized cellulose are also known to have good biocompatibility. The biocompatibility of methylcellulose (MC) in the peritoneum is not known, but the mixture of MC and HA has been reported to be biocompatible in intrathecal injection. Many work has been done on application of poly(lactic acid) (PLA) and poly(lactic-co-glycolic acid) (PLGA) in medical application (Kuo and Yeh, 2011).

To summarize, the materials for tissue engineering are based on optimized composition of several types of natural/synthetic, biocompatible, and biodegradable components:

- Collagen: low mechanical strength, relative rapid biodegradation, good cell adhesion, migration, and proliferation;
- Cellulose, or a cellulose derivative: good mechanical strength, slow degradation, poor cell adhesion;
- Glycosaminoglycans: The HA and CS: special function in assisting cell migration and cell differentiation during the healing process. Through its high water absorption, resists forces of compression, while collagen and cellulose withstand tensile forces;

- Poly(lactic acid) (PLA) and Poly(lactic-co-glycolic acid) (PLGA) with relatively high mechanical strength, but their hydrophobic surface is not favorable for cell seeding.

Preparation of Parallel-oriented Nano and Micro-diameter Fibrils

The natural ECM has fibrillar structure, with submicron to micron sized fibers (Wess, 2008). Thus, alternative methods of scaffolds fabrication that are more homogenous and biomimetic are needed. For the first time, there is an evidence that construction of nano- and micro-fibrillar biomaterials/scaffolds may be possible. Implants based on such materials are the next frontier of bioengineering. The basic building blocks of nano- and micro-fibrillar materials are likely to be nanofibers. One of the potential advantages of this method is that it should be capable of scaling up to produce nanofibers in quantities required to mimic the ECM. There has been significant increase, recently, in research on electrospinning of collagen fibers for tissue engineering (Powell et al., 2008; Yang et al., 2008).

Electrospinning and Coaxial Electrospinning

Several laboratory techniques for producing nanofibers are being developed. For application in tissue engineering, good progress has been made with the technique of electrospinning (Matthews et al., 2002). In the electrospinning apparatus, a pump feeds a solution of collagen into a long narrow tube, like a syringe needle. The flow through this tube results in formation of a longitudinally oriented solution filament, with collagen molecules oriented in direction of flow. A high DC voltage applied to the tube, leads to formation of an electric charge in the polymeric filament emerging from the tube. The charge generates forces of repulsion within the filament and neutralizes the effect of surface tension. At sufficiently high voltage, the emerging solution filament ceases to be held together by surface tension and splits into nano-fibrils. These are collected on an earthed drying—drum and on solvent evaporation become collagen nanofibrils. The diameter of the nanofibrils is related to the solution viscosity. A solvent composition will be found experimentally, to ensure stable nano-scale homogeneity in the polymer solution at the required viscosity. The choice of solvents is limited by the requirement for rapid evaporation from the nanofilaments. Since electric charge formation is fundamental to the process, the solution will have to meet strict requirements with respect to the dielectric constant and resistivity. There are several variables in the process, the effect of which will have to be investigated experimentally. It will be of paramount importance to confirm that no traces of solvents have been left in nanofibrils, which might be in a long term cytotoxic. A potential advantage of the electrospinning process is that, when proven to be successful, it can be scaled-up for larger scale production. Complex scaffold architecture is possible by using several spinnerets in parallel, extruding fibrils at different angles. Some information on the electrospinning methods in collagen nanofiber fabrication and its great potential for tissue regeneration is published (Zheng et al., 2010).

Electrospinning provides an important step in producing basic construction elements for the scaffolds. It also provides the preferred way of achieving two specific objectives: regulating the rate of biodegradation of the scaffolds and producing scaffolds for instant use. Moreover, coaxial electrospinning can be carried out using a solution of mixed collagen-based polymers (Thomas et al., 2007; Zhou et al., 2010).

Ultrasonic Fiber splitting
The alternative method of preparing parallel—oriented nano-diameter fibrils is using ultrasonic splitting of collagen micro-diameter fibers (Zhao and Feng, 2007). This technique has not been proposed before for use in tissue engineering, but if successful, it has advantages over electrospinning. Purified collagen fibers, dispersed in an aqueous medium are treated with powerful ultrasonic, which split the fibers into constituent, nano-diameter fibrils. The fibrils are parallel oriented by extruding the aqueous dispersion of the fibrils through a long, narrow tube onto a drying drum. An important advantage of this technique over electrospinning is that it avoids the process of dissolving collagen in an organic solvent. The mechanical and surface properties of the fibrils should be better and there will be no need for laborious removal of the last traces of an organic cytotoxic solvent. Although the technique is simple in principle, it requires considerable effort to develop. The technique can be used for producing scaffolds with modified rate of biodegradation and/or mechanical properties. This would be done by preparing nanofibrils from two or more biopolymers, extruding a dispersion of mixed nanofibrils, forming mixed fibril mats and cross-linking by chemoradiation reactions.

HYBRID SCAFFOLDS RESEMBLING STRUCTURE AND PROPERTIES OF NATURAL TISSUES

In order to perform its function well, after implantation a scaffold must have the best achievable biocompatibility, it should also have a microstructure resembling that of natural tissue (for cell seeding, ingrowth and proliferation), adequate mechanical strength and biodegradability matching the time-scale of tissue regeneration.

Restitution of skin loss is the oldest and still not totally resolved problems in the field of surgery. In severely burned patients, the prompt closure of full thickness wounds is critical for survival. Many different materials including collagen sponges and cellulose have been used to stimulate granulation tissue in wound base after deep burns and traumatic injures. Since spontaneous healing of the dermal defects does not occur, the scar formation for full thickness skin loss is inevitable, unless some skin substitutes are used. Recently, promising results have been obtained utilizing electrospun collagen scaffolds as skin substitutes in athymic mice (Powell et al., 2008). Electrospun collagen scaffolds have been shown to produce skin substitutes with similar cellular organization, proliferation and maturation to the current, clinically utilized model, and were shown to reduce wound contraction, which may lead to reduced morbidity in patient outcomes. The 3D biodegradable porous scaffolds that combined the advantages of natural Type I collagen and synthetic PLGA knitted mesh have been also developed for cartilage engineering tissue (Dai et al., 2010).

Collagen Hydrogels as Substitutes of Dura Mater
Other problem exists in the case of dura mater. Injury of dura mater may result from many causes, including cranio-cerebral trauma, destruction by tumor, surgical removal, and various congenital malformations. For almost 100 years various artificial

materials such as synthetic or collagenous tissues as well as collagen covered vicryl mesh have been employed for dural repair.

A commercially available substitute for dural repair is DuraGen® (Integra Life-Sciences Corporation, US) (Narotam et al., 1999). DuraGen® is made by acid precipitation, freeze drying, and thermal/chemical cross-linking of collagen in combination with chondroitin 6-sulfate. Another one, Promogran (Jonson &Jonson Medical) is designed for wound healing (Vin et al., 2002). Promogran is freeze dried composite of oxidized regenerated cellulose (45%) and collagen (55%). Recently, TissuDura® with equine collagen as dura mater has been developed (Parlato et al., 2011). Unfortunately, the great differences exist between natural (ECM) and manufactured synthetic implants. It is important to realize that up to now no ideal dural implant is available. Muscle fascia remains the most accepted autologous implant in general use.

Over 20 years ago the author of this review in collaboration with surgeons obtained collagen substitute of dura mater (Pietrucha, 1991). Collagen type I was derived from purified bovine Achilles tendons by pepsin digestion and acetic acid treatment to prepare dispersion. Whereas native collagen tissue possesses significant strength, this is lost when collagen products are made from collagen solution. These reconstituted products may therefore require treatment with cross-linking agent, so as to increase the strength for particular applications. Cross-linking agents (physical or mostly chemical treatments) are used to stabilize biological products. Cross-linking procedure can decrease biodegradation rate and immunogenicity. However, most cross-linking agents (i.e., aldehyde family) used to stabilize biological molecules produce toxic byproducts which are detrimental to the wound healing process. That is why reaction cross-linking of collagen and preparation hydrogels was carried out by irradiation with electron beam. However, even extensively radiation cross-linked collagen hydrogels have poor physical strength for the preparation of dural substitute. For that reason collagen, hydrogels were reinforced with poly(ethylene terephthalate) (PET) mesh. The key features which had to be achieved were: (a) an elastic hydrogel of defined stable structure, (b) optimized range of network pore sizes, (c) covalent bonding to the synthetic fiber net, and (d) freedom from harmful chemical residues.

All the key features were achieved by the use of high energy radiation under optimized and strictly controlled conditions. The radiation process produced a sterilized implant which could be stored for at least a year. It could be implanted without any additional treatment. The implant material underwent with good success three stages of evaluation: biochemical *in vitro*, pre-clinical on animals and clinical. This collagen composite material has been used in neurosurgery, plastic, and reconstructive surgery to replace, complete or strengthen the cerebral and spinal cord dura mater as well as other soft tissues. Satisfactory results have been also obtained in short-term and long-term (8 years) studies with dural substitute on over 100 patients in several independent clinics. The prostheses had been implanted in patients with posttraumatic CFC-leakage and with intracranial neoplasms as well as in the surgical treatment of meningoencephalocele, meningomyelocele, and various dystrophic defects on central nervous system of children (Figures 1–4) (Pietrucha, 2010a; Pietrucha and Polis, 2009; Pietrucha et al., 2000; Polis et al., 1993). Moreover, the biomaterial was used for closure of abdominal wall defect in new-born (Figure 5) (Chilarski and Pietrucha, 1996).

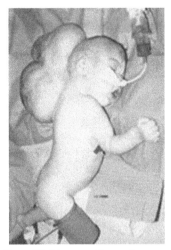

Figure 1. The large hernias on disqualified for operation so far.

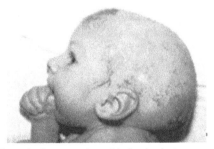

Figure 2. The child from Figure 1 after operation with dural substitute.

Figure 3. The child from Figure 1, 4 years after operation.

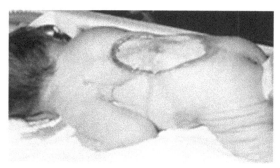

Figure 4. Myclomeningocele covered by substitute of dura mater.

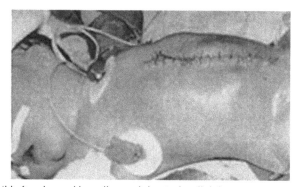

Figure 5. The child after closured by collagen abdominal wall defect.

After implantation as dural substitute, collagen gradually degrades and cells synthesize their own dura mater. The non-degradable PET mesh remains in organism as unnecessary ballast. It is very important that collagen biomaterials can be effectively cross-linked and simultaneously sterilized by radiation technique, which does not produce toxic by-products detrimental to wound healing.

Modified collagen hydrogels first appeared to be promising for skin. Unfortunately, their extensive contraction and their poor mechanical properties constituted major disadvantages toward their utilization as permanent graft. Recently, it has been shown that concentrated collagen hydrogels can be considered to be new candidates for dermal substitution because they are easy to handle, do not contract drastically, favor cell growth and can be quickly integrated *in vivo* (Helary et al., 2010; Yunoki et al., 2010).

The Effect Cross-linking on Properties of Scaffolds

In collaboration with research workers from Laval University, Quebec, Canada we have obtained very interesting results that show for the first time that activated poly(ethylene glycols) (PEGs) can be grafted onto collagen using radiation method (Doillon et al., 1994, 1999). Collagen was extracted from adult bovine hide using a series of acid dispersions and salt precipitations. In specific experimental conditions, *in vivo* tissue ingrowth was observed within a stable porous structure. It has been demonstrated that chemically activated or radiation modified monomethoxy poly(ethylene glycol) (MPEGs), derivatives conjugated to collagen decreased its biodegradability, immunogenicity and antigenicity while maintaining its biological activity. These biomaterials have been used as templates for tissue ingrowth assisting wound healing. The results obtained show unambiguously that some of the collagen-polymers compositions give rise to the appearance of formation of an ether link and modification of the protein secondary structure as determined by the analysis of the amide I and amide II infrared bands. Furthermore, using irradiation, it is possible to offer both a biodegradable implant with stable structure and a well-suitable implant for surgery in a sterile and ready–to-use state. Additional advance in this manner is that the collagen structure can be stabilized without any impurities in catalysts or chemical initiators.

The other type of collagen-based material is a new family of tri—component hydrogel consisting of water-soluble, substantially biocompatible synthetics, and bioactive collagen (Pietrucha and Banas, 2009; Pietrucha and Verne, 2010). The concept in the preparation of a copolymeric hydrogel is to incorporate in the system a slowly degradable, bio-inert poly(vinyl alcohol) (PVA) of high glass transition temperature, T_g, a highly biocompatible and hydrophilic MPEG of low T_g, and biodegradable, bioactive natural macromolecule collagen. These hydrogels were synthesized by using a two-stage process: multiple freeze-thaws followed by irradiation with an electron beam.

It has been shown that repetitive freeze-thawing of PVA solutions containing collagen and MPEG results in formation of physically cross-linked hydrogels. The mechanism by which these physical cross-links are formed may involve hydrogen bonding, formation of crystallites or a liquid-liquid phase separation process and chains entanglements. The immediate effect of irradiation of above polymer systems, in water and under oxygen-free atmosphere, is radiolysis of water and formation of polymers free radicals. The structure of these free radicals is shown in Figure 6.

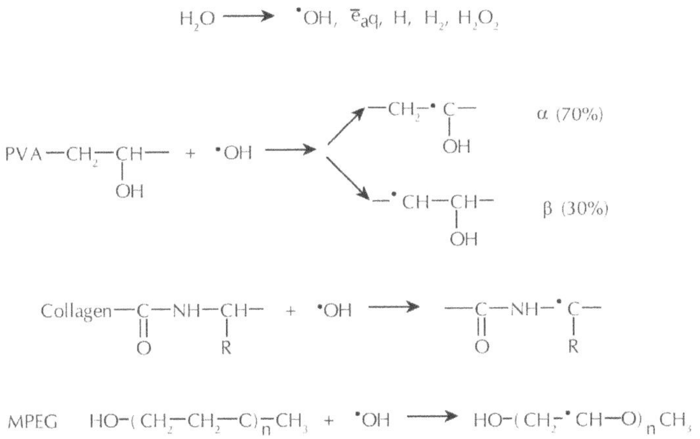

Figure 6. Schematic description of products radiolysis of H_2O, PVA, MPEG, and collagen.

The main species of water radiolysis reacting with the mentioned polymers are OH radicals. They attack the macromolecules and abstract hydrogen atoms. The structure of resulting macro radicals depends on the type of polymers. Mutual recombination of macroradicals induces intra- and inter-molecular cross-links. Then semi- and inter-penetrating INP 3D sterile polymer network without any additional chemical cross-linking agents are formed. Cross-linking was also found to improve the mechanical and morphological properties of the hydrogels. Small amounts of collagen and MPEG in structure of PVA lead to an unexpected improvement of porosity and strong me-chanical integrity of PVA-collagen-MPEG hydrogel. Beneficial effects of PEG on the stability of pores of PVA have been also observed by others (Bodugoz-Senturk et al., 2008). From the viewpoint of biomedical application it seems that addition of collagen and MPEG to PVA would produce a highly elastic and strong scaffold that would fa-cilitate high levels of cell attachment and growth. PVA-MPEG-Collagen hydrogels are promising materials to use as scaffolds in tissue engineering as well as a biomaterial for osteochondral defect repair or as artificial cartilage.

Work on many composite systems of collagen sponge-shape matrix with GAG has been reported in research papers, published in the last 10 years. Among them the author of this publication explored preparation of new composite of collagen with HA and CS to develop innovative biological scaffolds for tissue engineering (Pietrucha, 2005, 2007, 2010). In these studies, particular attention has been paid to the prepara-tion and characterization of cross-linked collagen-HA/CS matrices with potential for use in tissue engineering. To prepare collagen modified by HA/CS two complementary cross-linking methods, dehydrothermal (DHT) pre-treated and 1-ethyl-3-(3-dimethyl aminopropyl) carbodiimide (EDC) have been introduced.

Modifications in molecular organization of collagen-HA/CS scaffold induce varia-tions in the material application properties, and more particularly of their thermal, mechanical, and dielectric properties. The typical differential scanning calorimetry (DSC) and termogravimetric analysis (TGA) of collagen and collagen cross-linked by mixture EDC and NHS in the presence of HA are shown in Figure 7. The thermograms obtained by DSC for both collagen modified with HA and collagen alone evidenced

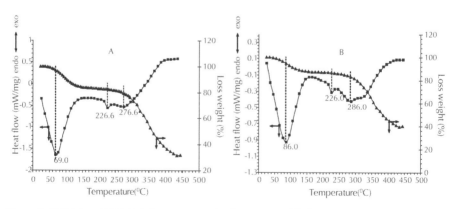

Figure 7. DSC (■) and TGA (▲) thermograms of collagen unmodified (A) and (B) modified by HA.

three different endothermal peaks: the first one related to temperature of thermal de-naturation (T_d) of collagen and the other two were connected with a complex phenom-enon of collagen modified by HA compared to unmodified collagen is significantly shifted to the region of thermal modification, finally leading to the destruction of ma-terials. However, the maximum of the peak for collagen modified by HA compared to unmodified is significantly shifted to the region of higher temperatures (from 69.5 to 85°C).

It is known that intra- and inter-molecular hydrogen bonds as well as hydrogen bound water are responsible for the stability of the triple helix conformation of col-lagen macromolecules. The mechanism of the EDC/NHS reactions with collagen/CS was presented previously in Figure 8 (Pieper et al., 1999).

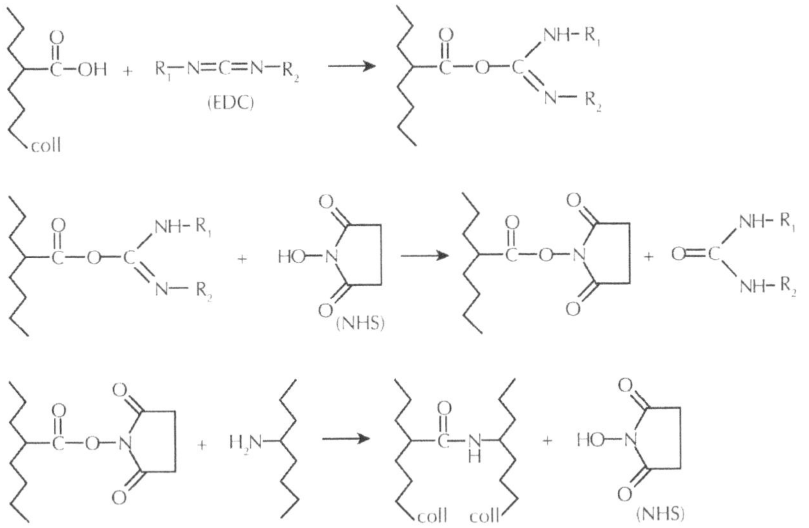

Figure 8. Scheme of cross-linking reactions of collagen by EDC and NHS.

These findings suggest that similar reactions may occur between collagen and HA, because the chemical structure of HA is similar to other GAGs except that HA is un-sulfated disaccharide. Hence, the most important reactions occurring in the collagen-HA system with the mixture of EDC and NHS can be schematically presented as in Figure 9.

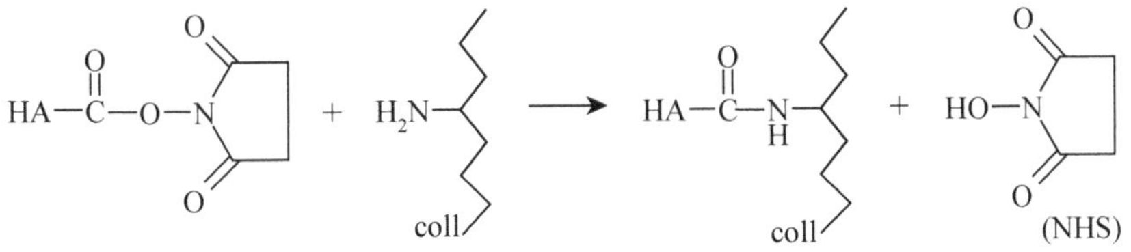

Figure 9. Scheme of covalent attachment of HA to collagen using EDC and NHS.

It has been suggested (Park et al., 2003) that the carboxyl group of HA reacts not only with amino groups of collagen but also with hydroxyl groups of collagen or HA resulting in the formation of ester linkages. However, it is worth noticing that an essential part of the scaffold is collagen therefore the most important reactions result in the formation of the collagen network. The EDC is not incorporated into the amide cross-links and it and other cytotoxic substitutes are completely rinsed out from the matrix. The architecture of scaffolds should also provide appropriate porosity spatial cellular organization. Spatial cellular organization determines cell-cell and cell-matrix interactions, and is crucial to the normal tissue and organ function. Otherwise, porosity and pore size of scaffolds play a vital role in tissue formation *in vitro* and *in vivo* (Pietrucha, 2010b). Results of distribution and pore size in collagen-based scaffolds prepared by mercury porosimetry are listed in Table 1.

Table 1. Characterizations of the porous structure of the collagen-based scaffolds.

Nr	Collagen-based scaffold	Pore size [μm]		
		1-10	**10-100**	**100-350**
		Pore volume [%]		
1.	Collagen unmodified	6.8	71.9	20.0
2.	Collagen cross-linked by EDC	8.0	76.6	14.0
3.	Collagen modified by HA with EDC	22.2	65.5	11.0

The addition of HA influenced the density and the pore size of collagen-HA matrices. That was verified by SEM (Figure 10).

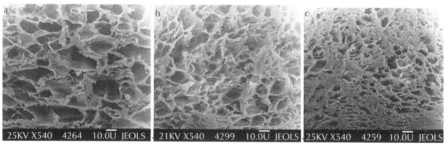

Figure 10. (a) The SEM of unmodified collagen scaffold, (b) collagen cross-linked by EDC, and (c) collagen with HA cross-linked by EDC.

It was observed that HA reduces the porosity of scaffolds with minimum about 68%. After cross-linking the pore size and porosity of the composite scaffold became smaller and a 3D network structure with incorporated HA was formed.

Generally, the cross-linking by different agents leads to the formation of 3D network structure with different pore size. This original approach might provide new insights into the structural differences between natural tissue and novel class of 3D network structure of collagen-based scaffold for tissue engineering. It has been shown (Miron-Mendoza et al., 2010; Murphy et al., 2010) that collagen translocation appeared to depend primarily on matrix stiffness, whereas cell spreading and migration is less dependent on matrix stiffness and more dependent on collagen matrix porosity.

In creation of functional scaffolds for tissue engineering their dielectric properties also play an important role. It is known that all the physiological processes are accompanied by the flow of electric current through such structural elements as intra- and extra-cellular fluids and charge accumulation at the interface, for example of collagen and water in the skin, tendon or bone. These conduction and polarization mechanisms in the tissues are possible thanks to their dielectric properties. So, in the tissue engineering the dielectric properties of scaffolds should be similar or even better than those of the regenerating tissues and organs. The results of the studies performed by the dielectric spectroscopy over a wide range temperatures provide new information on molecular interactions in cross-linked collagen (Marzec and Pietrucha, 2008; Pietrucha and Marzec, 2005, 2007). For collagen-HA or collagen-CS systems cross-linking is an effective way not only to decrease the biodegradation rate but also to optimize the dielectric properties of these systems. In fact, in the whole temperature range (22–230°C) the dielectric parameters are much higher for collagen-HA and collagen-CS scaffolds than for unmodified collagen because of the greater number of sites among which protons can jump and mobility of these carriers in the former.

To summarize, the result of the thermal and dynamic viscoelastic measurements supports strongly the conclusion that the collagen modified with HA/CS using EDC/NHS makes the materials stiffer and gives rise to a marked increase in their structure thermostability. The influence of cross-linking on the permittivity of collagen is significant over the entire temperature range. This may be ascribed to the possible simultaneous occurrence of two reactions: collagen-collagen cross-links and covalent grafting of HA to the collagen backbone, resulting in the formation of semi- interpenetrating polymer network (IPN). The results suggest that these collagen/HA and collagen-CS matrices as an analog of the natural 3D ECM may be useful *in vitro* investigation to support the attachment and proliferation of a variety of cells.

STRATEGY FOR NEW APPROACH TO TISSUE ENGINEERING

In the recently proposed projects, foundations are established for an entirely new approach to tissue engineering by constructing artificial scaffolds based on combination of cells, biodegradable multi-component matrix, and bioactive molecules (selected growth factors and genes) to recapitulate natural processes of tissue regeneration and development. Cooperative interplay of these components is imperative to achieve biologically functional engineered tissue.

Scaffolds with Bioactive Molecules

The researchers believe that combined nano- and micro-fibrous scaffolds, with high surface area to volume ratio, loaded with genes of growth factors will be a good system for treatment of chronic wounds. The chronic wounds are characterized with low proliferation activity of cells and increased of metalloproteinases activity. Delivery of the genes to chronic wounds and successful transfection allows the increased continuous synthesis of growth factors in wounds and acceleration of cells proliferation or angiogenesis that increases speed of healing. At application of the selected angiogenic growth factors for example vascular endothelial cell growth factor (VEGF) basic fibroblast growth factor (bFGF or FGF-2) and epidermal growth factor (EGF) may be useful for enhancing angiogenesis (Gerard et al., 2010; Lee and Shin, 2007) and *in vivo* wound healing of diabetic ulcers (Choi et al., 2008). The achievement of combining high loadability with controlled release of growth factors still represents a major challenge in the field tissue engineering (Pang et al., 2010).

Bacterial Collagen-like Protein

Predominantly, collagen-based scaffolds are constructed from bovine Type I collagen. It has weak immunogenic tendencies and of crucial importance, it bears great resemblance to human collagen. The Word Heath Organization (WHO) allows the use of bovine collagen in medical applications, providing that specified safety procedures are adhered to. However, when substances of animal origin are used it is not impossible for transmission of pathogens to occur, mainly BSE and TSE (animal transmittable spongio-encephalitis), from infected animal to man. The researches believe that, although the risk is small, it is appropriate for tissue engineering to develop scaffolds based on alternative to animal's collagen. Significant progress has been made in production of recombinant human collagen in yeasts to circumvent this and other issues. However, the use of yeast systems is complicated and difficulties in achieving large scale production. Recently, the new realistic prospect of a remedy has appeared from the latest advances in production of bacterial collagens (Peng et al., 2010). There is a conviction that bacterial collagens are an excellent source of collagen for use as biomaterials and as a scaffold for tissue engineering. The recombinant bacterial collagen can be produced in large quantities in *E. coli* without any animal contamination. In particular, it is non-immunogenic, supports cell attachment, can be made with additional functional domains and can be fabricated into fibrous structures and into formats such as sponges suitable for wound repair or other medical applications.

NEURAL TISSUE ENGINEERING

Damage of the central nervous system (nerves of spinal cord) can be caused by disease or trauma and often results in paralysis, which at present can not be cured. Every year, more than 11,000 individuals sustain spinal injury in the USA. These are mostly young men, median age 26. Of the quarter of a million people living in the USA after spinal cord injury, nearly half have irreversible loss of neurologic function below the level of injury (with either lower limbs or all four limbs paralyzed). (Madigan et al., 2009). It is known, that in Europe it is comparable number but complete data is not available. The aim of a lot researches is to pave the way for reversing this loss of neurologic functions, which is still perceived as "irreversible".

The new, realistic prospect of a remedy has emerged from the latest advances in tissue engineering. Tissue engineering progresses by applying newly developed scientific methods and concepts to techniques of regeneration of lost, damaged or diseased body tissue or organ. In the last few years, such new scientific methods and concepts have emerged in two relevant areas. One is the progress in the study and manipulation of stem cells. The other is the progress in research on repair of damaged spinal cord systems of animals (Okada et al., 2007; Yoshii et al., 2009).

The use of stem cells as a biological basis for tissue engineering has opened up an entirely new chapter in medicine and revealed the possibility of breakthrough in attempts to repair and regenerate damaged parts of nervous systems. Stem cells have capacity for self-renewal they can differentiate into multiple cell types and have capability of *in vivo* reconstitution of a given tissue. Both embryonic and adult (somatic) stem cells have shown great promise. Rapidly accelerating rates of progress are evident in more effective ways of controlling cell differentiation, in understanding the mechanisms of cell engraftment into host tissue and in growing cell cultures in bioreactors.

The future of use of stem cell therapy will depend on the availability of means of delivering the cells to the location of nerve damage. No applicable scaffold is currently available for implantation in the spinal cord and there is no evidence of an advanced state of development of such a scaffold (Tanaka and Ferretti, 2009). The requirement is for an implantable scaffold, biologically, chemically and physically compatible with the spinal cord environment. It will have to be designed to deliver preselected neural stem cells to the location of nerve damage and then to support, guide and co-ordinate the regeneration of the nerve and of the damaged tissue. In addition to stem cells, the scaffold will have to deliver therapeutic factors and cells identified as essential or helpful to reconstruction of nerves. Looking further ahead, a requirement may emerge for the scaffold to act also as a genetic delivery tool, for non-viral and viral gene delivery, with or without targeted genomic integration. The underlying therapeutic concept is that combination strategies will maximize the ability to recreate nerve connections and regenerate spinal cord tissue.

To summarise, a tissue engineering device for repair and regeneration of spinal cord nerves will consist of two components: stem cells which are in an advanced state of development and scaffold, which is yet to be developed. The aim of this program is to design and develop the required scaffold. The design stage will be preceded by research to establish the principal design parameters which are not yet known.

Scaffolds for Therapeutic Treatment of Injuries to Central and Peripheral Nervous Systems

The tissue engineering scaffolds for repair and regeneration of brain and spinal cord protective dura mater, designed and developed by the author of this chapter, have outstanding clinical success. But they were constructed, like virtually all clinically successful scaffolds for implanting into connective tissue. For this reason, they had structure resembling that of connective tissue. In connective tissue, the fibrillar collagen ECM is more plentiful than the cells and surrounds individual cells on all sides, determining the physical properties of the tissue. The tissue engineering scaffolds for connective tissue are constructed to have similar internal structure, where individual cells are surrounded by biopolymer on all sides, while providing conditions for cell diffusion and proliferation. The ECM is virtually absent from the spinal cord. The major part of the tissue of the spinal cord consists of large neurons which are responsible for the receptive, integrative and motor functions of the nerve system. The second component is neuroglial cells, smaller but more numerous than neurons. They are responsible for supporting and protecting neurons. The scaffold for implanting in spinal cord will have to be biologically and physically compatible with the environment of the spinal cord. It's structure will need to have directional orientation for both guiding the ends of the severed nerves and for guiding stem cells and neurotrophic factors to the location of the nerve repair process. A supply of stem cells from outside to the repair area is essential, because mature neuron cells do not divide and once the adult complement of neurons has been generated, no stem cells persist to generate more, or to repair damage. Two main forms of elongated nerve cells will need accommodation in the scaffold. The first is neuron, the nerve cell which may be 5–150 μm in diameter. The second is axon, a thin, about 15 μm in diameter, extension growing out of neuron; its function is to conduct signals towards distant target, which may be up to 1 m away. Axons are more frequently severed than neuronsand research on animals has shown that they are much easier to re-grow and regenerate. A newly generated or repaired axon will tend to grow along a directionally oriented fibril or pore channel of the scaffold. Structural support of axonal regeneration will be combined with targeted delivery systems for stem cells and also for therapeutic drugs and neurotrophic factors to regionalize growth of specific nerve cells. In relation to blood, the cerebral and spinal fluids are low in cellular nutrients. Scaffold permeability to a wide range of molecular sizes will be crucial for access of oxygen and nutrients and for removal of metabolic waste.

The author of this review has credible means, supported by experimental evidence, of adjusting the values of the known parameters of the scaffold. She has developed experimentally techniques for adjusting either or both of these properties by co-polymerization/grafting of collagen with minor proportions of other biopolymers. These include the internal surface area-to-volume ratio, the ranges of dimensions of interconnecting pores, diameters and orientation of fibrils and pore channels, as well as the equilibrium water content. Two additional properties of the scaffold may require adjustment: the rate of biodegradation after implanting and the mechanical strength. The results of these experiments will be accessible in another original publication.

Regeneration of peripheral nerve is also important topic in regenerative medicine. Ideally, a nerve guide material, first of all, should provide guidance and support for regeneration axons, exhibit biodegradability properties and have a shelf-life appropriate to the nerve trauma (Alluin et al., 2009; Parenteau-Bareil et al., 2010; Wang et al., 2009). Among the known nerves guides collagen-based composite materials such as CultiGuide®-composite poly-caprolactone and porous collagen-based beads, RevoNerv®-bioresorbable porcine collagen Type I+III nerve conduit and bovine collagen type I tube from Integra Life Sciences Cor. are developed. Although surgery techniques improved over the years, the clinical results of peripheral nerve repair remain unsatisfactory.

CONCLUSION

This review highlights current tissue engineering and novel therapeutic approaches to axonal regeneration following spinal cord injury. An important role in creation of functional, biological scaffolds for tissue engineering play theirs stiffness, thermostability, porosity and dielectric properties.

The results also showed that it is possible to produce ideal replacement for dura, the tough protective membrane of brain and spinal cord. The dura substitute is capable of regenerating the tough protective membrane, but also of bringing about regeneration of blood vessels. Regenerated blood vessels will enable restoration of natural functions of regenerated, replaced or supplemented tissues. The replacement composite scaffolds are effective in regenerating other types of tissues. Enhancing angiogenesis is effected by introducing specific bioactive growth factors. Matrix with incorporated bioactive molecules may become a way of introducing genes into selected locations in the body for "release-on-demand" therapy. Electrospinning is a process that can be used to fabricate collagen-based scaffolds economically at large scales. Furthermore using irradiation, it is possible to offer both a biodegradable implant with stable structure and a well-suitable implant for surgery in a sterile and ready-to-use state. The novel nano- and micro-structured collagen with minor proportions of other biopolymers will provide new possibilities in the field of regenerative medicine as a new path for applications of tissue engineering. Using several new technologies should open new horizons for tissue engineering and for an important branch of medical science.

KEYWORDS

- **Chondroitin sulfate**
- **Collagen**
- **Extracellular matrix**
- **Glycosaminoglycans**
- **Methylcellulose**
- **Tissues**
- **Transplantation**

Chapter 20

Pectic Polysaccharides of Fruits: Characterization, Fractional Changes and Industrial Application

Kausik Majumder

INTRODUCTION

Cell wall is a unique feature in a plant cell. This extra-protoplasmic entity is a fundamental distinguishing feature that makes a mark on a plant cell from the corresponding unitary biological material of the animal kingdom. The somatic cells of plants are characterized by three peripheral layers and this come to the vision of the scientist long before the underlying material that is the protoplasm was discovered. This structural barrier of a plant cell, which is looked upon as a secretory product of protoplasm is however, destined with a number of important functions to perform. In addition to providing the skeletal support, cell walls are known to participate in a number of physiological functions and have an important role in the carbon economy of the biosphere (Goodwin and Mercer, 1988).

Structurally, the cell wall is equipped with several complex organic compounds, chemically recognized as cellulose, hemicellulose, lignin, and the pectic substances. Among these polymers, pectic substances find their location in the middle lamella and also the primary wall of the cell wall. The proportion of the pectic material in the cell wall however, depends much on the type of layers present in it. For it is known that the amount of this component extracted from meristemetic or parenchymatous tissues may comprise of 15–30% of the wall material in contrast to 0.5–1.5%, that is extractable from tissues, which are highly lignified (Tarchevsky and Marchenko, 1991). This review aims to give insight into how the various structural changes that occur in the pectic polymers during fruit development and ripening. To the end, a brief note on industrial significance and application of this natural polymer are discussed.

STRUCTURE, CLASSIFICATION AND NOMENCLATURE OF PECTIC SUBSTANCES

Pectic polysaccharides or substances are polygalacturonate molecule in which $1\rightarrow4$, α-D-galacturonic acid chains are linked with branches of L-rhamnopyranosyl residues with neutral side chains of L-arabinose, D-galactose, and D-xylose (Saulnier and Brillouet, 1998). The carboxyl groups of galacturonic acid are partially esterified by methyl groups and partially or completely neutralized by sodium, potassium or ammonium ions (Kashyap et al., 2001). Some of the hydroxyl groups on C_2 and C_3 may be acetylated and the monomer is thought to have the C_1 confirmation (Pilnik and Voragen, 1970). The primary chain consists of α-D-galacturonic units linked α-$(1\rightarrow4)$, with 2–4% of L-rhamnose units linked β-$(1\rightarrow2)$ and β-$(1\rightarrow4)$ to the galacturonic units

(Whitaker, 1970). The most abundant component of the pectic polysaccharides is polyuronic acids. In the case of higher plants, these polymers are built up mainly by α-D-galacturonic acid residues. However, in the marine brown algae, the polyuronic acids are structurally different and are usually called alginic acids. Some pectic polysaccharide domains, such as homogalacturonate have a relatively simple primary structure. This polysaccharide are also called polygalacturonic acid, a (1→4) linked polymer of α-D-galacturonic acid residues. Another abundant pectic polysaccharide is rhamnogalacturonan (RG II), which has a long backbone of alternating rhamnose and galacturonic acid residues. This molecule is very large and is believed to carry long side-chains of arabinans and galactans. Much further up on the scale of molecular complexity is highly branched pectic polysaccharides, rhamnogalacturonan (RG II), which contains a homogalacturonan of at least ten different sugars in a complicated pattern of linkages (Carpita and McCann, 2001). The rhamnogalacturonans are negatively charged at pH ≥ 5. The side chains of arabinans, galactans, arabinogalactans, xylose or fucose are connected to the main chain through their C_1 and C_2 atom (vander Vlugt-Bergmans et al., 2000; Sathyanarayana and Panda 2003). Albersheim (1976) reported that there are three different polyuronic acids in the primary wall of suspension cultured sycamore cells. The two of which do confirm are a homogalacturonan composed of at least 25 D-galacturonic acid residues linked α (1–4) glycosidic linkages and a high molecular weight rhamnogalacturonan. It consists of segments composed of about 8 α- 1,4 linked D-galacturonic residues separated from one another by a trisaccharide nade up of L-rhamnopyranose (1→4), D-galacturonic (α,1→2), L-rhamnose. The terminal D-galacturonic residues of the pure galacturonan segments in linked to the L-rhamnose containing trisaccharide by an α,1→2 linkages cause the chain to zigzag. The polyuronic acids which does not conform to the usual pattern is also closed as a rhamnogalacturonan but appears to be a much branched structure containing in addition to D-galacturonic acid and L-rhamnose, a variety of other monosaccharides, some of which are quite rare.

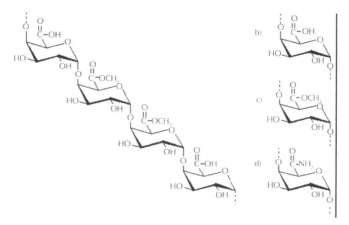

Figure 1. (a) A repeating segment of pectin molecule and functional groups; (b) carboxyl; (c) ester; and (d) amide in pectin chain (Sriamornsak, 2001).

Classification of the pectic substances is an area that grabbed much attention of the polymer biochemistry. Pectic polysaccharides are in fact, recognized in several forms and accordingly, they have been classified into distinct groups. The American Chemical Society classified pectic substances into four main groups as follows (Kertesz, 1951).

1. *Protopectin*: the term protopectin is applied to the water insoluble parent pectic substances, which occurs in plants and which upon restricted hydrolysis, yields pectinic acids

2. *Pectinic acid*: the term pectinic acid is used for colloidal polygalacturonic acid containing more than a negligible proportion of methyl ester groups

3. *Pectin*: the term pectin designates those water-soluble pectinic acids of varying methyl ester content and degree of neutralization which are capable of forming gels with sugar and acids under suitable condition

4. *Pectic acid*: the term pectic acid is applied to pectic substances, mostly composed of colloidal polygalacturonic acids and essentially free from methyl ester groups. Normal or acid salts of pectic acids are called pectates.

However, it needs mentioning in this context that classification and nomenclature of pectic substances in consideration to their chemical solubility in different solvents has also been emphasized. In this context, pectic substances are classified broadly into the following groups (Barnier and Thibault, 1982).

1. Water-soluble pectic substances (WSP)—are those pectic substances which are soluble in water and include pectin and colloidal pectinic acids of high methyl ester content.

2. Oxalate-soluble pectic substances (OXP)—include pectic acids and the colloidal pectinic acids of low methyl ester content

3. Acid-soluble (HP) and Alkali-soluble pectic substances (OHP)—are the groups includes protopectin of having methyl ester content.

FRACTIONAL CHANGES OF PECTIC POLYSACCHARIDES IN DEVELOPING FRUITS

Pectic substances are the major components of cell wall and middle lamella of plant tissues which undergo structural changes during development and ripening of the fruits and thereby contributing significantly to textural softening of ripening fleshy fruits (Proctor and Peng, 1989). Solubilization of pectic polysaccharides which occurs to different extent among fruits and fruit tissues, are involved with enzymic and physical mechanism (Gross, 1990), such as hydrolysis of polygalacturonate backbone, modification of side chains and disruption of ionic or hydrogen bonds (Liyama et al., 1994). The participation of cell wall modifying enzymes that is pectin methyl esterase and polygalacturonase in pectic metabolism during fruit ripening (Majumder and Mazumdar, 2002) causes molecular mass downshift, decreased methyl esterification, and glycosylation (Malis-Arad et al., 1983), which in turn alters the physical properties of the cell wall (Klein et al., 1995). The change-over of pectic substances from cell wall bound to soluble form has been well documented in a number of fruits, although the ripening phenomenon reflected by solubilization of pectic polymer is not uniform throughout tissues in a fruit. The reviewer would like to point out at the outset that researches done by the scientists on pectic compounds in fruits are much extensive and the studies have covered fruits of a large number of species with particular reference to those of economic importance. The pertinent works done on majority of such fruits have been brought in and discussed in the following under the respective headings.

Citruses (*Citrus* species) Fruits

Among the edible species belong to the genus *Citrus*, the mandarin orange (*Citrus reticulate* Blnco.) and the sweet orange (*Citrus sinensis* Osbeck) are known to have supreme importance as table or desert fruits all over the world. In Satsuma mandarin oranges, the water-soluble pectins in the pulp increased rapidly during advanced maturity condition and decreased gradually in ripening phase (Daito and Sato, 1984). The acid and water-soluble pectins as well as the total pectin had consistently steeped up in developing lemon (*Citrus limon* Burm.) pulp (Soni and Randhawa, 1969), although a regular increase of pectin content of lemon peel during ripening was also observed (Ramadan and Domah, 1986). The granulation of juice sometimes appears to be a great problem in many *Citrus* fruits. The changes of the different pectic fractions may be responsible for this granulation (El-zeftawi, 1978). The reduction of NaOH-soluble fraction of pectin accompanied with an increase in water-soluble pectin was also observed in process of juice granulation (El-zeftawi, 1978). In order to unfold the mechanism of hardening of juice-sacs during granulation, the stoichiometric relationship between pectic substances and the level of calcium ions in the healthy, gelated and granulated juice-sacs from *Citrus sulcata* fruits were investigated by Goto (1989) and the amount of calcium ions per unit amount of pectic substances in the EDTA-soluble fractions had higher in gelated and granulated juice-sacs than in the healthy juice-sacs. The crystallization of pectic substances as a result of increased levels of calcium ions in the middle lamella was a cause for hardening of the juice-sacs and collapsed juice vesicles of grape-fruit (Hwang et al., 1990) and was also associated with a modification of water-soluble and chelator-soluble pectin molecular weight.

Mango (*Mangifera indica* L.)

The pectic compounds and pectic degrading enzymes in mango pulp is a subject of study, undertaken by some Researchers. The level of pectin along with other constituents continued to increase in the fruits during their growth as opposed to the reverse trend for the acidity level in fruits (Gangwar and Tripathi, 1973). The water- and alkali-soluble pectic fractions were reported to decrease in contrast to ammonium oxalate-soluble pectin fraction in ripening stage of mango fruits, irrespective of cultivars variation (Tandon and kalra, 1984). The polygalacturonase activity markedly increased during ripening and the loss of fruit firmness is well correlated with this increased enzyme activity (El-Zoghbi, 1994). The rate of marked changes during fruit ripening could exert on important influence on storage life and export potentials of mango fruits. Pectic polymers become less tightly bound in the cell walls during ripening and this wall loosening phenomenon is involved with hydrolysis of galactose containing polysaccharides (Seymour et al., 1989). However, a close relation between changes in β-galactosidase activity, tissue softening and increased pectin solubility and degradation was also observed (Ali et al., 1995).

Guava (*Psidium guajava* L.)

Guava is not only a rich source of vitamin C but also stakes its claims to be one of the top ranking fruits in respect of pectin of high jelly-grade, for which a considerable part of guava produced is made use of in making fruit jelly in commerce, The amount of pectin content in fruits was found as maximum in the ripe stage (Pal and Selvaraj, 1979). However, there is a variation of pectin content among the cultivars of the fruit. A direct relationship on the increase of soluble pectin with an inverse trend between protopectin and softening in developing fruits has also been observed (El-Buluk et al., 1995; Esteves et al., 1984). Salmah and Mohamed (1987) observed a sigmoid curve in regards to the changes of total pectin content during development of fruits while, the water-soluble pectin followed a linear change (El-Buluk et al., 1995). However, pectin content is also affected by cropping season and higher amount of polysaccharides remain persistent with the winter crops (Dhingra et al., 1983). The fractional changes of pectic polysaccharides in developing fruits revealed that the extent of pectic solubilization was more pronounced in inner pericarp region as compared to outer and middle pericarp zone of the fruit tissue (Das and Majumder, 2010). A review on pectic polysaccharides of guava (Marcelin et al., 1990) discussed the metabolism of various forms of polygalacturonates and pectic degrading enzymes in guava fruits as well as importance of the levels of these substances for processing industries.

Apple (*Mallus pumilla* Mill)

Plenteous researches have in fact, being done on examining the pectic substances of apple fruits. Earlier reports (Efimova, 1981; Surinder Kumar et al., 1985; Tischenko, 1973) stated that pectin content decreased during ripening with an increase of polygalacturonase activity. Mangas et al. (1992) reported that the alcohol-insoluble solids decreased with ripening while water-soluble or the chelator-soluble pectin elevated in the final stage of fruit ripening. However, the galactose and arabinose residues were lost both from alcohol insoluble residues and different pectic fractions during ripening of fruits (Fischer and Amado, 1994; Fischer et al., 1994; Redgwell et al., 1997).

Prune Fruits

The rate of fruit softening in peach fruits was negatively correlated with the pectin associated sugars and the chelator-soluble pectin exhibited a decreasing galacturonic acid and rhamnose ratio during fruit ripening (Fishman et al., 1993; Maness et al., 1993; Selli and Sansavani, 1995). In plum fruits (*Prunus domestica* L.), the ethanol-insoluble solids elevated up to a certain level and then fell down and pectin content parallel; the change in weight of insoluble solids (Boothby, 1983). Batisse et al. (1994) reported that the oxalate and water-soluble pectic fractions increased while the acid-soluble part tended to decrease in ripening cherry fruits.

Other Fruits

The ratio of pectinic acid to that of the total pectin increased consistently with ripening of pear (*Pyrus* species) fruits (Yamaki and Matsuda, 1977). Ning et al. (1997) reported the increase of water-soluble pectin and decrease of hydrochloride-soluble pectin during ripening of Chinese pear fruits. The decrease of hydrochloric acid-soluble pectin that is protopectin was also reported in ripening strawberry (*Fragiria* sp.) fruits (Sato et al., 1986; Spayd and Morries, 1981). In kiwi (*Actinidia deliciosa*) fruits, the ripening and softening of fruits is accompanied with an increase of highly methylated pectin, a loss of protopectin and an increase of polygalacturonase activity (Fuke and Matsuoka, 1984; Miceli et al., 1995; Wang et al., 1995). Platt-Alolia et al. (1980) reported that the initial wall breakdown is involved with the degradation of pectin in middle lamella in ripening Avacado (*Persea americana* Mill.) fruits. Ripening of persimmon (*Diospyros kaki*) fruits is always accompanied with the reduction of water-soluble pectin and the concomitant increase of low molecular weight pectin (Inkyu et al., 1996; Matsui and Kitagawa, 1991). In blueberry (*Vaccinium cyanococcus* Camp.), a reduction of alkali-soluble pectin corresponding to an increase of water-soluble and chelator-soluble pectins are associated with fruit ripening (Proctor and Peng, 1990). The total pectin content was increased during ripening in litchi (*Litchi chinensis* Sonn.) (Singh and Singh 1995), banana (*Musa paradisiaca* L) (Tripathi et al., 1981) and water-melon (*Citrullus lanatus* Schrad.) (Vargas et al., 1991), while the amount of total pectin was reported to decrease in Figure (*Ficus carica* L.) (Tsantili, 1990), date (*Phoenix dactylifera* L.) (Bukhavev et al., 1987), bael (*Aegle marmelos* L.) (Pandey et al., 1986) and ber (*Zizyphus mauritiana* Lann.) (Abbas et al., 1988). However, there is a difference of pectin content in ripening pine-apple (*Ananas cosmosus* Merr.) (Trukhiiya and Shabel'skaya, 1983), between dry and wet season, as well as in the peel and pulp of tamarind (*Tamarindus indica* L.) (Hernandez-Unzon and Lakshminarayana 1982). The molecular weight of pectin polymers decreased during ripening of musk-melon (*Cucumis melo* Naad) fruits (Ranawala et al., 1992). Although the amount of pectic material extracted with NaOAc buffer increased consistently during ripening of olive (*Olea europaea* cv Koroneiki) fruits, the molecular weight distribution and the sugar composition of the pectic polymers remains unchanged (Vierhuis et al., 2000) The pectic fractions in the ripening Figure (*Ficus carica* L.) fruits has higher uronic acid content in addition to Ara, Gal and Rha (Owino et al., 2004). In cape-gooseberry (*Physalis peruviana* L.) fruit ripening, the increased level of polygalacturonase activity is correlated with the increase of water-soluble and oxalate-soluble pectic substances (Majumder and Mazumdar, 2002), while ethylene and auxin increase the solubilization of pectin, the increased accumulation of water-soluble pectic fraction (Majumder and Mazumdar, 2005), and gibberellin partially blocked the degree of pectin solubilization (Majumder and Mazumdar, 2001).

CHARACTERIZATION OF PECTIN DERIVED FROM DIFFERENT FRUITS

The examination of physical and chemical properties of pectic polysaccharides, extracted from fruits, peels of the fruits has an important area since pectin is extensively used in industries, particularly in fruit processing sector. The objective of the review in this line is to provide fundamental knowledge on the chemical composition and qualitative parameters of the fruit pectin which are useful for industrial purposes. A comparative characterization of the pectin isolated from the peels of lime, sour orange, sweet orange, and grapefruits showed that the yield of pectin (dry weight basis), jelly grade, setting time, esterification value, equivalent weight, methoxyl content, acetyl content, molecular weight, and intrinsic viscosity in the *Citrus* species had ranged as 14.5–17.2%, 180–225, 1.0–5.0 min, 56.1–63.2%, 859–1452, 7.40–8.62%, 0.32–0.46%, 67,000–92,600 and 3.2–4.4, respectively (Alexander and Sulebele, 1980). The albedo of Spanish lemons represents 61.3, 12.4, and 10.4% galacturonic acid for three different fractions respectively that is chelating-soluble, alkali-soluble and the residues (Ros et al., 1996). The de-esterification pectin of *Citrus unishiu* was further degraded by *endo*-polygalacturonase and the degradation products yields high and low molecular fractions of 39.5 and 1.8% respectively as esterification values (Matsuura, 1987). The labeling study on the chemistry of pectic substances in grapefruit provided evidence that (Baig et al., 1980) galacturonic acid was the only uronic acid with about 76% by weight of pectic polysaccharides. The remaining 24% was accounted for by neutral sugars viz. rhamnose, arabinose, xylose, mannose, galactose and glucose, mannose and glucose, which were in equal properties constituted 5% of the total neutral sugars and rhamnose, arabinose, xylose and galactose, and glucose constituted 16.3, 33.3, 13.4, and 31.5% respectively of the total neutral sugars components of the labeled polysaccharides. The water extract of cell wall polysaccharides of *Limonia acidissima* contains pectic polymers substituted with side chains comprising mainly of 1,5 – 1,3,5- linked arabinose together with 1,4→, 1,6→, 1,3,6→ linked galactose and lesser amounts of 1,2,4→ and 1,3→ linked galactose residues (Mondal et al., 2002). The citrus peel consists of homogalacturonans (HGs) stretches of similar length and similar number of galacturonic acids (Yapo et al., 2007). Low-pressure anion exchange and size-exclusion chromatography fractionations of xylose-rich pectin of yellow passion fruit (*Passiflora edulis* var *flavicarpa* Degener) reveals that HGs are the predominant polymer though rhamnogalacturonans accounts about 17% of the pectin (Yapo and Koffi, 2008).

The molecular weight, esterified carboxyl group contents, anhydrouronic acid content and the intrinsic viscosity of the pectin, extracted from mango are varied between the varities cultivated. The weight average molecular weights of pectin from Alphanso, Kitchener and Abu Samaka varieties of mango were found as 1.6×10^4, 2.1×16^4, 2.7×10^4, and the intrinsic viscosity are 0.60, 0.89, and 1.22 respectively (Saeed et al., 1975). However, the fruit peel of mango contains an appreciable amount of pectin and also it varied among the varieties of mango fruit. For example, the pectin yield, methoxyl content, anhydrouronic acid content, value of esterification, molecular weight, jelly-grade of three mango varieties viz. Dasheri, Langra, and Alphanso had ranged from 13 to 19, 8.07 to 8.99, 58.40 to 61.12, 83.38 to 85.93, 1,05,000 to 1,61,700, and 155 to 200%, respectively (Srirangarajan and Shrikhande 1977). Neukom et al. (1980) and O'beirne et al. (1982) reported that all pectic substances of apple fruits are not rhamnogalacturonan and the galacturonic acid rich polymer of chelator-soluble pectin are of lower in amount of arabinose, galactose and rhamnose residues. The analysis on the cellulosic residue pectin of apple fruits reveals that it has a highly ramified RG I backbone containing an equal number of galacturonic acid (1,4-GalAp) and rhamnose (sum of 1,2-Rha$p$ and 1,2,4-Rha$p$)residues (Oechslin et al., 2003). The arabinan-rich pectic polysaccharides from the cell walls of *Prunus dulcis* seeds has an arabinan glycosidic linkage composed of T-Ara$f$: (1→5)- Ara$f$: (1→3,5)-Ara$f$: (1→2,3,5)-Ara$f$ in the relative proportions of approximately 3:2:1:1 (Dourado et al., 2006). This polymer has also a lymphocyte stimulating effect. Cardoso et al. (2007) reported that the Arabinan-rich pectic polysaccharides accounts about 11–19% of the total pectic polymer, extracted from olive (*Olea europaea* L.) fruit pulp. The FTIR spectroscopic studies of the olive fruit pulp cell wall polysaccharides stated that the pectic polysaccharides are rich in uronic acid, arabinose, arabinose-rich glycoproteins, xyloglucans, and glucuronoxyalans (Coimbra et al., 1999).

The pectic substances extracted from the pulp of unripe fruits of papaya (*Carica papaya* L.) showed that D-glucose, D-galacturonic acid, and L-arabinose are the major constituent sugars with a trace amount of rhamnose (Biswas and Rao, 1969). The study on cherry fruits (*Prunus avisum*) revealed that the pectic fractions are composed of neutral sugarsviz. Rhamnose, fucose, ribose, arabinose, xylose, mannose, glucose, galactose, and 2-deoxyglucose, which account as 13.9–42.8% of the total carbohydrates (Barnier and Thibault, 1982). However, the neutral sugars of pectic substances of plum (*Prunus domestica* L.) fruits are completely esterified (Boothby, 1980). An appreciable amount of pectic material as of 0.85% on fresh weight basis was extracted from kiwi fruits (*Actinidia deliciosa* L.), which also exhibited a highly jellying temperature of 90°C, compared with 63°C for apple pectin. The cell wall polysaccharides of kiwifruit (*Actinidia deliciosa*) contain the pectic polysaccharides of the GTC- and KOH-soluble fractions which has more highly branched rhamnogalacturonan backbones than the CDTA- and Na$_2$CO$_3$-soluble polymers (Redgwell et al., 2001). The swollen peduncle of cashew nut (*Anacardium occidentalae* L.), known as cashew apple contains the pectic substances which has the methoxyl content of 3.5%, anhydrouronic acid of 45.10% and jelly-grade of 75. The acidic water soluble fraction of the polysaccharides of Indian gooseberry (*Phyllanthus emblica* L.) was reported to contain a pectin with D-galacturonic acid, D-arabinosyl, D-xylosyl, L-rhamnosyl, D-glucosyl, D-mannosyl, and D-galactosyl residues (Nizamuddin et al., 1982). The analysis on the cell walls of loquat (*Eriobotrya japonica* L.) fruit tissues reveals that the major component polysaccharides are the pectic polysaccharides, accounted up to 70% of the total cell wall polysaccharides (Femenia et al., 1997).

INDUSTRIAL SIGNIFICANCE OF PECTIC POLYSACCHARIDES

Over the years, pectin has been used in conventional food processing industrial processes such as preparation of jam, jelly, and so on. Pectin was first isolated in 1825 by Henri Braconnot, though the uses of pectin for making jams and marmalades was known long before. However, in 1930s, pectin was commercially extracted from dried apple pomace and citrus peels. They have also pharmaceutical and textile applications.

Fruit Processing Application

The most important use of pectin is based on its ability to form gels. High methoxyl pectin forms gels with sugar and acid. Oakenfull (1991) reported that hydrogen bonding and hydrophobic interactions are important forces in the aggregation of pectin molecules. Gel formation is caused by hydrogen bonding between free carboxyl groups on the pectin molecules and also between the hydroxyl groups of neighboring molecules. The rate of gel formation is also affected by the degree of esterification. A higher esterified pectin (i.e., pectin with a DE of above 72%) causes gel at lower soluble solids and more rapid setting than slow-set pectins (i.e., pectin with a DE of 58–65%). However, low methoxyl pectin requires the presence of divalent cations for proper gel formation. This unique property of this biopolymer has made it a very important food additive for the food processing industries (Pilnik and Voragen, 1970). Pectin is extensively used in the preparation of jam, jelly and marmalade. Pectin is also an indispensable ingredient in other food preparations like sweetmeat, salad dressing, ketchup, sauce, ice-cream, pudding, and so on. (Cruess, 1977). Pectin can also be used to stabilize acidic protein drinks, such as drinking yogurt, and as a fat substitute in baked goods.

Pharmaceutical Application

Pectin has a wide application in the pharmaceutical industry. Pectin has been reported to help in reducing blood cholesterol (Sriamornsak, 2001). Pectin acts as a natural prophylactic substance against poisoning with toxic cations. It has been shown to be effective in removing lead and mercury from the gastrointestinal tract and respiratory organs (Kohn, 1982). Pectin combinations with other colloids have been used extensively to treat diarrheal diseases, especially in infants and children. Pectin has a light antimicrobial action toward *Echerichia coli* (Thakur et al., 1997). Pectin hydrogels have been used in tablet formulations as a binding agent (Slany et al., 1981) and also have been used in controlled-release matrix tablet formulations (Naggar et al., 1992). Pectin has a promising pharmaceutical uses and is presently considered as a carrier material in colon-specific drug delivery systems (Sriamornsak, 2001). Pectin is used as a carrier of a variety of drugs for controlled release applications. Techniques on the manufacture of the pectin-based drug is a promising subject for the future pharmaceutical industries (Sriamornsak, 2001)

Other Uses

Pectin is commonly used as a demulcent in throat lozenges. Pectin is used in manufacturing of many healthcare products such as lotions, facial creams, and other types of makeup. In the cigar industry, pectin is considered an excellent substitute for vegetable glue and many cigar smokers and collectors will use pectin for repairing damaged tobacco wrapper leaves on their cigars. Among other uses, mention may be made to the preperation of adhesives sizing of textiles, hardening of steel and so on. (Cohn and Cohn, 1996).

CONCLUSION

A number of reports are in record on the chemistry, characterization, functional physiological properties, and applications of the pectic polysaccharides, a natural polymer. The foregoing discussion divulge that the long chain polymer of α-D-galacturonic acids, which find their existence outside the protoplasm of a plant cell have a number of important functions to perform. However, the function of the arabinan and galactan side-chains of pectin remains unknown. The basic properties of pectin have been known for nearly 200 years, but recently there has been tremendous progress in the fine structure of pectic polymer by application of the advanced techniques including high-throughput microarray analysis, enzymatic fingerprinting, mass spectrometry, NMR, molecular modeling and using of monoclonal antibodies. The phase diagrams of pectin-calcium systems showed that gelation is possible because of the strong interactions of calcium with pectins. An increase in the ionic strength, a neutral pH, and a decrease in the setting temperature and the degree of methylation lower the amount of calcium required for the gel formation. However, the complex physico-chemical properties of the pectic polymer are not yet well understood. Future studies would highly require to unfold the molecular mechanism of pectic polymer in sol-gel synthesis for more applicability in various novel applications. Moreover, new sophisticated designer pectin with specific functionalities would have a major impact on fruit processing and pharmaceutical industries.

KEYWORDS

- Cell wall
- Oxalate-soluble pectic substances
- Pectic polysaccharides
- Pectin
- Polyuronic acids
- Rhamnogalacturonan
- Water-soluble pectic substances

Chapter 21

Synthesis and Characterization of Alkyd Resin Microcapsules

Hariharan Raja, Swarupini Ganesan, Gokulnath Dillibabu, and Ravisankar Subburayalu

INTRODUCTION

Microencapsulation enables storing of reactive components in a hollow sphere which has been used for variety of applications such as pharmaceuticals, textiles, adhesives, cosmetics, nutrient retention, and advanced coatings. Recently, microcapsules in the form of microscopic spheres containing "Healants" or "coating repairing" compounds are involved in self-healing mechanism (Ashok Kumar et al., 2006; Dong Yang Wu et al., 2008). Upon damage-induced cracking, the microcapsules containing healing agents are ruptured by the propagating crack fronts resulting in the release of the healing agent into the cracks by means of capillary action. These microcapsules can be manufactured by various methods of polymerization techniques (Erin B. Murphy et al., 2010).

Poly(urea-formaldehyde) (UF) microcapsules prepared by *in situ* polymerization of urea and formaldehyde meet the demanding criteria required for use in self-healing materials (Suryanarayanaa, 2008). These criteria includes excellent bonding to the matrix, sequestration of the healing monomer from the surrounding environment, and rupture and release of the monomer into the crack plane upon matrix damage (Blaiszik B. J., 2009).

Here, we report on the development of urea-formaldehyde microcapsules filled with fast drying alkyd resin and xylene as a solvent for its use in self-healing. Not all resins are suitable for self-healing. Alkyd resin has fast drying property, film forming ability by atmospheric oxidation, self-healing property, surface transparency, and low flammability. Alkyd resin, being cost effective and commercially available resin it could be the suitable replacement for other resins which had been used for self-healing coatings.

EXPERIMENTAL PROCEDURE

Microcapsule Materials

Fast drying alkyd resin obtained from soya oil is used as a core material and mixed xylene is used as a solvent (to reduce the viscosity of the resin). Urea, formaldehyde, resorcinol and ammonium chloride were the wall forming materials. Polyvinyl alcohol as a surfactant and n-octanol as an anti-foaming agent supports to achieve high yield of microcapsules without any air filled capsules.

Encapsulation Technique

The encapsulation method was adapted from that of (Brown et al., 2003). Microcapsules containing a mixture of resin and solvent were prepared by *in situ* urea-formaldehyde microencapsulation procedure. 260 ml of deionised water is mixed with 10 ml of 5 wt% aqueous solution of polyvinyl alcohol in a 1000 ml beaker and is kept over a hot plate. The solution is agitated using REMI Mechanical stirrer consisting of 4 bladed low shear mixing propeller each of 40mm diameter, placed above the bottom of beaker. Under agitation, 5 g of urea, 0.5 g of ammonium chloride and 0.5 g of resorcinol were added to the solution. After agitation for 5 min, pH was adjusted to approximately 3.5 by using 5wt% solution of hydrochloric acid in deionised water. One to two drops of n-octanol was added to reduce foam.

A slow stream of 60 ml of alkyd resin and solvent containing 0.05 wt% cobalt octoate, 0.5 wt% lead octoate and 0.05 wt% manganese octoate driers were added slowly to form an emulsion and allowed to stabilize for 10 min under agitation. After stabilization, 12.67 g of 37 wt% aqueous solution of formaldehyde was added. The temperature is increased at a rate of 1°C/min upto 45°C under stirring at 200 rpm and maintained at this temperature for 4 hr. Contents were cooled to ambient temperature. Microcapsules from the suspension were recovered by sieve filtration through a stack of 3 sieves: 450, 650, and 1000 μm. A wet-sifting technique was performed by rinsing the set of sieves under a running faucet of deionised water, in order to remove better the urea-formaldehyde debris. The collected microcapsules were then dried at room temperature for approximately 24 hr as in Figure 1.

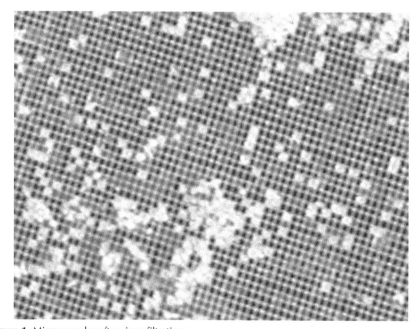

Figure 1. Microcapsules after sieve filtration.

Infrared Spectroscopy

In order to ascertain UF encapsulation of resin, spectra of original alkyd resin is compared with the spectra of soxhlet extracted resin. The purpose of Soxhlet apparatus is to extract core material from the outer membrane. The core material collected after soxhlet extraction was mixed with Potassium Bromide and palette was prepared for recording spectra using fourier transform infra red spectroscopy (FTIR).

Analysis of Shell Morphology

Scanning Electron Microscopy was performed to analyze surface morphology and shell wall thickness of the capsule. A dried microcapsule was placed on a conductive carbon tape attached to a mounting piece for imaging. Few microcapsules were ruptured with a razor blade to confirm the presence of alkyd resin inside the UF shell.

Thermal Analysis of Microcapsules

Microcapsules were analyzed using thermo gravimetric analyzer (SDT Q600V8.0, Ta instrument) in nitrogen environment with a sample weight of about 4.6170mg. Heating rate was maintained at 10°C/min in the temperature range of 30–800°C. Similarly samples were also analyzed by using differential scanning calorimeter(DSC Q10 V9.0, TA instruments) in oxygen environment with sample weight of about 6.2mg at heating rate of 10°C/min, between 30 and 400°C.

RESULTS AND DISCUSSION

Urea-formaldehyde capsules are formed by the reaction of urea and formaldehyde to obtain methylol ureas, which condenses under acidic conditions to form the shell Material. Alkyd Resin encapsulation takes place simultaneously during formation of cross-linked urea-formaldehyde polymer. When the pH is changed to acidic and heated to 55°C, urea and formaldehyde reacts to from poly (urea-formaldehyde).

FTIR Analysis

The FTIR spectra of original alkyd resin and soxhlet extracted resin are shown in Figure 2 which shows the characteristic peaks of ester stretching (COOR) at 1729.49 cm^{-1}, unsaturated fatty acid at 1266.23 cm^{-1} and aromatic CH deformation at 730cm^{-1} closely matches thus establishing that alkyd resin has been successfully encapsulated within urea-formaldehyde shell.

Thermal Analysis

Thermal analyses of microcapsules were carried out to prove encapsulation of alkyd resin inside UF Shell. The TGA of microcapsule is done and the curve is shown in Figure 3. The first endothermic peak at 229.2°C corresponds to UF decomposition temperature (Suryanarayanaa et al., 2008) and the endothermic peak at 350°C corresponds to the start of alkyd resin decomposition temperature which ends at 450.5°C. These results establish that microcapsules contain alkyd resin as core and UF resin as the shell. The weight loss of the sample was found to be 4.5 mg (98.11%)

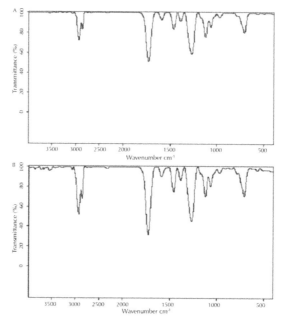

Figure 2. (A) FTIR spectra of original Resin and (B) Soxhlet extracted Resin.

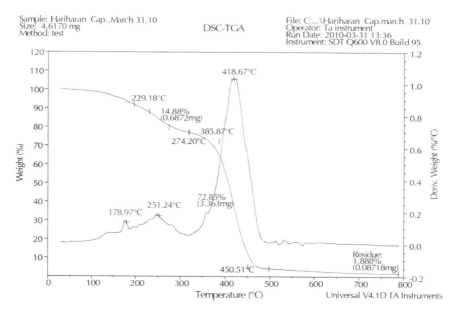

Figure 3. TGA curve.

The DSC curve of the microcapsule is shown in Figure 4. The endothermic peak at 115.7°C corresponds to melting of UF resin and at 199.4°C corresponds to melting of alkyd resin whereas at 147.4°C it corresponds to evaporation of solvent xylene. Deviations in the base line are said to be glass transition temperature

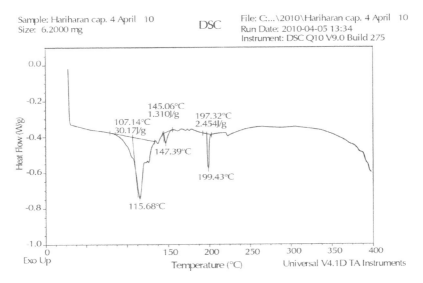

Figure 4. DSC curve.

Shell Wall Morphology

Scanning electron microscopy was performed to analyze capsule surface morphology and shell wall thickness. Figure 5 shows a resin-solvent filled microcapsule with rough exterior shell wall and size in the range of 450 μm.

Figure 5. SEM image of a microcapsule.

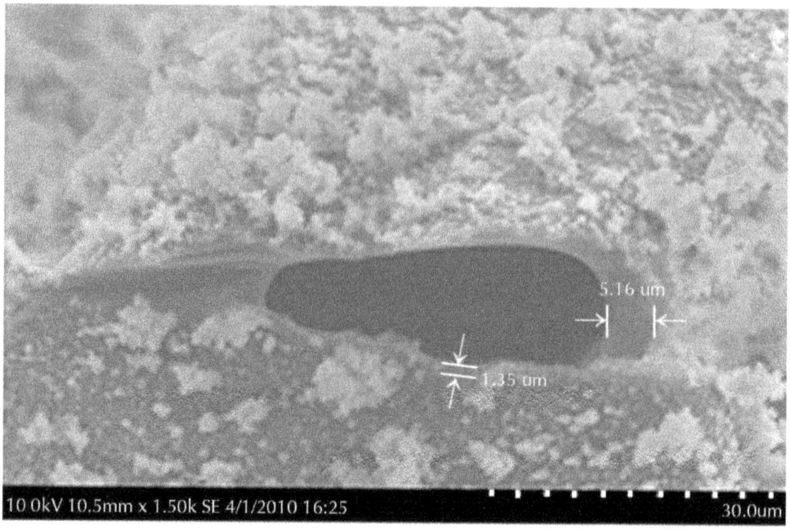

Figure 6. Shell wall morphology of a ruptured microcapsule.

Some microcapsules were ruptured with a razor blade to allow for viewing of the inner shell wall morphology. Figure 6 shows the shell wall morphology comprising of two distinct regions of thin continuous inner shell wall and rough exterior shell wall.

The thickness of UF shell wall was found as 5.16 μm. Continuous inner shell wall is formed due to reaction between urea and formaldehyde in aqueous phase and rough exterior is formed as colloidal UF particle coalesce and deposits along interface.

CONCLUSION

Alkyd resin along with solvent has been successfully encapsulated inside UF shell by *In situ* encapsulation technique. The SEM analysis shows that microcapsules were comprised of a thin continuous shell wall and a rough exterior shell wall. The FTIR spectra of original and soxhlet extracted resin proves the encapsulation of alkyd resin inside the shell. Moreover DSC and TGA analysis shows that the microcapsules contain both resin-solvent mixture as core and UF as shell material. These microcapsules incorporated in paints or coatings release the healing material when scratched which in turn heals the scratches and cracks.

KEYWORDS

- **Deionised water**
- **Morphology**
- **Octoate**
- **Polyvinyl alcohol**
- **Urea-formaldehyde**

Chapter 22

Environmental Recovery by Magnetic Nanocomposites Based on Castor Oil

Souza, Jr. Fernando Gomes, Oliveira, Geiza Esperandio, and Lopes Magnovaldo Carvalho

INTRODUCTION

In spite of the great efforts related to the petroleum consumption reduction, our society remains depending on this resource as an energetic source besides raw material for the production of several commodities. In addition, petroleum is principally carried through long distances, increasing the chances of accidents, mainly on the aquatic environments, as those shown in Figure 1. These accidents involving oil spill produce devastating impacts on the environment, since the lipophilic hydrocarbons of the petroleum make a strong interaction with the similar tissue of the higher organisms. This assimilation produces intoxication and even death. Unfortunately, some of the traditional cleanup process, which involves the use of dispersants and surfactants—see the reports about the Torrey Canyon (Bellamy et al., 1967) and the Alaska (Bragg et al., 1994) cases—presents a greater impact to the environment than the spill, since the absorption of the dispersed oil by organisms, leads to a much greater time to the bio recovery of the degraded environment (Barry, 2007).

Not considering the harmful side of the subject, the polymers can be engineered to produce chemical and physical properties which make them useful tools to the environment recovering process. Among these materials some are from petrochemical sources, such as poly(divinylbenzene) and poly(methyl methacrylate) (Queiros et al., 2006) and other ones from natural resources, such as the chitin, the chitosan and its derivates (Carvalho et al., 2006). In addition, the environmental recovery goal can be more easily reached through changes in the aliphatic/aromatic character of the used polymeric materials, allied to a magnetic induced behavior, making possible to create new magnetic materials able to remove petroleum from the water.

In this specific context, our group has dedicated some efforts to obtain polymer materials completely or partially "green" from renewable resources such as the cashew nut shell liquid (CNSL); the glycerol, byproduct of the production of biodiesel by transesterification process; and the lignin removed by the Kraft process. Resources of these chemicals are shown in Figure 2. These polymers are useful in the environmental recovery which involves the oil spill cleanup process. In addition, seeking for the raising of the clean up efficiency, magnetic nanoparticles are incorporated to these polymers, allowing for the use of electromagnetic devices to catch the composite impregnated with petroleum (Lopes et al., 2010; Souza et. al., 2010 a, 2010b).

Figure 1. (a) Oil spilled on the ocean (Atom, 2010), (b) and on the beach (Schambeck, 2010), (c) bird (Biagini, 2010), and (d) and dead whale (Marcelino, 2010) covered by oil.

Figure 2. (a) Cashew and cashew nut—a source of cardanol (Ramos, 2010; Juraci Jr., 2010), (b) flaxseed—a source of lignin (Tristro, 2010), (c) and castor beans—a source of castor oil and glycerin.

In this work, the experimental design technique (DoE) was used to find out the better way to produce maghemite nanoparticles, seeking to reach the smaller size associated with the larger magnetic force. The analyzed factors were the stirring speed, the digestion time of the precipitated and the temperature used in the thermal treatment. In addition, the DoE was also used to study the influence of the amount of the glycerol, castor oil and phthalic anhydride on the magnetic force and on the oil removal capability of the composite.

MAGNETIC RESINS PRODUCTION

Maghemite Nanoparticles Production

The synthesis of the maghemite particles were performed as shown in Figure 3. Initially, aqueous solutions of hydrochloric acid (2 M), ferric chloride (2 M), and sodium sulfite (1 M) were prepared. In a typical procedure, 30 ml of the ferric chloride solution and 30 ml of deionized water were added into a beaker under continuous agitation. Soon afterwards, 20 ml of the sodium sulfite solution was added to the beaker and also under continuous agitation. The reaction product was precipitated by slowly adding 51 ml of concentrated ammonium hydroxide into the beaker under continuous agitation. After 30 min, the medium was filtrated and the obtained particles were washed several times with water and finally dried at 60°C in an oven. Magnetite was converted into maghemite through annealing at 200°C along 1 hr (Souza Jr. et al., 2009a). The maghemite nanoparticles are shown in Figure 4. These maghemite nanoparticles were chosen because they present a good magnetic behavior. It means that these maghemite nanoparticles have residual magnetization close to zero, after removing the applied magnetic field. It is possible to notice that in the presence of the magnetic field the nanoparticles were aligned (see Figure 4(a)) and after removing the applied magnetic field, they lose the alignment (see Figure 4(b)).

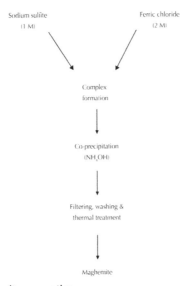

Figure 3. Scheme of maghemite preparation.

Figure 4. (a) Maghemite nanoparticles in the presence and (b) absence of the magnetic field.

The influence of temperature, stirring speed and time of reaction on the magnetic force and size of particles was evaluated using a three factor experimental design at two levels plus a central point (2^3 + 0). The lowest levels used for the stirring speed, the time of reaction and the annealing temperature were equal to 1,000 rpm, 15 min, and 150°C, respectively. The highest levels used for the previously mentioned factors were equal to 3,000 rpm, 45 min, and 250°C, respectively. The central point experiments were performed in triplicate, adjusting the factors at 2,000 rpm, 30 min, and 200°C, respectively.

The AFM was used for determination of the characteristic diameter of the produced maghemite particles. The AFM was performed in a DI Nanoscope IIIa microscope (LNLS, Brazil—AFM#8421/08 and AFM#9637/10), at non-contact mode, NSC-10-50, 20N.m^{-1} and 260 kHz.

The WAXS/SAXS measurements were performed using the beam line of the Brazilian Synchrotron Light Laboratory (LNLS, Brazil D11A-SAXS1-8507). This beam line is equipped with an asymmetrically cut and bent silicon (111) monochromator (λ = 1.743Å), which yields a horizontally focused X-ray beam. A linear position-sensitive X-ray detector (PSD) and a multichannel analyzer were used to determine the SAXS intensity I(q) as function of the modulus of the scattering vector q = ($4\pi/\lambda$) sin θ, 2θ being the scattering angle. All SAXS scattering patterns were corrected for the parasitic scattering intensity produced by the collimating slits, for the non-constant sensitivity of the PSD, for the time varying intensity of the direct synchrotron beam and for differences in sample thickness. Thus, the SAXS intensity was determined for all samples in the same arbitrary units, so that they can be directly compared to each other. Since the incident beam cross-section at the detection plane is small, no mathematical deconvolution of the experimental SAXS function was needed (Souza Jr. et al., 2009b, 2007).

The SAXS results, shown in Figure 5, indicate that all different maghemites, obtained of each experiment of experimental design, present similar dimensions around (31.3 ± 0.7) nm. It shows that size of nanoparticles do not depend on these parameters. This result is according to AFM results, shown in Figure 6. The probability density function (PDF) of the maghemite nanoparticles showed that, with 95% of confidence, the diameters of the particles ranges between 14.1 and 46.0 nm. In addition, the most probable observed value is equal to 27.9 nm.

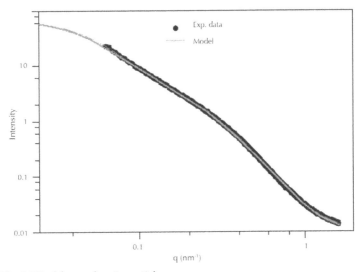

Figure 5. The SAXS of the maghemite particles.

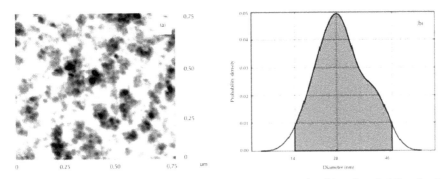

Figure 6. (a) The AFM micrograph of the maghemite nanoparticles (b) and probability density function (PDF) of diameters of maghemite.

Magnetic force was determined using a very simple experimental setup, constituted by an analytical balance and a digital caliper, shown in Figure 7. Two removable supports were used: the first one was placed on the plate of the balance, while the second and highest one was settled outside the balance. The distance between the lowest and highest tops was equal to 25.9 ± 0.2 mm. Tested materials were placed on the support located on the plate of the balance, as shown in Figure 7. Balance was unset and a neodymium N42 magnet—external Gauss strength and energy density equal to 13.2 kG and 42 MGOe, respectively - was placed on the highest support. Then, the apparent variation of mass was recorded and magnetic force calculation (opposite to gravitational one) was calculated according to

$$MF = \Delta m.g \qquad (1)$$

In Equation (1) MF is the magnetic force, Δm is the apparent variation of mass in the presence of the magnetic field and g is the acceleration of gravity.

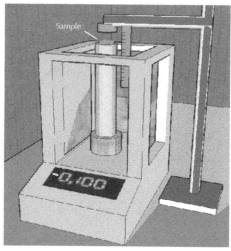

Figure 7. Experimental setup to determinate the magnetic force (Lopes et al., 2010; Marcelino, 2010).

The calculated effects of the factors over the magnetic force are shown in Table 1. This table also showed the standard deviation and confidence degree (p). The results that present $p < 0.05$ are statistically significant and they are marked in bold. As one can see, only time and temperature influence the magnetic force. How bigger this factors, higher the nanoparticles magnetic force. It can be easier observed in Figure 8, which presents the magnetic force as a temperature and time function. This Figure makes obvious that the biggest magnetic force is achieved when the time of preparation and the annealing temperature are equal to 60 min and 250°C, respectively.

Table 1. Calculated effects of the studied factors over the magnetic force of the maghemites.

Factor	Effect	Standard Deviation	p
Average	1.8253E+2	1.5016E+1	1.1990E-3
Curvature	−1.3612E+2	5.7509E+1	9.8775E-2
(1)Stirring Speed	8.6621E+1	3.0033E+1	6.3312E-2
(2)Time	1.0003E+2	3.0033E+1	4.4698E-2
(3)Temperature	2.1055E+2	3.0033E+1	5.9610E-3
1 by 2	-8.0325E+1	3.0033E+1	7.5398E-2
2 by 3	8.1612E+1	3.0033E+1	7.2708E-2

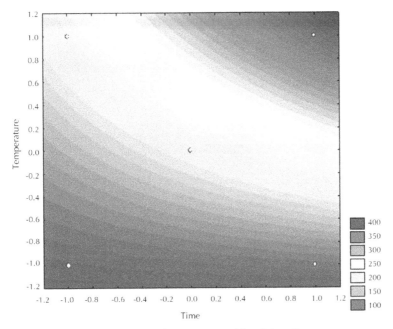

Figure 8. Magnetic force as a function of the temperature (T) and time (t).

Alkyd Resin Production

The polymeric resin was prepared using the castor oil obtained from castor beams. The castor oil possesses large amounts of triglycerides, Figure 9. These triglycerides react with an alcohol, in a basic medium, to produce biodiesel and glycerin, as shown in Figure 10.

(a)   (b)

Figure 9. (a) Castor oil from castor beams (behind) (Proquinor, 2010), (b) triglycerides structure.

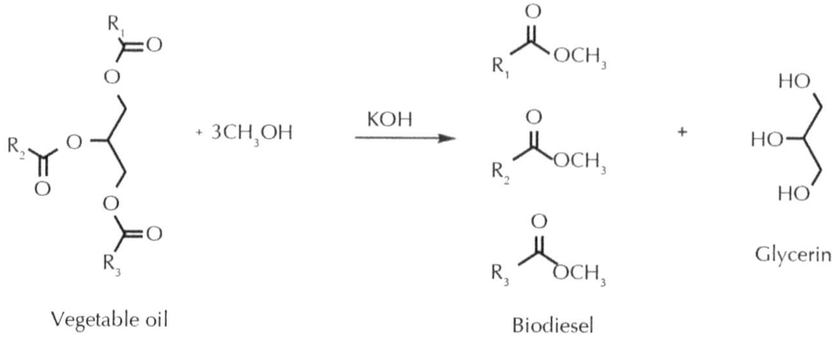

Figure 10. Biodiesel production reaction.

The resin was prepared using castor oil and phthalic anhydride (1:1 in moles). These reactants were poured into a three-necked flask under continuous stirring. Soon afterwards, the medium was acidified with 0.2 ml sulfuric acid to each gram of the castor oil. The system was heated until 230°C and kept at this temperature along 1 hr under constant nitrogen flow. The highly viscous organic medium obtained was removed from the flask using a spatula. The polymerization reaction is shown in Figure 11, while the Figure 12% an optical micrography of the alkyd resin obtained through this reaction.

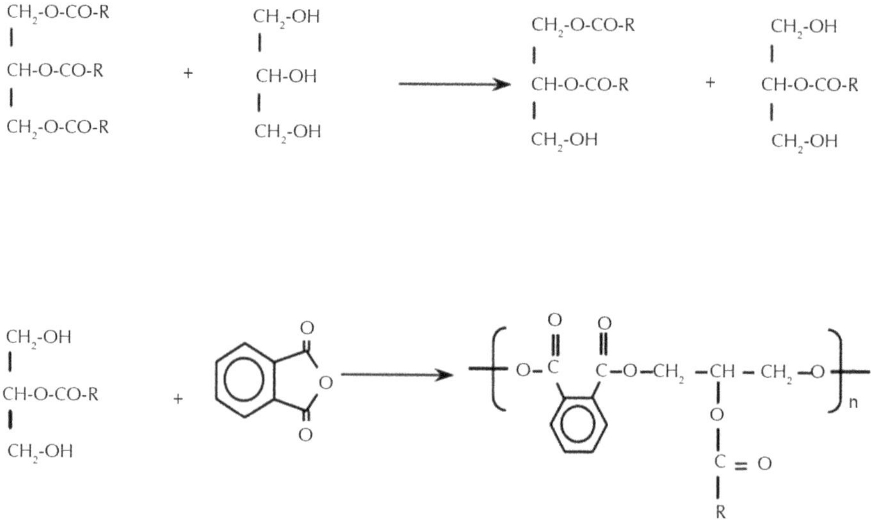

Figure 11. Alkyd resin polymerization reaction.

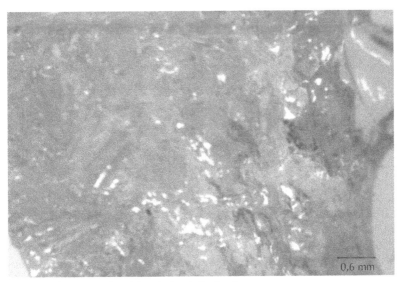

Figure 12. Optical micrography of the obtained alkyd resin.

The resin preparation was carried out using an experimental design, where the studied factors were the amount of the castor oil amount and the phthalic anhydride. The glycerin amount was always kept constant. This DoE was performed using a two factor experimental design at two levels plus a central point ($2^2 + 0$), yielding eleven experiments, where the lowest and highest levels were performed in duplicates and the central point in triplicate.

The obtained resins were analyzed by FTIR spectroscopy. The obtained spectra do not show significant differences among the resins and two of them are presented in Figure 13. The characteristic bands of resins are showed in the Table 2.

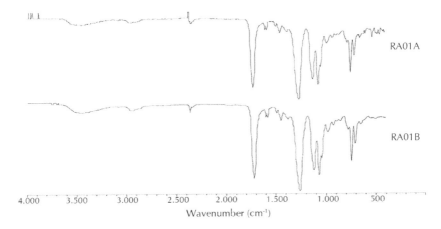

Figure 13. The FTIR spectra of alkyd resins.

Table 2. Characteristic bands of alkyd resins.

Region (cm⁻¹)	Phenomenon
1720	C=O axial deformation; carbonyl group
1600–1580	Aromatic ring–Axial deformation doublet
1266	C-O axial deformation
1070	Aromatic ring "in plane" deformation
741–705	Aromatic def. associated with the presence of polyester

All bands are typical of the alkyd resins (Souza Jr. et al., 2010b; Lopes et al., 2010). The spectrum of the pure resins present characteristic bands at 1720 cm⁻¹, related to the C=O axial deformation of the carbonyl group; at 1,600 and 1,580 cm⁻¹, related to axial deformation of the aromatic ring; at 1,266 cm⁻¹ related to the C-O axial deformation; at 1,070 cm⁻¹, related to the "in plane" deformation of the aromatic ring and at 741–705 cm⁻¹, related to the aromatic deformation associated with the presence of polyester (Souza Jr. et al., 2010b).

Composite Production

To prepare the composite, each one of the obtained alkyd resin was milled and fused. Afterwards, maghemite nanoparticles were mixed with the resin to obtain a composite containing 5 wt% of maghemite. This proceeding is schematically shown in Figure 14.

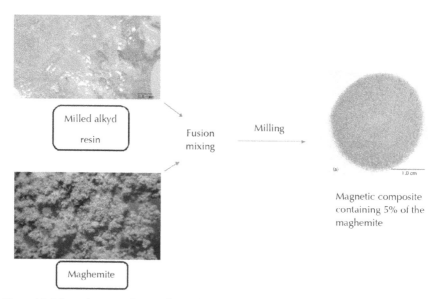

Figure 14. Schematic proceeding to obtain the magnetic composite containing 5 wt% of maghemite nanoparticles.

The obtained composites were analyzed by FTIR spectroscopy and their spectra were compared to pure alkyd resin ones, as shown in Figure 15. The characteristic bands of the composites are presented in Table 3.

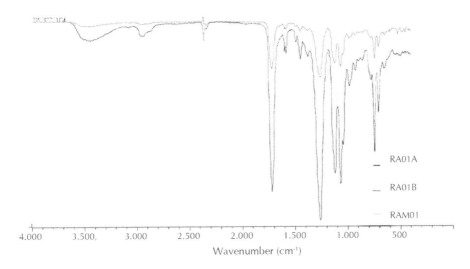

Figure 15. The FTIR spectra of composite (RAM01) and pure alkyd resin (RA01A and RA01B).

Table 3. Characteristic bands of composite and alkyd resin.

Region (cm⁻¹)	Phenomenon
3420	O-H stretch related to hydrogen bond
2900	C-H axial deformation of the C-H (CH₂ and CH₃ groups)
1727	C=O axial deformation.
1215	C-O axial deformation
1045	Axial deformation of the C-O-C bonding

The spectra of the biocomposites do not present any significant differences, when compared to the spectrum of the matrix, probably indicating the absence of significant chemical interactions between the matrix and the filler (Lopes et al., 2010).

The XRD results, shown in Figure 16, suggest that the mono-domains of the particles are not changed, since the crystallite size (calculated using the 311 plane @ 2θ = 37.2°) of the pure maghemite and of the particles present inside the resin, obtained using the Scherrer equation, are equal to (17 ± 2) nm and (21 ± 2) nm, respectively.

As previously said, each one of the resins obtained from the DoE was used to prepare the composites. Therefore, the effect of the studied factors—castor oil and phthalic anhydride amounts over magnetic force of the composites was evaluated. The Table 4 and the Figure 17 show that the increasing of castor oil amount increases the magnetic force. On the other hand, the same effect is caused by the decreasing of phthalic anhydride amount. In this case, as previously mentioned, the results that present $p < 0.05$ are marked in bold.

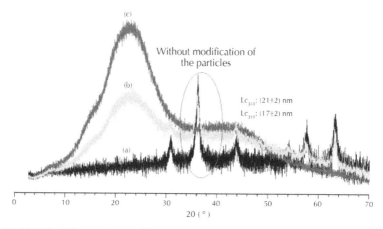

Figure 16. (a) XRD of the maghemite, (b) composite containing 5 wt% of maghemite, (c) and resin.

Table 4. Calculated effects of the studied factors over the magnetic force of the composites

Factors	Effect	Standard Error	P
Average	1.8286E+0	2.1838E-2	0.0000E+0
Curvature	-1.2414E-1	6.6717E-2	8.1268E-2
(1)Anhydride	-2.5334E-1	4.3677E-2	2.7000E-5
(2)Castor oil	5.2404E-1	4.3677E-2	0.0000E+0
1 by 2	6.0748E-2	4.3677E-2	1.8331E-1

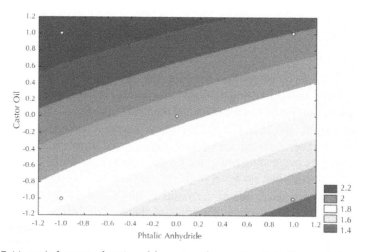

Figure 17. Magnetic force as a function of the castor oil amount and phtalic anhydride amount.

The magnetic tests were performed as described to maghemite nanoparticles. The Figure 17 makes clear that, among the tested conditions, the one which used the lowest amount of the phtalic anhydride and the highest amount of the castor oil produced the highest magnetic force. It is also important to notice that the composites containing 5 wt% of the maghemite presented a magnetic force equal to (2.2 ± 0.1) mN/g. This magnetic force is enough to the proposed cleanup process (Souza Jr. et al., 2010b).

The obtained magnetic force results seem to be related with the superstructure of the composite materials. The SAXS results of the composites showed that the condition of the best magnetic force (lowest level of the anhydride and highest level of the castor oil) presents superstructures composed by 23 particles of radius equal to eight nm, Figure 18(a). On the other hand, the composite prepared according the condition which yields the lowest magnetic force (using the highest level of the anhydride and the lowest level of the castor oil), as shown in Figure 18(b), presents no obvious spatial correlation. This result may be related to the presence of rigid polymer structures which make more difficult the formation of the superstructures, decreasing the magnetic force.

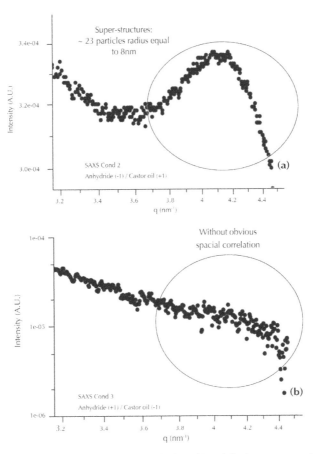

Figure 18. (a) The SAXS of the composites with the highest (b) and the lowest magnetic force among tested materials.

Oil Removal Capability of Composites

The oil removal tests were performed according to the analytical procedure established in our laboratory (Lopes et al., 2010; Souza Jr. et al., 2010b). In spite of different magnetic forces, all the composites were used in the cleanup processes. These tests were performed according to Figure 19. The first one shows the oil spilling, Figure 19(a). The second one shows the deposition of the nanocomposite on the oil, Figure 19(b). The third picture shows the magnet, Figure 19(c), and the fourth shows the removal of the petroleum, Figure 19(d).

Table 5 shows the oil amount removed by each one of the obtained nanocomposites. It is possible to notice that the increase of the oil mount on the water produces a larger amount of oil removed from the medium. Therefore, from this table, it is possible to infer that, in the best case, the 1 g of the composite is able to remove (18.03 ± 0.80) g of the petroleum from the water.

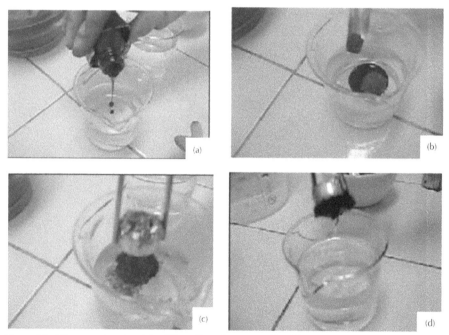

Figure 19. Oil removal testing (a) spilling oil on the water, (b) adding magnetic nanocomposite, (c) approaching the magneto, and (d) removing oil from the water.

Table 5. Oil removal capability with different petroleum amounts.

Composite	Removed mass using one gram of the resin (g/g)		
	Petroleum (8)	Petroleum (12)	Petroleum (16)
RAM1	9.3 ± 0.4	12.7 ± 0.1	16.9 ± 0.4
RAM2	8.6 ± 0.2	12.5 ± 0.2	16.7 ± 0.8
RAM3	9.0 ± 0.3	12.8 ± 0.0	17.0 ± 0.3

Table 5. *(Continued)*

Composite	Removed mass using one gram of the resin (g/g)		
	Petroleum (8)	**Petroleum (12)**	**Petroleum (16)**
RAM4	8.6 ± 0.2	13.2 ± 0.5	17.9 ± 0.4
RAM5A	9.0 ± 0.3	14.6 ± 0.6	17.3 ± 0.5
RAM5B	9.3 ± 0.7	13.0 ± 1.0	17.1 ± 0.5
RAM5C	9.7 ± 0.2	13.6 ± 1.1	18.0 ± 0.8

The effect of the amount of the oil, phthalic anhydride, and castor oil on the petroleum removal capability of the resins was evaluated using the DoE technique. The obtained results are shown in Table 6.

Table 6. Oil removal answers to the composites from different alkyd resin.

Factors	Effect	Standard Error	p
Average	1.3128E+1	7.6107E-2	0.0000E+0
(1) Petroleum(L)	8.1358E+0	1.8307E-1	0.0000E+0
Petroleum(Q)	−7.9000E-2	1.5854E-1	6.2018E-1
(2) Anhydride(L)	2.9614E-1	1.9773E-1	1.3974E-1
Anhydride.(Q)	4.7381E-1	1.5102E-1	2.6970E-3
(3) Castor oil (L)	6.2100E-2	1.9773E-1	7.5461E-1

The obtained results show that the phthalic anhydride amount presents a quadratic effect on the oil removal capability, while the petroleum amount presents a linear effect.

Figure 20 shows the oil removal capability as a function of the petroleum and anhydride amount. It is easier to observe that the increase of the anhydride amount leads to the increase of the oil removal capability possibly associated with a best aromatic/aliphatic balance. In addition, the increase of the petroleum amount leads to the increase of the oil removal capability.

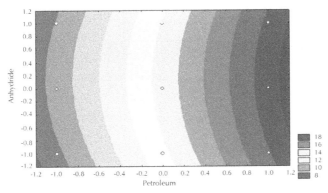

Figure 20. Oil removal capability as a function of the petroleum and phtalic anhydride amounts.

This information leads to one conclusion: the maximum oil removal capability was not reached yet. This result is very encouraging since several commercial products are able to remove only 5 g of the petroleum from the water (Souza Jr. et al., 2010b).

CONCLSION

A new magnetic and green nanomaterial was produced using castor oil. This material possesses a good magnetic force and a considerable oil removal capability, which allows its use in oil spill cleanup processes. Thus, the prepared material contributes to the environment encouraging nobler uses to some of the available renewable resources besides reducing the environmental anthropogenic impact on areas degraded by oil spill accidents, being, possibly, an important tool against environmental disasters related to the oil spill on the water.

KEYWORDS

- **Cashew nut shell liquid**
- **Magnetic force**
- **Nanocomposites**
- **Petroleum**
- **Probability density function**

ACKNOWLEDGMENT

The authors thank to CNPq, CAPES and FAPERJ for the financial support. The authors also thank to Brazilian Synchrotron Light Laboratory by the support on SAXS (D11A-SAXS1 #9077/10 and #9078/10) and AFM (AFM#8421/08 and AFM#9637/10) experiments.

Chapter 23

Magnetic Biofoams Based on Polyurethane Applied in Oil Spill Cleanup Processes

Oliveira, Geiza Esperandio, Souza Jr., Fernando Gomes, and Lopes, Magnolvaldo Carvalho

INTRODUCTION

Humanity is still very dependent on crude oil. In spite of the large efforts related to the reduction of the petroleum consumption, our society remains depending on this resource as an energetic source besides raw material for the production of several commodities. In addition, petroleum is, principally, carried through far distances, increasing the chances of accidents. These accidents involving oil spill produces devastating impacts on the environment, such as the ones shown in Figure 1, since the lipophilic hydrocarbons of the petroleum present a strong interaction with the similar tissue of the higher organisms. This assimilation produces intoxication and even death. Unfortunately, the traditional cleanup process, which involves the use of dispersants and surfactants (Bellamy et al., 1967; Bragg et al., 1994), often presents a larger impact to the environment than the spill. This larger environmental impact is related to the absorption of the dispersed oil by organisms, which leads to a greater time for the biorecovery of the degraded environment (Barry, 2007).

Figure 1. (a) Several impacts caused by spilled oil: Man removing oil spilled on the beach (Porto, 2010); (b) Aerial view of oil spill in the ocean (Bouza, 2010); (c) Bird covered in oil spilled on environmental accident (CPDA NEWS, 2010); and (d) dead fish on oil spill (Caetano, 2010).

Therefore, seeking to overcome the humanity addiction on crude oil, energy from alternative resources is being researched by scientific community around the world. Among them, the biodiesel production is very promising and it is encouraged by the Brazilian Government. The Brazil possesses many sources to produce biodiesel, such as those shown in Figure 2. Nowadays, all diesel from petroleum produced in Brazil receive 4% of biodiesel, and the resulting mixing is named B4. Furthermore, there is the prediction, that diesel will receive 5% of biodiesel (B5) in 2013.

Figure 2. Several Brazilian's resources to the biodiesel production (Flexor, 2010).

However, the biodiesel production yields large amounts of glycerin as byproduct, as one can see in Figure 3. Therefore, innovative uses for this new glycerin source must be researched, avoiding the collapse of the animal glycerin chain. Among them, the production of resins is interesting due to the likeness between these polymers and the petroleum, promoting these resins as potential spill cleanup agents. Therefore, the polymers can be engineered to produce chemical and physical properties which make them useful tools to the environment recovering process. Among these materials there are the ones from petrochemical sources, such as poly(divinylbenzene) and poly(methyl methacrylate) (Queirós et al., 2006) and other ones from natural resources, such as the chitin, the chitosan, and its derivates (Carvalho et al., 2006). In addition, the environmental recovery goal can be more easily reached through changes in the aliphatic/aromatic character of the used polymer materials, allied to a magnetic induced behavior, making possible to create new magnetic materials able to remove petroleum from the water.

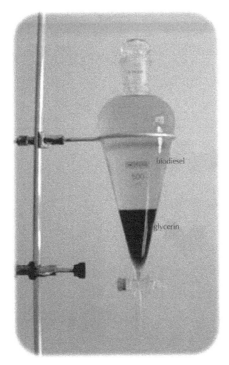

Figure 3. Produced biodiesel and glycerin in laboratory scale (Deboni, 2010).

In this specific context, our group has dedicated great efforts to obtain polymer materials completely or partially "green" from renewable resources such as the cashew nut shell liquid (CNSL); the glycerol, byproduct of the production of biodiesel by the transesterification reaction; and the lignin removed by the Kraft process. These polymers are useful in the environmental recovery which involves the oil spill clean-up process. In addition, seeking for the raising of the clean up efficiency, magnetic nanoparticles are incorporated to these polymers, allowing the use of electromagnetic devices to catch the composite impregnated with petroleum (Lopes et al., 2010; Souza Jr. et al., 2010 a, b).

MAGNETIC BIOFOAMS

Biofoams Production

Magnetic biofoams can be prepared using different contents of water and toluene diisocyanate. In our experiments, we used the compositions shown in Table 1. The optical micrographs of the produced biofoams are shown in Figure 4. It can be noticed that the sample PU-1 has large pores connected, which presents a most open aspect. Moreover, it is possible to infer that the sample PU-2 also has large pores, but one can see that these pores are smaller and posses thicker walls. The increase of the pore walls becomes more visible as more water is used, as observed in the sample PU-3.

Table 1. Composition of the biofoams.

Sample	Castor Oil (mL)	Toluene Diisocyanate (mL)	Water (mL)
PU-1	2.0	2.0	0.25
PU-2	2.0	2.0	0.35
PU-3	2.0	2.0	0.45
PU-4	2.0	1.0	0.25

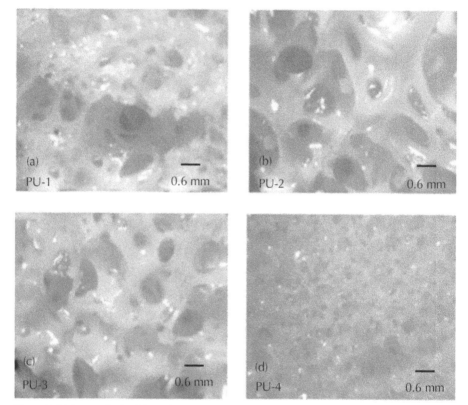

Figure 4. Biofoams obtained from the compositions shown in Table 1: (a) PU-1, (b) PU-2, (c) PU-3, (d) PU-4.

Nanoparticles Production

Nanoparticles can be produced from several materials. Among them, iron oxide has shown great versatility. To obtain magnetic biofoams, it was used maghemite nanoparticles, which is a kind of iron oxide nanoparticles. These nanoparticles were prepared through the homogeneous precipitation technique. In a typical procedure, the ferric chloride solution and the deionized water were added into a beaker under continuous agitation. Soon afterwards, the sodium sulfite solution was added to the beaker, still under continuous stirring. The reaction product was precipitated by the slow addition of concentrated ammonium hydroxide into the beaker under continuous agitation. After 30 min, the medium was filtrated and the obtained particles were washed several times with water and finally dried at 60°C in an oven. Magnetite was converted into maghemite through annealing at 200°C along one hour. The maghemite nanoparticles are shown in Figure 5. These maghemite nanoparticles were chosen because they present a good magnetic behavior. It means that, after removing the applied magnetic field, these maghemite nanoparticles have residual magnetization close to zero. It is possible to notice that in presence of the magnetic field the nanoparticles were aligned (Figure 5a) and after removing the applied magnetic field, they lose their alignment.

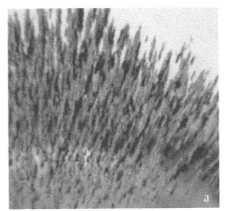 

Figure 5. (a) Maghemite nanoparticles in the presence and (b) absence of the magnetic field.

Magnetic Biofoam Production and Characterization

The opened porous volumes of the prepared resins are presented in Table 2. Generally speaking, it can be seen that the increase of water amount cause a weakening of the walls, evidenced by porosity decrease from PU-1 to PU-4. This is due to fewer available diisocyanate to react with the hydroxyl groups of castor oil. It caused the collapse of several cells, leading to the decrease of the internal volume. Among the prepared bioresins, the values presented in Table 2 show that PU-1 was the most porous material ($2.6 \pm 0,2$ cm^3 of opened pores volume). It means that this material has the thinnest walls, been perfect to prepare particulated material useful to the environmental recovered processes. So, the PU-1 foam was chosen to prepare the composite containing the maghemite nanoparticules, sample PU-5. This magnetic biofoam was prepared through fusion and mixture of polyurethane resin and maghemite nanoparticles, generating a material containing 5wt% of the maghemite, shown in Figure 6. It can be noticed that PU-5 sample presented pores smaller than the ones presented by the PU-1 one. The nanoparticules presence generated a decrease in the open porous volume. It is related to an increase of the thickness of the walls. However, the reddish composite remains easily processable.

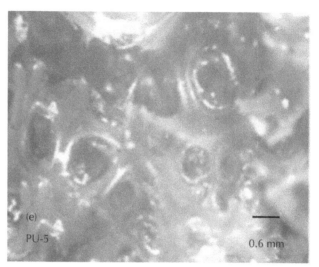

Figure 6. Magnetic biofoam contained 5%wt of maghemite nanoparticules.

Table 2. Opened porous volumes of the obtained biofoams (PU-1 to PU-4) and biocomposite (PU-5.)

Sample	Volume (cm³)
PU-1	2.6 ± 0.2
PU-2	2.0 ± 0.2
PU-3	1.2 ± 0.1
PU-4	0.7 ± 0.1
PU-5	2.0 ± 0.2

The FTIR spectra of the prepared materials can be seen in Figure 7. It can be noticed that all spectra are similar, indicating that the biocomposites are chemically comparable, independent on water and toluene diisocyanate amounts. Moreover, the biofoam PU-4 has more intense characteristic bands.

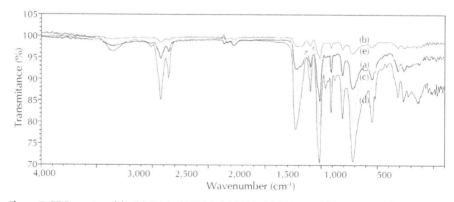

Figure 7. FTIR spectra of the (a) PU-1, (b) PU-2, (c) PU-3, (d) PU-4, and (e) PU-5 samples.

The characteristic bands of the biofoams are the axial deformation related to hydrogen bonds of NH group, at 3,300cm^{-1}. The doublet, around 2,900 cm^{-1}, corresponds to the axial deformation of C-H bond of the CH_2 and CH_3 groups. The intense characteristic band that appears at 1,727cm^{-1}, is correlated to the axial deformation of the C=O bond conjugated with the amide I band. Characteristic bands at 1,600 and 1,512 cm^{-1} are due to angular deformations of N-H bonds, named amide II band. The axial deformation of the C-N bond appears at 1,411cm^{-1}, while, the angular deformation of the N-H bond is at 1,300 cm^{-1}. The characteristic band at 1,215 cm^{-1} corresponds to axial deformation of the (C=O)-O group. Finally, the characteristic band, around 700 cm^{-1}, is related to angular deformation out-of-plane of the N-H bond.

The diffractograms of the maghemite nanoparticles, bioresins and nanocomposite samples, presented in Figure 8, show that maghemite nanoparticles were not changed by the bioresins. The crystallinity values are shown in Table 3 and they were calculated using the Ruland method. (Souza Jr., F. G. et al., 2010b) It can be noticed that crystallinity of pure foams is always low, independent of their preparing method. In addition, these values are, statically, equals. It can be seen that the bioresins presented crystallinity values around 10%, while, the maghemite nanoparticles presented 70% of crystallinity. On its turn, the biocomposite showed an intermediate crystallinity around 20%.

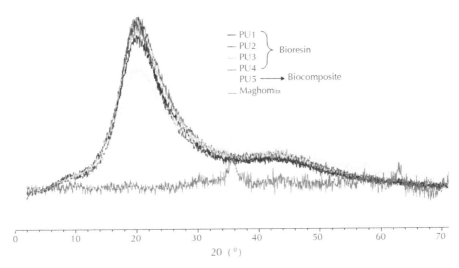

Figure 8. Diffractograms of the maghemite nanoparticles, bioresins (PU1-PU4) and nanocomposite (PU5) samples.

Table 3. Crystallinity degree of the maghemite nanoparticles, bioresins (PU1-PU4), and nanocomposite (PU5) samples.

Sample	Crystallinity (%)
PU-1	9 ± 2
PU-2	12 ± 1
PU-3	13 ± 1
PU-4	12 ± 4
PU-5	21 ± 3
Maghemite	70 ± 1

Figure 9 shows the SAXS analysis of the maghemite nanoparticles inside the bio-foam matrix. The Figure detail shows the Gaussian deconvolution of the SAXS data. This Gaussian deconvolution allowed several calculations, of which main results are shown in Table 4. These SAXS results showed two main kinds of scattering centers. These centers are related to superstructures, which present diameters equal to 60 and 38 nm, spaced from each other by 23 and 15 nm, respectively.

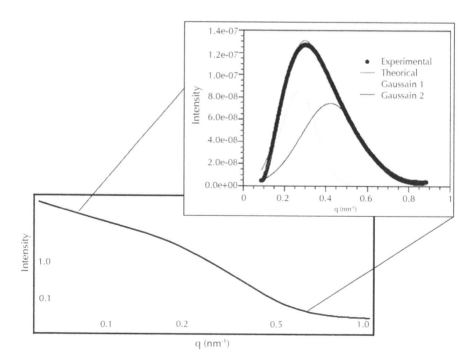

Figure 9. SAXS analysis of the maghemite nanoparticles inside the biofoam matrix (the detail of the figure show the Gaussian deconvolution).

Table 4. Main results obtained from SAXS analysis.

Peak	Center (nm⁻¹)	FWHM (nm⁻¹)	ds (nm)	Lc (nm)
1	0.27	0.21	23.3±0.2	59.8±0.6
2	0.42	0.33	15.0±0.1	38.1±0.4

MAGNETIC FORCE OF MAGNETIC BIOFOAMS

The magnetic force of composite material was measured by weighting in two distinct moments: in the absence of the magnetic field and in the presence of the magnetic field. In both cases it is used a known distance between the magnet and the sample. There is an exponential increase in the magnetic force when the magnet gets close to the maghemite and magnetic biofoam samples. The empiric model to describe this behavior is

$$MF = F_\infty + a_1 . e^{da_2}$$ Eq.1

In the Eq. 1, MF is the magnetic force; F_∞ is the force exercised by the magneto over the sample with the distance tending to infinite; a_1 is the amplitude, a_2 is the decay constant and finally d is the distance between the magneto center and the sample.

Using this model is possible to determinate the previously mentioned parameters and the correlation between the experimental data and the model, shown in Table 5. This model allows extrapolating the magnetic force to obtain its value when the distance between the magneto and the samples is zero, this value is named initial magnetic force or MF_0, which corresponds to the sum of F_∞ and a_1. The values found to MF_0 for the magnetic biofoam and the maghemite are equal to $(2.0 \pm 0.4) \times 10^{-2}$ and $(9 \pm 8) \times 10^{-3}$ N, respectively. It indicates that the maghemite nanoparticles, dispersed into the polymer matrix, are attracted by magneto with a force around 110% higher, in comparison with the pure maghemite nanoparticles, shown in Figure 10. It is possibly due to high dispersion degree that nanoparticles own in the polymer matrix, which allows the increasing of the attraction capability. These results are in agreement with the SAXS ones.

Table 5. Parameter values and correlations of the model.

Parameter	PU-5	Maghemite
F_∞	$(3.4 \pm 0.4) \times 10^{-5}$	$(3 \pm 2) \times 10^{-6}$
a_1	$(2.0 \pm 0.4) \times 10^{-2}$	$(9 \pm 8) \times 10^{-3}$
a_2	$(7.6 \pm 0.3) \times 10^{-1}$	$(1.2 \pm 0.3) \times 10^{0}$
R^2	0.9997	0.9959

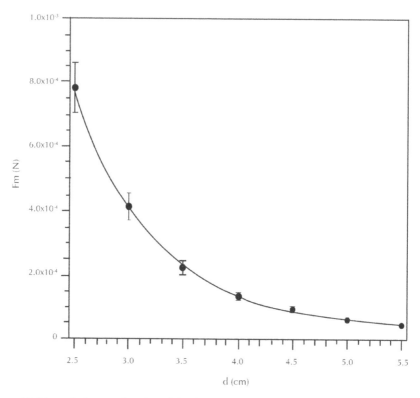

Figure 10. Magnetic force as function of the distance between the magnet and the sample to the magnetic biofoam (PU-5).

OIL REMOVAL CAPABILITY OF THE MAGNETIC BIOFOAM

As proposed, the magnetic biofoam can be used to remove spilled oil over the water. The obtained results are shown in Figure 11, which allowed performing the linear regression of the data. The results are shown in the empirical model described by the Eq. 2.

$$RO = -(1.81 \pm 2.19) \times 10^{3} + (4.1 \pm 0.1) \times 10^{0} \times MB \qquad \text{Eq.2}$$

In the Eq. 2, RO is the removed oil amount and MB is the amount of the magnetic biofoam used in the test.

The results shown in Figure 11 and in the Eq. 2 allow to infer that, inside the tested interval, each gram of the magnetic biofoam is able to remove (4.1 ± 0.1)g of the petroleum from the water. Thus, a small portion of maghemite inside polyurethane biofoam produces a strong magnetic attraction, enough to remove, completely, the mass of biofoam soaked by petroleum. Therefore, these materials may be used in environmental recovery processes related to the oil spill on the water, besides providing nobler uses to some of the renewable resources available. It produces a new perspective to environmental recovery of degraded aquatic environments, besides it presents a new alternative to use the castor oil.

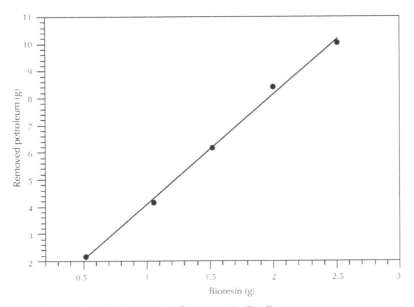

Figure 11. Oil removal capability test using biocomposite (PU-5).

CONCLUSION

A new magnetic and more ecologic material was obtained using castor oil and maghemite nanoparticles. This material joins magnetic properties of the maghemite with low density characteristic of the foams. This specific polyurethane foam, obtained from castor oil and toluene diisocyanate, shows a good oil removal capability. It makes possible to use this material for cleaning up spilled oil in aquatic environments.

ACKNOWLEDGMENTS

The authors thank to CNPq, CAPES and FAPERJ for the financial support. The authors also thank to LNLS by the support on SAXS experiments (D11A-SAXS1 #7086/08, #8519/09, #8507/09, #9077/10 and #9078/10).

KEYWORDS

- **Biofoams**
- **Lipophilic hydrocarbons**
- **Maghemite nanoparticles**
- **Toluene diisocyanate**
- **Transesterification reaction**

Chapter 24

Self-assembly of Lipid A-phosphates: Supramolecular Liquid Crystals

Henrich H. Paradies, Chester A. Faunce, Hendrik Reichelt, and Kurt Zimmermann

INTRODUCTION

Self-assembly takes place through both a reversible non-covalent interaction and a unique recognition involving a variety of building blocks. The components involved in the formation of two-dimensional (2-D) and three-dimensional (3-D) nano-or μm scaled periodicities such a cubic, cylindrical, or mesophases are van der Waals and electrostatic forces, hydrogen bonding as well as π-π stacking forces (Gazit, 2010). The various observed phases have been investigated extensively for example surfactant water (Seddon, 1996), block copolymers (Matsen and Bates, 1996), and thermotropic materials (Tschierske, 2001). Lyotropic materials can provide templates for porous inorganic materials (Attard et al., 1997) and lyotropics such as double chained N-cationic lipids (Clancy and Paradies, 1997; Clancy et al., 1995). These materials are also capable of forming complexes with DNA and neutral lipids as well as being useful as carriers in gene therapy and other pharmaceutical formulations (Koltover et al., 1998; Paradies, 1992; Thies et al., 1996). Other important materials in self-assembly are dendrons (Rosen et al., 2009) and nanostructural soft matter which show a tapered shape and account for the formation of bulk phases such as lyotropic materials (Bates and Fredrickson, 1999) or block copolymers (Anderson et al., 1988). Recently, Ungar et al. (2003) showed that a self-organized supramolecular dendrimer nanostructure possessed a noncubic phase and established a relationship between the chemical structure and the self-assembly composed of tapered dendrons. These materials reveal σ and BCC phases with an increase in temperature. Subsequently, dodecahedral quasicrystals were formed, displaying wedge-shaped dendrimer molecules that self-assembled into virtually spherical particles. It was established that each spherical particles from the branched compounds contained on average 11.6 molecules, on a tetragonal lattice with 30 particles per unit cell. Similar hierarchical structures, though not identical, have been found for colloidal crystals of lipid A-phosphates from the bacterial *E. coli* source (Faunce and Paradies, 2010). Understanding the characteristics of jammed lipid A-phosphate packings provides basic insights into the structural arrangements of lipid A-phosphate liquid crystals and bulk properties of these supramolecular liquid crystals, glasses, and selected aspects of their biological actions. This contribution also covers recent advances in understanding jammed packings of polydisperse sphere mixtures, non-spherical particles for example ellipsoids within the assembly and polyhedra for example by "*E.coli Autovaccines*" (Zimmermann et al., 2003)

The most conserved component in lipopolysaccharides (LPS) from gram-negative bacteria is lipid A. Lipid A is linked to a core of oligopolysaccharides (Raetz and Whitefield, 2002), and in its di-phosphorylated form consists of a β-1,6-linked D-glucosamine disaccharide carrying six saturated fatty-acid residues and two negatively charged phosphates at the reducing and non-reducing end of the glucosamine (Figure 1). Approximately 1,00,000 patients die in the United States every year from infections acquired while in the hospital or nursing homes due to the presence of LPS from gram-negative bacteria. The lipophilic part of LPS, lipid A-diphosphate is associated with lethal endotoxicity, pyrogenicity, specific immune response. It is also responsible for triggering a cascade of cellular mediators, for example tumor necrosis factor (TNF) α, interleukins, leukotrienes, thromboxane A2 from monocytes and macrophages. A detailed insight into the spatial-structure and the packing of toxic and non-toxic lipid A phosphates will aid in the understanding of the following: self-assembly, physical interactions with gram-negative bacteria, biofilm associations, LPS components, antibiotic resistance, membrane components, cationic antimicrobial (CAM) peptides, membrane fusion, and divalent cations, particularly Mg^{2+} and Ca^{2+} ions (Ernst et al., 1999; Faunce et al., 2005). From specular neutron scattering studies it is observed that these divalent counterions modulate the mechanical properties of interacting LPS membranes (Schneck et al., 2009).

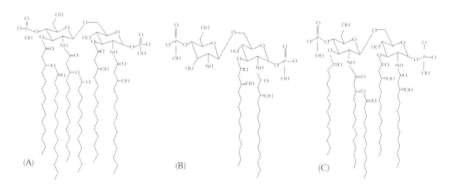

Figure 1. Chemical structures of lipid A-diphosphate (**A**) and two antagonistic lipid A-diphosphate molecules, (B and C). Lipid A-diphosphate from *E. coli* is a 1,4-di-phosphorylated β-1,6-linked D-glucosamine disaccharide with four residues of amide-and esterified R-(-)-3-hydroxy fatty acids (∗ denotes the chiral centers in the hydroxy fatty-acid esters, apart form the chiral and epimeric carbons in the disaccharide moieties which are not marked). The antagonistic lipid A-diphosphate molecules shown in (B) and (C) contain the same disaccharide as in (A); however, they differ in the number anchored carbohydrate positions and the number of chiral fatty-acid chains but the chain lengths is the same (C_{14}). The corresponding monophosphate of lipid A is only phosphorylated at the reducing end of the disaccharide.

Until recently the self-organization of charged particles, like lipid A-diphosphate, its analogues and their effects on complex fluids received little attention. In addition, charged particle dispersions in nano-size the regime such as lipid A-diphosphate can influence the stability of the system. This is a result of a segregation of domains near large uncharged colloidal particles, especially in systems with charged small spheres or large asymmetry in shape and/or charge.

The present contribution is concerned with nm scaled ordered assemblies of a unique composition of nontoxic lipid A-phosphate analogs which can be obtained from non-pathogenic and non-replicating Gram-negative bacteria. The lipid A-phosphate analogs are acquired according to a strongly defined protocol under non-denaturing conditions, by applying purification steps using preparative HPLC. The structural characterization is carried out using a MALDI-TOF mass spectroscopy, NMR and FT-IR. Where the starting materials consist of inactivated, non-replicating gram-negative commensal bacteria for example Enterobacteriaceae such as *E. coli* grown and harvested *in vitro* or in reactors for up-scaled production of the desired and specified material. These preparations reveal strong attenuating immune-regulating responses on the interleukin,

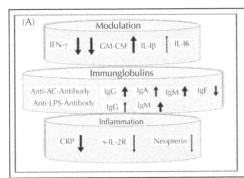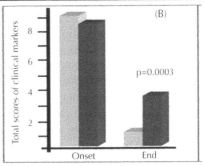

Figure 2. (A) Changes in determined clinical parameters during the administration of "*E. coli autovaccines*" to patients (n = 78) suffering from chronic sinusitis over a period of time of between 4 and 6 weeks (Rusch et al., 2001; Zimmermann et al., 2003). The various levels of the stimulation of the cellular and humoral immunological defense are shown: IFN-γ: interferon-γ; GM-CSF: granulocyte-macrophage colony factor; IL-6: interleukin-6; IL-1β: interleukin-1β; CRP: C-reactive protein; s-IL-2R: soluble interleukin-2 receptor; IgX: the various immunoglobulins, (B) Randomized double-blind and placebo controlled clinical study encompassing 114 patients suffering from chronic sinusitis before and after treatment with "*E. coli autovaccines*" (p = 0.0003).

Table 1. Endotoxicity and TNF induction in murine bone marrow macrophage cultures of various lipid A-phosphates. Results obtained according to the protocol of Sayers et al. (1987).

Source	Concentration µg/mL	TNF units (-) interferon-γ	TNF units (+) interferon-γ (100 units)
Lipid A-diphosphate	1.0	$1,700 \pm 20$	$3,900 \pm 35$
+ compound B	0.1	850 ± 20	$2,750 \pm 30$
+compound C	0.01	120 ± 25	$1,300 \pm 20$
+compound B & C	0.01	10 ± 5	900 ± 20
Lipid-4 phosphate	10.0	350 ± 22	890 ± 22
+Compound B	1.0	100 ± 20	310 ± 20
+Compound C	10.0	500 ± 22	790 ± 20
+Compound B & C	1.0	50 ± 9	100 ± 18

Table 1. *(Continued)*

Source	Concentration µg/mL	TNF units (-) interferon-γ	TNF units (+) interferon-γ (100 units)
Mixture compounds B & C	100	4.0 ± 0.5	6.0 ± 0.8
	50	1.0 ± 0.3	2.0 ± 0.4
	10	0.4 ± 0.5	7.0 ± 0.5
	5	0.2 ± 0.6	4.0 ± 0.5
	1	< 0.2	1.0 ± 0.5
LPS, from *E. coli* positive control	0.1	850 ± 20	3,100 ± 28
	0.01	140 ± 20	1,000 ± 21
	0.001	8 ± 2	100 ± 20
E.coli autovaccine	0.01	≤ 0.4	≤ 3.0
	0.001	< 0.2	< 3.0
	0.0001	< 0.2	< 2.0

Antibody and on cellular levels in humans (Figure 2) because of their chemical entities and unique supramolecular assemblies on the µm and nm scale. The different lipid A-diphosphates molecules which are associated with endotoxicity and pyrogenicity and the corresponding nontoxic lipid A-phosphate molecules, to which the lipid A-monophosphate belongs, were investigated as models for potential "*E. coli auto-vaccines*" (Zimmermann et al., 2003). The activity of the auto-vaccines modulates the innate immunity (Wehkamp et al., 2007), environmental changes in bacteria populations, changes in nutrition and chronic inflammation (Zimmermann et al., 2003). *Note:* The fecal microbial composition of an irritable bowel syndrome patient differs significantly from that of a healthy subject.

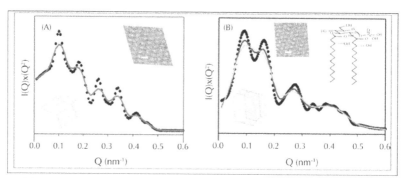

Figure 3. (A) SAXS profiles I(Q) versus Q, with Q = (4π/λ)×sinθ/2), of BCC (*Im3m*) type colloidal crystals (λ = 1.54 nm) with a = 37.6 nm. The black dotted scattering profile is for the lipid A-diphosphate phase at φ = 3.5 x 10⁻⁴, I = 0.5 mM NaCl, the solid-redline is the profile for an equimolar mixture of the antagonistic molecule depicted in (B) of Figure 1 for lipid A-diphosphate with a = 35.5 nm, (B) SAXS profiles of colloidal crystals of the FCC type (*Fd3m*), the black-dotted line corresponds to lipid A-diphosphate with a = 57.5 nm. The red-solid line is for the colloidal mixture of lipid A-diphosphate and antagonistic lipid A-diphosphate (Figure 1(B)) both are at φ = 5.4 x 10⁻⁴, I = 0.5 mM NaCl. The green solid SAXS profile represents the results from a mixture of lipid A-diphosphate with the corresponding monosaccharide of lipid A-diphosphate (inset) with two fatty-acid chains but at φ = 3.4 x 10⁻⁴, I = 0.5 mM NaCl. *Insets:* Crystal morphologies and corresponding TEM images.

Colloidal clusters from dispersions with low polydispersity in shape, size and charge of lipid A-diphosphate and antagonists of lipid A-diphosphate were investigated using small-angle X-ray scattering, light scattering (LS), and electron microscopy. These aqueous dispersion and colloidal crystalline phases were tailored with different chain lengths (C_{12-16}) with even or odd numbers of covalently attached chiral (R-[-])-fatty-acid chains, or the number of chains (Figure 1(A)). The clusters were successfully prepared by self-assembly at very low ionic strength (I $\cong$ 10^{-4} M, at 25°C) and physically analyzed and studied for their clinical efficacy. The colloidal crystalline phases of lipid A-diphosphate were identified by the presence of resolution-limited Bragg peaks that were indexed according to their structures which were as follows: a BCC (*Im3m*) lattice (a = 37.5 ± 2.0 nm) at ϕ = 3.5 × 10^{-4}, an FCC (*Fd3m*) lattice (a = 57.5 ± 2.5 nm) at ϕ = 5.0 × 10^{-4} with I = 5.0 mM NaCl and a second FCC (*Fm3m*) lattice (a = 55.0 ± 2.5 nm) at ϕ = 2.5–4.0 × 10^{-4} with 1.0–10.0 μM HCl (Figure 3).

The peaks observed, for lipid A-diphosphate dispersion (ϕ = 3.5 × 10^{-4}), which were assigned a BCC lattice ($h^2 + k^2 + l^2$: 2, 4, 6, 8 and 10, and $h + k + l = 2n$) with a = 37.6 nm. A possible space group was *Im3m* (Figure 2(A)). The molecular weight calculations for this phase of lipid A-diphosphate were made using the (110) diffraction planes, $d = d_{110} = a/(2)^{1/2}$. For this BCC lattice type, the estimated molecular weight was approximately 10.1 × 10^6 g/mol (formula monomer molecular weight $\cong$ 1,950 g/mol). Under these experimental conditions the observed cubic phases were all non-lamellar and none belonged to the space group *Pn3m*, with a spacing ratio of 1, 1:$\sqrt{2}$, 1:3; 1:$\sqrt{4}$, 1:6, and so on, or to the *Pm3n* structure with diffraction lines of the following ratios: $\sqrt{2}$, $\sqrt{4}$, $\sqrt{5}$, $\sqrt{6}$, $\sqrt{8}$ and $\sqrt{10}$, and so on. Both the *Im3m* and *Pm3n* structures were possibilities and the overall structure consisted of disconnected clusters of lipid A-diphosphate separated by a continuous film of water. The body-centered cubic phase was generally present at higher water contents and low ionic strengths. The structure could be modeled by the packing of clusters of identical rigid spheres which were embedded in water. The closest packing of the spheres took place in the array and the radius of the closely packed rigid sphere was $R_{sp} = (\sqrt{3}/4 \times a) = 16.4$ nm. In contrast to the *Im3m* structure, the phase represented by the space group *Pm3n* was thought to contain two types of clusters.

Interestingly, when employing colloidal dispersions prepared from lipid A-diphosphate and the antagonistic molecules from either 1(B) or 1(C), using the same protocol and at low ionic strength (I $\cong$ 10^{-4} M), like results were obtained. For low volume fractions of lipid A-diphosphate (Figure 3(A)), very similar SAXS-diffraction spectra, high resolution transmission images (TEM) and LS profiles were observed (Figure 4). These assemblies were consistent with a colloidal assembly for a BCC lattice (*Im3m*) with a = 35.5 nm. However, a mixture of equimolar concentrations of the two antagonistic molecules 1(B) and 1(C), revealed a SAXS-powder diffraction pattern and a LS profile that could be indexed for a much larger face-centered (*Fd3m*) unit cell, with a = 58.0 nm. For dilute solutions the LS profile conformed to a hydrodynamic sphere of radius of R = 34.7 nm, similar values were found for lipid A-diphosphate dispersions at high volume fraction ϕ = 5.4 × 10^{-4}. An investigation of another aqueous preparation which contained lipid A-diphosphate and each of the antagonistic molecules B and C in Figure 1 was carried out at volume fractions 2.5 £ ϕ ≤ 3.4 × 10^{-4}. Here, a primitive cubic lattice was revealed with $a = 26.8$ nm, and the space group *Pm3n*. A different colloidal crystals phase was found for the volume-fraction range $4.0 \times 10^{-4} \leq \phi \leq 6.5 \times 10^{-4}$ and successfully indexed as an FCC structure with $a = 39.6$ nm (*Fd3m*), different from one found previous (Figure 3(B) and (C)). According to chemical analyses, colloidal dispersions and the colloidal crystals contained added molecules; however, in different molar ratios as identified by MALDI-TOF-MS, and ion-spray-MS. The dispersions containing these molecules shown in Figure 1 (B) and (C) tested in an *in vitro* assay were the most potent ones. A significant reduction in endotoxicity and the induction of TNF was observed in the presence or absence of γ-interferon in the murine bone marrow macrophage assay (Table 1). These molecules also acted strongly on the main inflammatory cytokines.

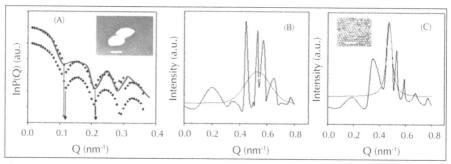

Figure 4. (A) Light scattering (LS) profiles of intensity I(Q) versus. Q ((λ = 637.8 nm) for the liquid phase of the aqueous lipid A-diphosphate dispersions (dark crosses). The crosses with the red solid line represent the scattering profile of an equimolar mixture of lipid A-diphosphate and the antagonistic molecule depicted in Figure 1(B) as well as lipid A-diphosphate. The two dispersions are at φ = 3.5 x 10⁻⁴, I = 0.5 mM NaCl. The profiles conform to a polydispersity of σ = 5.2% and R = 35 nm. The arrows show the positions of the first, second, and third minima in the RGD approximation. *Inset:* Colloidal crystals observed on SEM images, bar = 1 μm, (B) Experimental X-ray powder diffraction profiles of the *Pm3n* cubic phase with a = 26.5 nm (see text); (³⁄₄) corresponds to the effective colloidal structure factor, $S_{eff}(Q)$, calculated from the scattered light intensities revealing a strong (200) peak. The ionic strength was 0.154 M NaCl (0.9 w/w%), and (**C**) An experimental X-ray powder diffraction profiles of the FCC phase with a = 39.6 nm with a possibly space group of *Fd3m*. The solid-red line (³⁄₄) represents $S_{eff}(Q)$ the (222) peak at of 0.154 M NaCl. *Inset:* A TEM image of the assembly and bar = 10 nm.

Sharp peaks were recorded from SAXS experiments on lipid A-diphosphate at a volume fraction $\phi = 5.0 \times 10^{-4}$ with an ionic strength $I = 0.5$ mM NaCl, indicating the presence of long-range order (Figure 2(B)).

It was possible to assign virtually all of the peaks to the reflections of an FCC lattice ($h^2 + k^2 + l^2$: 3, 8, 11, 12, and 16) as indicated in Figure 2, with $a = 57.5$ nm, corresponding to the space group *Fd3m*. The systematically absent reflections were the $h + k, k + l$ and $h + l \neq 2n$, for the general reflections (*hkl*) and $k + l \neq 4n$ for the *(0kl)* zone. The absence of these reflections reinforced the argument that the lattice type was face-centered cubic, therefore, two space groups were possible, namely *Fd3* and *Fd3m*. The two space groups were also centrosymmetric and belong to the Laue classes *m3* and *m3m*. Note: *Fd3m* corresponds to special positions of *Fd3*.The observed and calculated *hkl* sets of Bragg reflections were consistent with a combination of different sites in the *Pm3n* (*a* and *d*), or of sites *a* and *d* in the *Fd3m* (Figure 5). This could be attributed to two different space-filling packings: (i) two dodecahedra on site *a* and six tetrakaidecahedra on site *d*, forming a *Pm3n* lattice; (ii) or sixteen dodecahedra on site *d* and eight hexadecahedra on site *a*, forming an *Fd3m* lattice. The space-filling network of lipid A-diphosphate consisting of slightly distorted polyhedra was similar to known basic frameworks of gas-hydrates and sodium silicon sodalites. The final geometry of the "spheres" that is whether they were rounded or faceted in shapes, was established. The equilibrium separation distance between the interfaces of the "spheres" suggested that they repel each other as a result of electrostatic, steric, van der Waals forces as well as water layer surrounding the spheres.

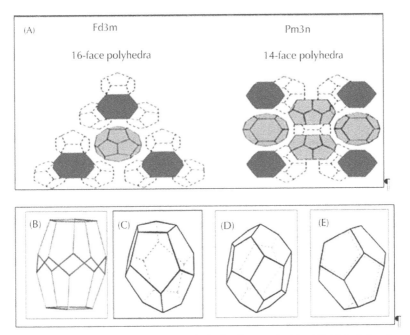

Figure 5. (A) Packing of Fd3m and Pm3n tetrahedrally close-packed (tcp) lipid A-diphosphate and lipid A-monophosphate structures for two cubic crystalline phases. For both structures, aqueous bilayer compartmentalizes the hydrophobic portion of the lipid A-phosphates into tetrahedrally networks. This network is a combination of a pentagonal dodecahedron (blue) with 14-face polyhedra (green) in Pm3n and with 16-face polyhedra (green) in Fd3m, (B) possible stacking of tetrakaidecahedra along the [001] and along the [100], and [010] in cubic (Pm3n) assuming in-plane orientation of the fatty acid chains. (C)–(E) possible polygon representations assuming polyballs of different lipid A-phosphates as multiple components, shown as Wigner-Seitz cells as derived from X-ray diffraction and electron microscopy. Multiplicity and Wyckoff positions, coordination number are 2b and 12 in (C), 4f and 15 in (D), and 8i and 12 in (E).

The cubic *Pm3n* structure, which was seen in the lipid A-monophosphate clusters Faunce and Paradies 2009), materialized as a result of a space-filling combination of two polyhedra, a dodecahedra and a tetrakaidodecahedra. This is in contrast to the tetrakaidecahedra (*Im3m*) or rhombodo-decadecahedra (*Fd3m*) packings observed for the lipid A-diphosphate assemblies (Faunce et al., 2005). The lipid A-diphosphate structures may be reconciled as credible structures of compressed emulsions (De Geyer at al., 2000). Thus, it may be argued that favorable space filling packings will be achieved when satisfactory geometrical conditions were fulfilled (Plateaus's law). For the lipid A-phosphates this resulted from minimizing the surface area between the aqueous films and so achieving a high homogeneous curvature. However, in the case of the lipid A-monophosphate, rhombodo-decadecahedra ($Fd3m$) packing seems to be suppressed a result of instability in the mean curvature between the tetrahedral and the octahedral nodes. Tetrakaidodecahedra packing shows only tetrahedral nodes. However, the tetrahedral angle (109.47°) can only be restored between all of the edges if the hexagonal faces of the truncated octahedron change and generate change and generate planer surfaces with no mean curvature and form Kelvin's minimal polyhedra (Weaire and Phelan, 1994).

Furthermore in case of the lipid A-monophosphate, the flat-faced lipid A mono-phosphate cluster observed in position "c" implies space-filling combinations with at least two different polyhedra. This was shown by Rivier and Aste (1996) for tcp ensembles, as well as in a study of lipid A-diphosphate (Faunce et al., 2003a). The structure reported by Rivier and Aste (1996), face and edge angles was close to the ideal values of 120° and 109.47°. In the $Pm\overline{3}n$ and $Fd3m$ cellular networks there were two examples of 24 tcp structures. One structure with A15 tcp packing was comprised of two dodecahedra and eight tetradecahedra showing a $Pm\overline{3}n$ symmetry. The other structure with C15 tcp packing consisted of 16 dodecahedra and eight hexakaideca-hedra with a space group $Fd3m$. However, a similar spatial arrangement may be as-sumed for lipid A-monophosphate clusters and other comparable assemblies (Faunce and Paradies, 2009), producing two types of cells with different volumes by reduction in the total surface area (Ziherl and Kamien, 2001). In the case of lipid A-monophos-phate this would cause an ideal structure to coarsen, by either particle diffusion, where small nanocrystals form larger nanocrystals or by growth of the larger nanocrystals at the expense of the smaller ones (Ostwald ripening). However, unlike the reported foam systems, the lipid A-monophosphate system revealed both the A15 and the C15 structures to be stable and which and possessed polyhedral cells each of different vol-umes. These cells corresponded to an A15 structure, displaying space-filling packing and polyhedra with very little distortion.

GLASS PHASES

Glass phase were found recently, which could play a crucial role in the preparation of pharmaceuticals, adjutants and vaccine formulation (Reichelt et al., 2008). The glassy phase occurred in almost all of the humoral-biochemical, cellular activities, and studies; nevertheless, the phase went unnoticed because the solutions were buff-ered. Furthermore, a loss of ergodicity was observed with either an increase in volume fraction or ionic strength and by the addition of HCl at 20°C. All of these factors hindered particle motion as a result of cluster formation with the neighboring lipid A-diphosphate particles (attractive glass). It was also possible, that at a temperature ~40°C and at a high-volume fraction $\phi \cong 5.8 \times 10\text{--}4$, the system experienced a re-entrant liquid-glass transition.

Therefore, if either of these two conditions occurs independently or simultaneously the system evolves to form a vitreous state known as a repulsive glass. The distinction between the glass and liquid transitions observed in different regions of the phase diagram was made on the basis of the height and width of the S(Q) peaks. The structure factors of the glass phase, which appear like an "enlarged" FCC structure, displayed a measured peak width (FWHM) ranging between 0.06 nm^{-1} and 0.073 nm^{-1} for the first peak of $S_{eff}(Q)$ for lipid A-diphosphate. The width of liquid phase peak was greater than 0.013 nm^{-1} with the glass phase showing a structure-factor peak well above Hansen and Verlet (1969) criteria for the volume-fraction range $1.5 \times 10^{-4} \leq \phi \leq 3.5 \times 10^{-4}$ and 1.0–10.0 μM HCl. The dynamic-light scattering and shear-viscosity results also agreed well with the calculated values obtained from the mode coupling theory (Bosse et al., 1978). A dramatic reduction in the low shear viscosity occurred for this sample if the surface charges were neutralized or protonated. Also observed for this system was a substantial decrease in stress at the onset of shear thickening.

After weeks or months crystallization took place in the system if a polydispersity of 5.5% existed in the charge or the sphere radii. At high-volume fractions and a polydispersity of ~7.5%, a glass phase formed. For a system like lipid A-diphosphate, which contained a single species interacting with a spherical symmetric potential, a glass phase will normally form if quenched at a sufficiently rapid rate. Such a system was heavily dependent on deionization time, requiring several days (or weeks) to form a solid phase from a liquid-stock solution. The fundamental time steps in the crystallization of atomic materials were the inverse of the phonon frequency $\cong 10^{-13}$ s. For a system of lipid A-diphosphate with a polyball morphology, an interparticle spacing of 55 nm, a Stokes-Einstein diffusion coefficient $D_0 = 4.0 \times 10^{-8}$ cm^{2}s^{-1} and in an aqueous solution at 20°C, the value calculated for $a^2/D_0 \cong 1.0–1.5$ s. The deionization process on a time scale of between 3 and 4 days would be equivalent to a 10^{-4} s quench from the liquid to the solid phase. It was much easier to attain the crystalline state when the liquid-phase boundary was approached through an increase in the volume fraction and/or by slowly changing the ionic strength. Phonon frequencies in colloidal systems are small because of the large particle mass and long spacing distances (0.05–1.5 μm). The Phonon frequencies scale with $^{m-0.5} \times a^{-1}$ and are 10^5 times smaller for colloidal crystals than for normal solids. The links of the onset of melting to the thermally driven rms displacements, δr, of particles about their mean positions roughly approach $0.15 \times a$ when the crystal melts (Lindemann criterion): This results in a movement of ~6.5 nm, a distance considerably less than the spacing between the surfaces of the colloidal spheres and less than the screening length. Therefore, if either of these two conditions occurred independently or simultaneously the system would evolve to form a vitreous state known as a repulsive glass. Constructing a phase diagram for for example lipid A-diphosphate, for behavior at a low, moderate and high volume fractions (φ) in the pH range 5.6–7.0, revealed additional phases. The disclosed phases which formed in an acidic pH were; another cubic structure and two glassy forms. In the experiments, the bare lipid A-diphosphate charge was not constant when the particle number density, n, changed, because of the formation of distinct cubic structures (FCC, BCC, and S.C.). Three main effects could be responsible for the observation: (i) counter ion condensation, (ii) self screening and (iii) many-body effects appearing as macro ion shielding. Furthermore, for lipid A-diphosphate assemblies in the presence of Na^+, K^+, and Ca^{2+}, and Mg^{2+}, condensation of counterions at the particle surface might occur for large surface charge densities but without a change in the particle number density, n. For sufficiently large particle number densities, the lipid A-mono and diphosphate initiated crystallization in distinct cubic lattices (Faunce and Paradies, 2007, 2009, 2011; Faunce et al. 2003b, 2007). Crystallization occurred with an increased in effective particle charge, Z_{eff} and in effective temperature $T^* = k_B \times T/V$ $(\overline{d})$ at low ionic strength. *Note*: The elasticity charges were significantly lower than the conductivity charges, indicating the presence of macroion shielding.

A decrease in crystal stability was observed at pH 5.5 with the occurrence significant deviations in large bare charges. This was seen for lipid A-diphosphate complexes with ERI-1 (Aschauer et al., 1990) and other nontoxic lipid A-analogues (Christ et al., 2003), when examining the charge from conductivity measurements. The system first freeze and then re-melted through modifying the charge by adding OH^-, or after complex formations with the following: single chained N-cationic lipids (Paradies and Habben, 1993; Thies et al., 1996) and double chained N-cationic surfactants (Alonso et al., 2009; Thies et al., 1996), nontoxic lipid A-phosphates or CAM peptides (defensins). High packing density and volume fractions of $\phi = 0.3$–0.5 were essential in generating short-range order. It was not possible to achieve this condition for lipid A-phosphate dispersions because the nearest neighbors were generally located at a distance of one molecular diameter. Dispersions of particles exhibiting long-range repulsive interactions also undergo a less apparent order-disorder transition, providing the ionic strength is very low. If the repulsion between the particles is large and low polydispersity is achieved, the transition from liquid-like to ordered solid-like behavior occurs over a very narrow volume-fraction regime ϕ. Consequently, it was possible for interactions to take place between colloidal lipid A-phosphate particles as well as assemblies comprised of for example lipid A-diphosphate and non-toxic lipid A-phosphate analogues. These may have different chain lengths, numbers of chains and disaccharides. The tuning of these systems was made possible by modifying either the particle surfaces or the properties of the matrix in which they were suspended. As long as at pH of between 5.5 and 7.0 and temperatures of between 5 and 25°C were maintained various lipid A-phosphate crystalline cubic and trigonal assemblies formed for example $Im\overline{3}m$, Fdm, $R\overline{3}m$, $Pm\overline{3}m$, and $Ia\overline{3}d$, and $R32$.

BIOMINERALIZATION IN THE PRESENCE OF LIPID A-PHOSPHATES

With the presence of lipid A-diphosphate, *in vitro* crystallization of various forms of $CaCO_3$ took place for example calcite, vaterite or aragonite. This demonstrated that lipid A-diphosphate induced and stabilized the metastable vaterite phase above a volume fraction $\phi = 7.8 \times 10^{-4}$, at low ionic strength, a pH 5.8 and at ambient temperature. The morphology of the vaterite phase did not change with an increase in volume fraction. The calcite phase formed at significantly lower volume fraction value, $\phi = 5.8 \times 10^{-4}$, at pH 8.5, and in the presence of 1 mM Ca^{2+}. An elemental analysis of the supersaturated bicarbonate solution showed that the Mg^{2+} concentration was too low with respect to the Ca^{2+} concentration to allow the nucleation of significant amounts of aragonite. In the light of the considerable influence of Ca^{2+} and Mg^{2+} ions on the structure and activities of lipid A-diphosphate, the experiments were repeated in the presence of 150 µM Mg^{2+} in the same crystallization environment. Even at this Mg^{2+} concentration no changes in the above findings were observed. However, recrystallization experiments in the presence of higher amounts of Mg^{2+} (150 mM) revealed aragonite to be the major polymorph of the precipitated $CaCO_3$ (Faunce and Paradies, 2008).

The crystalline $CaCO_3$ particles that formed remained well separated between the interfaces of the lipid A-diphosphate assemblies and the aqueous surrounding. The CD spectrum between 210 and 500 nm did not change with time or $[Ca^{2+}]$ at either pH. This indicated that no changes in the secondary structures of the template occurred upon crystallization. Accordingly, no contamination of lipid A-diphosphate by the $CaCO_3$ crystals was detected from MALDI-TOF-MS and LC-MS analyses. Interparticle interactions at a well-defined screening length $k^{-1/2} \sim 9.5$ nm (surface charge density) between the template, crystals and water may have arisen. This would be from a balance between long-range attractive and repulsive forces at the above screening length. Roughness-induced capillary effects (Hashmi et al., 2005) and irregular meniscus defects may explain the prevalence of the observed vaterite cluster at high-volume fractions, in the subphase. Loose aggregates of vaterite formed at high-volume fractions by suppressing the growth of calcite.

The development of a specific well-formed mineral layer by the various polymorphs of $CaCO_3$ with lipid A-diphosphate as a template may cause direct resistance to viral and bacterial invasion and/or penetration with respect to CAM activity. The CAM activity was not changed in the presence of 10 µM to 5 mM $CaCl_2$ at pH 8.5, thus underlining the importance of the physical behavior of lipid A-diphosphate at high pH. However, it was significantly altered in the presence of 10 mM $CaCl_2$ but revealed no further increase in CAM activity.

Lipid A-phosphate Assemblies at pH 8.0-9.0

The solution structure of lipid A-diphosphate at an alkaline pH that is pH 8.5–9.0 was of considerable importance for a number of reasons, which are as follows: (i) potential production of vaccines; (ii) compounds known to bind to LPS or lipid A-diphosphate at pH 8.0; (iii) chain length-dependent agglutination of oligosaccharide clustering at alkaline pH by multivalent anion binding (Christ et al., 1994), (iv) inactivation of endotoxins by cationic surfactants in the textile, cleaning and disposal processes in hospitals and nursing homes, and (v) neutralizing the endotoxic structure of lipid A-diphosphate with cationic complexes (Paradies et al., 2003). The solutions of lipid A-diphosphate at pH 8.5 showed significantly lower solubility than organic salts with for example cetylpyridinium, cetyltrimethyl-ammonium, triethylamine, dihexadecyl-, or didecyldimethylammonium than one would expect. The physical behavior of dispersions of lipid A-diphosphate at alkaline pH was highly warranted, therefore, preliminary results of this system at pH 8.0 at 25°C presented. The influence of a variable charge on lipid A-phosphates, particularly, for lipid A-diphosphate, was more subtle to study and difficult to access experimentally because of the chemical instability at an alkaline pH over certain time periods. Counterion condensation played a significant role as shown for example micelles by Bucci et al. (1991), Yamanaka et al. (1999)

and Wette at al. (2010) for charged silica particles supported the idea that a counterion condensation took place at the particle surface, over large surface charge densities with an increase in pH to 8.5 or 9.0. Changes in the NaOH concentrations from μM to mM raised the pH but did not necessarily raise the particle surface charge density. However, this concentration adjustment resulted in a change form a spherical to a rod-like assembly (Faunce and Paradies, 2008). Two current models explain the concentration-dependent findings at pH 8.0. At $C \cong C^*$ the orientational entropy and the electrostatic repulsion terms of the free-energy expression for a dispersion of charged rods favor anti-parallel alignment. The anti-parallel alignment of the rods, gives rise to a cubic lattice-like interparticle structure. However, for $C \gg C^*$, a hexagonal packing of parallel rods was anticipated to be an appropriate model, Where C^* denotes the critical concentration of lipid A-diphosphate, $C^* = $ particle/l^3, which amounts to 1.5×10^{12} particles per mL. Both models would fulfill the conditions at $C \gg C^*$ because the separation between rod-like particles must be considerably less than the rod length, the rods may be considered to be infinitely long. The rods were charged and consequently the surfaces must be as far apart as possible. For both models, $Q_{max} L = 2\pi\gamma \, (C/C^*)^{1/2}$ scales with $\gamma = 0.58$ for a cubic structure, and for a hexagonal structure with $\gamma = 0.93$. The experimental values obtained of 0.57 for $C \cong C^*$ and 0.90 for $C \gg C^*$ were close to the theoretical values. If the correlation of nearest-neighbor rods of lipid A-diphosphate were considered, then parallel alignment of nearest-neighbor rods occurred, with a local ordering parameter $S = 0.071$ at $C = 2.7 \, C^*$. Accordingly, monodisperse rod-like particles with a standard deviation of $< 5.5\%$ and an L/d range from 20–25 were seen on the TEM and SEM images (Figure 6).

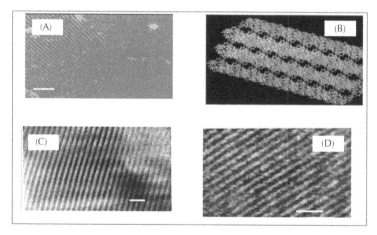

Figure 6. The TEM images of various lipid A-diphosphate assemblies at pH 8.0 in the presence of 100 μM NaOH. (A) Rod-like lipid A-diphosphate assemblies comprising of lipid A-diphosphate (Figure 1(A)) and nontoxic but antagonistic lipid A-diphosphate (Figure 1(C)). Scale bar is 10 nm, (B) A 3-D assembly of an antagonistic lipid A-diphosphate analog (Figure 1(B)) modeled from TEM images and arranged in a hexagonal packing (R3), (C) TEM image of a 3-D assembly with two antagonistic lipid A-phosphates (Figure. 1(B) and 1(C)) and endotoxic lipid A-diphosphate. Scale bar is 5 nm, and (D) Single lipid A-diphosphate (Figure 1(A)) particles forming a highly ordered chain-like structure at pH 8.5 in the N-phase. The scale bar is 100 nm. (D)

Before the lipid A-diphosphate specimens were dried for TEM or SEM inspection, the initial concentration of the dispersions was $C = 0.5C^*$. From $C = 0.5\ C^*$ to $C = 0.85\ C^*$ an isotropic phase appeared as rod-like clusters, some of which form individual particles, while others were aligned side by side and/or head to tail. With increasing particle density more clusters appear, grew, and finally formed Sm structures. Below an ionic strength of $I = 5$ mM NaOH at pH 8.5 the nematic region increases at the expense of the I-phase. With an increase in the particle-number density it is possible to observe the Sm-phase and the formation of the N-phase, which is induced with a dilution of the Sm phase. The N-phase forms in a lipid A-diphosphate suspension with an ionic strength of $I = 1.5 \times 10^{-4}$ M NaCl, but at pH 5.6, spherical particles of diameters of 90–100 nm are noticed, and at higher concentrations ($C \cong 11.0\ C^*$) colloidal crystals are observed, which have a cubic FCC structure with $a = 57.3$ nm. Individual elongated lipid A-diphosphate particles may be seen, oriented in a direction approximately normal to the arrow. When the particle-number density is increased, lateral cluster growth takes place and the contours becoming clearer and cluster layering becomes more apparent. These simulations predicted a critical minimal L/d ratio of 4.5–5.0 for the stable I-phase where the I-N-triple point was also located. The Sm layer period and in-layer particle separation were found to be 2.0 and 2.8 nm, respectively ($C = 5C^*$). The packing-particle fraction η for this Sm phase was estimated to be 0.38, smaller than the 0.45 predicted if electrostatic interactions are taken into account. This can only be explained (adding NaOH) as being due to an increase in the number of free counterion upon charging the particles for example self-screening implying like charge attraction and many-body effects which might be attributed to macroion shielding but not in enhancing aggregation.

Moreover, high-resolution electron microscopy and electron diffraction on lipid A-diphosphate rods (length of the order of µm and diameter of several nanometers) revealed that the rods were held to the truncated polyhedral with a 5-fold symmetry (see also Figure 7 and legend). It was possible to show that most of the lipid A-diphosphate particles were orientated in the [001] direction with respect to the substrate for one of the five deformed tetrahedral subunits that is the 5-fold axis was parallel to the surface of the substrate.

CUBIC MIMICRY OF THE SELF-ASSEMBLY AND DENSE PACKING OF LIPID A-PHOSPHATE

Both 2- and 3-D assemblies of lipid A-mono or diphosphate constructed from single or multiple chemical entities of known chemical composition ("coded subunits") were quite complex. These assemblies were very unlikely to be loose or unordered combinations of lipid A molecules or of their analogs. The "coded subunit" addresses the chemical appearances of single or identical chemical structures of lipid A-phosphate. However, they were non-identical and differed in chemical structure for example number of chiral fatty acid chains, length of the hydrocarbon chain (number of carbons), inserted double bonds in the hydrocarbon chain, and hexose compositions (Christ et al., 2003). The assemblies were characterized by being comprised of two phosphates residues: one at the reducing end and the other one at the non-reducing end for the diphosphate. In the case of the monophosphate, the phosphate was located only at the reducing end of the disaccharide. From known chemical and physical data it appeared that stoichiometric ratios existed between the different chemical entities of lipid A-phosphate, but they have not yet been elucidated. However, on the basis of the experimental X-ray diffraction, LS, and quantitative electron microscopy data the structures were found to be lipid A-phosphate "quasicrystals" (Faunce and Paradies, 2006; Faunce et al., 2003b, 2011; Torquato and Stillinger, 2010). These quasi crystals can also demonstrate non-crystallographic packing of non-identical lipid A-phosphate spheres. A spatial packing of these spheres in for example cuboctahedron or icosahedron represent reasonable physical models. Other possible spatial arrangements (Torquato and Stillinger, 2010) may also exist. Many aperiodically ordered materials are known to possess 5-, 7-, 8-, 10-, and 12-fold symmetry most of these were prepared by supercooling multi component liquids or melts (Lee et al., 2010), which show a σ phase (Kasper and Frank, 1959). A BCC phase was also discovered following an increase in temperature which gave rise to subsequent dodecagonal quasicrystals which form spherical particles. It is possible that the distinct formation of discrete (self)-assemblies, formed by large and thermodynamically stable quasicrystals may add information on for example local icosahedral ordering. This would also be the case for single component lipid A-phosphates (identical subunits) or for example, a two component system where another component acts as a copolymer or a non-identical subunit. The equilibrium structures of such systems with respect to lipid A-phosphates built from coordinated spherical or ellipsoidal polyhedral, as described by Frank and Kasper in 1959, may also be controlled by components of the molecular shape and branching, and applicable to the various subunits of lipid A-phosphates.

From X-ray diffraction traces and the electron diffraction pattern obtained for for example lipid A-diphosphate (identical subunits), the material crystallizes to form a BCC cubic lattice ($a = 36.1$ nm). This BCC lattice was associated with the spatial accommodation of identical spherical (or slightly ellipsoidal) particles in the first approximation. According to the form-factor scattering at high Q, there was a spherical domain of size R = 7.31 nm. Assuming identical spherical lipid A-diphosphate morphologies and a mass density of 1.02 g/cm^3 (Faunce and Paradies, 2008), there were on the average 28–30 lipid A-diphosphate clusters present in the initial BCC lattice, according to the Bragg scattering. Therefore, each spherical lipid A-diphosphate cluster contains approximately 514.5 lipid A-diphosphate molecules or 18.4 molecules per aggregate. The spatial domain of R = 7.31 nm can incorporate almost 32.9 hexagons when applying the previously determined unit cell dimensions of $a = 3.65$ nm, $c = 1.97$ nm and $\alpha = 120°$ (space group R32), with eight lipid A-diphosphate molecules per unit cell, or three molecules per rhombohedral unit cell with $a_{rh} = 2.25$ nm and $\alpha_{rh} = 66.7°$. Real space electron density maps constructed form X-ray diffraction and selected area electron diffraction data and using the Rietveld method, the space group R32 (Faunce et al., 2003b), for a cluster domain R = 7.31 nm. This was for a set of 30 cluster in the BCC unit cell, where 12 have the coordination number 12, (Wyckoff positions 2b and 8i), and 16 have the coordination number 14 (8i and 8j), and four with the coordination number 15 (4g). In a comparison between the simulated and experimentally I(Q) versus Q pattern, good agreement was revealed between the modeled and experiment results (Figure 7). The resulting lipid A-diphosphate model consists of a complex periodic structure built from fused dodecagonal cells with layers originating from hexagonal cylinders which developed as columns of spherical lipid A-diphosphate.

The I(Q) versus Q simulations (Figsure 3 and 4) showed noticeable differences in the shape and the magnitude of the maxima and the minima of the main peaks with respect to the Wyckoff positions. These peaks are related are related to the Wigner-Seitz cell volumes; an effect especially related to the low Q region. Interestingly the different Wigner-Seitz polyhedra which make up the overall volume of the Frank-Kasper phase unit cells as depicted in Figure 5 range from between 90 and 98% of the average cell volumes. The lipid A-diphosphate arrangement constructed from a mixture of an icosahedron and truncated octahedron (BCC) symmetry follows the Landau theory (De Gennes and Prost 1993; Witten and Pincus 2004) and appears to add further packing frustration thus reducing the system entropy.

Another packing possibility exists, particularly for the aforementioned *E. coli autovaccines*, assuming an assembly of equal spheres where each sphere was surrounded by its first coordination polyhedron, which might consists of 12 spheres at the vertices of a cuboctahedron. A second layer of spheres packed over the first layer may require 42 spheres. In general the nth layer consists of ($10n^2 + 2$) spheres of for example lipid A-phosphate molecules, where each sphere may be an identical or non-identical subunit (Figure 7). The self-assembled lipid A-phosphate chemical formulas for the various "identical" component lipid A-phosphates are shown in (Figure 1). Discrete combinations of for example lipid A-diphosphate with compound B or C are revealed in and they were positioned over the first layer. The spheres were in contact along the 5-fold axes. Spherical lipid A-diphosphate subunits within the icosahedron a maximum packing was reached at a volume fraction ($\phi = \pi\sqrt{18} = 0.74$). The packing maximum was attained by stacking single or multiple lipid A-phosphates layers in FCC, HCP arrays or in random sequences. The ellipsoidal multiple lipid A-phosphate molecules pack more efficiently than spherical multiple lipid A-phosphate molecules, revealing dense packing, generating both random and colloidal crystalline arrangements. At the least a spherical aspect ratios $x = a/b = \pm \sqrt{3}$ of these ranges, the ellipsoidal centers lie on an FCC lattice with the ellipsoidal axis parallel to any face-centered square plane, however, rotated by 90° from Figure 1(C and D) and conform to BCC or FCC cubic lattices (Batista and Miller, 2010). This was dependent upon the volume fractions ϕ and ionic strength in aqueous dispersions although with different unit cell dimensions. It was possible to accommodate the lipid A-phosphate anions by lowering the symmetry from cubic to rhombohedral and finally to a monoclinic structure (Faunce et al., 2011). Here the strategic outline follows laws from nature, using the geometric and energetic characteristics of icosahedral symmetry (Williams, 1979).

Furthermore, an icosahedron composed of 12 spherical (identical or non-identical subunits) lipid A-phosphate clusters can be constructed and arranged around a central sphere. The sphere was surrounded by a second icosahedral shell twice the size of the first. The shell contained 42 spheres one layer to the next one. This implies in light of multiple lipid A-phosphate molecules packed as deformable spheres that the adoption of an HCP arrangement was only possible for the high volume fractions mentioned previously. When multiple component lipid A-particles were constrained by their neighbors to form a permanent ellipsoidal cluster, denser colloidal crystal of the lipid A-phosphate assemblies developed (Batista and Miller, 2010).

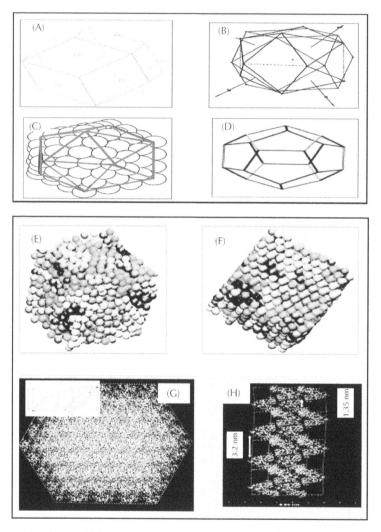

Figure 7. (A) A cubic lipid A-diphosphate colloidal crystal of single component self-assembled cluster with a unit cell dimensions of $a = 36.5$ nm. This cubic quasicrystal can be subdivided into rhombohedral unit cells (one unit cell is shown in G) used to construct the icosahedral quasicrystal packing, In (B) a process is shown by which a cuboctahedron (in shape as noticed in Figure 3 (E) and (H)) of 12 rods of various lipid A-phosphates, jointed at their ends (thin lines) in the self-assembly can be moved by rotations of its triangular faces about their normal to yield a regular icosahedron (thick lines and dotted lines), (C) Icosahedral packing of equal spheres for lipid A-diphosphate encountering multiple lipid A-phosphates (shown in the last row as colored spheres representing different single lipid A-phosphates present in the *E. coli autovaccines*), and only the third layer is shown. On each triangular face the layers of spheres succeed each other in cubic close-packing sequence by applying a distorted tetrahedron. This arrangement follows the suggestions proposed by Mackay (1962), (D) Shows an idealized 3-D colloidal crystal composed of identical and non-identical lipid A-phosphates

Figure 7. (Caption Continued)

subunits (*E. coli autovaccines*) arranged in icosahedral cubic symmetry. A layer packing can be arranged which corresponds to a Penrose local isomorphism class, where a decomposition may include addition of rhombohedra above and below layer packing, (E) Model packing of lipid A-diphosphates for non-identical lipid A-phosphates in a random order for a cubic icosahedron (yellow, blue, red, and green), in (F) a similar packing is depicted for a BCC cubic lattice (as in A), but in an ordered array of the non-identical lipid A-phosphate subunits, (G) Rhombohedral packing (a = 1.55 nm, α = 67°, space group or R32) of identical lipid A-phosphate molecules (Figure 1(A)). The inset shows the shape of a cubic crystal (a = 36.5 nm, red), in which a rhombohedral unit cell has been fitted (blue), and (H) A 3-D packing of the lipid A-diphosphate (Figure 1(A)) is shown similar as in Figure 6(B), derived form rhombohedral symmetry assuming that the lipid A-diphosphate anion is completely orientationally ordered, and lowering the symmetry from cubic to rhombohedral and finally to monoclinic (P2$_1$ or C2). A rope model of lipid A diphosphate (Figure 1(A)) is shown, based on an image of lipid A-diphosphate aligned along the 2-fold axis screw axis, the b-axis of the unit cell with 2a = 3.78 nm, 4b = 7.11 nm, c = 3.94 nm, and β = 62.5°. Atoms are presented as stick models: Phosphorus violet, oxygens red, carbons black, and hydrogens white.

The polyhedral shapes of the lipid A-phosphates, *E.coli autovaccines* and stoichiometric mixtures of lipid A-diphosphate and antagonistic lipid A-diphosphates (Figure 1) in general, may be attributed to the deformable nature of the fluid clusters. The clusters were composed of a small high electron density core (460 e/nm^3) approximately 1.95 nm in diameter. These were surrounded by a large, flexible, aliphatic fatty-acid chain shell 2.70 nm in diameter. The polyhedra deformations conform to the different symmetry sites in *Pm3n*, or icosahedrons and truncated octahedron structures. This leads to a space-filling arrangement consisting of dodecahedra and tetrakaidodecahedra networks (Figure 5). For clusters that exhibit flattened surfaces, the film curvature will be distributed along the cell edges and at all of the vertices.

CONCLUSION

It was possible to construct different Wigner-Seitz polyhedra that make up the overall volume of the Frank-Kasper type unit cells with complexes comprised of lipid A-diphosphate, antagonistic and non-toxic lipid A-phosphate analogs depending on volume fraction and nature of the counterions (Faunce et al., 2005). They form by spontaneous self-assembly and appear to obey the principles of thermodynamically reversible self-assembly but once self assembled strongly resists disassembly. Base on the principles outlined in this contribution, lipid A-phosphate assemblies can be designed which form large unit cells by containing more than hundred lipid A-phosphates. The range of lipid A-phosphate structures may also be increased further by employing various different ("non-identical subunits") and identical subunits of lipid A-phosphate in analogy with block copolymers.

The rational design of such assemblies including those of biocompatible quantum dots for biological imaging, nucleation of polymorphic inorganic minerals, production of suitable aerosols for immunization, and structure-function relationships will be impacted by a theoretical and practical understanding of these spherical assemblies, rod-like assemblies and the mixtures thereof. Given the theoretical and practical importance of this system, we expect that the attention given to it will substantially increase our knowledge on *LPS*, innate immunity, mineral nucleation, the driving forces for the ordered assemblies as well as their interactions with CAM, and the resistance against antibiotics where gram-negative bacteria are involved. Furthermore, the structure of the lipid A-diphosphate rod as prepared at pH 8.0 can be explained as truncated large dodecahedra. Due to the low ionic strength and pH 8.0, at the decahedral site nucleation started and the growth is suppressed and can only take place along the direction, that is the 5-fold axis, resulting in well ordered long rods.

KEYWORDS

- **A-diphosphate**
- **Cationic antimicrobial**
- **Gram-negative**
- **Lipopolysaccharides**
- **Tumor necrosis factor**

Chapter 25

Bone Mechanical Stimulation with Piezoelectric Materials

J. Reis, C. Frias, F. Silva, J. Potes, J. A. Simxes, M. L. Botelho,
C. C. Castro, and A. T. Marques

INTRODUCTION

Fukada and Yasuda were the first to describe bone piezoelectrical properties, in the 1950s. When submitting dry bone samples to compressive load, an electrical potential was generated, an occurrence explained by the direct piezoelectric effect (Fukada and Yasuda, 1957). The nature of the piezoelectric effect is closely related to the occurrence of electric dipole moments in solids. In connective tissues such as bone, skin, tendon and dentine, the dipole moments are probably related to the collagen fibbers, composed by aligned strongly polar protein molecules (ElMessiery, 1981; Fukada and Yasuda, 1964; Halperin et al., 2004). The architecture of bone itself, with its aligned concentric lamellae, concurs for the existence of potentials along bone structure (El-Messiery, 1981).

Bone piezoelectric constants, that is the polarization generated per unit of mechanical stress, change according to moisture content, maturation state (immature bone has lower piezoelectric constants when comparing to mature bone) and architectural organization (samples from osteossarcoma areas show lower values due to the unorganized neoplastic changes) (Marino and Becker, 1974).

Early studies concentrated on dry bone and because collagen's piezoelectricity was described as nearly zero with 45% moisture content, there were doubts that wet bone could, in fact, behave as a piezoelectric material, but further studies confirmed it in fact does (Fukada and Yasuda, 1957; Marino and Becker, 1974; Reinish and Nowick, 1975). Some of the published studies reinforce the importance of fluid flow as the main mechanism for stress generated potentials in bone, and piezoelectricity's role was, and still is, quite unknown (Pienkowski and Pollack, 1983).

More recently, bone piezoelectrical properties have rouse interest, in the context of bone physiology and electro-mechanics. It has been associated to bone remodeling mechanisms, and to streaming potential mechanisms (Ahn and Grodzinsky, 2009; Ramtani, 2008). Piezoelectricity explains why, when under compression, collagen reorganizes its dipole and shows negative charges on the surface, which attract cations like calcium. Conversely, if tensed, collagen yields predominance of positive charges, thus obviously influencing the streaming potential and mineralization process (Noris-Suárez et al., 2007).

The commercially available biomaterials for bone replacement and reinforcement do not take into account the bone natural piezoelectricity and the mechanism of streaming potential, and thus its use is accompanied by a break in bone natural electro physiologic mechanisms.

On the other hand, so far orthopedic implants only perform fulfill primary functions such as mechanical support, eliminating pain and re-establishing mobility and/ or tribologic/articular contact. Arthroplasty is liable to cause intense changes on strain levels and distribution in the bone surrounding the implant, namely stress shielding. Metal stiffness is much higher than that of bone, so the rigid stems tend to diminish the amount of stress transmitted to the surrounding bone and produce stress concentration in other areas, depending on geometry and fixation technique. Stress shielding leads to bone resorption, which in turn may cause implant instability and femoral fracture, and make revision surgery more challenging (Beaulé et al., 2004; Huiskes et al., 1992; Mintzer et al., 1990; Sumner and Galante, 1992). Ideally, the bone implant should present sensing capability and the ability to stimulate bone, maintaining physiological levels of strain at the implant interface.

The work here summarized explores *in vitro* and *in vivo* use of a piezoelectric polymer for bone mechanical stimulation.

Piezoelectric Materials for Mechanical Stimulation of Bone Cells—The *in vitro* Study

Osteocytes and osteoblasts are essential for mechanosensing and mechanotransduction, and cell response depends on strain and loading frequency (Kadow-Romacker et al., 2009; Mosley et al., 1997). We explored the use of piezoelectric materials as a mean of directly straining bone cells by converse piezoelectric effect.

The MCT3T3-E1 cells were cultured under standard conditions and on the surface of Polyvinylidene Fluoride (PVDF) films, subjected to static and dynamic conditions, as described by Frias et al. (2010).

Polymeric piezoelectric films (PVDF) were used as substrate for cell growth. These thin films consisted of a 12 x 13 mm active area, printed with silver ink electrodes on both surfaces in a 15 x 40 mm die-cut piezoelectric polymer substrate, polarized along the thickness. In dynamic conditions the substrates were deformed by applying a 5 V current, at 1 Hz and 3 Hz for 15 min.

To guarantee adhesion of osteoblasts to the device surface and electric insulation, the surface was uniformly covered with an electric insulator material. The chosen material for covering was an acrylic, poly (methyl methacrylate) (PMMA), (PERFEX®, International Dental Products, USA), used alone in the first three layers and a in forth layer along with 4% of Bonelike® (250–500 µm) particles added (kindly offered by INESC Porto). The coating was performed by dip-coating at constant velocity of 0, 238 mm/sec. Impedance was measured both in saline and culture medium, in non-coated and coated devices, and electric insulation achieved. The coating procedure aimed improvement of cell adhesion and electrical insulation. Electrically charged particles are known to improve osteoblast proliferation and it was important to prevent cell damage and other means of stimulation other than the mechanical (Dekhtyar et al., 2008; Kumar et al., 2010; Nakamura, 2009).

To estimate the magnitude of stress/strain, finite numerical models were applied and theoretical data was complemented by optic experimental data. The finite numerical method estimated displacement varying from 6.44 to 77.32 nm in uncoated films, with strain levels around 2.2 μ strains along the surface. The Electronic Speckle Pattern Interferometry (ESPI) method showed the displacement in coated films was lower, and the maximum substrate displacement was 0.6 μm, in the central area of the coated devices; displacement was minimum in the encastre (clamped) region.

Piezoelectric substrates (standing on culture dishes, TPP) and controls (standard culture dishes, TPP) were seeded with 16×10^4 cells, with a total volume of 100 μl of cell suspension. Cells were allowed to adhere to the substrate, then the rest of culture medium added (n = 6); and cells grown in both static and dynamic piezoelectric substrates. The MCT3T3-E1 cells were cultured in standard conditions, using α-MEM medium (Cambrex), 2 mM L-Glutamine (Cambrex), 10% of bovine fetal serum (Gibco), 0.5% gentamicin and 1% amphotericin B (Gibco).

The statistical analysis was done using software Origin Pro 8 (OriginLab Corporation, USA).

Normal distribution of the results was verified using the Kolmogorov–Smirnov test, homogeneity of variance assessed through the Levene test and differences between groups tested using one-way ANOVA (at a level of 0.05).

Cell viability and metabolic activity was accessed through the resarzurin method, after stimulation of dynamic group; viable cells reduce resarzurin, producing resorufin, a highly fluorescent product. Previous studies indicated PVDF affects negatively adherent cell lines' viability (Hung et al., 2006; Tabary et al., 2007).

The assessment of cell viability and proliferation evidenced a material's poorer performance than control standard culture vessels, in spite of the coating procedure (Table 1). The results are expressed as percentage of the value of controls (considered as 100%) ± standard error of the mean and show higher viability values on mechanically stimulated substrates, although the differences are not statistically significant.

Table 1. Cell viability 24 hr and 48 hr after seeding and daily stimulation of the dynamic group, results are expressed in percent related to controls (standard cell culture dish), assumed as 100%. Means and Error bars show Means ± Standard Error of the Mean.

Proliferation and Viability	Static	Dynamic
24 hours post-seeding	49.9 ± 5.25	59.7 ± 15.7
48 hours post-seeding	76.4 ± 16.9	83.4 ± 25.1

Nitric oxide (NO) is a messenger molecule produced in response to mechanical stimulation of osteoblasts and osteocytes, with a large variety of biological functions (Smalt et al., 1997; Van't Hof, 2001). In this study, culture medium samples were collected immediately after stimulation and NO measured, using NO Assay Kit (Biochain), based on the Griess reaction, after sample deproteinization, and according to the manufacturer's instructions (Figure1). Culture medium NO measurements in the samples subjected to mechanical stimulation were of 3.7 ± 0.65 and 3.2 ± 0.54 μmol/ml, respectively at 24 and 48 hr post-seeding. The nitric oxide values in static conditions were significantly lower, 2 ± 0.35 µmol/ml, 24 hr post-seeding and 1.7 ± 0.3 µmol/ml, 48 hr post-seeding (Figure 1).

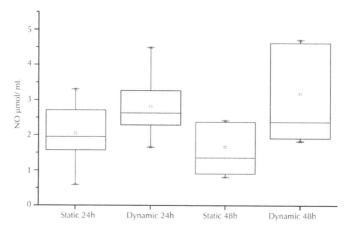

Figure 1. Nitric oxide measurement (µmol/ml) in culture medium in static versus dynamic conditions, 24 and 48 hr after seeding MC3T3 on the devices, and immediately after stimulation at 1 and 3 Hz. NO values are significantly higher in the dynamic group.

The NO measurement results of culture under dynamic conditions versus static conditions, suggest osteoblasts detect and respond in a reproducible way to small displacements and strain levels.

Changes induced by mechanical stimulation on the cytoskeleton were qualitatively assessed through indirect immunofluorescence. Primary antibodies against actin, laminin and tubulin were used; stains show stronger fluorescence on mechanical stimulated cells, clearer images of the cytoskeleton elements and nucleus delimitation and prominent cytoplasmatic extensions (Figures 2 to 4).

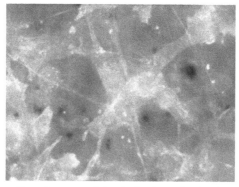

Figure 2. MC3T3 cells on the active area of the device immediately after mechanical stimulation. Indirect immunofluorescence using primary antibody against actin (Actin, pan Ab-5, Thermo Scientific, used at 1:50) and secondary antibody (Chromeo™ 488 conjugated Goat anti-Mouse IgG, Active Motif 1:500); (400X, microscope Olympus BX41, Olympus Cell A Imaging Software).

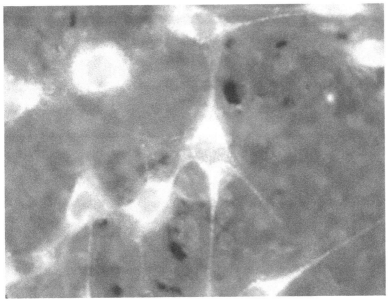

Figure 3. MC3T3 cells on the active area of the device immediately after mechanical stimulation. Indirect immunofluorescence using primary antibody against actin (Laminin, Ab-1, Thermo Scientific, used at 1:50) and secondary antibody (Chromeo™ 488 conjugated Goat anti-Rabbit IgG, Active Motif 1:500); (400X, microscope Olympus BX41, Olympus Cell A Imaging Software).

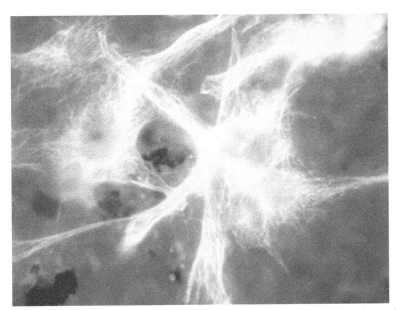

Figure 4. MC3T3 cells on the active area of the device immediately after mechanical stimulation. Indirect immunofluorescence using primary antibody against actin (Tubulin β, Thermo Scientific, used at 1:50) and secondary antibody (Chromeo™ 488 conjugated Goat anti-Rabbit IgG, Active Motif 1:500); (400X, microscope Olympus BX41, Olympus Cell A Imaging Software).

The *in vitro* study showed evidence of effective bone cell mechanical stimulation, and the concept was further explored *in vivo*. The *in vivo* implantation of piezoelectric actuators for tissue mechanical stimulation is innovative and a potential use in the development of smart implants.

Piezoelectric Materials for Bone Mechanical Stimulation–the *in vivo* Study

The actuator device was developed, composed of a micro-board containing a ultra-low power 16-bit microcontroller (eZ430-RF2500, Texas Instruments, USA), powered by lithium battery and encapsulated in polymethylmetacrilate (PMMA) and a set of six actuators composed of PVDF and silver electrodes, electrically insulated by dip-coating as previously described. A similar, but static, control device was also developed, sterilized and implanted.

The sterilization of the device posed a challenge in itself. The devices included a 16-bit processor, which corrupted its memory when submitted, in a Co-60 source, to 25 kGy at a dose rate of 2 kGy/hr. On the other hand, it was not possible to sterilize by moist or dry heat since the PVDF actuators depolarize at temperatures equal or above 60°C. An alternative sterilization method, which ensured absence of toxic residues, was developed. The methodology of its development and validation was based on ISO 11737-1 and ISO 14937, as described elsewhere (Reis et al., 2010).

The actuator device was implanted in the left hind limb and the control static device was implanted in the right hind limb of a 4 year old merino ewe, with 45 kg body-weigh, under general inhalatory anaesthesia. Two osteotomies were made on the medial surface of the tibial proximal physis using an especially metal designed guide to make two regular and well orientated osteotomies using an oscillating saw. The bone was continuously irrigated with a sterile saline solution during the process of low speed drilling and cutting. The same procedure was followed with a different design guide for the distal femoral physis, where four osteotomies were done. The portion of the devices containing the microprocessor and the power supply were left in the subcutaneous space.

One week after implantation calcein (Sigma, USA) was injected subcutaneously (15 mg/kg) and 1 week prior to sacrifice the same procedure was done with alizarin complex one (25 mg/kg) (Sigma). Thirty days after implantation the ewe was sacrificed by intravenous sodium pentobarbital injection. The present study was authorized by competent national authorities and conducted accordingly to FELASA's guidelines for animal care. Proper analgesia procedures began before the surgery and were maintained through a week.

Both hind limbs were dissected, the implanted materials and surrounding tissue removed and fixed in 4% paraformaldehyde for 2 weeks. Bone samples were cut transversally to the long axis of the bone, each including a piezoelectric film and the surrounding bone

Specimens were dehydrated through an ascending ethanol series. Soft tissues (local lymph nodes and samples of the fibrous capsule surrounding the implants) were routinely processed and embedded in paraffin. Undecalcified bone samples of each of the implants were included in resin (Technovit® 9100, Heraeus Kulzer, Germany) according to the manufacturer's instructions and 80 μm thick sections cut with a saw microtome (Leica 1600, Germany) parallel to the piezoelectric film long axis. A minimum of five sections of each resin block was cut. Sections were then appropriately processed for routine staining (Giemsa Eosin), mounted for fluorescence microscopy.

The prepared slides were evaluated qualitatively. For histomorphometric studies the interface between the bone and implant was divided in four distinct areas: A1, A2, A3, and A4, from cortical towards bone surrounding the free extremity of the piezoelectric film (Figure 5).

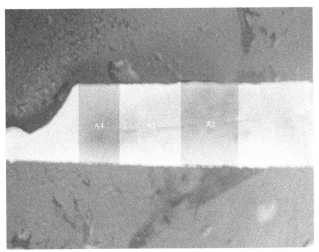

Figure 5. Tibia section the osteotomy where the piezoelectric film was placed. The figure shows example of bone section prior to inclusion and how the areas for histomorphometry were distributed; A1 corresponds to the film encastre (clamped) region.

Pictures were taken from the bone surrounding both sides of the film in areas A1 to A3 and A4.

For immunohistochemistry, bone sections were decalcified in formic acid 5% for 3 weeks, dehydrated in ethanol, cleared in xylene and embedded in paraffin wax and 3 μm sections cut. After deparaffinization and rehydration, immunohistochemistry sections were treated with 3% hydrogen peroxide for 10 min.

Primary antibodies for Proliferating Cell Nuclear Antigen (PCNA) (NeoMarkers, USA, Mouse Monoclonal Antibody, Ab-1, Clone PC10, Cat. #MS-106-P0), Osteopontin (NeoMarkers, Rabbit Polyclonal Antibody, Cat. #RB-9097) and Osteocalcin (Abcam, Mouse monoclonal [OC4-30], Cat. ab13418) were diluted to 1:200, 1:50, and 1:40, respectively. Prior to immunostaining the sections were pretreated for antigen retrieval at 100°C in 10 mM citrate buffer, pH 6, for 20 min in microwave oven, followed by cooling for 30 min at room temperature. For double staining of PCNA and osteopontin, immunohystochemistry was done using a double staining kit Pic TureTM (Zymed Laboratories Inc, USA.), according with the manufacturer's instructions. Slides were counterstained with Mayer's hematoxilin and Clearmount used to mount the slides.

Osteocalcin immunohystochemistry was performed with resource to kit Pic TureTM-MAX Polymer (Invitrogen, USA.). Slides were counterstained with Mayer's hematoxilin, dehydrated, mounted with Entellan® (Merck, Germany), and covers lipped. For all sections positive controls were made simultaneously. As negatives controls adjacent sections were incubated: (a) without primary antibody and, (b) with rabbit/rat normal serum (similar concentration as that of primary).

After one month implantation period, there were statistically significant differences. Total bone area around the actuators was significantly higher, when comparing to static controls (39.91 ± 14.08% vs 27.20 ± 11.98%) (Figure 6).

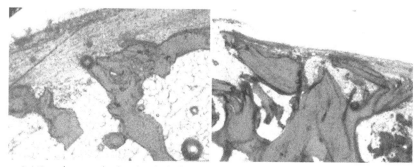

Figure 6. Microphotograph of undecalcified sections, Giemsa-Eosin stain, of A3 area of static control (on the right) and actuator (on the left). Both were implanted in the same position in the tibia. A fibrous capsule was present on the bone/film interface. Scale bar represents 200 μm.

The increment of bone occupied area was due to new bone formation, as evidenced in Figure 7. In actuators the area occupied by woven bone and osteoid was 64.89 ± 19.32% of the total bone area versus. 31.72 ± 14.54% in static devices. With the aid of the fluorochrome labeling, we measured bone mineral deposition rate in the distal third of the piezoelectric devices. Bone deposition rate was significantly higher around actuated devices (4.44 ± 1.67 μm/day) than around static devices (2.70 ± 0.95 μm/day).

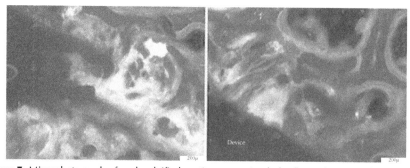

Figure 7. Microphotograph of undecalcified sections, unstained; the fluorochromes calcein (green) and alizarin complexone (red) signal the areas of newly formed bone around one of the actuators placed in the femur; picture on the left shows A4; picture on the right shows A3. Scale bar represents 200 μm.

Newly formed bone and increase in total bone area were unevenly distributed along the length of the actuators; significant differences when comparing actuators to static controls arose from the two distal thirds of the devices. No differences in total bone area and new bone area were found in the actuators encastre region (clamped region). These findings are in agreement with the previous Finite Numerical Method and ESPI method studies on the displacement of the piezoelectric films under the experimental conditions.

Immunohistochemistry shows a marked elevation in osteopontin detection around actuators, in A3 and A4 (Figure 8). Since it is known that OPN production is increased in association with mechanical loading (Harter et al., 1995; Perrien et al., 2002), the increased expression we found around actuators' areas of higher deformation, when comparing to static controls, is most likely associated with mechanical stress. No marked differences were found in PCNA detection.

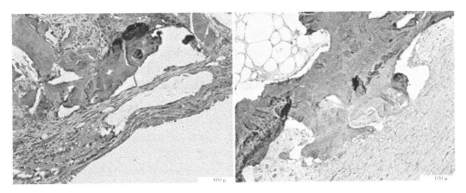

Figure 8. Microphotograph of decalcified sections, double Fast-Red and DAB immunohistochemistry staining for osteopontin and PCNA, respectively. Picture shows A3 areas of actuator (on the left) and static control (on the right), evidencing much more extensive osteopontin labelling around actuator. Scale bar represents 100 μm.

Osteocalcin detection was also increased around actuators, when compared to controls (Figure 9). Osteocalcin is a non-collagenous protein and a major constituent of bone matrix; it is produced by osteoblasts and binds strongly to hydroxyapatite; osteocalcin is considered a sensitive marker of bone formation, and it has been described as rising as consequence of mechanical stimulation-induced cell differentiation (Mikuni-Takagaki, 1999; Pavlin et al., 2001).

We observed that all the devices were separated from neighbouring bone by a fibrous capsule with an average thickness of 292 μm. This is probably due to the material itself, since the fibrous capsule was obvious both in actuators and static devices, with no statistical significant differences in capsule thickness between the two groups (289.59 ± 131.20 μm in actuated films vs 293.93 ± 84.79 μm). It would be mandatory to develop and test a material with improved biocompatibility to evaluate accurately the bone material interface.

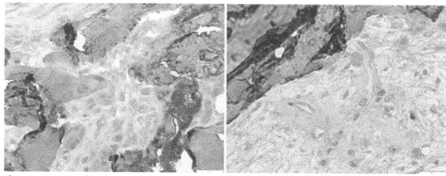

Figure 9. Microphotograph of decalcified sections, DAB immunohistochemistry staining for osteocalcin, respectively. Picture shows A3 areas of actuator (on the left) and static control (on the right), evidencing more extensive osteocalcin labelling around actuator. Scale bar represents 20 µm.

The results are very clear in evidencing qualitative and quantitative statistically significant differences when comparing static and actuated films but it would be necessary to enlarge the animal study. However, considering the limitations evidenced by the material itself, we feel this could be ethically questionable unless alternative electrodes and materials with piezoelectric properties are developed.

CONCLUSION

The huge potential of piezoelectric materials as a mean to produce direct mechanical stimulation lies also on the possibility of producing stimuli at a high range of frequencies and in multiple combinations, in order to avoid routine loading accommodation.

The use of piezoelectric material based actuators to produce bone mechanical stimulation seems promising in theory and the present *in vitro* and *in vivo* studies were a first step towards the validation of the concept.

Taking into account what is already known on bone physiology, and particularly, bone mechanotransduction, developing materials for bone regeneration that are able to respect bone electrophysiology seems like a logical move towards better clinical results whenever treatment of bone defects is being considered.

KEYWORDS

- **Actuator device**
- **Nitric oxide**
- **Osteoblasts**
- **Osteocytes**
- **Piezoelectric effect**
- **Polymeric piezoelectric films**

Chapter 26

Role of Water in Dielectric Properties of Starch Membranes

Perumal Ramasamy

INTRODUCTION

There are different kinds of polarization namely electronic, atomic, orientational, and ionic polarization. These polarizations vary in their frequencies. The behavior of the orientation polarization in time-dependent fields can, as a good approximation, be characterized with respect to their relaxation times. This behavior is generally denoted as *dielectric relaxation*. Dielectric spectroscopy can be used to probe the dynamics of local motions that are of dimensions less than 1 nm, segmental motions that are of dimensions 1–10 nm and dynamics of chain contour that are of dimension ~ 10–100 nm. Some of the useful properties that can be measured using dielectric measurements are: (i) Dielectric constant, (ii) Loss factor, (iii) Dissipation factor, (iv) Relaxation time, (v) Relaxation strength, and (vi) Conductivity. Dielectric spectroscopy can be used to study a wide variety of materials such as water, glass forming liquids, clusters, ice and porous materials and colloids. It can be used to study biological materials like lipids, proteins, cells, DNA, RNA, and tissues.

Biological materials have the advantage of being environmentally friendly materials. Dielectric relaxation spectroscopy (DRS) has been found to be a very useful tool in studying the polymer dynamics of polysaccharides (Butler and Cameron, 2000; Einfiled et al., 2001, 2003;.Majumder et al., 2004; Moates et al., 2000; Smits et al., 2001; Viciosa et al., 2004). In this chapter the usefulness of dielectric spectroscopy for understanding the dynamics of molecules in starch samples containing varying water contents will be discussed. Starch is emerging as a useful biomaterial in the field of energy science. It is an abundantly available biomaterial that is used for energy storage in plants. Application of starch as membranes for solid state batteries would greatly help the environment as it is a biodegradable material. Making of membranes usually involves solvents like water. Water also acts as a plasticizer. The incorporation of water in membranes can affect the transport of ions in the membranes. It is therefore essential to learn about the role of water in membranes that contain water. Also, temperature will affect the transport of charges in the membranes. Hence, in this chapter, the modifications in the dielectric properties of starch films as a function of temperature and water content will be discussed.

MATERIALS AND METHODS

Amioca starch (corn starch) was used for this study. Samples were made by hot pressing starch between two brass plates of diameter 30 mm at 1900 bar. The thickness of the discs obtained were ~ 0.4–0.5 mm. In order to make samples with 0% humidity, the discs were annealed at 100°C in oven for 24 hr. Discs were also placed inside sealed containers having $MgNO_3$ and NaCl salt for at least 2 weeks to achieve a relative humidity of 51% and 75%, respectively. The samples were quickly transferred (less than 10 min) from the sealed containers to the sample holder in the DRS machine so as to minimize absorption of atmospheric moisture.

Dielectric relaxation spectra were collected isothermally using a Novocontrol GmBh Concept 40 broadband dielectric spectrometer in the frequency range $0.1–10^6$ Hz. Temperatures were controlled within 0.2°C. The diameter of the top electrode was 15 mm while the diameter of the bottom electrode was 30 mm. In order to better resolve the spectra due to high values for dielectric loss caused by high conduction loss, the dielectric loss was calculated using the expression (1) (Wubbenhorst and Turnhout, 2002).

$$\varepsilon''der = -(\pi/2)(\partial\varepsilon'(\omega)/\partial\ln\omega) \qquad (1)$$

The relaxation positions were determined using WINFIT software.

Dielectric Constant as a Function of Water Content

The variation of ε' with respect to the frequency for various temperatures for starch stored at 0% humidity is as shown in Figure 1a and 1b. From Figure 1a and 1b, it is observed that the dielectric constant reaches a maximum of ~10 at 100°C. From Figure 1a, it is observed that a relaxation (a broad peak) exists at temperature range –100–0°C. Another relaxation is observed at temperatures close to 100°C.

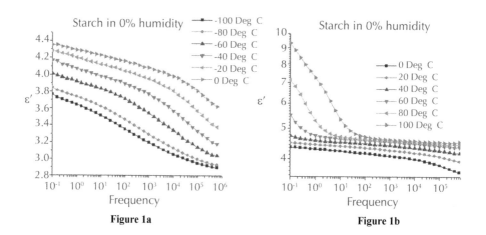

Figure 1a Figure 1b

The variation of ε' with respect to the frequency for various temperatures for starch stored at 51% humidity is as shown in Figure 2a and 2b. A relaxation is observed at temperatures –100–0°C. Another relaxation is observed at higher temperatures. The dielectric constant at low frequencies is observed to increase rapidly as temperature increases. The value of the dielectric constant is large at low frequencies due to electrode polarization.

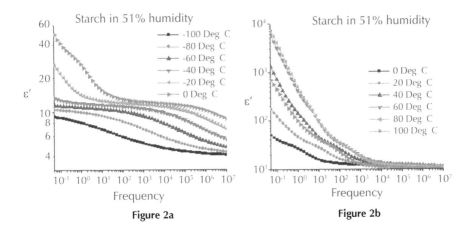

Figure 2a

Figure 2b

The variation of ε' with respect to the frequency for various temperatures for starch stored at 75% humidity is as shown in Figure 3a and 3b. The value of the dielectric constant for any given temperature or frequency for starch in 75% humidity is larger than that for dry starch or for starch in 51% humidity. The dielectric values increases rapidly with increasing temperatures at low frequencies. It has very high values (~ 10^5) at low frequencies and high temperatures due to electrode polarization. Also, the curves have almost the same value at high temperatures indicating the loss of water at higher temperatures.

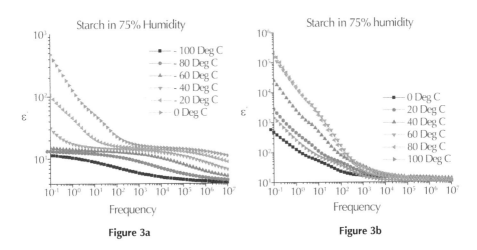

Figure 3a

Figure 3b

Dielectric Loss as a Function of Water Content

The variation of the dielectric loss (ε'') as a function of frequency for various temperatures for starch in 0% humidity is shown in Figure 4a and 4b. At low temperature, it is observed that a very prominent relaxation is observed with its frequency maxima increasing in its intensity and position as the temperature increases from –100 to 0°C.

The value of the maximum value for the dielectric loss is small at low temperatures (~ 0.2 at 0°C). As the temperature increases yet another relaxation is observed. When the sample is held at high temperature at 150°C, the dielectric loss still exist showing that the starch is not disintegrated at such temperature.

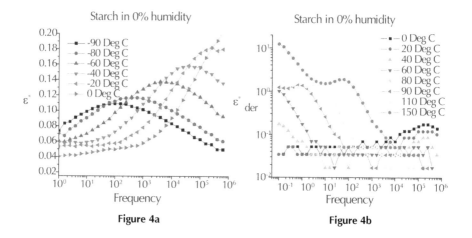

Figure 4a

Figure 4b

The variation of the dielectric loss (ε") as a function of frequency for various temperatures for starch in 0% humidity is shown in Figure 5a and 5b. In this case, the dielectric loss values are higher than that for starch in 0% humidity. At low temperatures, a broad relaxation is observed. The position of the frequency maxima increases with increasing temperatures. As the sample approaches 0°C yet another relaxation is observed. ε"$_{der}$ was used instead of ε" to better resolve the spectra due to high values for dielectric loss caused by high conduction loss. The spectra had nearly the same values at high temperatures (~ 80°C).

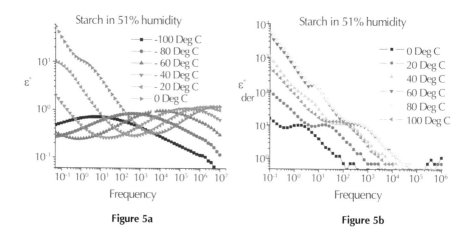

Figure 5a

Figure 5b

The variation of the dielectric loss (ε'') as a function of frequency for various temperatures for starch in 0% humidity is shown in Figure 6a and 6b. Two relaxations are observed one at temperatures $-100-0°C$ and another at $-20-100°C$. Here at $-20°C$, the beginning of the presence of the second relaxation can be observed. The values of the dielectric loss are found to be highest for starch in 75% humidity. The values for dielectric loss at 60 and 80°C are nearly the same, while that for 100°C is found to be lesser. This indicates the evaporation of the water from the sample. It is nearly the same as that at 20°C this indicates that relaxations are very much dependent upon the availability of water in the local environment of the starch molecules.

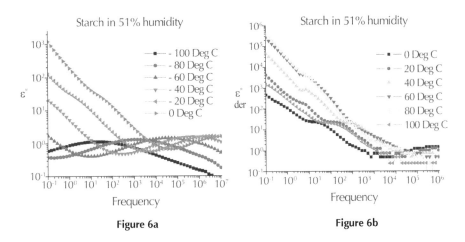

Figure 6a   Figure 6b

Relaxation Frequencies

The relaxation frequencies were determined by estimating the frequency maxima from the dielectric loss spectra. The relaxation frequencies for starch with different water contents is shown in Figure 7. At low temperatures ($-100-0°C$), it was observed that dry starch, starch in 51% humidity and starch in 75% humidity had nearly the same values. They also showed an Arrhenius behavior. At low temperatures, the relaxation frequencies increased with increase in temperature as the water content increased. The effect was found to be more at higher temperatures ($0-100°C$). It was observed that the relaxation frequencies were highest for samples with 75% humidity. This indicates that the presence of water facilitates the response of the molecules to the applied field. It is also observed that for starch with 51 and 75% humidity, the frequencies increases with increasing temperatures and then decreases after attaining a peak at ~80°C. This indicates the stiffening of the membranes due to evaporation of water. The activation energies can be calculated using the Arrhenius relation. It is found that the activation energy for dry starch and starch in 51 and 75% humidity is ~46 kJ mol⁻¹. This is indicative of presence of bound water. The corresponding dielectric strength values were found to be low (<10) indicating that it is a local relaxation process. At higher temperatures, it was observed that the activation energies were ~86 kJ mol⁻¹. Also, the dielectric strength values for water containing starch was found to be ~15–20 indicative of Maxwell–Wagner–Sillars (MWS) relaxation that arises from the charge carriers accumulated at the interphase between amorphous and crystalline regions.

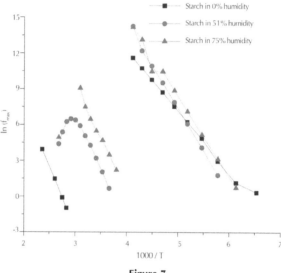

Figure 7

Conductivity Measurements

The conductivity of the samples at 20°C is shown in Figure 8. It is observed that at high frequencies the conductivity of dry starch is greater than that of starch with water in it. It is also observed that the conductivities for starch with 51 and 75% humidity coincide at high frequencies (> 10^5 Hz). The conductivity at low frequency (~ 1 Hz) increases as the water content increases.

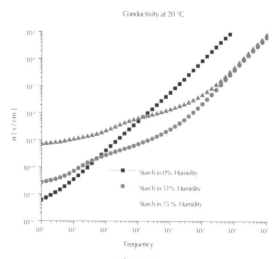

Figure 8

The DC conductivity is given by the conductivity values at 0 Hz. The conductivity for dry starch at 80°C was found to be ~10^{-12} S/cm. Figure 9 shows the variation in conductivity at 0.05 Hz for starch in 51 and 75% humidity. It is observed that the conductivity increases with increasing temperatures and it reaches a maximum at ~80°C. The profiles look similar for starch containing 51 and 75% humidity. At low temperatures ~ −80°C, the conductivities are nearly the same. The conductivity reaches maximum value at ~80°C for starch in 75% humidity. The decrease in the conductivity at temperatures >80°C is due to evaporation of water at higher temperatures.

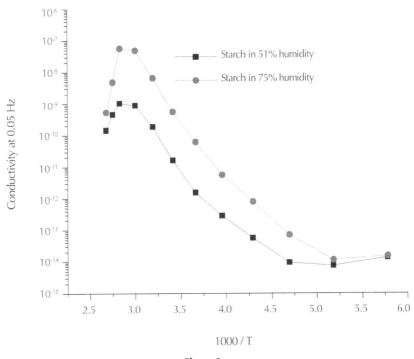

Figure 9

CONCLUSION

The dielectric characteristics of starch are highly modified in the presence of water. In the absence of water the dielectric spectra shows a local relaxation at low temperatures regions. Inclusion of water shows the presence of yet another relaxation at higher temperatures. The relaxation observed at higher temperatures is indicative of Maxwell-Wagner-Relaxation that arises from the charge carriers accumulated at the interphase between amorphous and crystalline regions. This indicates that the addition of water gives rise to both crystalline and amorphous regions in the starch films. The relaxation frequencies increase with the increase in the water content in the samples indicating that the presence of water increases the flexibility of the polymer chains. The conductivities also increase as the water content increases. The decrease in the conductivity at temperatures greater than 80°C, the decrease in the relaxation frequencies at temperature greater than 80°C and the reduction in the dielectric loss for samples having high water content at temperatures greater than 80°C are explained as a result of the evaporation of water from the starch samples at temperatures close to boiling point of water leading to stiffening of the starch membrane.

ACKNOWLEDGMENT

The author would like to gratefully thank Prof. James Runt of Department of Materials Science and Engineering, Pennsylvania State University for providing the opportunity to carry out this research as a postdoctoral researcher in his laboratory. The author also gratefully thanks Dr Georgios Polizos of Oak Ridge National Lab for his valuable help in dielectric measurements.

KEYWORDS

- **Amioca starch**
- **Dielectric loss**
- **Dielectric relaxation spectroscopy**
- **Maxwell-Wagner-Sillars**
- **Relaxation frequencies**

Chapter 27

Tribo Performance of T-BFRP Composite Subjected to Dry/Wet Contact Conditions

Umar Nirmal, Jamil Hashim, Dirk Rilling, P.V. Brevern, and B. F. Yousif

---

INTRODUCTION

Recently, new and more stringent environmental regulations coupled with the depletion of oil resources have evoked a concern among researchers to find a substitute for synthetic fibers in polymeric composites (Yousif and El-Tayeb, 2008a). As an alternative, natural fibers are becoming an attractive alternative due to their advantages over the synthetics such as recyclability, biodegradability, renewability, low cost, light weight, high specific mechanical properties, and low density (Corbiere et al., 2001; Gowda et al., 1999; Joshi et al., 2004; Yousif and El-Tayeb, 2008a; Wambua et al., 2003). Nowadays, applications of natural fiber reinforced polymeric composites can be found in housing construction material, industrial and automotive parts (Baiardo et al., 2004; Huda et al., 2008; Liu et al., 2004; Nishino et al., 2003).

It is known from the literature that, untreated oil palm (Yousif and El-Tayeb, 2007a, 2008a, 2010), sugarcane (El-Tayeb, 2008a, 2008b), banana (Pothan et al., 2003) and coir (Yousif, 2008) fibers have very poor interfacial adhesion strength with the matrix by nature. The poor interfacial adhesion is due to foreign impurities/substances which prevent the matrix to bond firmly with the fibers. Interestingly, betelnut fibers have many tiny hairy spots termed *trichomes* which protrude from the outer layer of the fiber surface (Nirmal and Yousif, 2009). The presence of *trichomes* may results in high interfacial adhesion with the polymer matrix and may prevents pulling out processes during tribological and single fiber pullout tests (SFPT).

From the tribological point of view, few works have been pursued on jute (Thomas et al., 2009), cotton (Hashmi et al., 2007), oil palm (El-Tayeb, 2008b; Yousif and El-Tayeb, 2007a, 2008a), sugarcane (El-Tayeb, 2008a, 2008b), coir (Yousif, 2008) and bamboo (Tong et al., 1998, 2005) fibers regarding their usage for tribo-polymeric composites. For instance, wear and frictional characteristics of oil palm fiber reinforced polyester composite (Yousif and El-Tayeb, 2007a, 2008a) revealed that oil palm fibers enhanced the wear performance of polyester by three to 4-folds. This was due to the presence of oil palm fibers at the surface of the composite forming a mixed layer of broken fiber and polyester debris which protected the polyester regions during the sliding.

Considering fiber orientation, the effect of sugarcane fiber has been studied on tribo-characteristics of polyester composites (El-Tayeb, 2008a). It has been found that fiber mats oriented parallel to the sliding direction showed lower wear performance than fibers oriented anti-parallel under the same test conditions. This was because in the parallel orientation, the path ahead of the wear debris is exposed, thus easing the fragmentation of fibers and removal of abrasive particles (El-Tayeb, 2008a). In anti-parallel orientation, abrasive particles were moving through different interfaces alternately, that is were more hindrance in the path of abrasive particles which constitutes resistance and traps wear debris which in turn, reduces wear.

Contact conditions (dry/wet) have an equal important role which controls the tribo performance of polymeric composites (Borruto et al., 1998; El-Tayeb, 2008b; Pothan et al., 2003; Sınmazcelik and Yılmaz, 2007; Sumer et al., 2008; Wu and Cheng, 2006; Yamamoto and Takashima, 2002; Yousif, 2008; Yu et al., 2008). It has been reported that tribo performance of some polymeric composites were improved under wet contact condition compared to dry (Wu and Cheng, 2006; Yamamoto and Takashima, 2002). It is known that increased interface temperature during adhesive dry loading conditions caused high damaged on the composite surface during sliding especially at the resinous regions due to thermo-mechanical loading conditions (Yousif and El-Tayeb, 2010). As such, the cooling effect introduced by water prevents the pullout of oil palm fibers from the polyester matrix as opposed to dry contact, that is wear is only controlled by mechanical loading (Yousif and El-Tayeb, 2008b, 2010).

In previous work by the participating authors (Nirmal and Yousif, 2009; Yousif et al., 2008), untreated betelnut fiber reinforced polyester (UT-BFRP) composite was used to study the wear and frictional behavior of the composite under dry contact condition. The work revealed that the average wear and friction coefficient of the composite were reduced by 98 and 73% compared to neat polyester namely when the fibers were oriented parallel to the sliding direction.

Thus, through the author's knowledge, there is no work reported on polymeric composites based on treated betelnut fibers under dry and wet contact conditions. Hence, the current work aims to study the effect of treated betelnut fibers on the tribo-behavior of polyester composites. The interfacial adhesion strength of the treated fiber with the polyester was determined using single fiber pullout test. The sliding wear and frictional characteristics of the developed composite were evaluated using a Block-On-Disc (BOD) machine under dry/wet contact conditions. The tests were conducted at different applied loads (5–200 N) and sliding distances (0–6.71 km) against a smooth stainless steel counter face with sliding velocity; 2.8 m/s.

MATERIALS PREPARATION

Preparation of Betelnut Fibers

The preparation of betelnut fibers was explained in a past publication done by the author (Nirmal and Yousif, 2009). The length and diameter of individual fiber were in the range of 30–50 mm and 150–200 μm respectively. However, the prepared fibers were soaked in a 6% Natrium Hydroxide (NaOH) solution mixed with tap water at temperature of 26 ± 5°C for 48 hr. The fibers were rinsed and left to dry at room temperature before being put in an oven for 5 hr at 45°C.

One can see from Figures 1(a) and (b) that significant modifications occurred when betelnut fiber was treated. Very rough fiber surface can be seen on the treated one, Figure 1(b). Moreover, the *trichome* in Figure 1(b) seems to be rougher than in Figure 1(a). This could improve the interaction between the betelnut fibers with the polyester matrix. In previous works (Yousif and El-Tayeb, 2008a, 2010), the interfacial adhesion of oil palm fibers was highly improved when the fiber was treated with 6% NaOH. For the current work, the effect of treatment on the interfacial adhesion property of betelnut fiber and its effect on the tribological behavior of the polyester composite will be explained.

The prepared fine fibers (Nirmal and Yousif, 2009) were arranged and pressed into uniform mats and the mats were then cut into the dimensions of the composite fabrication mould. The density of the fibers in mat sheets was determined to be about 200 ± 10 g/m^2. Figure 1(c) shows a micrograph of a randomly oriented treated betelnut fiber mat. The average distance of the fiber in the mat was about 83 ± 5 μm.

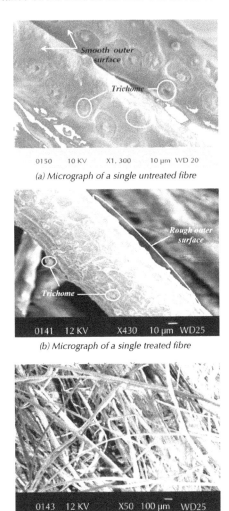

(a) Micrograph of a single untreated fibre

(b) Micrograph of a single treated fibre

(c) Micrograph of treated fibre mat

Figure 1. Micrographs of betelnut fiber. (a) Micrograph of a single untreated fiber; (b) Micrograph of a single treated fiber; (c) Micrograph of treated fiber mat.

Fiber Pullout Test

The SFPT were conducted on universal test system (100 Q Standalone) to determine the interfacial adhesion characteristics of treated betelnut fiber with the polyester matrix. Figure 2 shows the schematic drawing of the pullout test. Further detail on the sample preparation and the test procedure were explained in the past publication done by the author (Nirmal and Yousif, 2009). The loading speed was 1 mm/min. It should be mentioned here that the tensile properties of single betelnut fiber were studied for dry and wet fibers. Under wet conditions, the fibers were soaked in tap water (hardness 120–130 mg/l) for 24 hr and then tested.

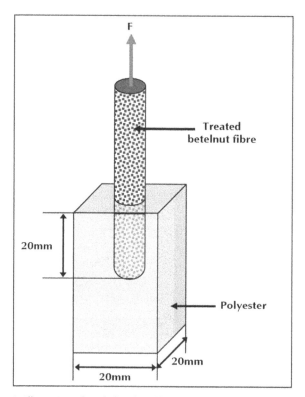

Figure 2. Schematic illustration of single betelnut fiber pullout test.

The pullout result for single fiber (dry/wet) is presented in Figure 3(a). The Figure shows that both trends (under dry/wet) are the same. The maximum stress for the dry fiber is about 280 MPa which is almost similar to the single fiber strength. Similarly, the wet fiber reached to about 250 MPa. This indicates that there is no pullout of fiber took place during the test. Moreover, the strength is also the same as the single tensile result. This shows that the interfacial adhesion of the treated fiber under dry/wet conditions is very high preventing the pulling out process. The microscopy of the pullout samples are shown in Figures 3(b) and (c) which explain the above results.

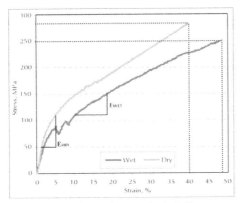

(a) Stress / Strain diagram of a single fibre

(b) Micrograph of fibre breakage after pull-out for dry test

(c) Micrograph of fibre breakage after pull-out for wet test

Figure 3. Stress/Strain diagram and corresponding micrographs for Single fiber pull-out test under dry/wet conditions (a) Stress/Strain diagram of a single fiber; (b) Micrograph of fiber breakage after pull-out for dry test; (c) Micrograph of fiber breakage after pull-out for wet test.

The main reason of higher interfacial adhesion of the fiber is due to the presents of *trichomes* and rough surface of the fiber after treating with 6% NaOH. This is a promising result which has not been reported before on natural fibers such as oil palm, sugarcane, coir and jute fibers (El-Tayeb, 2008a, 2008b; Navin and Dwivedi, 2006; Pothan et al., 2003; Thomas et al., 2009; Yousif, 2008; Yousif and El-Tayeb, 2007a, 2008a, 2010).

Preparation of Composite

Unsaturated polyester (Butanox M-60) mixed with 1.5% of Methyl Ethyl Ketone Peroxide (MEKP) as catalyst was selected as a resin for the current work. Treated betelnut fiber reinforced polyester (T-BFRP) composite was fabricated using hand lay-up technique. In composite preparation, a metal mould (100 × 100 × 12 mm) was fabricated. The inner walls of the mould were coated with a thin layer of wax as release agent. The first layer of the composite was built by pouring a thin layer of polyester. A prepared mat was placed carefully on the polyester layer. Steel roller was used to arrange the mat and eliminate trapped bubbles. This process was repeated until the composite block was built containing 13 layers of fiber mats and 14 layers of polyester. The prepared blocks were pressed at approximate pressure of 50 kPa in order to compress the fiber mats and to force out the air bubbles. The blocks were cured for 24 hr and then machined into specimens in the size of 10 × 10 × 20 mm.

TRIBOLOGICAL EXPERIMENTAL PROCEDURE

Figure 4 shows a schematic drawing of BOD machine which was used for the current work. Under wet contact condition, water system was adopted at the machine. Water was supplied to the counterface by a pump at a flow rate of 0.4 l/min. Water flowing to the counterface was collected by a container. A filter was placed in the water flow and cleaned from wear debris after each test. Accutec B6N-50 load cell was adapted to the BOD load lever to measure the frictional forces between the specimens and counterface while a weight indicator was integrated in order to capture the frictional forces simultaneously.

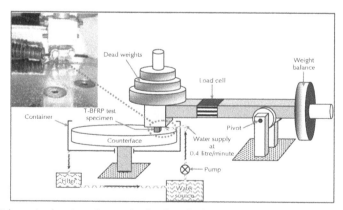

Figure 4. Schematic drawing of a newly developed Block-On-Disc (BOD) tribological machine operating under dry/wet contact conditions.

The tests were performed at a sliding velocity of 2.8 m/s, different sliding distances (0–6.72 km) and different applied loads (5–200 N). All specimens after the wet test were dried in an oven at temperature of 40°C for 24 hr. The specific wear rate was computed using Equation (1) where the weight lost of the specimens was determined using Setra weight balance (± 0.1 mg). Figure 5 illustrates the sliding direction with respect to the fibers mats under dry/wet contact conditions.

$$W_S = \frac{\Delta V}{F_N . D} \tag{1}$$

where;

W_s = Specific wear rate [mm³/N m]

ΔV = Volume difference [mm³]

F_N = Normal applied load [N]

D = Sliding distance [m]

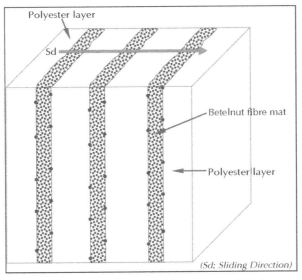

Figure 5. Schematic illustration of T-BFRP composite showing the sliding direction.

DISCUSSIONS AND RESULTS

Wear performance of T-BFRP Composite

Specific wear rate of T-BFRP composite as a function of sliding distance at different applied loads are presented in Figure 6 under dry/wet contact conditions respectively.

Under dry contact condition; Figure 6(a), specific wear rate (Ws) of the composite has less influence by sliding distance especially at higher range of applied loads. However, at an applied load of 5 N, there is an increase in Ws until 5 km of sliding distance that is a steady state reached after 5 km of sliding distance. On contrary, Figure 6(b) shows similar trends of specific wear rate. One can see that the curves are divided into two regions; "running in" and "steady state". From the Figure, as sliding distance builds up, specific wear rate gradually reduces until a steady state transition (6.72 km). Surprisingly, the steady state specific wear rate was much shorter ($\approx$ 4.2 km) as compared to the dry test ($\approx$ 5 km); cf. Figure 6(a). The presence of water helped to cool the interface that is reducing the thermo mechanical loading of the composite during the sliding. This enhanced the wear (low values of specific wear rate) namely under wet contact conditions. From Figure 6(b), one can see that superior improvement on Ws was achieved compared to the dry tests; cf. Figure 6(a). It is suggested that introducing water at the interface served two main purposes; as a cleaning and cooling agent (Baley et al., 2006; Bijwe et at., 2002). As such, in wet contact conditions, the specific wear rate of the composite was low by about five times compared to the dry tests.

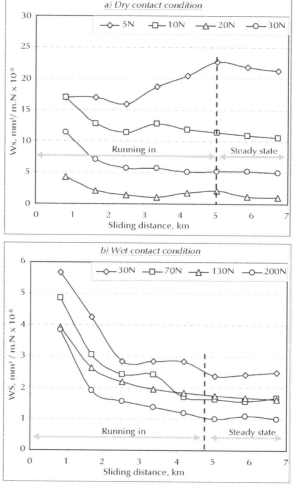

Figure 6. Specific wear rate (Ws) of T-BFRP composite *vs.* sliding distance at different applied loads and 2.8 m/s sliding velocity under dry/wet contact conditions.

Frictional Performance of T-BFRP Composite

The frictional performance of T-BFRP composite at different applied loads against sliding distances is presented in Figure 7 under dry/wet contact conditions. In general, Figure 7(a) shows that T-BFRP composite exhibits lower friction coefficient values approximately in the range of 0.4–0.7 at all applied loads. Figure 7(b) however shows a tremendous drop in friction coefficient values as compared to the dry test. One can see that the friction coefficient values were in the range of 0.01 ~ 0.08 respectively. The drastic reduction in friction coefficient under wet contact condition is due to the presence of water at the interface which assisted to wash away the generated wear debris and to reduce the interaction between asperities in contact during sliding. Similar results were reported on polyester composites based on glass fiber (Yousif and El-Tayeb, 2007b, 2008b).

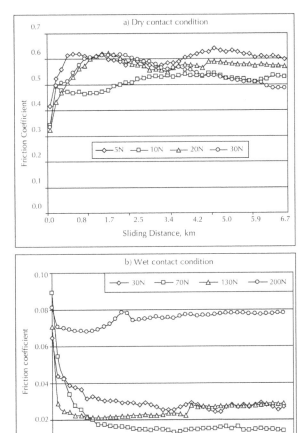

Figure 7. Friction coefficient of T-BFRP composite *vs.* sliding distance at different applied loads and 2.8 m/s sliding velocity under dry/wet contact conditions.

Worn Surfaces of the Composite Morphology

Dry Contact Condition

Figure 8(a) shows evidence of fiber debonding micro-cracks associated with generated fine debris. At longer sliding distance (5 km), Figure 8(b), the wear mechanism was predominant by plastic deformation, detachment and debonding of fibers. The Figure shows the end of fibers which is covered by polyester associated with plastic deformation indicating high intimate contact between asperities (composite and counterface) leading to higher friction coefficient values, cf. Figure 7(a). Due to the side force being anti parallel to the sliding direction, there was evidence of softened polyester (marked SP) causing higher material removal when the sliding escalates. It was reported that a high friction coefficient is possible when the contact of rubbing was between neat polyester and stainless steel (Yousif et al., 2008). Moreover, the softened polyester regions had modified the roughness of the counterface (cf. Figure 10(b)) compared to the virgin one (cf. Figure 10(a)).

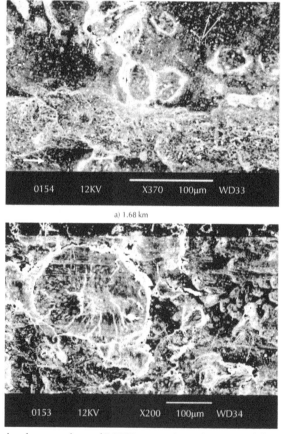

a) 1.68 km

Figure 8. Micrographs of worn surfaces of T-BFRP composite under 30 N at different sliding distances for dry contact condition (a) 1.68 km; (b) 5.0 km.
(Crack: crack, De: debonding, Dt: detachment, Fd: fine debris, Pd: plastic deformation, R: resinous, Sp: softened polyester)

Wet Contact Condition

From Figure 9(a), when the composite is subjected to low applied load (70 N) and longer sliding distance (6.72 km), the fibers were squeezed parallel to the sliding force causing debonding of fibers. The SEM image also concludes that the fibers were torn apart. However, the fibers were still in good shape that is no delamination. Consequently at higher applied loads (200 N) and shorter sliding distance (1.68 km); cf. Figure 9(b); the wear was initiated by debonding of fibers especially the ones close to the resinous regions associated with torn fibers which eventually formed wear debris during the sliding. The wear debris could have left very fine grooves on the worn surfaces of the composite as evidenced in Figure 9(b) marked 'Fg'. When the wear escalates to 6.72 km of sliding distance; Figure 9(c), the predominant wear mechanism is due to debonding and delamination of fiber mats. The Figures also confirm that there were no signs of fine grooves evidenced on the worn surfaces as the water had washed away the generated wear debris during longer sliding distance, that is 6.72 km. This may be the main reason why Ws was significantly lower at higher applied loads; 200 N which is confirmed by Figure 6(b).

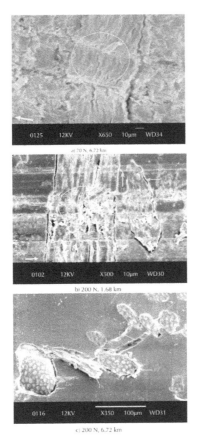

Figure 9. Micrographs of T-BFRP composite under 70 N and 200 N at different sliding distances for wet contact condition (a) 70 N, 6.72 km; (b) 200 N, 1.68 km; (c) 200 N, 6.72 km
(De: debonding, Dl: delamination, Dt: detachment, Fg: fine grooves, Fd: fine debris, Tf: torn fiber)

Effect of Sliding on Surface Roughness

Before test, the average roughness profile of the stainless steel counterface was Ra = 0.052 μm; Figure 10(a). After test under both dry/wet contact conditions, there were slight modifications on the counterface roughness. The roughness profiles of the counterface are presented in Figures 10(b) and (c). The roughness of the wear track was measured in the presence of film transfer. The film transfer was removed by acetone, where the polyester is soluble in acetone and the results are displayed in Figure 11.

From Figure 11, one can see that the average roughness values were slightly lower when the T-BFRP composite that was subjected to wet contact condition as compared to the dry test. As discussed previously, water played an important role to wash away trapped/generated wear debris between the contacting interface and thus lowering the Ra values in wet contact conditions. For dry tests, the higher roughness is due to the trapped wear debris from the fibrous and resinous regions on the counterface which contributed to increase the Ra values for all three orientations. From Figure 11, it can be said that the counterface roughness increased for both dry and wet contact conditions after testing the composite in the three orientations. However in dry contact condition; after cleaning the counterface, the roughness decreased noting that the counterface roughness is still higher than the virgin one. This indicates the presence of rough film transfer during the sliding. Interestingly, under wet contact conditions, there were not many changes in the Ra values of the counterface. It can be observed that the wear track roughness after testing before cleaning and after cleaning was not highly remarkable. This could have been because of water introduced at the interface which washed away all trapped wear particles by the T-BFRP composite test specimen during the sliding. In spite to this, the reduction of counterface surface roughness under wet contact condition was about 21% as compared to the dry test.

The optical microscopy images of the virgin counterface and after the test are shown in Figure 12 for dry/wet contact conditions. In Figure 12(b), composite experienced film transfer on the counterface. However, there was much worn polyester debris from the resinous region of the composite which caused greater surface roughness on the counterface due to the fact that the worn polyester debris are brittle by nature. When the composite was subjected to wet contact condition, the counterface was polished with the presence of water during sliding. As a result, there was no evidence of film transfer which is confirmed by Figures 12(d) and (e). Therefore, this can be the reason why the specific wear rate under wet contact condition for the three orientations was significantly lower compared to the dry test.

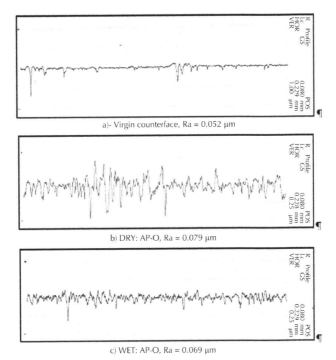

a)- Virgin counterface, Ra = 0.052 µm

b) DRY: AP-O, Ra = 0.079 µm

c) WET: AP-O, Ra = 0.069 µm

Figure 10. Roughness average profiles of the virgin counterface and after testing at 30 N applied load, (a) Virgin counterface, Ra = 0.052 µm; (b) DRY: AP-O, Ra = 0.079 µm; (c) WET: AP-O, Ra = 0.068 µm. 3.36 km sliding distance and 2.8 m/s sliding velocity under dry/wet conditions

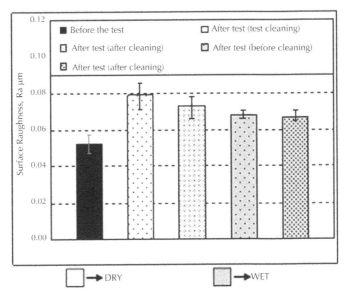

Figure 11. Roughness averages (Ra) of the counterface before and after the test under dry/wet contact conditions

Before Testing

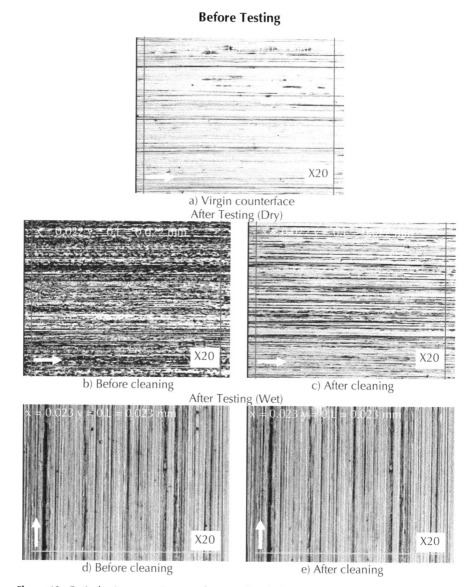

Figure 12. Optical microscopy images of counterface before and after testing the composite at applied load of 30 N and sliding distance of 3.36 km at sliding velocity of 2.8 m/s under dry/wet contact conditions; (a) Virgin counterface; (b) Before cleaning; (c) After cleaning; (d) Before cleaning; (e) After cleaning.

CONCLUSION

After conducting the experimental work and discussing the results, few points can be drawn as follows:

1. The 6% NaOH fiber treatment enhanced the wear resistance of the T-BFRP composite under dry/wet contact conditions compared to the untreated ones which was conducted previously by the participating authors (Nirmal and Yousif, 2009).

2. The presence of treated betelnut fibers in the matrix improved the wear and frictional performance of polyester, that is the average wear and friction coefficient was reduced by about 54 and 95% respectively under wet contact conditions compared to the dry.

3. The effect of introducing water at the interface served two main purposes; as a cleaning and cooling agent. As such, the Ws of the T-BFRP composite under wet test were lower by about five times compared to the dry tests.

4. Significant improvement on wear and frictional performance of the T-BFRP composite was achieved under wet contact conditions compared to dry. This was due to the tremendous reduction in the thermo mechanical loading during the sliding in wet contact conditions. In addition, higher loads up to 200 N can be applied under wet contact conditions.

5. The wear mechanism under dry contact conditions was predominated by micro-cracks, plastic deformation, debonding and detachment of fibers. Under wet contact conditions, the wear mechanism was predominant by debonding, delamination and detachment of fibers associated with loose and torn fibers.

6. The counterface surface roughness was increased after testing the T-BFRP composite under dry/wet contact conditions. For dry contact conditions, there was evidence of film transfer on the counterface meanwhile for wet contact conditions, there was no evidence of film transfer but instead the continuous rubbing by the T-BFRP composite on the counterface modified the initial surface roughness of the counterface.

KEYWORDS

- **Betelnut fiber**
- **Block-On-Disc**
- **Natural fiber**
- **Single fiber pullout tests**
- **Tribological**

Chapter 28

Properties and Nano-observation of Natural Rubber with Nanomatrix Structure

Yoshimasa Yamamoto and Seiichi Kawahara

INTRODUCTION

"Nanomatrix structure" is a novel phase separated structure for a multi component system, which may provide outstanding properties such as high elastic property and high proton conductivity (Akabori et al., 2009; Kawahara et al., 2008, 2007, 2003; Pukkate et al., 2007; Suksawad et al., 2009, in press; Yusof et al., 2008). It is defined to consist of dispersoid of a major component and matrix of a minor component. The nanomatrix structure may be formed by covering particles with a nanolayer followed by coagulation of the resulting nanolayer-covered particles. In this case, the particles are required to chemically link to the nanolayer, in order to stabilize the nanomatrix structure in equilibrium state. Especially, for the polymeric materials, the chemical linkages may be formed by gaft copolymerization of a monomer onto polymer particles in latex stage, since the particles in the latex are dispersed in water. In this regard, natural rubber has attracted much attention, since natural rubber is isolated from *Hevea brasiliensis* in latex stage (Steinbchel et al., 2001). However, natural rubber contains proteins on the surface of the rubber particle (Creighton, 1984; Steinbchel et al., 2001). Since, the proteins trap radicals to inhibit the graft copolymerization of vinyl monomer, it is important to remove the proteins from natural rubber latex (Tuampoemsab et al., 2007). This chapter will discuss the deproteinization of natural rubber latex, formation of nanomatrix structure, and properties and morphology of natural rubber with nanomatrix structure.

DEPROTEINIZATION OF NATURAL RUBBER LATEX

Removal of proteins from natural rubber (NR) may be essentially concerned with methods on how to control interactions between the rubber and proteins in the latex stage (i.e., chemical and physical interactions). The former is cleaved with proteolytic enzyme such as alkaline protease (Eng et al., 1993) and the latter is denatured with urea, which may change conformation of the proteins (Creighton, 1984). In the previous work (Eng et al., 1992; Fernando et al., 1985; Fukushima et al., 1998), the removal of proteins was mainly made in the latex stage by enzymatic deproteinization to remove proteins present on the surface of the rubber particle as a dispersoid. After the enzymatic deproteinization, the nitrogen content of NR was reduced to less than 0.02 wt%, which was about 1/20 of that of the untreated NR. Despite the significant decrease in the nitrogen content, however, problems still exist, that is, both a long incubation time necessary for the enzymatic deproteinization (i.e., more than 24 hr), and remaining proteins, peptides, or amino acid sequences. It is, thus, quite important to establish a novel procedure to remove the proteins from NR rapidly and efficiently.

The structure of NR has been proposed to consist of α-terminal, two *trans*-1,4-isoprene units, long sequence of *cis*-1,4-isoprene units, and ω-terminal, aligned in this order (Tanaka, 1983, 1989a, b). The α-terminal was inferred to be a modified dimethylallyl group that can form hydrogen bonds between proteins, while the ω-terminal was comprised of a phospholipid that may form chemical crosslinks with ionic linkages. According to the proposed structure of NR, it is expected that there is little possibility to form chemical linkages between NR and the proteins.

In previous work (Allen et al., 1963; Tangpakdee et al., 1997), NR coagulated from fresh NR latex, just after tapping from *Hevea brasiliensis*, was found to be soluble in toluene, cyclohexane, and tetrahydrofuran. In contrast, the rubber from latex preserved in the presence of ammonia contained about 30–70% gel fraction, which was insoluble in the solvents. The formation of the insoluble fraction would be concerned with the interactions of rubber and proteins, because the gel fractions are reported to be soluble in the solvents after the enzymatic deproteinization (Eng et al., 1994; Tangpakdee et al., 1997). If the interactions are physical but not chemical, it is possible to remove the proteins from the rubber after denaturation of the proteins with urea. Here, the removal of the proteins from fresh NR latex and preserved high-ammonia latex was investigated with urea in the presence of surfactant.

Deproteinization of NR with Urea

Total nitrogen content, X, of both untreated and deproteinized rubbers, is shown in Table 1. The total nitrogen content of HANR was reduced to 0.017 wt% after enzymatic deproteinization (E-DPNR), as reported in the previous study (Tanaka et al., 1997). On the other hand, it was reduced to 0.020 wt% after the treatment with urea, being similar to the nitrogen content of E-DPNR. This implies that most proteins present in NR are attached to the rubber with weak attractive forces. To remove further the proteins, the treatment with urea was carried out after the enzymatic deproteinization of HANR latex. The nitrogen content of the resulting rubber, EU-DPNR, was 0.008 wt%, less than that of E-DPNR and U-DPNR. This suggests that the most of proteins are removed by denaturation with urea, whereas the residue must be removed with proteolytic enzyme in conjunction with urea.

To assure the difference in the role between the proteolytic enzyme and urea, the treatment of fresh NR latex was made as well as HANR latex. The total nitrogen content of both untreated and deproteinized rubbers, which were prepared from fresh NR latex, is also shown in Table 1. The total nitrogen content of fresh NR was reduced to 0.014 wt% after enzymatic deproteinization (fresh E-DPNR), and 0.005 wt% after enzymatic deproteinization followed by the treatment with urea (fresh EU-DPNR), as in the case of HANR. On the other hand, after treatment with urea (fresh U-DPNR), the total nitrogen content of fresh NR was 0.004 wt%, being the least among the deproteinized rubbers. This may be explained in that most proteins present in fresh NR are attached to the rubber with weak attractive forces, which are able to be disturbed with urea. Thus, it is possible to expect that almost all proteins present in fresh NR are removed rapidly from the rubber by urea treatment.

Table 1. Nitrogencontent and incubation time for HANR, Fresh NR, and Deproteinized NR.

Specimens	Incubation time (min)	X (wt %)
HA-NR	0	0.300
E-DPNR[a]	720	0.017
U-DPNR[b]	60	0.020
EU-DPNR[c]	780	0.008
Fresh NR	0	0.450
Fresh E-DPNR	720	0.014
Fresh U-DPNR	60	0.004
Fresh EU-DPNR	780	0.005

[a] E-DPNr; enzymatically deproteinized HA-NR.
[b] U-DPNR:urea-treated HA-NR.
[c] EU-DPNR: urea-treated E-DPNR.

A plot of total nitrogen content versus time, t, required for the deproteinization of HANR and fresh NR with urea at 303 K is shown in Figure 1. The nitrogen content of HANR and fresh NR decreased suddenly to 0.022 and 0.005 wt% after 10 min, respectively, after adding urea. The difference in the nitrogen content between HANR and fresh NR may be attributed to the amount of the proteins that are weakly attracted to the rubber, as mentioned above. It is quite important to note that the nitrogen content of fresh NR decreases to and reaches a definite value of 0.004 wt% within 10 min, expressing an advantage of urea compared to proteolytic enzyme in view of the rapid, efficient deproteinization. The dependence of the total nitrogen content on temperature is also shown for fresh NR in Figure 1. The nitrogen content decreased to 0.004 wt% within 10 min at all temperatures, ranging from 303 to 363 K. This may be in part due to the ability of urea to form hydrogen bonds with the proteins and detach themselves from the rubber particles with urea, based upon the previous work (Creighton, 1984). Consequently, urea is proved to be more effective to remove the proteins from fresh NR, rather than the proteolytic enzyme.

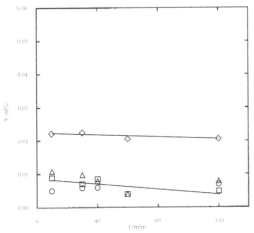

Figure 1. Nitrogen content of HA-NR at 303 K (◊), andfresh U-DPNR at 303 K (○), 333 K (□), and 363 K (△) versustime for incubation.

Concentration of Urea for Deproteinization

It is quite important to determine the amount of urea necessary to remove the proteins from NR. Figure 2 shows the relationship between the total nitrogen content of fresh NR latex versus concentration of urea after treatment with urea. The nitrogen content was dependent upon the concentration of urea and was found to be the lowest (i.e., 0.005 wt%, at urea concentration of 0.1 wt%). The higher nitrogen content of 0.007 wt% at 0.05wt% urea may be due to a lesser amount of urea that interacts with the proteins present in the rubber. In contrast, the higher nitrogen content at higher concentration of urea may be expected to be due to the residual urea that interacts with the rubber, because the nitrogen content increased linearly as the concentration of urea increased.

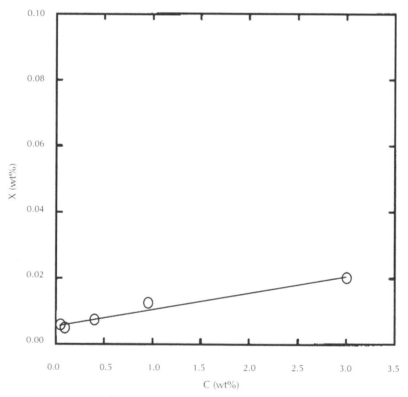

Figure 2. Nitrogen content of fresh U-DPNR versus urea concentration.

Figure 3(A) shows FTIR spectra for fresh U-DPNR, in which the peak at 3,320 cm⁻¹is identified to mono- or dipeptides, as reported in the previous study (Eng et al., 1994). In the spectrum of Figure 3(A)a–d, a small peak was observed at 3,450 cm⁻¹in addition to the peak at 3,320 cm⁻¹, the intensity of which increased as the concentration of urea increased. A mixture of synthetic *cis*-1,4-polyisoprene with urea showed a peak at 3,450 cm⁻¹characteristic of urea in addition to the peak at 3,320 cm⁻¹, which is identified to an antioxidant as a mixture, as shown in Figure 3(B). Thus, we esti-mated the concentration of residual urea present in fresh U-DPNR from the intensity of the peak at 3,450 cm^{-1}, using a calibration curve made with the mixture of synthetic *cis*-1,4-polyisoprene with urea.

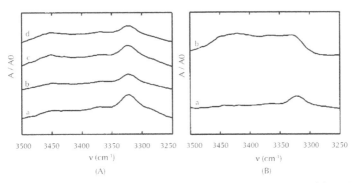

Figure 3. (A) FTIR spectrums of fresh U-DPNR at various urea concentrations: (a) 0.05 wt% urea, (b) 0.1 wt% urea, (c) 0.5wt% urea, and (d) 1.0 wt% urea; (B) FTIR spectrums of synthetic *cis*-1,4-polyisoprene and its mixture (a) synthetic*cis*-1,4-polyisoprene, and (b) synthetic *cis*-1,4-polyisoprene mixed with urea.

A plot of the concentration of the residual urea versus urea concentration is shown in Figure 4. The residual urea increased as the urea concentration increased. Further-more, the nitrogen content estimated from the concentration of the residual urea was similar to that shown in Figure 2. Thus, the increase in nitrogen content of the depro-teinized rubber was proved to be due to not only the mono- or dipeptides but also the residual urea. In the present study, a suitable condition of urea for the deproteinization was determined to be 0.1 wt%.

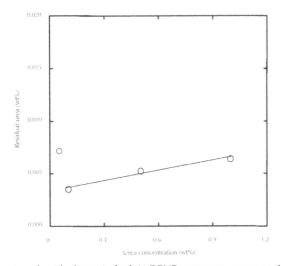

Figure 4. Concentration of residual urea in fresh U-DPNRversus urea concentration.

FORMATION OF NANOMATRIX STRUCTURE

Nanomatrix structure was formed by graft-copolymerization of styrene onto deproteinized natural rubber. To form the nanomatrix structure, 1.5 mol/kg-rubber of styrene is fed into natural rubber latex after adding an initiator (Akabori et al., 2009; Kawahara et al., 2008, 2007, 2003; Pukkate et al., 2007; Suksawad et al., 2009, in press; Yusof et al., 2008). The resulting graft-copolymer, DPNR-*graft*-PS, was characterized through ¹H-NMR spectroscopy. A typical ¹H-NMR spectrum of DPNR-*graft*-PS, is shown in Figure 5. Signals characteristic of *cis*-1,4-isoprene units appeared at 1.76, 2.10, and 5.13 ppm, which were assigned to methyl, methylene and unsaturated methyne protons of isoprene units, respectively. Broad signals at 6–7 ppm were assigned to phenyl proton of styrene units, whose intensity was dependent upon the feed of styrene as a monomer. Thus, we estimated a content of styrene units in the DPNR-*graft*-PS and conversion of styrene from a ratio of signal intensity of phenyl proton to methyl proton and the feed of styrene.

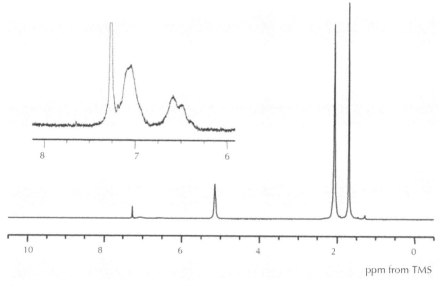

Figure 5. Typical 1H-NMR spectrum for DPNR-graft-PS.

The estimated content of styrene units and conversion of styrene are shown in Figure 6. Both of the content and conversion were dependent upon the feed of styrene; for instance, the higher the feed of styrene, the higher is the content of styrene units. In contrast, a locus of convex curve was drawn for the conversion, in which a maximum is shown at 1.5 mol/kg-rubberstyrene-feed. This suggests that the most significant feed of styrene is 1.5 mol/kg-rubber in the case of graft-copolymerization of styrene onto DPNR with *tert*-butyl hydroperoxide of 3.3×10^{-5} mol/g-rubber with tetraethylene pentamine at 30°C.

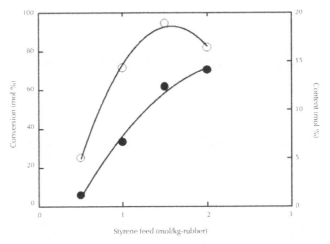

Figure 6. Conversion and content of styrene for DPNR-graft-PS.

To estimate the grafting efficiency, free polystyrene, which was a mixture present in the graft-copolymer, was removed by extraction with acetone/2-butanone 3:1 mixture. The grafting efficiency, ν, was estimated as follows:

$$\nu = \frac{\text{Mole of PS linked to DPNR}}{\text{Mole of PS produced by reaction}} \times 100$$

The estimated value of grafting efficiency of styrene is shown in Figure 7. The grafting efficiency was significantly dependent upon the feed of styrene and it showed a maximum at 1.5 mol/kg-rubber feed of styrene, as in the case of conversion of styrene. This may be explained to be due to deactivation and chain transfer of the radicals. At 1.5 mol/kg-rubber feed of styrene, almost all polystyrene, thus produced, was proved to link up to natural rubber as a grafting polymer.

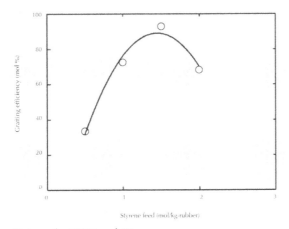

Figure 7. Grafting efficiency for DPNR-graft-PS.

MORPHOLOGY OF NANOMATRIX STRUCTURE

Recently, a serious problem on TEM images was suggested by Jinnai and co-workers, on the basis of computer simulation and observation of the phase separated structure of a block-copolymer through transmission electron micro-tomography, TEMT (Kato et al., 2007). The transmitted electron beam throughout an ultra-thin section made an apparent two-dimensional image, different from a real object, because of a thickness of the section (Kaneko et al., 2005; Kawaseet al., 2007). In other words, the TEM image is just a projection of three-dimensional entity; hence, structural information along a beam-through direction, perpendicular to the plane of the specimen, that is, Z-direction, may be lost. To obtain the real image of the phase separated structure, therefore, we must perform the three dimensional observation for the nanomatrix structure. The three dimensional image is made by stacking tomograms reconstructed from a series of tilted transmitted two dimensional images, which are taken by TEM. Although a resolution along the Z-direction is known to be less than that in the specimen plane, we may recognize the real structure through the Z-direction. Furthermore, we have to take a notice of a diameter of the dispersoid of the nanomatrix structure, which is about a micro-meter in diameter. Since, the thickness of the ultra-thin section is less than several hundred nm, the three-dimensional image is not obtained for the nanomatrix structure as long as we adapt TEMT; that is, the TEMT image is similar to the two-dimensional TEM image. To observe the nanomatrix structure in a range of nano-metric scale and micro-metric scale, thus, we have to apply not only TEMT but also recently proposed dual beam electron microscopy, that is field-emission scanning electron microscopy equipped with focused ion beam, FIB-SEM (Beach et al., 2005; Brostow et al., 2007; Kato et al., 2007). Using these techniques, we shall investigate a relationship between the properties and three-dimensional structure for the nanomatrix structured material.

After alignment of the observed images of surfaces of DPNR-*graft*-PS, which were sliced with the focused ion beam at each 102 nm interval, a three-dimensional image was made for the nanomatrix structure, as shown in Figure 8. Since the surfaces were completely flat, without damage or void, we successfully constructed the three-dimensional image of the nanomatrix structure in micro-metric scale. At this low magnification, the brighteach other in all directions. The natural rubber particles were positively confirmed to be dispersed into the nanomatrix of polystyrene.

The TEMT observation at high magnification makes possible to investigate the matrix, precisely. The three-dimensional TEMT image is shown in Figure 9. A curvature of the natural rubber particles was well-shown in the image, in spite of a limitation of thickness of the ultra thin section, that is about 100 nm. The natural rubber particles of various dimensions were randomly dispersed in the polystyrene matrix, reflecting a broad distribution of the particle dimension, that is 50 nm–3 μm in volume mean particles diameters (Kawahara et al., 2001). The three-dimentional image reveals not only the connected matrix but also it defect, at which rubber particles are fused to each other.

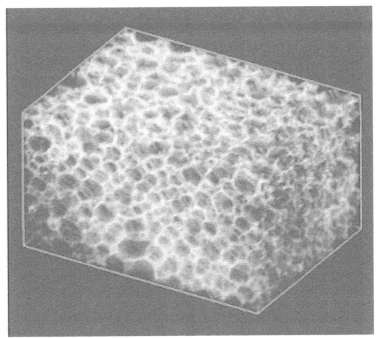

Figure 8. FIB-SEM image of DPNR-graft-PS with the nanomatrix structure before annealing. After aligning the observed surfaces sliced with the focused ion beam at each 102 nm interval, a three-dimensional image in micro-metric scale was made for the rubber with nanomatrix structure. Box size of the three-dimensional image is 6045 nm, 4123 nm and 5000 nm in X, Y, and Z directions, respectively.

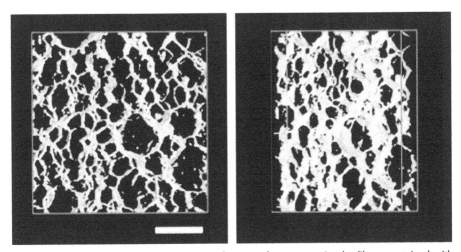

Figure 9. Three-dimensional TEMT image. To focus on the nanomatrix, the film was stained with RuO_4 and the matrix was colored in the image; hence, transparent domains represent natural rubber and light yellow domains represent polystyrene. A mirror image indicates a three-dimensional picture. Scale bar shows 1 mm. Box size of the three-dimensional image is 3740 nm, 3740 nm, and 472 nm in X, Y, and Z directions, respectively.

In order to focus on the defect to understand a reason why the dramatic increase in the plateau modulus occurs after annealing the rubber with the nanomatrix structure, we made the three-dimensional image of the nanomatrix at extremely high magnification. Figure 10 and Figure 11 show the three-dimensional TEMT images of nano-matrix structure before and after annealing the rubber with nanomatrix structure at 130°C, respectively. The image of a through view in Figure 10 apparently showed the connected nanomatrix before annealing. However, the edge views revealed that a part of the nanomatrix was disconnected; that is, many lumps of granular polystyrene gathered to form the nanomatrix structure. A distance between the polystyrene-granules was less than several nm, suggesting a flocculation of the granules. In contrast, after annealing, the granules were partly fused to each other (Figure 11). Thus, the annealing may bring about the connected nanomatrix.

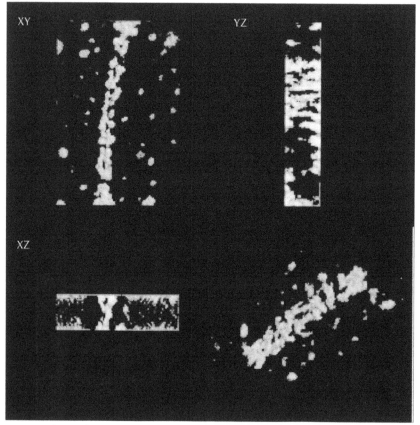

Figure 10. Orthogonal cross-sectional views and three-dimensional TEMT image of the rubber with the nanomatrixstructure. Box size is 289 nm, 434 nm, 65 nm, where dark domains represent natural rubber and bright domains represent polystyrene. In a XY through view, the nanomatrix is confirmed to be about 15 nm in thickness, which is disconnected to each other in XZ and YZ edge views. The three-dimensional TEMT image reveals that a part of the matrix was connected to each other, whereas the other was disconnected, before annealing.

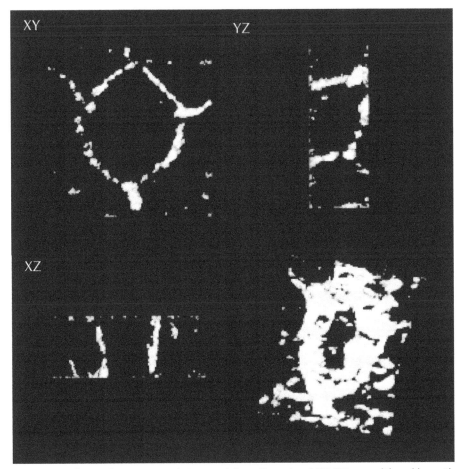

Figure 11. Orthogonal cross-sectional views and three-dimensional TEMT image of the rubber with the nanomatrixstructure after annealing at 130°C. Box size is 362 nm, 362 nm, 139 nm, where dark domains represent natural rubber and bright domains represent polystyrene. In a XY through view, the nanomatrix of about 15 nm in thickness is shown to be connected each other. XZ and YZ edge views reveal that the thin nanomatrix is connected to each other after annealing. We may see the fused nanomatrix in three-dimensional TEMT image.

PROPERTIES OF NATURAL RUBBER WITH NANOMATRIX STRUCTURE

Elastic Property of Natural Rubber

A storage modulus at plateau region (plateau-modulus) versus frequency for natural rubber, the rubber with the nanomatrix structure (DPNR-*graft*-PS) and the rubber with the island-matrix-structure is shown in Figure 12. The value of the plateau-modulus of natural rubber was about 10^5 Pa, as in the case of that reported in the literature (Roberts, 1988). When the nanomatrix structure was formed in natural rubber, the plateau-modulus of the rubber increased about 10 times as high as that of natural rubber. In contrast, the plateau-modulus increased a little, about 1.3 times, when the island-matrix-structure was formed in natural rubber. The 1.3 times increase in the plateau-modulus is the same as an ideal increase of 1.3 times estimated from the moduli of natural rubber and polystyrene by the following expression (Takayanagi, 1966);

$$\sigma = E\gamma$$

$$E = \left\{ \frac{1-\lambda}{E_{PS}} + \frac{\lambda}{\phi E_{NR} + (1-\phi)E_{PS}} \right\}$$

where λ and φ represent volume fractions of the fragments in series and parallel models, respectively, and E_i is a storage modulus of i component, that is, ENR=10^5 Pa and EPS=10^9 Pa. The large increase in the plateau-modulus is an advantage of the nanomatrix structure, compared to the island-matrix structure, in which a ratio of natural rubber to polystyrene is the same, that is, about 90:10. We find a further increase in the plateau-modulus of the rubber with nanomatrix structure after annealing it at 130°C, above glass transition temperature of polystyrene. The 35 times increase in the plateau-modulus is a remarkable virtue characteristic of the nanomatrix structure.

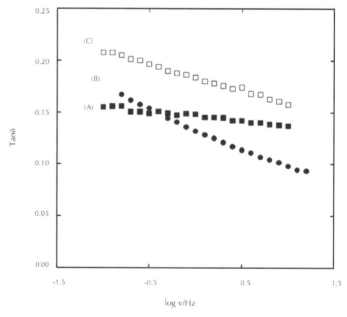

Figure 12. Storage modulus at plateau region versus frequency for (A) natural rubber, (B) DPNR-graft-PS (PS: ca 10%) with island-matrix structure, (C) DPNR-graft-PS (PS: ca 10%) with the nanomatrix structure annealed at 30°C and (D) DPNR-graft-PS (PS: ca 10%) with the nanomatrix structure annealed at 130°C. The plateau modulus characteristic of the rubbery material increased a little with the island-matrix-structure and dramatically with the nanomatrix structure, respectively. As for the nanomatrix structured material, the plateau modulus increased about 35 times as high as that of natural rubber, after annealing it at 130°C.

Figure 13 shows a plot of loss tangent, tanδ, versus frequency at plateau regain. The value of tanδ of natural rubber was in a range of 0.1 to 0.15, as in the case of literature value (Roberts, 1988). When we form the nanomatrix structure, the tanδ of the product increased dramatically at high frequency region but a little at low frequency region. After annealing the product at 130°C, the tanδ increased significantly over whole frequency range at plateau region. This may be explained to be due to the well-connected nanomatrix at 130°C but less-connected one at 30°C. Thus, the product was proved to accomplish not only high storage modulus but also outstanding loss modulus, where tanδ is a ratio of loss modulus to storage modulus. The difference in the storage modulus and loss tangent at the plateau region must come from a change in the morphology.

Based upon the three-dimensional observation at low and extremely high magnifications, the 10 times increase in the storage modulus may be associated with the flocculation of the polystyrene-granules to form the nanomatrix structure, which is distinguished from the isolated dispersoid in the island-matrix structure. The less than several nano-metric spaces between the granules in the nanomatrix enable them to interfere to increase the storage modulus. The further increase in the plateau-modulus, that is 35 times, may be explained to be due to the connected nanomatrix after annealing. In an ordinary circumstance, the minor hard-component in the multi component system, consisting of hard and soft components, is known to become the island phase in the island-matrix-structure for polymer blends or block- and graft-copolymers. In this case, little increase in the modulus is anticipated for the multi component systems. In contrast, it is difficult to form co-continuous structure for the natural rubber/polystyrene 90:10 blend. Only by forming the nanomatrix of the minor component, we will create tough, functional materials, which we have never seen.

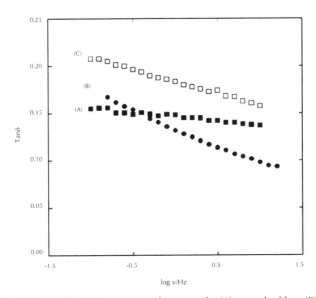

Figure 13. Loss tangent at plateau region versus frequency for (A) natural rubber, (B) DPNR-graft-PS (PS: ca 10%) with the nanomatrix structure annealed at 30°C and (C) DPNR-graft-PS (PS: ca 10%) with the nanomatrix structure annealed at 130°C. The loss tangent increased at high frequency region after forming the nanomatrix structure, and it increased significantly after annealing it at 130°C.

Electric Property of Natural Rubber

Figure 14 shows TEM images for (a) DPNR-*graft*-PS-1.5, (b) DPNR-*graft*-PS-4.5 and (c) DPNR-*graft*-PS-5.5. The ultra thin sections of the DPNR-*graft*-PS-1.5, the DPNR-*graft*-PS-4.5 and the DPNR-*graft*-PS-5.5 were stained with OsO_4, in which the bright domains represent polystyrene and the dark domains represent natural rubber. As for the DPNR-*graft*-PS-1.5, the natural rubber particles of about 1 μm in average diameter, ranging from 50 nm to 3 μm in volume mean particle diameter, were well dispersed in polystyrene matrix of about 15 nm in thickness. In contrast, for the DPNR-*graft*-PS-4.5 and the DPNR-*graft*-PS-5.5, the natural rubber particles of about 1 μm in average diameter were dispersed intouncontinuous matrix of condensed polystyrene particles of about 60 nm in diameter that are densely close to each other in the matrix. On the other hand, Figure 14 (d) shows TEM imagesfor thesulfonated DPNR-*graft*-PS-5.5, which was prepared bysulfonation of the DPNR-*graft*-PS-5.5 (32 wt% polystyrene) with 0.8 N chlorosulfonic acid. The ultra thin section of

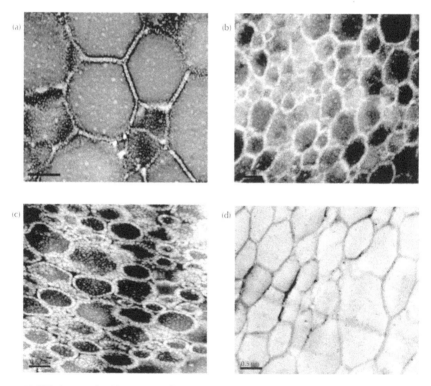

Figure 14. TEM images for (a) DPNR-*graft*-PS-1.5, (b) DPNR-*graft*-PS-4.5 (c) DPNR-*graft*-PS-5.5 and (d) DPNR-*graft*-PS-5.5 sulfonated with 0.8 N chlorosulfonic acid 30 °C for 5 h. The ultra thin sections of about 100 nm in thickness for (a) DPNR-*graft*-PS-1.5, (b) DPNR-*graft*-PS-4.5 (c) DPNR-*graft*-PS-5.5 were stained with OsO_4 at room temperature for 3 min, in which gloomy domains represent natural rubber and bright domains represent the PS. On the other hand, the ultra thin section of about 100 nm in thickness for (d) sulfonated DPNR-*graft*-PS-5.5 was stained with RuO_4 for 1 min, in which bright domains represent natural rubber and gloomy domains represent the PS.

hesulfonated DPNR-*graft*-PS was stained with RuO$_4$, in which the bright domains represent the DPNR and the gloomy domains represent polystyrene. The sulfonation of DPNR-*graft*-PS-5.5made connecting the PS particles to form completely continuous nanomatrix, since the PS particles were dissolved into chloroform before sulfonation with chlorosulfonic acid, but the DPNR particles were not due to three-dimensional network structure. Thickness of the connected nanomatrix is about 80 nm, reflecting that several PS particles of about 60 nm in diameter fuse to each other after sulfonation. A volume fraction of the resulting sulfonated PS is estimated by image-analysis to be about 35 vol% in the sulfonated DPNR-*graft*-PS-5.5. The estimated value of the volume fraction is quite similar to 36 vol% of fed styrene for graft copolymerization, as expected from more than 90 mol% conversion and about 80 mol% grafting efficiency of styrene.

Figure 15 shows solid state ^{13}C CP/MAS NMR spectra for (a) DPNR-*graft*-PS-5.5 and (b) DPNR-*graft*-PS-5.5 sulfonated with 0.8 N chlorosulfonic acid. As for the DPNR-*graft*-PS-5.5 (Figure 15 (a)), signals at 23.4 (5), 26.5 (6) and 32.4 (7) ppm in aliphatic region are assigned to methyl and methylene carbon atoms of *cis*-1,4-isoprene units, whereas the signals at 125.2 (9) and 134.8 (10) ppm in olefinic region are assigned to C=C of *cis*-1,4-isoprene units. Furthermore, three signalssignals appeared at 40.0 (8), 127.1 ppm (11) and 145.7 ppm (12), respectively, which are assigned to aliphatic carbon of methylene group and aromatic carbons of styrene unit, respectively.

After sulfonation, signals characteristic of DPNR and PS disappear, whereas broad signals spectrum appear to overlap to each other in aliphatic region and new signal appears in olefinic region, as shown in Figure 15 (b). The new signal at 139.5 ppm (3) in Figure 15 (b), appears after sulfonation, which overlapped with the signals at 127.2 (2) and 149.4 (4) ppm. According to the previous literatures (Kawahara et al., 2003), the signal at 139.5 ppm (3) is assigned to aromatic carbon substituted with sulfonic acid group, whereas the broad signals in aliphatic region ranging from 10 ppm to 40 ppm were assigned to methyl and methylene carbons of cyclized *cis*-1,4-isoprene units (Patterson et al., 1987). Therefore, the overlapped broad signals, marked as no.5 in Figure 15 (b), are explained to be due to the cyclization of DPNR, which is supported by a dramatic increase in glass transition temperature, T_g, and gel content (Table 2).

Electrochemical properties of the sulfonated DPNR-*graft*-PS are summarized in Table 2. A conversion of the sulfonation, estimated from a sulfur content (mol%) of the sulfonated DPNR-*graft*-PS-5.5, is 75 mol% at 0.8 N chlorosulfonic acid. The value of the ion exchange capacity (IEC) of the sulfonated DPNR-*graft*-PS-5.5 is 2.4 meq/g at 75 mol% sulfur content, which is larger than the value of the IEC of sulfonated PS, itself, and Nafion®117. On the other hand, the value of the proton conductivity of the sulfonated DPNR-*graft*-PS-5.5 was 9.5×10^{-2} Scm^{-1} at 75 mol% sulfur content, which was about third as large as that of the sulfonated PS. The large values of the IEC and the proton conductivity may be attributed to the formation of the nanomatrix structure as a channel for the ion-transpotation, since the well-connected nanomatrix of the sulfonated PS is formed between the cyclized DPNR particles.

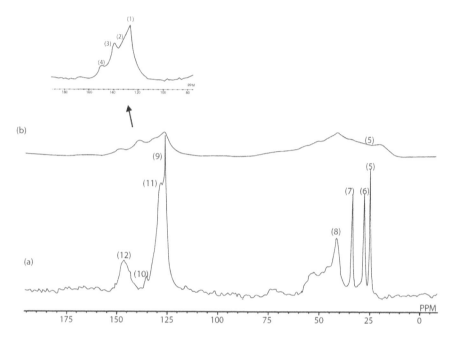

Figure 15. ^{13}C CP/MAS NMR spectra for (a) DPNR-*graft*-PS-5.5 and (b) DPNR-*graft*-PS-5.5 sulfonated with 0.8 N ClSO$_3$H.

The water uptake of the sulfonated DPNR-*graft*-PS-5.5 at 75 mol% sulfur content was 1/15 times as low as that of the sulfonated PS. It was comparable to the water uptake of Nafion®117. This may be explained to be due to an effect of the nanomatrix. The nanomatrix structure was found to play an important role in not only the excellent proton conductivity but also less water uptake, which were necessary for the proton conductive polymer electrolyte.

Table 2. T_g, sulfur content, and electrochemical properties of sulfonated DPNR-*graft*-PS, sulfonated PS and Nafion®117.

Sample name	T_g °C	Gel content wt%	Sulfur content wt%	IEC meq/g	Conductivity S/cm	Water uptake wt%
Sulfonated DPNR-*graft*-PS 0.8 N	140.6	90.4	75	2.43	9.5×10^{-2}	24.3
Sulfonated PS	–	–	–	1.50	3.2×10^{-2}	340.0
Nafion®117	169.2	90.6	–	1.02	8.0×10^{-2}	18.6

CONCLUSION

Natural rubber with the nanomatrix structure was found to possess the outstanding viscoelastic properties and the excellent proton conductivities. Natural rubber with the nanomatrix structure was prepared by graft-copolymerization of styrene onto

DPNR. The increases in storage modulus (G′) for the DPNR-*graft*-PS-1.5 and the DPNR-*graft*-PS-4.5, which were about 10 and 100 times as high as that of natural rubber, were attributed to the formation of nanomatrix structure. After sulfonation of the DPNR-*graft*-PS-5.5, the values of the proton conductivity and water uptake was 9.5×10^{-2} and 24.3 wt%, which were higher than those of the sulfonated PS itself and corresponded to the commercial proton conductive polymer electrolyte.

KEYWORDS

- **Deproteinization**
- **Deproteinization**
- **Nanolayer**
- **Nanomatrix**
- **Proteolytic enzyme**

References

1

An, X., Su, Z., and Zeng, H. (2003). Preparation of highly magnetic chitosan particles and their use for affinity purification of enzymes. *J. Chem. Technol. Biotechnol.* **78**, 596–600.

Arai, S. and Akiya, F. (1978). Desalination reverse osmotic membranes and their preparation, U.S. Patent, 4111810.

Arrascue, M. L., Garcia, H. M., Horna, O., and Guibal, E. (2003). Gold sorption on chitosan derivatives. *Hydrometallurgy* **71**(1–2), 191–200.

Asfari, Z., Bohmer, V., Harrowfield, J., and Vicens, J. (2001). Calixarenes 2001. Kluwer Academic Publishers, Dordrecht.

Atia, A. A. (2005). Studies on the interaction of mercury(II) and uranyl(II) with modified chitosan resins. *Hydrometallurgy* **80**, 13–22.

Atia, A. A., Donia, A. M., and Shahin, A. E. (2005). Studies on the uptake behavior of a magnetic Co_3O_4-containing resin for Ni(II), Cu(II) and Hg(II) from their aqueous solutions. *Sep. Purif. Technol.* **46**, 208–213.

Atia, A. A., Donia, A. M., and Yousif, A. M. (2003). Synthesis of amine and thio chelating resins and study of their interaction with zinc(II), cadmium(II) and mercury(II) ions in their aqueous solutions. *React. Funct. Polym.* **56**(1), 75–82.

Aydin, A., Imamoglu, M., and Gulfen, M. (2008). Separation and recovery of gold(III) from base metal ions using melamine–formaldehyde–thiourea chelating resin. *J. Appl. Polym. Sci.* **107**(2), 1201–1206.

Babel, S. and Kurniawan, T. A. (2003). Low-cost adsorbents for heavy metals uptake from contaminated water: A review. *J. Hazard Mater.* **B97**(1), 219–243.

Bailey, S. E., Olin, T. J., Bricka, R. M., and Adrian, D. D. (1999). A review of potentially low costs sorbents for heavy metals. *Water Res.* **33**(11), 2469–2479.

Brugnerotto, J., Desbrie`res, J., Roberts, G., and Rinaudo, M. (2001b). Characterization of chitosan by steric exclusion chromatography. *Polym J.* **42**, 9921–9927.

Brugnerotto, J., Lizardi, J., Goycoolea, F. M., Arg elles-Monal, W., Desbrières, J., and Rinaudo, M. (2001a). An infrared investigation in relation with chitin and chitosan characterization. *Polym J.* **42**, 3569–3580.

Butelman, F. (1991). U.S. Patent 5021561.

Butewicz, A., Campos Gavilan, K., Pestov, A. V., Yatluk, Y., Trochimczuk, A. W., and Guibal, E. J. (2010). Palladium and platinum sorption on a thiocarbamoyl-derivative of chitosan. *J. of Applied Polymer Science.* **116**(6), 3318–3330.

Chang-hong, P., Yi-feng, C., and Mo-tang, T. (2003). Synthesis and adsorption properties of chitosan-crown ether resins. *J. Cent. South Univ. Technol.*, **10**(02), 103–107.

Chanthateyanonth, R., Ruchirawat, S., and Srisitthiratkul, C. (2010). Preparation of New Water-Soluble Chitosan Containing Hyperbranched-Vinylsulfonic Acid Sodium Salt and Their Antimicrobial Activities and Chelation with Metals. *J. Appl. Polym. Sci.* **116**, 2074–2082.

Chiou, M. S. and Li, H. Y. (2003). Adsorption behavior of reactive dye in aqueous solution on chemical cross-linked chitosan beads, *Chemosphere* **50**, 1095.

Cimerman, Z., Galic, N., and Bosner, B. (1997). The Schiff bases of salicylaldehyde and aminopyridines as highly sensitive analytical reagents. *Anal. Chim. Acta*, **343**, 145–153.

Deans, J. R. and Dixon, B. G. (1992). Uptake of Pb^{2+} and Cu^{2+} by novel biopolymers. *Water Research*, **26**(4), 469–472.

Ding, S. M., Zang, X. Y., Feng, X. H., Wang, Y. T., Ma, S. L., Peng, Q., and Zhang, W. A. (2006). Synthesis of *N,N'*-diallyl dibenzo 18-crown-6 crown ether cross-linked chitosan and their adsorption properties for metal ions. *React Funct Polym.* **66**(6), 357–363.

Donia, A. M., Atia, A. A., and Elwakeel, K. Z. (2005). Selective separation of mercury(II) using a synthetic resin containing amine and mercaptan as chelating groups. *React. Funct. Polym.* **65**, 267–275.

Donia, A. M., Atia, A. A., El-Boraey, H. A., and Mabrouk, D. (2006a). Uptake studies of copper(II) on glycidyl methacrylate chelating resin containing Fe_2O_3 particles. *Sep. Purif. Technol.* **49**, 64–70.

Donia, A. M., Atia, A. A., El-Boraey, H. A., and Mabrouk, D. H. (2006b). Adsorption of Ag(I) on glycidyl methacrylate/*N,N*-methylene *bis*-acrylamide chelating resins with embedded iron oxide. *Sep. Purif. Technol.* **48**, 281–287.

Donia A. M., Atia A. A., and Elwakeel K. Z. (2008). Selective separation of Hg(II) using magnetic chitosan resin modified with Schiff's base derived from thiourea and glutaraldehyde. *J. Hazard Mater.* **151**, 372–379.

Eikebrokk, B. and Saltnes, T. (2002). NOM removal from drinking water by chitosan coagulation and filtration through lightweight expanded clay aggregate filters, *J. Water Supply Res. Technol. Aqua.* **51**, 323.

Emara, A. A. A., Tawab, M. A., El-ghamry, M.A., and Elsabee Maher, Z. (2011). Metal uptake by chitosan derivatives and structure studies of the polymer metal complexes. *Carbohyd Polym.* **83**(1), 192–202.

Favere, V. T. (2008). Chitosan cross-linked with a metal complexing agent: Synthesis, characterization and copper(II) ions adsorption. *React. Funct. Polym.* **68**, 572–579.

Fujiwara, K., Ramesh, A., Maki, T., Hasegawa, H., and Ueda, K. (2007). Adsorption of platinum(IV), palladium(II), and gold(III) from aqueous solutions onto L-lysine modified cross-linked chitosan resin. *J. Hazard Mater.* **146**(1–2), 39–50.

Gerente, C., Lee, V. K. C., Le Cloirec, P., and McKay, G. (2007). Application of chitosan for the removal of metals from wastewaters by adsorption - Mechanisms and models review. *Crit. Rev. Env. Sci. Tec.* **37**(1), 41–127.

Gotoh, T., Matsushima, K., and Kikuchi, K. (2004). Adsorption of Cu and Mn on covalently cross-linked alginate gel beads. *Chemosphere* **55**, 57–64.

Guibal, E. (2004). Metal ion interactions with chitosan—A review. *Separ. Purif. Technol.* **38** (1), 43–74.

Guibal, E., Van Vooren, M., Dempsey, B. A., and Roussy, J. (2006a). *Sep. Sci. Technol.* **41**(11), 2487–2514.

Guibal, E., Van Vooren, M., Dempsey, B., and Roussy, J. (2006b). A review of the use of chitosan for removal of particulate and dissolved contaminants. *Separ. Sci. Technol.* **41**(11), 2487–2514.

Gutsche, C. D. (1998). In *Calixarenes Revisited*, J. F. Stoddart (Ed.). The Royal Society of Chemistry, Cambridge.

Hall, L. D. and Yalpani, M. (1980). Enhancement of the metal-chelating properties of chitin and chitosan. *Carbohydr. Res.* **83**, C5–C7.

Hasan, M., Ahmad, A. L., and Hameed, B. H. (2008). Adsorption of reactive dye onto cross-linked chitosan/oil palm ash composite beads. *Chem. Eng. J.* **136**, 164–172.

Heux, L., Brugnerotto, J., Desbrie`res, J., Versali, M. F., and Rinaudo, M. (2000). Solid state NMR for determination of degree of acetylation of chitin and chitosan in A. Niger. *Biomacromolecules.* **1**(4), 746–751.

Ho, Y. S. and McKay, G. (1998). Sorption of dye from aqueous solution by peat. *Chem. Eng. J.* **70**, 115–124.

Hosseini, M., Mertens, S. F. L., Ghorbani, M., and Arshadi, M. R. (2003). Asymmetrical Schiff bases as inhibitors of mild steel corrosion in sulphuric acid media. *Mater. Chem. Phys.* **78**, 800–808.

Hubicki, Z., Leszczyn´ska, M., Łodyga, B., and Łodyga, A. (2007). Recovery of palladium(II) from chloride and chloride-nitrate solutions using ion-exchange resins with S-donor atoms. *Desalination* **207**(1), 80–86.

Igwe, J. C. and Abia, A. A. (2007). Adsorption isotherm studies of Cd (II), Pb (II) and Zn (II) ions bioremediation from aqueous solution using unmodified and EDTA-modified maize cob. *Ecletica Quimica* **32**(1), 33–42.

Ikeda, M., Gotanda, T., Imamura, Y., and Hirakawa, C. (1999). Method for microbially decomposing organic compounds and method for isolating microorganism, U.S. Patent, 5919696,

Juang, R. S. and Chiou, C. H. (2001). Feasibility of the use of polymer-assisted membrane filtration for brackish water softening, *J. Membr. Sci.* **187**, 119.

Jung, B. O., Kim, C. H., Choi, K. S., Lee, Y. M., and Kim, J. J. (1999). Preparation of amphiphilic chitosan and their antimicrobial activities. *J. Appl. Polym. Sci.* **72**(13), 1713–1719.

Kałedkowski, A. and Trochimczuk, A. W. (2006). Chelating resin containing hybrid calixpyrroles: New sorbent for noble metal cations. *React. Funct. Polym.* **66**(9), 957–966.

Kartal, S. N. and Imamura, Y. (2005). Removal of copper, chromium, and arsenic from CCA-treated wood onto chitin and chitosan. *Bioresource Technol.* **96**, 389–392.

Katarina, R. K., Oshima, M., and Motomizu, S. (2009). High-capacity chitosan-based chelating resin for on-line collection of transition and rare-earth metals prior to inductively coupled plasma-atomic emission spectrometry measurement. *Talanta.* **79**, 1252–1259.

Katarina, R. K., Takayanagi, T., Oshima, M., and Motomizu, S. (2006). Synthesis of a chitosan-based chelating resin and its application to the selective concentration and ultratrace determination of silver in environmental water samples. *Anal. Chim. Acta.* **558**, 246–253.

Katarina, R. K., Takayanagi, T., Oshita, K., Mitsuko Oshima, M., and Motomizu, S. (2008). Sample Pretreatment Using Chitosan-based Chelating Resin for the Determination of Trace Metals in Seawater Samples by Inductively Coupled Plasma–Mass Spectrometry. *Analytic. Sci.* **24**, 1537–1544.

Kavitha, D. and Namasivayam, C. (2007). Experimental and kinetic studies on methylene blue adsorption by coir pith carbon. *Bioresource Technology* **98**(1), 14–21.

Kittur, F. S., Prashanth, K. V. H., Udaya Sankar, K., and Tharanathan, R. N. (2002). Characterization of chitin, chitosan and their carboxymethyl derivatives by differential scanning calorimetry. *Carbohydr Polym.* **49**, 185–193.

Kotrba, P. and Ruml, T. (2000). Bioremediation of heavy metal pollution exploting constituents, metabolites and metabolic pathway of livings. *Collect. Czech. Chem. Commun.* **65**, 1205–1217.

Krishnapriya, K. R. and Kandaswamy, M. (2010). A new chitosan biopolymer derivative as metal-complexing agent: Synthesis, characterization, and metal(II) ion adsorption studies *Carbohydr Res.* **345**(14), 2013–2022.

Kyzas, G. Z., Kostoglou, M. and Lazaridis, M. K. (2009). Copper and chromium(VI) removal by chitosan derivatives—Equilibrium and kinetic studies. *Chem Eng J.* **152**(2–3), 440–448.

Laus, R., Costa, T. G., Szpoganicz, B., and Favere, V. T. (2010). Adsorption and desorption of Cu(II), Cd(II) and Pb(II) ions using chitosan cross-linked with epichlorohydrin-triphosphate as the adsorbent. *J. Hazard Mater.* **183**(1–3), 233–241.

Lee, S. T., Mi, F. L., Shen, Y. J., and Shyu, S. S. (2001). Equilibrium and kinetic studies of copper(II) ion uptake by chitosan-tripolyphosphate chelating resin. *Polymer* **42**(5), 1879–1892.

Li, F., Bao, C., Zhang, J., Sun, Q., Kong, W., Han, X., and Wang, Y. (2009). Synthesis of Chemically Modified Chitosan with 2,5-Dimercapto-1,3,4-thiodiazole and Its Adsorption Abilities for Au(III), Pd(II), and Pt(IV). *Inc. J. Appl. Polym. Sci.* **113**, 1604–1610.

Malkondu, S., Kocak, A., and Yilmaz, M. (2009). Immobilization of Two Azacrown Ethers on Chitosan: Evaluation of Selective Extraction Ability Toward Cu(II) and Ni(II). *Journal of Macromolecular Science, Part A: Pure and Applied Chemistry* **46**(8), 745–750.

McKay, G., Blair, H. S., and Findon, A. (1989). Equilibrium studies for the sorption of metal ions onto chitosan. *Indian Journal of Chemistry* **28**(23), 356.

McKay, G., Blair, H. S., and Gardner, J. (1983). Rate studies for the adsorption of dyestuffs onto chitin, *J. Colloid Interface Sci.* **95**, 108.

McKay, G., Ho, Y. S., and Ng, J. C. Y. (1999). Biosorption of copper from wastewaters: A review. *Sep. Purif. Meth.* **28**, 87–125.

Memon, S., Tabakci, M., Roundhill, D. M., and Yilmaz, M. (2006). Synthesis and evaluation of the Cr(VI) extraction ability of amino/nitrile calix[4]arenes immobilized onto a polymeric backbone. *React. Funct. Polym.* **66**, 1342–1349.

Mi, F. L., Shyu, S. S., Lee, S. T., and Wong, T. B. (1999a). Kinetic study of chitosan-tripolyphosphate complex reaction and acid-resistive properties of the chitosan-tripolyphosphate gel beads prepared by in-liquid curing method. *J. Polym. Sci. B: Polym. Phys.* **37**(14), 1551–1564.

Mi, F. L., Shyu, S. S., Wong, T. B., Jang, S. F., Lee, S. T., and Lu, K. T. (1999)b. Chitosan–polyelectrolyte complexation for the preparation of gel beads and controlled release of anticancer drug. II. Effect of pH-dependent ionic cross-linking or interpolymer complex using tripolyphosphate or polyphosphate as reagent. *J. Appl. Polym. Sci.* **74**(5), 1093–1107.

Monier, M., Ayadb, D. M., Weia, Y., and Sarhan, A. A. (2010a). Adsorption of Cu(II), Co(II), and Ni(II) ions by modified magnetic chitosan chelating resin. *J. Hazard Mater.* **177**(1–3), 962–970.

Monier, M., Ayad, D. M., Wei, Y., and Sarhan, A. A. (2010b). Preparation and characterization of magnetic chelating resin based on chitosan for adsorption of Cu(II), Co(II), and Ni(II) ions. *React. Funct. Polym.* **70**(4), 257–266.

Negm, N. A. and Ali, H. E. (2010). Modification of heavy metal uptake efficiency by modified chitosan/anionic surfactant systems. *Eng. Life Sci.* **10**(3), 218–224.

Ng J. C. Y., Cheung W. H., and McKay G. (2002). Contact time optimization of two-stage batch adsorber systems using the modified film-pore diffusion model. *J. Colloid Interf. Sci.* **255**(1), 64–74.

Ngah, W. S. W. and Fatinathan, S. (2010). Adsorption characterization of Pb(II) and Cu(II) ions onto chitosan-tripolyphosphate beads: Kinetic, equilibrium and thermodynamic studies. *J. of Environmental Management* **91**(4), 958–969.

Ni, C. and Xu, Y. (1996). Studies on Syntheses and Properties of Chelating Resins Based on Chitosan. *J. Appl. Polym. Sci.* **59**(3), 499–504.

Onsoyen, E. and Skaugrud, O. (1990). Metal recovery using chitosan. *J. Chem. Tech. Biotechnol.* **49**(4), 395–404.

Oshita, K., Sabarudin, A., Takayanagi, T., Oshima, M., and Motomizu, S. (2009.) Adsorption behavior of uranium(VI) and other ionic species on cross-linked chitosan resins modified with chelating moieties. *Talanta.* **79**, 1031–1035.

Paik, I. (2001). Application of chelated minerals in animal production. Asian-Australasian *Journal of Animal Science* **14**, 191–198.

Pearson, R. G. (1963). Hard and soft acids and bases. *J. Am. Chem. Soc.* **85**, 3533–3539.

Peng, C., Wang, Y., and Tang, Y. (1998). Synthesis of cross-linked chitosan-crown ethers and evaluation of these products as adsorbents for metal ions. *J. Appl. Polym. Sci.* **70**(3), 501–506.

Peplow, D. (1999). Environmental Impacts of Mining in Eastern Washington, Center for Water and Watershed Studies Fact Sheet, University of Washington, Seattle.

Prasad, M., Saxena, S., and Amritphale, S. S. (2002). Adsorption Models for Sorption of Lead and Zinc on Francolite Mineral. *Ind. Eng. Chem. Res.* **41**(1), 105–111.

Rabea, E. I., Badawy, M. E. T., Stevens, C. V., Smagghe, G., and Steurbaut, W. (2003). Chitosan as Antimicrobial Agent: Applications and Mode of Action. *Biomacromolecules* **4**(6), 1457–1465.

Radwan, A. A., Alanazi, F. K., and Alsarra, I. A. (2010). Microwave Irradiation-Assisted Synthesis of a Novel Crown Ether Cross-linked Chitosan as a Chelating Agent for Heavy Metal Ions (M^{+n}). *Molecules.* **15**(9), 6257–6268.

Ramesh, A., Hasegawa, H., Sugimoto, W., Maki, T., and Ueda, K. (2008). Adsorption of gold(III), platinum(IV) and palladium(II) onto glycine modified cross-linked chitosan resin. *Bioresour Technol.* **99**, 3801–3809.

Reddad, Z., Gerente, C., Andres, Y., and Cloirec, P. (2002). Adsorption of several metal ions onto a low-cost biosorbent: kinetic and equilibrium studies. *Environ. Sci. Technol.* **36**(9), 2067–2073.

Repo, E., Warchol, J. K., Kurniawan, T. A., and Sillanpaa, M. (2010). Adsorption of Co(II) and Ni(II) by EDTA- and/or DTPA-modified chitosan: Kinetic and equilibrium modeling. *Chem. Eng. J.* **161**(1–2), 73–82.

Roberts, G. A. F. (1992). *Chitin Chemistry*, MacMillan, London, UK.

Rodrigues, C. A., Laranjeira, M. C. M., De-Favere, V. T., and Stadler, E. (1998). Interaction of Cu(II) on N-(2-pyridylmethyl) and N-(4-pyridylmethyl) chitosan. *Polymer* **39**(21), 5121–5126

Ruiz, M., Sastre, A., and Guibal, E. (2002). Pd and Pt recovery using chitosan gel beads: II. Influence of chemical and physical modification on sorption properties. *Sep Sci Technol.* **37**(10), 2385–2403.

Ruiz, M., Sastre, A. M., and Guibal, E. (2003). Osmium and iridium sorption on glutaraldehyde cross-linked chitosan gel beads. *Solv Extr Ion Exch.* **21**(2), 307–329.

Sag, Y. and Aktay, Y. (2002). Kinetic studies on sorption of Cr(VI) and Cu(II) ions by chitin, chitosan and Rhizopus Arrhizus. *Biochem. Eng. J.* **12**, 143–153.

Sashiwa, H., Yajima, H., and Aiba, S. I. (2003). Synthesis of chitosan-dendrimer hybrid and its biodegradation. *Biomacromolecules* **4**(5), 1244–1249.

Shimizu, Y., Nakamura, S., Saito, Y., and Nakamura, T. (2008). Removal of Phosphate Ions with a Chemically Modified Chitosan/Metal-Ion Complex. *J. Appl. Polym. Sci.* **107**, 1578–1583.

Tabakci, M. and Yilmaz, M. (2008). Synthesis of a chitosan-linked calix[4]arene chelating polymer and its sorption ability toward heavy metals and dichromate anions. *Bioresour. Technol.* **99**(14), 6642–6645.

Tabakci, M., Ersoz, M., and Yilmaz, M. (2006). A calix[4]arene-containing polysiloxane resin for removal of heavy metals and dichromate anion. *J. Macromol. Sci. Pure Appl. Chem.* **43**, 57–69.

Tabakci, M. and Yilmaz, M. (2008). Synthesis of a chitosan-linked calix[4]arene chelating polymer and its sorption ability toward heavy metals and dichromate anions. *Bioresource Technol.* **99**(14), 6642–6645.

Tan, I. A. W., Hameed, B. H., and Ahmad, A. L. (2007). Equilibrium and kinetic studies on basic dye adsorption by oil palm fibre activated carbon. *Chem. Eng. J.* **127**, 111–119.

Tang, L. and David, N. (2001). Chelation of chitosan derivatives with zinc ions. II. Association complexes of Zn^{2+} onto **O,N**-carboxymethyl chitosan. *J. Appl. Polym. Sci.* **79**(8), 1476–1485.

Tang, X. H., Tan, S. Y., and Wang, Y. T. (2002). Study of the synthesis of chitosan derivatives containing benzo-21-crown-7 and their adsorption properties for metal ions. *J. Appl. Polym. Sci.* **83**(9), 1886–1891.

Tien, C. (1994). *Adsorption Calculations and Modeling*, Butterworth-Heinemann, Newton, MA, p. 244.

Tsigos, L., Martinou, A., Kafetzopoulos, D., and Bouriotis, V. (2000). Chitin deacetylases: New, versatile tools in biotechnology. *Trends Biotechnol.* 18(7), 305–312.

Unlu, N. and Ersoz, M. (2006). Adsorption characteristics of heavy metal ions onto a low cost biopolymer sorbent from aqueous solutions. *J. Hazard. Mater.* **B136**, 272–280.

Uysal, M. and Ar, I. (2007). Removal of Cr(VI) from industrial wastewaters by adsorption, part 1: determination of optimum conditions. *J. Hazard. Mater.* **149**, 482–491.

Varma, A. J., Deshpande, S. V., and Kennedy, J. F. (2004). Metal complexation by chitosan and its derivatives: A review. *Carbohydrate Polymers* **55**, 77–93.

Vasconcelos, H. L., Camargo, T. P., Goncalves, N. S., Neves, A., Laranjeira, M. C. M., and Fávere, V. T. (2008). Chitosan cross-linked with a metal complexing agent: Synthesis, characterization and copper(II) ions adsorption. *Reactive and Functional Polymers* **68**, 572–579.

Vicens, J. and Bohmer, V. (1991). Calixarenes: A Versatile Class of Macrocyclic Compounds Topics in Inclusion Science. Kluwer Academic Publishers, Dordrecht.

Volesky, B. (2001). Detoxification of metal-bearing effluents: Biosorption for the next century. *Hydrometallurgy* **59**(2–3), 203–216.

Wagner, M. and Nicell, J. A. (2002). Detoxification of phenolic solutions with horseradish peroxidase and hydrogen peroxide. *Water Res.* **36**, 4041.

Wan Ngah, W. S. and Liang, K. H. (1999). Adsorption of Gold (III) ions onto chitosan and N-Carboxymethyl Chitosan: Equilibrium Studies. *Ind. Eng. Chem. Res.* **38**(4), 1411–1414.

Weber, T. W. and Chakravorti, R. K. (1974). Pore and solid diffusion models for fixed-bed adsorbers. *AIChE J.* **20**, 228.

Weber, W. J. and Morris, J. C. (1962). Advances in water pollution research. In: *Proc. Int. Conf. Water pollution symposium*, Vol. 2, Pergamon press, Oxford, pp. 231–266.

Wong, Y. C., Szeto, Y. S., Cheung, W. H., and McKay, G. (2004). Adsorption of acid dyes on chitosan—Equilibrium isotherm analyses. *Process Biochem.* **39**, 693.

Wong, Y. C., Szeto, Y. S., Cheung, W. H., and McKay, G. (2003). Equilibrium studies for acid dye adsorption onto chitosan, *Langmuir* **19**, 7888.

Wu, F. C., Tseng, R. L., and Juang, R. S. (2001). Kinetic Modeling of Liquid-Phase Adsorption of

Reactive Dyes ond Metal Ions on Chitosan. *Water Res*. **35**, 613–618.

Xing, R., Liu, S., Guo, Z., Yu, H., Wang, P., Li, C., Li, Z., and Li, P. (2005). Relevance of molecular weight of chitosan and its derivatives and their antioxidant activities *in vitro*. *Bioorg. Med. Chem*. **13**(5), 1573–1577.

Yilmaz, M., Memon, S., Tabakci, M., and Bartsch, R. A. (2006). In: *New Frontiers in Polymer Research*, R. K. Bregg (Ed.). Nova Science Publishers, Hauppauge NY, pp. 125–171.

YoneKura, L. and Suzuki, H. (2003). Some polysaccharides improve zinc bioavailability in rats fed a phytic acid-containing diet. *Nutrition Research* **23**(3), 343–355.

Yoshida, H., Okamoto, A. and Kataoka, T. (1993). Adsorption of acid dye on cross-linked chitosan fibers—Equilibria. *Chem. Eng. Sci*. **48**, 2267.

Zalloum, H., Al-Qodah, Z., and Mubarak, M. S., (2009). Copper Adsorption on Chitosan-Derived Schiff Bases. *J. Macromol. Sci. A: Pure Appl Chem* **46**, 1–12.

2

Altman, G. H., Diaz, F., Jakuba, C., Calabro, T., Horan, R. L., Chen, J., Lu, H., Richmond, J., and Kaplan D. L. (2003). Silk-based biomaterials. *Biomaterials* **24**(3), 401–416.

Arai, T., Freddi, G., Colonna, G., Scotti, E., Boschi, A., Murakami, R., and Tsukada, M. (2001). Absorption of metal cations by modified B. mori silk and preparation of fabrics with antimicrobial activity. *Journal of Applied Polymer Science* **80**, 297–303.

Arai, T., Freddi, G., Innocenti, R., and Tsukada, M. (2004). Biodegradation of *Bombyx mori* silk fibroin fibers and films. *Journal of Applied Polymer Science* **91**, 2383–2390.

Artan, M., Karadeniz, F., Karagozlu, M. Z., Kim, M. M., and Kim, S. K. (In press). Anti-HIV-1 activity of low molecular weight sulfated chitooligosaccharides. Carbohydrate Research.

Cunniff, P., Fossey, S., Auerbach, M., Song, J., Kaplan, D., Adams, W., Eby, R., Mahoney, D., and Vezie, D. (1994). Mechanical and thermal properties of dragline silk from the spider Nephila

clavipes. *Polymers for Advanced Technologies* **5**(8), 401–410.

Davidenko, N., Carrodeguas, R., Peniche, C., Solís, Y., and Cameron, R. (2010). Chitosan/apatite composite beads prepared by in situ generation of apatite or Si-apatite nanocrystals. *Acta Biomaterials* **6**, 466–476.

Di Martino, A., Sittinger, M., and Risbud, M. (2005). Chitosan: a A versatile biopolymer for orthopaedic tissue-engineering. *Biomaterials* **26**, 5983–5990.

Disa, J., Alizadeh, K., Smith, J., Hu, Q., and Cordeiro, P. (2001). Evaluation of a combined calcium sodium alginate and bio-occlusive membrane dressing in the management of split-thickness skin graft donor sites. *Annals of Plastic Surgery* **46**, 405.

Engelberg, I. and Kohn, J. (1991). Physico-mechanical properties of degradable polymers used in medical applications: A comparative study. *Biomaterials* **12**(3), 292-304.

Fan, H., Liu, H., Wong, E. J. W., Toh, S. L., and Goh, J. C. H. (2008). *In vivo* study of anterior cruciate ligament regeneration using mesenchymal stem cells and silk. *Biomaterials* **29**, 3324-3337.

Felt, O., Carrel, A., Baehni, P., Buri, P., and Gurny, R. (2000). Chitosan as tear substitute: a wetting agent endowed with antimicrobial efficacy. *Journal of Ocular Pharmacology and Therapeutics* **16**, 261-270.

Fini, M., Motta, A., Torricelli, P., Giavaresi, G., Nicoli Aldini, N., Tschon, M., Giardino, R., and Migliaresi, C. (2005). The healing of confined critical size cancellous defects in the presence of silk fibroin hydrogel. *Biomaterials* **26**, 3527–3536.

Foldvari, M. (2000). Non-invasive administration of drugs through the skin: Challenges in delivery system design. *Pharmaceutical Science and Technology Today* **3**, 417–425.

Freddi, G., Romano, M., Massafra, M. R., and Tsukada, M. (1995). Silk fibroin/cellulose blend films: Preparation, structure, and physical properties. *Journal of Applied Polymer Science* **56**, 1537.

Freddi, G., Tsukada, M., and Beretta, S. (1999). Structure and physical properties of silk fibroin/

polyacrylamide blend films. *Journal of Applied Polymer Science* 71, 1563.

Fuchs, S., Motta, A., Migliaresi, C., and Kirkpatrick, C. (2006). Outgrowth endothelial cells isolated and expanded from human peripheral blood progenitor cells as a potential source of autologous cells for endothelialization of silk fibroin biomaterials. *Biomaterials* 27, 5399–5408.

Gombotz, W. R. and Wee, S. F. (1998). Protein release from alginate matrices. *Advanced Drug Delivery Reviews* 31, 267–285.

Gotoh, Y., Minoura, N., and Miyashita, T. (2002). Preparation and characterization of conjugates of silk fibroin and chitooligosaccharides. *Colloid and Polymer Science* 280(6), 562–568.

Goy, R. C., de Britto, D., and Assis, O. B. G. (2009). A Review of the antimicrobial activity of chitosan. *Polímeros: Ciência e Tecnologia* 19(3), 241–247.

Guan, Y., Liu, X., Zhang, Y., and Yao, K. (1998). Study of phase behavior on chitosan/viscose rayon blend film. *Journal of Applied Polymer Science* 67, 1965–1972.

Gupta, K. C. and Ravi Kumar, M. N. V. (2000). Drug release behavior of beads and microgranules of chitosan. *Biomaterials* 21, 1115.

Hasegawa, M., Isogai, A., Onabe, F., and Usuda, M. (1992). Characterization of cellulose-chitosan blend films. *Journal of Applied Polymer Science* 45, 1873–1879.

Hirano, S. (1978). A facile method for the preparation of novel membranes from. N-acyl and N-acrylidone chitosan gels. *Agricultural and Biological Chemistry* 42, 1938.

Horan, R., Antle, K., Collette, A., Wang, Y., Huang, J., Moreau, J., Volloch, V., Kaplan, D., and Altman G. (2005). *In vitro* degradation of silk fibroin. *Biomaterials* 26, 3385–3393.

Hirano, S., Ohe, Y., and Ono, H. (1976). Selective N-acylation of chitosan. *Carbohydrate Research* 47, 315.

Hirano, S., Tobetto, K., Hasegawa, M., and Matsuda, N. (1980). Permeability properties of gels and membranes derived from chitosan. *Journal of Biomedical Material Research* 14, 477.

Hu, K., Lv, Q., Cui, F., Feng, Q., Kong, X., Wang, H., Huang, L., and Li, T. (2006). Biocompatible fibroin blended films with recombinant human-like collagen for hepatic tissue engineering. *Journal of Bioactive and Compatible Polymers* 21, 23.

Japanese Chitin and Chitosan Society. (1994). *Chitin and Chitosan Handbook*. Gihodo, Tokyo, Gihodo, pp. 460–483.

Je, J. Y., Park, P. J., and Kim, S. K. (2004). Free radical scavenging properties of hetero-chitooligosaccharides using an ESR spectroscopy. *Food and Chemical Toxicology* 42, 381–387.

Jeon, Y. J. and Kim, S. K. (2000). Production of chitooligosaccharides using an ultrafiltration membrane reactor and their antibacterial activity. *Carbohydrate Polymer* 41, 133–141.

Jeon, Y. J., Park, P. J. and Kim S. K. (2001). Antimicrobial effect of chitooligosaccharides produced by bioreactor. *Carbohydrate Polymer* 44, 71–76.

Jeon, Y. J., Shahidi, F., and Kim, S. (2000). Preparation of chitin and chitosan oligomers and their applications in physiological functional foods. *Food Reviews International* 16, 159–176.

Jiang, L., Li, Y., Wang, X., Zhang, L., Wen, J., and Gong, M. (2008). Preparation and properties of nano-hydroxyapatite/chitosan/carboxymethyl cellulose composite scaffold. *Carbohydrate Polymer* 74, 680–684.

Karageorgiou, V., Meinel, L., Hofmann, S., Malhotra, A., Volloch, V., and Kaplan, D. (2004). Bone morphogenetic protein-2 decorated silk fibroin films induce osteogenic differentiation of human bone marrow stromal cells. *Journal of Biomedical Material Research Part A* 71, 528–537.

Karthik, A., Latha, S., Thenmozhi, N., and Sudha, P. (2009). Removal of heavy metals chromium and cadmium using chitosan impregnated polyurethane foam. *Ecoscan* 3, 157–160.

Kean, T., Roth, S., and Thanou, M. (2005). Trimethylated chitosans as non-viral gene delivery vectors: Cytotoxicity and transfection efficiency. *Journal of Controlled Release* 103, 643–653.

Kifune, K., Yamaguchi, Y., and Kishimoto, S. (1988). Wound healing effect of chitin surgical dressing. *Transactions Society of Biomaterials* 11, 216.

Kim, H., Lee, H., Oh, J., Shin, B., Oh, C., Park, R., Yang, K., and Cho, C. (1999). Polyelectrolyte complex composed of chitosan and sodium alginate for wound dressing application. *Journal of Biomaterials Science, Polymer Edition* **10**, 543–556.

Kim, J. H., Kim, J. Y., Lee, Y. M., and Kim, K. Y. (1992). Properties and swelling characteristics of cross-linked poly (vinyl alcohol)/chitosan blend membrane. *Journal of Applied Polymer Science* **45**, 1711.

Kim, M. M. and Kim, S. K. (2006). Chitooligosaccharides inhibit activation and expression of matrix metalloproteinase-2 in human dermal fibroblasts. *FEBS Letters* **580**, 2661–2666.

Kim, U., Park, J., Joo Kim, H., Wada, M., and Kaplan, D. (2005). Three-dimensional aqueous-derived biomaterial scaffolds from silk fibroin. *Biomaterials* **26**, 2775–2785.

Kofuji, K., Ito, T., Murata, Y., and Kawashima, S. (2000). The controlled release of a drug from biodegradable chitosan gel beads. *Chemical and Pharmaceutical Bulletin* **48**, 579.

Kojima, K., Okamoto, Y., Miyatake, K., Kitamura, Y., and Minami, S. (1998). Collagen typing of granulation tissue induced by chitin and chitosan. *Carbohydrate Polymer* **37**(2), 109–113.

Kong, L., Gao, Y., Cao, W., Gong, Y., Zhao, N., and Zhang, X. (2005). Preparation and characterization of nano-hydroxyapatite/chitosan composite scaffolds. *Journal of Biomedical Material Research Part A* **75A**, 275–282.

Kweon, H., Ha, H. C., Um, I. C., and Park, Y. H. (2001). Physical properties of silk fibroin/chitosan blend films. *Journal of Applied Polymer Science* **80**, 928.

Li, C., Vepari, C., Jin, H., Kim, H., and Kaplan, D. (2006). Electrospun silk-BMP-2 scaffolds for bone tissue engineering. *Biomaterials* **27**, 3115–3124.

Liang, C. X. and Hirabayashi, K. (1992). Mechanical properties of fibroin-chitosan membrane. *Journal of Applied Polymer Science* **45**, 1937.

Liu, B., Liu, W., Han, B., and Sun, Y. (2007). Antidiabetic effects of chitooligosaccharides on pancreatic islet cells in streptozotocin-induced diabetic rats. *World Journal of Gastroenterology* **13**, 725.

Liu, D., Hsieh, J., Fan, X., Yang, J., and Chung, T. (2007). Synthesis, characterization and drug delivery behaviors of new PCP polymeric micelles. *Carbohydrate Polymer* **68**, 544–554.

Lv, Q. and Feng, Q. L. (2006). Preparation of 3-D regenerated fibroin scaffolds with freeze drying method and freeze drying/foaming technique. *Journal of Material Science: Material in Medicine* **17**(12), 1349–1356.

Mi, F. L., Sung, H. W., and Shyu, S. S. (2002). Drug release from chitosan–alginate complex beads reinforced by a naturally occurring cross-linking agent. *Carbohydrate Polymer* **48**, 61–72.

Minoura, N., Aiba, S., Gotoh, Y., Tsukada, M., and Imai, Y. (1995). Attachment and growth of cultured fibroblast cells on silk protein matrices. *Journal of Biomedical Material Research* **29**, 1215–1221.

Minoura, N., Tsukada, M., and Nagura, M. (1990). Fine structure and oxygen permeability of silk fibroin membrane treated with methanol. *Polymer* **31**, 265.

Moore, G. K. and Roberts, G. A. F. (1981). Reactions of chitosan: 2. Preparation and reactivity of N-acyl derivatives of chitosan. *International Journal of Biological Macromolecules* **3**, 292–296.

Moreau, J., Chen, J., Kaplan, D., and Altman, G. (2006). Sequential growth factor stimulation of bone marrow stromal cells in extended culture. *Tissue Engineering* **12**, 2905–2912.

Motta, A., Migliaresi, C., Faccioni, F., Torricelli, P., Fini, M. and Giardino, R. (2004). Fibroin hydrogels for biomedical applications: Preparation, characterization and *in vitro* cell culture studies. *Journal of Biomaterial Science, Polymer Edition* **15**, 851–864.

Moy, R. L., Lee, A., and Zalka, A. (1991). Commonly used suture materials in skin surgery. *American Family Physician* **44**(6), 2123–2128.

Muzzarelli, R. A. A., Jeuniaux, C., and Gooday, G. W. (1986). *Chitin in Nature and Technology.* Plenum, New York.

Muzzarelli, R., Baldassarre, V., Conto, F., Ferrara, P., Biagini, G., Gazzanelli, G., and Vasi, V. (1988). Biological activity of chitosan: Ultrastructural study. *Biomaterials* **9**, 247–252.

Nazarov, R., Jin, H. J., and Kaplan, D. L. (2004). Porous 3-D scaffolds from regenerated silk fibroin. *Biomacromolecules* **5**, 718–726.

Niamsa, N., Srisuwan, Y., Baimark, Y., Phinyo-cheep, P., and Kittipooms, S. (2009). Preparation of nanocomposite chitosan/silk fibroin blends films containing nanopore structures. *Carbohydrate Polymer* **78**, 60–65.

Nishimura, K., Nishimura, S., Seo, H., Nishi, N., Tokura, S., and Azuma, I. (1986). Macrophage activation with multi porous beads prepared from partially deacetylated chitin. *Journal of Biomedical Material Research* **20**, 1359–1372.

Nishimura, S. I., Kohgo, O., and Kurita, K. (1991). Chemospecific manipulation of a rigid polysaccharide: Syntheses of novel chitosan derivatives with excellent stability in common organic solvents by regioselective chemical modification. *Macromolecules* **24**, 4745.

Nunthanid, J., Laungtana-anan, M., Sriamornsak, P., Limmatvapirat, S., Puttipipatkhachorn, S., Lim, L. Y., and Khor, E. (2004). Characterization of chitosan acetate as a binder for sustained release tablets. *Journal of Controlled Release* **99**, 15–26.

O'Donoghue, J., O'Sullivan, S., Beausang, E., Panchal, J., O'Shaughnessy, M., and O'Connor, T. (1997). Calcium alginate dressings promote healing of split skin graft donor sites. *Acta Chirurgiae Plasticae* **39**, 53.

Oh, H., Lee, J., Kim, A., Ki, C., Kim, J., Park, Y., and Lee, K. (2007). Preparation of silk sericin beads using LiCl/DMSO solvent and their potential as a drug carrier for oral administration. *Fibers Polymer* **8**, 470–476.

Park, S. J., Lee, K. Y., Ha, W. S., and Park, S. Y. (1999). Structural changes and their effect on mechanical properties of silk fibroin/chitosan blends. *Journal of Applied Polymer Science* **74**, 2571.

Patel, V. R. and Amiji, M. M. (1996). Preparation and characterization of freezedried chitosan poly(ethylene oxide) hydrogels for site-specific antibiotic delivery in the stomach. *Pharmaceutical Research* **13**, 588–593.

Paul W. and Sharma C. (2004). Chitosan and alginate wound dressings: A short review. *Trends in Biomaterials and Artificial Organs* **18**, 18–23.

Perez-Rigueiro, J., Viney, C., Llorca, J., and Elices, M. (2000). Mechanical properties of single-brin silkworm silk. *Journal of Applied Polymer Science* **75**(10), 1270–1277.

Pins, G., Christiansen, D., Patel, R., and Silver, F. (1997). Self-assembly of collagen fibers. Influence of fibrillar alignment and decorin on mechanical properties. *Biophysical Journal* **73**(4), 2164–2172.

Pusateri, A., McCarthy, S., Gregory, K., Harris, R., Cardenas, L., McManus, A., and Goodwin, C. Jr. (2003). Effect of a chitosan-based hemostatic dressing on blood loss and survival in a model of severe venous hemorrhage and hepatic injury in swine. *Journal of Trauma* **54**, 177.

Puttipipatkhachorn, S., Nunthanid, J., Yamamoto, K., and Peck, G. E. (2001). Drug physical state and drug–polymer interaction on drug release from chitosan matrix films. *Journal of Controlled Release* **75**, 143–153.

Qi, L. and Xu, Z. (2006). *In vivo* antitumor activity of chitosan nanoparticles. *Bioorganic and Medicinal Chemistry Letters* **16**, 4243–4245.

Qin, Y. (2008). Alginate fibres: An overview of the production processes and applications in wound management. *Polymer International* **57**, 171–180.

Rajapakse, N., Kim, M. M., Mendis, E., Huang, R., and Kim, S. K. (2006). Carboxylated chitooligosaccharides (CCOS) inhibit MMP-9 expression in human fibrosarcoma cells via down-regulation of AP-1. *Biochimica et Biophysica Acta (BBA)—General Subjects* **1760**, 1780–1788.

Ribeiro, A. J., Ferreira, S. C., and Veiga, F. (2005). Chitosan-reinforced alginate microspheres obtained through the emulsification/internal gelation technique. *European Journal of Pharmaceutical Science* **25**, 31–40.

Roh, D., Kang, S., Kim, J., Kwon, Y., Young Kweon, H., Lee, K., Park, Y., Baek, R., Heo, C., and Choe, J. (2006). Wound healing effect of silk fibroin/alginate-blended sponge in full thickness skin defect of rat. *Journal of Material Science: Material in Medicine* **17**, 547–552.

Rujiravanit, R., Kruaykitanon, S., Jamieson, A. M., and Tokura, S. (2003). Preparation of cross-linked chitosan/silk fibroin blend films for drug delivery system. *Macromolecular Bioscience* **3**, 604–611.

Sakabe, H., Ito, H., Miyamoto, T., Noishiki, Y., and Ha, W. S. (1989). *In vivo* blood compatibility of regenerated silk fibroin. *Sen-I Gakkaishi* **45**, 487.

Santin, M., Motta, A., Freddi, G., and Cannas, M. (1999). *In vitro* evaluation of the inflammatory potential of the silk fibroin. *Journal of Biomedical Material Research* **46**, 382.

Shu, X. Z. and Zhu, K. J. (2002). The release behavior of brilliant blue from calcium–alginate gel beads coated by chitosan: the preparation method effect. *European Journal of Pharmaceutics and Biopharmaceutics* **53**, 193–201.

Singh, D. K. and Ray, A. R. (1994). Graft copolymerization of 2-hydroxyethylmethacrylate onto chitosan films and their blood compatibility. *Journal of Applied Polymer Science* **53**, 1115–1121.

Singh, D. K. and Ray, A. R. (1997). Radiation-induced grafting of N,N'-dimethylaminoethylmethacrylate onto chitosan films. *Journal of Applied Polymer Science* **66**, 869–877.

Sudha, P. and Celine, S. (2008). Removal of heavy metal cadmium from industrial wastewater using chitosan coated coconut charcoal. *Nature Environment and Pollution Technology* **7**, 601–604.

Suzuki, K., Mikami, T., Okawa, Y., Tokoro, A., Suzuki, S., and Suzuki, M. (1986). Antitumor effect of hexa-N-acetylchitohexaose and chitohexaose. *Carbohydrate Research* **151**, 403–408.

Teng, S., Lee, E., Yoon, B., Shin, D., Kim, H., and Oh, J. (2009). Chitosan/nano-hydroxyapatite composite membranes via dynamic filtration for guided bone regeneration. *Journal of Biomedical Material Research A* **88**(3), 569–580.

Ueno, H., Mori, T., and Fujinaga, T. (2001). Topical formulations and wound healing applications of chitosan. *Advanced Drug Delivery Reviews* **52**, 105–115.

Van Ta, Q., Kim, M. M., and Kim, S. K. (2006). Inhibitory effect of chitooligosaccharides on matrix metalloproteinase-9 in human fibrosarcoma cells (HT1080). *Marine Biotechnology* **8**, 593–599.

Venkatesan, J. and Kim, S. K. (2010). Chitosan composites for bone tissue engineering—an overview. *Marine Drugs* **8**, 2252–2266.

Venkatesan, J., Qian, Z. J., Ryu, B., Ashok Kumar, N., and Kim, S. K. (in press). Preparation and characterization of carbon nanotube-grafted-chitosan—Natural hydroxyapatite composite for bone tissue engineering. *Carbohydrate Polymer*.

Vepari, C. and Kaplan, D. L. (2006). Covalently immobilized enzyme gradients within three-dimensional porous scaffolds. *Biotechnology and Bioengineering* **93**, 1130–1137.

Vepari, C. and Kaplan, D. L. (2007). Silk as a biomaterial. *Progess in Polymer Science* **32**, 991–1007.

Vollrath, F. and Knight, D. P. (2001). Liquid crystalline spinning of spider silk. *Nature* **410**(6828), 541–548.

Wang, L., Khor, E., Wee, A., and Lim, L. (2002). Chitosan alginate PEC membrane as a wound dressing: Assessment of incisional wound healing. *Journal of Biomedical Materials Research* **63**, 610–618.

Wang, Y., Kim, H. J., Vunjak-Novakovic, G., and Kaplan, D. L. (2006). Stem cell-based tissue engineering with silk biomaterials. *Biomaterials* **27**, 6064–6082.

Wong Po Foo, C. and Kaplan, D. (2002). Genetic engineering of fibrous proteins: spider dragline silk and collagen. *Advanced Drug Delivery Reviews* **54**, 1131–1143.

Xiao, C., Gao, S., Wang, H., and Zhang L. (2000). Blend films from chitosan and konjac glucomannan solutions. *Journal of Applied Polymer Science* **76**, 509.

Yalpani, M. and Hall, L. D. (1984). Some chemical and analytical aspects of polysaccharide modifications. 3 Formation of branched- chain, soluble chitosan derivatives. *Macromolecules* **17**, 272.

Yamaura, K., Kuranuki, N., Suzuki, M., Tanigami, T., and Matsuzawa, S. (1990). Properties of mixtures of silk fibroin/syndiotactic-rich poly(vinyl alcohol). *Journal of Applied Polymer Science* **41**, 2409-2425.

Yang, E. J., Kim, J. G., Kim, J. Y., Kim, S., Lee, N., and Hyun, C. G. (2010). Anti-inflammatory effect of chitosan oligosaccharides in RAW 264.7 cells. *Central European Journal of Biology* **5**, 95–102.

Yang, Y., Ding, F., Wu, J., Hu, W., Liu, W., Liu, J., and Gu, X. (2007). Development and evaluation of silk fibroin-based nerve grafts used for peripheral nerve regeneration. *Biomaterials* **28**, 5526-5535.

Yannas, I. and Burke, J. (1980). Design of an artificial skin. I. Basic design principles. *Journal of Biomedical Material Research* **14**, 65–81.

Yao, K. D., Liu, J., Cheng, G. X., Lu, X. D., Tu, H. L., and Lopes Da Silva, J. A. (1996).Swelling behavior of pectin/chitosan complex films. *Journal of Applied Polymer Science* **60**, 279–283.

Zhen-ding, S., Wei-qiang, L., and Qing-ling, F. (2009). Preparation and cytocompatibility of silk fibroin/chitosan scaffolds. *Frontiers of Materials Science in China* **3**(3), 241–247.

Zong, Z., Kimura, Y., Takahashi, M., and Yamane, H. (2000). Characterization of chemical and solid state structures of acylated chitosans. *Polymer* **41**, 899–906.

3

Acosta, R. I., Rodriguez, X., Guiterrez, C., and Motctezuma, G. (2004). Biosorption of chromium (VI) from aqueous solutions onto fungal biomass. *Bioinorganic Chemistry and Appllications* **2**(1–2), 1–7.

Agboh, O. C. and Quin, Y. (1997). Chitin and chitosan fibers. *Advances in polymer technology* **8**, 355.

Aksu, Z. (2005). Application of biosorption for the removal of organic pollutants: A review. *Process Biochemistry* **40**(3–4), 997–1026.

Alka Boricha, G. and Murthy, Z. V. P. (2010). Preparation of N,O-carboxymethyl chitosan/cellulose acetate blend nanofiltration membrane and testing its performance in treating industrial wastewater. *Chemical Engineering Journal* **157**(2–3), 393–400.

Alves, N. M. and Mano, J. F. (2008). Chitosan derivatives obtained by chemical modifications for biomedical and environmental applications. *International Journal of Biological Macro molecules* **43**, 401–414.

Amaike, M., Senoo, Y., and Yamamoto, H. (1998). Sphere, honeycomb, regularly spaced droplet and fiber structure of polyion complexes of chitosan and gellan. *Macromolecule Rapid Communications* **19**, 287.

Arai, K., Kinumaki, T., and Fujita, T. (1968). Toxicity of chitosan. *Bulletin of the Tokai Regional Fisheries Research Laboratory*, **56**, 89–94.

Avadi, M. R., Zohuriaan-Mehr, M. J., Younesi, P., Amini, A., Shafiee, A., and Rafiee-Tehrani, M. (2003). Optimized synthesis and characterization of N-triethyl chitosan. *Journal of Bioactive and Compatible Polymers* **18**, 469–479.

Avadi, M. R., Sadeghi, A. M. M., Tahzibi, A., Bayati, K. H., Pouladzadeh, M., Zohuriaan-Mehr, M. J., and Rafiee-Tehrani, M. (2004). Diethyl methyl chitosan an antimicrobial agent: Synthesis, characterization and antibacterial effects. *Journal of European Polymer*, **40**, 1355–1361.

Baba, Y. and Hirakawa, H. (1992). Selective adsorption of palladium(II), platinum(IV) and mercury(II) on a new chitosan derivatives processing pyridyl group. *Chemistry Letters* **44**, 1905.

Baba, Y., Masaak, and Kawano, Y. (1998). Synthesis of chitosan derivative recognizing planar metal ion and its selective adsorption of copper(II) over iron(III). *Reactive and Functional Polymers* **36**, 167.

Babel, S. and Kurniawan, T. A. (1999). Low-cost adsorbents for heavy metals uptake from contaminated water: A review. *Journal of Hazardous Materials* **97**, 219–243.

Bailey, S. E., Olin, T. J., Bricka, R. M., and Adrian, D. D. (1999). A review of potentially low-cost sorbents for heavy metals. *Water Resource* **33**, 2469–2479.

Baker, R. W. (2004). *Membrane Technology and Applications*, J. Wiley & Sons, Chictester, U.K.

Baraka, Hall. and Heslop, P. J. (2007). Melamine-formaldehyde-NTA chelating gel resin; synthesis, characterization and application for copper(II) ion removal from synthetic wastewater. *Journal of Hazardous Materials* **140**, 86–94.

Barcellos, I. O., Andreaus, J., Battisti, A. M., and Borges J. K. (2008). Nylon 6, 6/chitosan blends as adsorbent of acid dyes for the reuse of treated wastewater in polyamide in dyeing. *Polymeros* **18**(3), 215–221.

Becker, T., Schlaak, M., and Strasdeit, H. (2000). Adsorption of nickel(II), zinc(II) and cadmium(II) by new chitosan derivatives. *Reactive and Functional Polymers* **44**(3), 289–298.

Beppu, M. M., Arruda, E. J., Vieira, R. S., and Santos, N. N. (2004). Adsorption of Cu(II) on porous chitosan membranes functionalized with histidine. *Journal of Membrane Science* **240**, 227–235.

Berger, J., Reist, M., Mayer, J. M., Felt, O., Peppas, N. A., and Gurny, R. (2004). Structure and interactions in chitosan hydrogels formed by complexation or aggregation for biomedical applications. *European Journal of Pharmaceutics and Biopharmaceutics* **57**, 69–84.

Bhattacharya, A. and Misra, B. N. (2004). Grafting; A versatile means to modify polymers techniques, factors and applications. *Programme Polymer Science* **29**, 767–814.

Boddu, V., Abburi, K., Talbott, J. L., and Smith, E. D. (2003). Removal of hexavalent chromium from wastewater using a new composite chitosan biosorbent. *Environmental Science and Technology* **37**, 4449–4456.

Bolto, B. A. (1995). Soluble polymers in water purification. *Programme Polymer Science* **20**, 987–1041.

Bose, P., Bose, M., and Kumar, S. (2002). Critical evaluation of treatment strategies involving adsorption and chelation for wastewater containing copper, zinc, and cyanide. *Advances in Environmental Research* **7**, 179–195.

Cao, Z., Ge, H., and Lai, S. (2001). Studies on synthesis and adsorption properties of chitosan cross-linked by glutaraldehyde and Cu(II) as template under microwave irradiation. *European Polymer Journal* **37**, 2141.

Chao, A. C., Shyu, S. S., Lin, Y. C., and Mi, F. L. (2004). Enzymatic grafting of carboxy groups on to chitosan to confer on chitosan the property of a cationic dye adsorbent. *Bioresource Technology* **91**, 157–162.

Chaufer, B. and Deratani, A. (1988). Removal of metal ions by complexation-ultra filtration using water-soluble macro-molecules: Perspective of application to wastewater treatment. *Nuclear and Chemicals Waste Management* **8**, 175.

Chun Xiu Liu and Renbi Bai (2006). Adsorptive removal of copper ions with highly porous chitosan/cellulose acetate blend hollow fiber membranes. *Journal of Membrane Science* **284**(1–2), 313–322.

Copello, G. J., Varela, F., Martinez Vivot, R., and Diaz, L. E. (2008). Immobilized chitosan as biosorbent for the removal of Cd(II), Cr(III) and Cr(VI) from aqueous solutions. *Bioresource Technology* **99**, 6538–6544.

Crini, G. (2005). Recent developments in polysaccharide-based materials used as absorbents in wastewater treatment. *Programme Polymer Science* **30**, 38–70.

Crini, G. (2006). Non-conventional low-cost adsorbents for dye removal; A review. *Bioresource Technology* **60**, 67–75.

Crini, G. and Badot, P. M. (2008). Kinetic and equilibrium studies on the removal of cationic dyes from aqueous solution by adsorption on to a cyclodextrin polymer. *Programme Polymer Science* **33**, 399–447.

De Smedt, S. C., Demeester, J., and Hennink, W. E. (2000). Cationic polymer-based gene delivery systems. *Pharmaceutical Research* **17**, 113–126.

Denkbas, E. B., Kilicay, E. Birikeseven, C., and Ozturik, E. (2002). Magnetic chitosan micro spheres; preparation and characterization. *Reactive and Functional Polymers* **50**, 225.

Dinesh Karthik, A., Latha S., Thenmozhi, N., and Sudha P. N. (2009). Removal of heavy metals chromium and cadmium using chitosan impregnated polyurethane foam. *The Ecoscan* **3**(1–2), 157–160.

Divya Chauhan and Nalini Sankararamakrishnan (2008). Highly enhanced adsorption for decontamination of lead ions from battery wastewaters using chitosan functionalized with xanthate. *Bioresource Technology* **99**, 9021–9024.

Dragan J., Susana V., Tatjana T., Antorio N., Peter J., Mara Rosa J., and Pilac, E. (2005). Chitosan/acid dye interactions in wool dyeing system. *Carbohydrate Polymers* **60**, 51–59.

Dzul Erosa, M. S., Saucedo Medina, T. I., Navarro Mendoza, R., Avila Rodriguez, M., and Guibal, E. (2001) Cadmium sorption on chitosan derivatives. *Hydrometallurgy* **61**(3), 157–167.

Elangovan, R., Philip, L., and Chandraraj, K. (2008). Biosorption of chromium species by aquatic weeds: kinetics and mechanism studies. *Journal of Hazardous Materials* **152**, 100–112.

Evans, J. R., Davids, W. G., MacRae, J. D., and Amirbahman, A. (2002). Kinetics of cadmium uptake by chitosan-based crab shells. *Water Resource* **36**, 3219.

Fan, T., Liu, Y., Feng, B.,Zeng, G., Yang, C., Zhou, M., Zhou, H., Tan, Z., and Wang, X. (2008). Biosorption of cadmium(II), zinc(II) and lead(II) by *Penicillium simplicissimum*: isotherms, kinetics and thermodynamics. *Journal of Hazardous Materials* **160**, 655–661.

Mi, F. L., Shyu S. S., Wu, Y. B., Lee, S. T., Shyong, J. Y., and Huang, R. N. (2001). Fabrication and chacterization of a sponge-like asymmetric chitosan membrane as a wound dressing. *Biomaterials* **22**, 165.

Al-Momani, F., Touraud, E., Degorce-Dumas, J. R., Roussy, J., and Thomas, O. (2002). Biodegradability enhancement of textile dyes and textile wastewater by VUV photolysis. *Journal of Photochemical and Photobiology A* **153**, 191–197.

Forgacs, E., Cserhati, T., and Oros, G. (2004). Removal of synthetic dyes from wastewater: A review. *Environmental International* **30**, 953–971.

Geckeler, K. E. and Volchek, K. (1996). Removal of hazardous substances from water using ultra filtration in conjunction with soluble polymers. *Environmental Science and Technology* **30**, 725.

Gibbs, G., Tobin, J. M., and Guibal, E. (2003). Sorption of acid green 25 on chitosan; influence of experimental parameters on uptake kinetics and sorption isotherms. *Journal of applied polymer science* **90**, 1073–1080.

Gotoh, T., Matsushima, K., and Kikuchi, K. I. (2004). Preparation of alginate-chitosan hybrid gel beads and adsorption of divalent metal ions. *Chemosphere* **55**, 135–140.

Guibal, E., Jansson-Charrier, M., Saucedo, I., and Le Cloirec, P. (1995). Enhanced of metal ion sorption performances of chitosan; effect of structure on the diffusion properties. *Langumuir II*, 591–598.

Guibal, E., Milot, E. C., and Roussy, J. (2000). Influence of hydrolysis mechanisms on molybdate sorption isotherms using chitosan. *Separation Science and Technology* **35**, 1020–1038.

Guibal, E., Milot, C., and Roussy, J. (1999b). Molybdate sorption by cross-linked chitosan beads; dynamic studies. *Water Environment Resource* **71**, 10–17.

Guibal, E., Saucedo, I., Jansson-Charrie, M. R., Delanghe, B., and Le Cloirec, P. (1994). Uranium and vanadium sorption by chitosan and derivatives. *Water Science and Technology* **30**(9), 183.

Guillen, M. G., Sanchez, A. G., and Zamora, M. E. M. (1992). A derivative of chitosan and 2, 4-pentanodione with strong chelating properties. *Carbohydrate Research* **233**, 255–259.

Guillen, M. G., Sanchez, A. G., and Zamora, M. E. M. (1994). Preparation and chelating properties of derivatives of chitosan and 1,3-dicarbonyl compounds. *Carbohydrate Research* **258**(1–2), 313–319.

Gupta, V. K., Shrivastava, A. K., and Jain, N. (2001). Biosorption of chromium(VI) from aqueous solution by green algae Spirogyra species. *Water Resource* **35**, 4079–4085

Hejazi, R. and Amiji, M. (2003). Chitosan-based gastrointestinal delivery systems. *Journal of Controlled Release* **89**, 151.

Heras, A., Rodriguez, N. M., and Ramoz, V. M. (2001). *N*-methylene phosphonic chitosan; A novel soluble derivative. *Carbohydrate Polymers* **44**, 1–8.

Hiroyuki., Y. and Takeshi, T. (1997). Adsorption of direct dye on cross-linked chitosan fiber; breakthrough. *Water science and Technology* **35**(7), 29–37.

Hoang vinh Tran, Lam Dai Tran, and Thin Ngoc Nguyen. (2010). Preparation of chitosan/magnetite composite beads and their application for removal of Pb(II) and Ni(II) from aqueous solutions. *Material Science and Engineering C* **30**(2), 304–310.

Holme, K. R. and Hall, L. D. (1991). Novel metal chelating chitosan derivative; attachment of iminodiacetate moieties via a hydrophilic spacer group. *Canadian Journal of Chemistry*, **69**(2), 585–590.

Hong-Mei, K., Yuan-Li, C., and Peng-Sheng, L. (2006). Synthesis, characterization and thermal sensitivity of chitosan-based graft copolymers. *Carbohydrate Research* **341**, 2851–2857.

Hua-Yue, Z., Ru, S., and Ling, K. (2010). Adsorption of an anionc azo dye by chitosan/kaolin/$\dot{Y}$-Fe$_2$O$_3$ composites. *Applied Clay Science* **48**(3), 522–526.

Inger Vold, M. N., Kjell Varum, M., Eric Guibal, and Olav Smidsrod (2003). Binding of ions to chitosan; selectivity studies. *Carbohydrate Polymers* **54**, 471–477.

Inoue, K., Baba, K. Y., and Yoshizuka, K. (1993). Adsorption of metal ions on chitosan and cross-linked copper(II) complexed chitosan. *Bulletin of Chemical Society of Japan* **66**, 2915–2921.

Inoue, K., Hirakawa, H., Ishikawa, Y., Yamaguchi, T., Nagata, J., Ohto, K., and Yoshizuka, K. (1996). Adsorption of metal ions on gallium(III)-templated oxine type of chemically modified chitosan. *Seperation Science and Technology* **31**(16), 2273–2285.

Jenkins, D. W. and Hudson, S. M. (2001). Review of vinyl graft copolymerization featuring recent advances toward controlled radical-based reactions and illustrated with chitin/chitosan trunk polymers. *Chemical Reviews* **101**, 3245–3273.

Jha, N., Leela, I., and Prabhakar Rao, A. V. S. (1988). Removal of cadmium using chitosan. *Journal of Environmental Engineering* **114**, 962.

Ji-Yan, L., Xue-Qing, L., and Yuan-Fang, Z. (2010). Adsorption performance of chitosan/poluurethane porous composite material for Cu(2+) and Cd(2+) in water. *Environment Science and Technology*.

Kang, D. W., Choi, H. R., and Kweon, D. K. (1996). Selective adsorption capacity for metal ions of amidocimated chitosan bead-g-PAN co-polymer. *Polimoros* **20**, 989–295.

Karthikeyan, T., Rajagopal, S., and Miranda, L. M. (2005). Chromium (VI) Adsorption from aqueous solution by Hevea Brasilinesis sawdust activated carbon. *Journal of Hazardous Materials B* **124**, 192–199.

Katrina, C. M., Kwok., Vinci, K. C. Lee., Claire Genente, and Gordon McKay. (2009). Novel model development for sorption of arsenate on chitosan. *Journal of chemical Engineering* **151**, 122–133.

Kozlowski, C. A. and Walkowiak, W. (2002). Removal of chromium (VI) from aqueous solutions by polymer inclusion membranes. *Water Resource* **36**, 4870–4876.

Krajeswska, B. (1996). Pore structure of gel chitosan membranes.III.Pressure driven mass transport measurements. *Polymer Gels Networks* **4**, 55.

Krajeswska, B. (2001). Diffusion of metal ions through gel chitosan membranes. *Reactive Functional Polymers* **47**, 37.

Krajeswska, B. and Olech, A. (1996). Pore structure of gel chitosan membranes. I. solute diffusion measurements. *Polymer Gels and Network* **4**, 33–43.

Krajeswska, B. and Olech, A. (1996). Pore structure of gel chitosan membranes II. Modelling of the pore size distribution from solute diffusion measurements. Gaussian distribution-Mathematical limitations. *Polymer Gels Networks* **4**, 45.

Kratochvil, D., Pimentel, P., and Volesky, B. (1998). Removal of trivalent and hexavalent chromium by seaweed biosorbent. *Environmental Science and Technology* **32**, 2693–2698.

Krishnapriya, K. R. and Kandaswamy, M. (2009). Synthesis and characterization of a cross-linked chitosan derivative with a complexing agent and its adsorption studies toward metal (II) ions. *Carbohydrate Research* **344**, 1632–1638.

Kubota, N. and Kikuchi, Y. (1999). Macromolecular complexes of chitosan. In: S. Dumitriu (Ed), Polysaccharides structural diversity and functional versatility. New York, Deckker, pp. 595–628

Kumar, M., Tripathi, B. P., and Shahi, V. K. (2009). Cross-linked chitosan/polyvinyl alcohol blend beads for removal and recovery of Cd(II) from waste water. *Journal of Hazardous Materials* **172**(2–3), 1041–48.

Kurita, K. (2001). Controlled funtionalization of polysaccharide chitin. *Programme Polymer Science* **26**, 1921–1974.

Kurita, K., Kojima, T., Munakata, T., Akao, H., Mori, T., and Nishiyama, Y (1998). Preparation of non natural branched chitin and chitosan. *Chemistry Letters* **27**, 317–318.

Lasko, C. L. and Hurst, M. P. (1999). An investigation in to the use of chitosan for the removal of soluble silver from industrial wastewater. *Environment Science and Technology* **33**, 3622.

Lee, S. T., Mi, F. L., Shen, Y. J., and Shyu, S. S. (2001). Equilibrium and kinetic studies of copper (II) ion uptake by chitosan–tripolyphosphate chelating resin. *Polymer* **42**, 1879–1892.

Li, E. T., dunn, Q., Grandmaison, E. W., and Goosen, M. F. A. (1992). Application and properties of chitosan. *Journal of Bioactive and Compatible Polymers* **7**, 370–397.

Li, H. B., Chen, Y. Y., and Liu, S. L. (2003). Synthesis, characterization and metal ion adsorption property of chitosan-calixaranes. *Journal of Applied Polymer Science* **89**, 1139.

Liu, X. D. S., Tokura, Nishi, N., and Sakairi, N. (2003). A novel method for immobilization of chitosan onto non-porous glass beads through a 1,3-thiazolidine linker. *Polymer* **44**, 1021–1026.

Liu, X. D., Tokura, S., Haruki, M., Nishi, N., and Sakairi, N. (2002). Surface modification of nonporous glass beads with chitosan and their adsorption property for transition metal ions. *Carbohydrate Polymers* **49**, 103–108.

Malette, W., Quigley, H., Adickes, E., Muzzarelli, R. A. A., Jeuniaux C., and Gooday Editors G.W. (1986). *Chitin in Nature and Technology*, Plenum Press, New York, pp. 435–442.

Martel, B., Devassine, M., Crini, G., Weltrowski, M., Bourdonnaeu, M., and Morcellet, M. J. (2001). Preparation and sorption properties of a ß-cyclodextrin-linked chitosan derivative. *Journal of applied polymer science* **39**, 169–176.

Masitah, H., Bassim, H. H., and Latif, A. A. (2009). Thermodynamics studies on removal of reactive blue 19 dyes on cross-linked chitosan/oil palm ash composite beads. *Energy and Environment*, 50–54.

Mckay, G., Blair, H. S., and Gardner J. R. (1989). Adsorption of dyes on chitin. I. Equilibrium studies. *Journal Applied Polymer Science* **27**, 3043.

Metwally, E., Elkholy, S. S., Salem, H. A. M., and Elsabee, M. Z. (2009). Sorption behaviour of ^{60}Co and ^{154}Eu radio nuclides on to chitosan derivatives. *Carbohydrate polymers* **76**, 622–631.

Milot, C. (1998). Adsorption de Molybdate VI sur Billes de Gel de Chitosan-Application au Traitement d'Efflyents Molydiferes, Ph.D., Unite doctorale "Mecanique des Materiaux, Structures et Genie des Procedes-Laboratoire d'accueil: Laboratoire Genie de I'Environment Industriel (Ecole des Mines d'Ales), p. 214.

Ming-shen, C., Pang-Yen, H., and Hsing-Ya, L. (2004). Adsorption of anionic dyes in acid solutions using chemically cross-linked chitosan beads. *Dyes and Pigments* **60**, 69–84.

Mohan, D. and Pittman, C. U. Jr. (2006). Activated carbons and low cost adsorbents for remediation of tri and hexavalent chromium from water. *Journal of Hazardous Materials B* **137**, 762–811.

Muizzarelli, R. A. A. (1974). Chitosan membranes. *Ion Exchange Membrane* **1**, 193.

Muzzarelli, R. A. A. (1977). *Chitin*, Pergamon Press, Oxford, N.Y., pp. 139–150.

Muzzarelli, R. A. A. (1985). Removal of uranium from solutions and brines by a derivative of chitosan and ascorbic acid. *Carbohydrate Polymers* **5**(2), 85–89.

Muzzarelli, R. A. A. (1989). Carboxy methylated chitins and chitosan. *Carbohydrate Polymers* **8**(1), 1–21.

Muzzarelli, R. A. A. and Tanfani, F. (1982). *N*-(*O*-Carboxybenzyl) chitosan, *N*-carboxymethyl chitosan and dithiocarbamate chitosan; New chelating derivatives of chitosan. *Pure and Applied Chemistry* **54**(11), 2141–2150.

Muzzarelli, R. A. A., Tanfani, F., and Emanuelli, M. (1984). Chelating derivatives of chitosan obtained by reaction with ascorbic acid. *Carbohydrate Polymers* **4**(2), 137–151.

Muzzarelli, R. A. A, Tanfani, F., Emanuelli, M., and Bolognini, L. (1985). Aspartate glucan, glycine glucan and serine glucan for the collection of cobalt and copper from solutions and brines. *Biotechnology and Bioengineering* **27**(8), 1151–1121.

Nair, K. G. R. and Madhavan, P. (1984). Chitosan for removal of mercury from water. *Fishery Technology* **21**, 109.

Nomanbhay, S. M. and Palanisamy, K. (2005). Removal of heavy metal from industrial wastewater using chitosan coated oil palm shell charcoal. *Electronic Journal of Biotechnology* **8**, 43–53.

Onsoyen, E. and Skaugrud. (1990). Metal recovery using chitosan. *Journal of Chemical Technology and Biotechnology* **49**, 395–404

Pearce, C. I., Lloyd, J. R., and Guthrie, J. T. (2003). The removal of colour from textile wastewater using whole bacterial cells: a review. *Dyes Pigments* **58**, 179–196.

Peng, C., Wang, Y., and Tang, Y. (1998). Synthesis of cross-linked chitosan-crown ethers and evaluation of these products as adsorbents for metal ions. *Journal of Applied Polymer Science* **70**(3), 501–506.

Peter, M. G (2005). In: *Biopolymers for Medical and Pharamaceutical Applications*, A.

Steinbuchel and R. H. Marchessault (Eds.), Wiley-VCH, Weinheim, pp. 419–512.

Ramani, S. P. and Sabharwal, S. (2006). Adsorption behaviour of Cr(VI) onto radiation cross-linked chitosan and its possible application for the treatment of wastewater containing Cr(VI). *Reactive and Functional Polymers* **66**, 902–909.

Ravi Kumar, M. N. V. (2000). A review of chitin and chitosan applications. *Reactive and Functional Polymers* **46**, 1–27.

RaZiyeh, S., Mokhate, A., Niyar, M., Hajir, B., and Shooka Khorram, F. (2010). Novel biocompatible composite (chitosan-zinc oxide nanopaticles); preparation, characterization and dye adsorption properties. Colloids and Surfaces B. *Bio interfaces* **1**, 86–93.

Rinaudo, M. (2006). Chitin and chitosan; properties and applications. *Programme Polymer Science* **31**, 603.

Rio, S. and Delebarre, A. (2003). Removal of mercury in aqueous solution by fluidized bed plant fly ash. *Fuel* **82**, 153–159.

Roberts, G. A. F. (1992). *Chitin Chemistry*. The Macimillan Press Ltd., Hong Kong.

Robinson, T., Chandran, B., and Nigam, P. (2002). Studies on desorption of individual textile dyes and a synthetic dye effluent from dye-adsorbed agricultural residues using solvents. *Bioresource Technology* **84**, 299–301.

Rodrigues, C. A., Laranjeira, M. C. M., De Favere, V. T., and Stadler, E. (1998). Interaction of Cu(II) on *N*-(2-pyridyl methyl) and *N*-(4-pyridyl methyl) chitosan. *Polymer* **39**(21), 5121–5126.

Rojas, G., Silva, J., Flores, J. A., and Rodriguez, A. (2005). Adsorption of chromium onto cross-linked chitosan. *Seperation and Purification Technology* **44**, 31–36.

Roman, O. M., (2001). Progress and New Perpectives on Integrated Membrane operations For Sustainable Industrial Growth. *Industrial and Engineering Chemistry Research* **40**, 1277.

Rorrer, G. L., Hsien, T. Y., and Way, J. D. (1993). Synthesis of porous-magnetite chitosan beads for removal of cadmium ions from wastewater. *Industrial and Engineering Chemistry Research* **32**, 2170.

Ruey-Shin, J. and Huey-Jen, S. (2002). A simplified equilibrium model for sorption of heavy metal ions from aqueous solutions on chitosan. *Water Research* **36**, 2999–3008.

Ruey-Shin, J. and Ruey-Chang, S. (2000). Metal removal from aqueous solutions using chitosan-enhanced membrane filtration. *Journal of Membrane Science* **165**, 159–167.

Ruiz, M., Sastre, A. M., and Guibal, E. (2000). Osmium and Iridium sorption on chitosan derivatives. *Reactive Functional Polymers* **45**, 155.

Sahin, Y. and Ozturk, A. (2005). Biosorption of chromium (VI) ions from aqueous solution by the bacterium Bacillus thuringiensis. *Process Biochemistry* **40**, 1895–1901.

Sakaguchi, T., Hirokoshi, T., and Nakajima, A. (1981). Adsorption of uranium by chitin phosphate and chitosan phosphate. *Agricultural and Biological Chemistry* **45**, 2191–2195.

Sandipan, C., Sudipta, C., Bishnu, C. P., and Arun G. K. (2007). Adsorption removal of congo red, carcinogenic textile dye by chitosan hydrogels; Binding mechanism, equilibrium and kinetics. *Colloids and Surfaces A: Physicochemical Engineering Aspect* **299**, 146–152.

Sashiwa, H. and Shigemasa, Y. (1999). Chemical modification of chitin and chitosan 2; Preparation and water soluble property of *N*-acylated or *N*-alkylated partially deacetylated chitins. *Carbohydrate Polymers* **39**, 127–138.

Saucedo, I., Guibal, E., Roulph, C., and Le Cloirec, P. (1992). Sorption of uranyl ions by a modified chitosan; Kinetics and equilibrium studies. *Environmental Technology* **13**(12), 1101–1116.

Schmuhl, R., Krieg, H. M., and Keizer, K. A. (2001). Adsorption of Cu(II) and Cr(VI) ions by chitosan: Kinetics and equilibrium studies. *Water SA* **27**, 1–7.

Singh, V., Sharma, A. K., Tripathi, D. N., and Sanghi, R. (2009). Poly(methyl methacrylate) grafted chitosan; An efficient adsorbent for anionic azo dyes. *Journal of Hazardous Materials* **161**, 955–966.

Sreenivasan, K. (1998). Synthesis and preliminary studies on a beta-cyclodextrins-coupled chitosan as a novel adsorbent matrix. *Journal of Applied Polymer Science* **69**(6), 1051–1055.

Steenkamp, G. C., Keizer, K., Neomagus, H. W. J. P., and Krieg, H. M. (2002). Copper II removal from polluted water with alumina/chitosan composite membranes. *Journal of Membrane Science* **197**, 147–156.

Sudha, P. N., Celine, S., and Jayapriya, S. (2008). Removal of heavy metal cadmium from industrial wastewater using chitosan coated coconut charcoal. *Nature Environment and Pollution Technology* **6**(3), 421–424.

Tang, L. G. and Hon, D. N. S. (2001). Chelation of chitosan derivatives with zinc ions.III.Association complexes of Zn^{2+} on to O, *N*-carboxy methyl chitosan. *Journal of Applied Polymer Science* **79**(12), 1476–1485.

Tang, X. H., Tan, S. Y., and Wang, Y. T. (2002). Study of the synthesis of chitosan derivatives containing benzo-21-crown-7 and their adsorption properties for metal ions. *Applied Journal of Polymer Science* **83**, 1886.

Tan, T. W., He, X. J., and Du, W. X. (2001). Adsorption behaviour of metal ions on imprinted chitosan resins. *Chemical Technology and Biotechnology* **76**, 191.

Thanou, M., Verhoef, J. C., and Junginer, H. E. (2001). Oral drug absorption enhancement by chitosan and its derivatives. *Advanced Drug Delivery Reviews* **52**, 117–126.

Tokura, S., Nishimura, S. I., and Nishi, N. (1983). Studies on chitin IX. Specific binding of calcium ions by carboxy methyl-chitin. *Polymer Journal* **15**(3), 597–603

Van der Bruggen, B. and Vandecasteele, C. (2003). Removal of pollutants from surface water and groundwater by nanofiltration: overview of possible applications in the drinking water industry. *Environment Pollution* **122**, 435–445.

Van der Lubben, I. M., Verhoef, J. C., Borchard, G., and Junginger, H. E. (2000). In: *Chitosan per os from dietary supplement to drug carrier*, R. A. A. Muzzarelli (Ed.), ATEC, Grottammare, Italy, pp. 115–126.

Van der Oost, R., Beyer, J., and Vermeulen, P. E. (2003). Fish bioaccumulation and biomarkers in environmental risk assessment: A review. *Environmental Toxicology and Pharmacology* **13**, 57–149.

Van Luyen, D. and Huong, D. M. (1996). Chitin and Derivatives. In: *Polymeric Materials Encyclopedia*, J. C. Salamone (Ed.), CRC Press, Boca Raton (Florida), vol. 2, pp. 1208–1217.

Vandana, S., Ajit, K. S., and Rashmi, S. (2009). Poly(acrylamide) functionalized chitosan; An efficient adsorbent for azo dyes from aqueous solutions. *Journal of Hazardous Materials* **166**(1), 327–335,

Varma, A. J., Deshpande, S. V., and Kennedy, J. F. (2004). Metal complexation by chitosan and its derivatives: A review. *Carbohydrate Polymers* **55**, 77–93.

Vieira, R. S. and Beppu, M. M. (2006). Interaction of natural and cross-linked chitosan membranes with Hg (II) ions. *Colloids Surface A: Physiochemical Engineering and Aspects* **279**, 196–207.

Vincent, T. and Guibal, E. (2000). Non-dispersive liquid extraction of Cr (VI) by TBP/aliquat 336 using chitosan-made hollow fibers. *Solvent Extraction and Ion Exchange* **18**(6), 1241–1260.

Vincent, T. and Guibal, E. (2001). Cr(VI) extraction using aliquat 336 in a hollow fiber module made of chitosan. *Industrial and Engineering Chemistry Research* **40**, 1406.

Wan Ngah, W. S., Endud, R., and Mayanar, C. S. (2002). Removal of copper(II) ions from aqueous solution onto chitosan and cross-linked chitosan beads. *Reactive and Functional Polymers* **50**, 181–190.

Wan, L., Wang, Y., and Quian, S. (2002). Study on the adsorption properties of novel crown ether cross-linked chitosan for metal ions. *Applied Journal of Polymer Science* **84**, 29.

Wu Jun, X., Yong-liang, M., Yang Gang, S., and Bai-ye Zhan, G. (2010). Preparation and performances of polysilicate aluminum ferric-chitosan. *Bio Informatics and Biomedical Engineering (ICBBE)*, 4th International Conference, pp. 1–4.

Weltrowski, M., Martel, B., and Morcellet, M. (1996). Chitosan *N*-benzyl sulfonate derivatives as sorbents for removal of metals-ions in an acidic medium. *Journal of Applied Polymer Science* **59**(4), 647–654.

Wu, Y., Zhang, S., Guo, X., and Huang, H. (2008). Adsorption of chromium (III) on ligin. *BioresourceTechnology* **99**, 7709–7715.

Wurzburg, O. B. (Ed.). (1986). *Modified Starches: Properties and Uses*, CRC Press, Boca Raton.

Xiao, W., Yong, S C., Won S. L., and Byung, G. M. (2006). Preparation and properties of chitosan/polyvinyl alcohol) blend foams for copper adsorption. *Polymer International* **55**(11), 1230–1235.

Xie, W. M., Xu, P. X., Wang, W., and Lu, Q. (2001). Antioxidant activity of water-soluble chitosan derivatives. *Bioorganic Medical Chemistry Letters* **11**, 1699–1703.

Xie, W. M., Xu, P. X., Wang, W., and Lu, Q. (2002). Preparation and antibacterial activity of water- soluble chitosan derivative. *Carbohydrate Polymers* **50**, 35–40.

Yang, T. C. and Zall, R. R. (1984). Absorption of metals by natural polymers generated from seafood processing. *Industrial and Engineering Chemistry Product Research and Development* **23**, 168–172.

Yang, Z. and Cheng, S. (2003). Synthesis and characterization of macro cyclic polyamine derivative of chitosan. *Journal of Applied Polymer Science* **89**(4), 924–929.

Yang, Z., Wang, Y., and Tang, Y. (1999). Preparation and adsorption properties of metal ions of cross-linked chitosan azacrown ethers. *Journal of Applied Polymer Scince* **74**(13), 3053–3058.

Yasar Andelib aydin. and Nuran Deveci Aksoy. (2009). Adsorption of chromium on chitosan; Optimization, kinetics, and thermodynamics. *Chemical Engineering Journal* **151**, 188–194.

Yi, Y., Wang, Y., and Liu, H. (2003). Preparation of new cross-linked with crown ether and their adsorption for silver ion for antibacterial activities. *Carbohydrate Polymer* **53**, 425–430.

Yimin, Qin. (1993). The chelating properties of chitosan fibers. *Journal of Applied Polymer Science* **49**, 727–731.

Yoshida, H., Okamoto, A., and Katako, A. (1993). Adsorption of acid dye on cross-linked chitosan; equilibria. *Chemical Engineering Science* **48**, 2267.

Yoshida, H., Okamoto, A., Yamasaki, H., and Kataoka, T. (Eds.) (1992). Breakthrough curve for adsorption of acid dye on cross-linked chitosan fiber in; Proceedings of the international conference on fundamentals of adsorption. Kyoyo, Japan, May 17–22. *International adsorption Society*, pp. 767–774.

Zhao, Binyu Y. Z., Yue, T., Wang, X.,Wen, Z., and Liu, C. (2007). Preparation of porous chitosan gel beads for copper(II) ion adsorption. *Journal of Hazardous Materials* **147**, 67–73.

Zohuriaan-Mehr, M. J. (2005). Advances in chitin and chitosan modification through graft copolymerization; A comprehensive review. *Iranian Polymer Journal* **14**(3), 235–265.

4

Dweib M. A., Hu B., O' Donnell A., Shenton H. W., and Wool R. P. (2004). All natural composite sandwich beams for structural applications. *Composite Structures* **63**, 147–157.

Dweib M. A., Hu B., Shenton III H.W., and Wool R. P. (2006). Bio-based composite roof structure: Manufacturing and processing issues. *Composite Structures* **74**, 379–388.

Herrmann A. S., Nickel J., and Riedel U. (1998). Construction materials based upon biologically renewable resources from components to finished parts. *Polymer Degradation and Stability* **59**, 251–261.

Idicula M., Abderrahim Boudenne, Umadevi L., Laurent Ibos, Yves Candau, and Sabu Thomas. (2006). Thermo physical properties of natural fiber reinforced polyester composites. *Composites Science and Technology* **66**, 2719–2725.

Jacob M. and Thomas S. (2008). Bio fibers and Bio composites. *Carbohydrate Polymers* **71**, 343–364.

Kumar G. C. M. (2008). A Study of Short Areca Fiber Reinforced PF Composites in Proceedings of the World Congress on Engineering 2008, Vol II, July 2–4, London, U.K.

Mohanty A. K., Misra M., and Drzal L. T. (2001). Surface modifications of natural fibers and performance of the resulting biocomposites: An overview. *Composite Interfaces* **8**(5), 313–343.

Mohanty, A. K, Misra M., and Drzal, L. T. (2005). *Natural fibers, Biopolymers, and Biocomposites*. CRC Press/Taylor & Francis Group, Boca Raton FL.

Reddy N. and Yang Y. (2005). Biofibers from agricultural byproducts for industrial applications. *Trends in Biotechnolgy* **23**(1), 22–27.

Reddy N. and Yang Y. (2005a). Structure and properties of high quality natural cellulose fibers from cornstalks. *Polymer* **46**(15), 5494–5500.

Reddy N. and Yang Y. (2009). Natural cellulose fibers from soybean straw. *Bioresource Technology* **100**(14), 3593–3598.

Sanadi A. R., Caulfield D. F., and Jacobson R. E. (1997). Agrofiber/thermoplastic composites. Paper and composites from agro-based resources. R. M. Rowell, R. A. Young, and J. K. Rowell (Eds.), Chapter 12, 377–402, CRC Lewis Publishers, Boca Raton, FL.

Sheikh S. (2002). Performance of concrete structures retrofitted with fibre reinforced polymers. *Engineering Structures* **4**, 869–879.

Sreekumar P. A., Redouan Saiah, Jean Marc Saiter, Nathalie Leblanc, Kuruvilla Joseph, Unni Krishnan G., and Sabu Thomas (2009a). Dynamic mechanical properties of sisal fiber reinforced polyester composites fabricated by resin transfer molding *Polymer Composites* **30**(6), 768–775.

Sreekumar P. A., Selvin P. Thomas, Jean marc Saiter, Kuruvilla Joseph, Unni Krishnan G., and Sabu Thomas. (2009). Effect of fiber surface modification on the mechanical and water absorption characteristics of sisal/polyester composites fabricated by resin transfer molding. *Composites Part A: Applied Science and Manufacturing* **40**(11), 1777–1784.

Uomoto T., Mutsuyoshi H., Katsuki F. and Misra S. (2002). Use of fiber reinforced polymer composites as reinforcing material for concrete. *ASCE Journal of Materials in Civil Engineering*, **14**(3), 191–209.

Wool R. P. and Sun X. S. (2005). *Bio-based polymers and composites*. Elsevier Press, Amsterdam.

5

Ahn, W. S., Park, S. J., and Lee, S. Y. (2000). Production of poly(3-hydroxybutyrate) by fed-batch culture of recombinant Escherichia coli with a highly concentrated whey solution .Applied and Environmental *Microbiology* **66**(8), 3624–3627.

Alias, Z. and Tan, K. P. I. (2005). Isolation of palm oil-utilising, polyhydroxyalkanoate (PHA)-producing bacteria by an enrichment technique. *Bioresource Technology* **96**, 1229–1234.

Anderson, A. J. and Dawes, E. A. (1990). Occurrence, metabolism, metabolic role and industrial uses of Bacterial polyhydroxyalkanoates. *Microbiology Reviews* **54**, 450–472.

Anil Kumar, P. K., Shamla, T. R., Kshama, L., Prakash, M. H., Joshi, G. J., Chandrashekar, A., Kumari, K. S. L., and Divyashree, M. S. (2007). Bacterial synthesis of poly(hydroxybutyrate-co-hydroxyvalerate) using carbohydrate-rich mahua (*Madhuca* sp.) flowers. *Journal of Applied Microbiology* **103**, 204–209.

Ashby, R. D., Solaiman, D. K. Y., and Foglia, T. A. (2002). The synthesis of short- and medium-chain-length poly(hydroxyalkanoate) mixture from glucose- or alkanoic acid-grown *Pseudomonas oleovorans*. *Indian Journal of Microbiology and Biotechnology* **28**, 147–153.

Ballistreri, A., Giuffrida, M., Guglielmino, S. P. P., Carnazza, S., Ferreri, A., and Impallomeni, G. (2001). Biosynthesis and structural characterization of medium-chain-length poly(3-hydroxyalkanoates) produced by *Pseudomonas aeruginosa* from fatty acids. *Bio Macromolecules* **29**, 107–114.

Beccari, M., Majone, M., Massanisso, P., and Ramadori, R. A. (1998). Bulking Sludge with High Storage Response Selected Under Intermittent Feeding. *Water Research*. **32**, 3403–3413.

Bertrand, J. L., Ramsay, B. A., Ramsay, J. A. and Chavarie, C. (1990). Biosynthesis of poly-β-hydroxyalkanoates from pentoses by *Pseudomonas pseudovora*. *Applied and Environmental Microbiology* **56**, 3133–3138.

Bitar, A. and Underhill, S. (1990). Effect of ammonium supplementation on poly-β-hydroxybutyrate production by *A. eutrophus* in batch culture. *Biotechnology Letter* **12**, 563–568.

Bonartseva, A. P., Myshkina1, V. L., Nikolaeva, D. A., Furina1, E. K., Makhina1, T. A., Livshits, V. A., Boskhomdzhiev, A. P., Ivanov, E. A., Iordanskii, A. L., and Bonartseva, G. A. (2007). Biosynthesis, biodegradation, and application of poly(3- hydroxybutyrate) and its copolymers natural polyesters produced by diazotrophic bacteria. *Current Research and Educational Topics and Trends in Applied Microbiology A*. Méndez-Vilas (Ed.), p. 295.

Bormann, E. J. and Roth, M. (1999). The production of polyhydroxybutyrate by *Methylobacterium rhodesianum* and *Ralstonia eutropha* in media containing glycerol and casein hydrolysates. *Biotechnology Letter* **21**, 1059–1063.

Byrom, D. (1994). Polyhydroxyalkanoates In: *Plastics from Microbes: Microbial synthesis of polymers and polymer precursors*, D. P. Mobley (Ed.), Hanser, Munich, Germany, pp. 5–33.

Chien, C. C., Chen, C. C., Choi, M. H., Kung, S. S., and Wei, Y. H. C. (2007). Production of poly-β-hydroxybutyrate (PHB) by *Vibrio* spp. isolated from marine environment. *Journal of Biotechnology* 132, 259–263.

Cho, K. S., Ryu, H. W., Park, C. H., and Goodrich, P. R. (1997). Poly(hydroxybutyrate-co-hydroxyvalerate) from swine waste liquor by *Azotobacter vinelandii* UWD. *Biotechnology Letter* 19, 7–10.

Choi, J and Lee, S. Y. (1999). Process analysis and economic evaluation for Poly-β-hydroxybutyrate production by fermentation. *Bioprocess Engineering* 17, 335–342.

Choi, J. I., Lee, S. Y., and Han, K. (1998). Cloning of the *Alcaligenes latus* Polyhydroxyalkanoate Biosynthesis Genes and Use of These Genes for Enhanced Production of Poly(3-hydroxybutyrate) in *Escherichia coli. Applied and environmental Microbiology* 64(12), 4897–4903.

Dawes, E. A. (1988). Polyhydroxybutyrate: An intriguining polymer. *Bioscience Reports* 8, 537–547

Dawes, E. A. and Senior, P. J. (1973). The role and regulation of energy reserve polymers in microorganisms. *Advances in Microbial Physiology* 10, 135–266.

Dionisi, D., Majone, M., Ramadori, R., and Beccari, M. (2001). The storage of acetate under anoxic conditions. *Water Research* 35, 2661–2668.

Doi, Y., Kawaguchi, Y., Koyama, N., Nakamura, S., Hiramitsu, M., Yoshida, and Kimura, U. (1992). Synthesis and degradation of polyhydroxyalkanoates in *Alcaligenes eutrophus. FEMS Microbiological Reviews* 103, 103–108.

Fuchtenbusch, B. and Steinbuchel, A. (1999). Biosynthesis of polyhydroxyalkanoates from low-rank coal liquefaction products by *Pseudomonas oleovorans* and *Rhodococcus ruber. Applied Microbiology and Biotechnology* 52, 91–95.

Haas, R., Jin, B., and Zepf, F. T. (2008). Production of poly(3-hydroxybutyrate) from waste potato starch. *Bioscience Biotechnology and Biochemistry* 72, 253–256.

Hahn, J. J., Eschenlauer, A. C., Sletyr, U. B., Somers, D. A., and Srienc, F. (1999). Peroxisomes as sites for synthesis of polyhydroxyalkanoates in transgenic plants. *Biotechnology Progress* 15, 1053–1057.

Hassan, M. A., Shirai, Y., Kubota, A., Karim, M. I. A., Nakanishi, K., and Hashimoto (1998). The effect of oligosaccharides on glucose consumption by Rhodobacter sphaeroides in polyhydroxyalkanoate production from enzymatically treated crude sago starch. *Journal of Fermentation Bioengineering* 86(1), 57–61.

Haung, R. and Reusch, R. N. (1996). Poly(3-hydroxybutyrate) is associated with specific proteins in the cytoplasm and membranes of Escherichia coli. *Journal of Biological chemistry* 271(36), 22196–22202

Haywood, G. W., Anderson, A. J., Chu, L., and Dawes, E. A. (1988a). Characterization of two 3-ketothiolases possessing differing substrate specificities in the polyhydroxyalkanoate synthesizing organism *Alcaligenes eutrophus. FEMS Microbiological Letter* 52, 91–96.

Haywood, G. W., Anderson, A. J., Chu, L., and Dawes, E. A. (1988b). The role of NADH- and NADPH-linked acetoacetyl-CoA reductases in the poly-3-hydroxybutyrate synthesizing organism *Alcaligenes eutrophus. FEMS Microbiological Letter* 52, 259–264.

Haywood, G. W., Anderson, A. J., and Dawes, E. A. (1989). The importance PHB-synthase substrate specificity in polyhydroxyalkanoate synthesis by *Alcaligenes eutrophus. FEMS Microbiological Letter* 57, 1–6.

Haywood, G. W., Anderson, A. J., Williams, D. R., Dawes, E. A., and Ewing, D. F. (1991). Accumulation of a poly (hydroxyalkanoate) copolymer containing primarily 3-hydroxyvalerate from simple carbohydrate substances by *Rhodococcus* sp. NCIMB 40126. *International Journal of Biological Macromolecules* 13, 83–88.

Huang, T. Y., Duan, K. J., Huang, S. Y., and Chen, C.W. (2006). Production of polyhydroxybutyrates from inexpensive extruded rice bran and starch by *Haloferax mediterranei. Journal of Industrial Microbiology and Biotechnology* 33, 701–706.

Huisman, G. W., Wonink, E., Meima, R., Kazemier, B., Terpstra, P., and Witholt, B. (1991). Metabolism of poly (3-hydroxyalkanoates)

(PHAs) by *Pseudomonas oleovorans,* Identification and sequences of genes and function of the encoded proteins in the synthesis and degradation of PHA. *Journal of Biological Chemistry* **266**, 2191–2198.

Jau, M. H., Yew, S. P., Toh, P. S., Chong, A. S., Chu, W. L., Phang, S. M., Najimudin, N., and Sudesh, K. (2005). Biosynthesis and mobilization of poly(3-hydroxybutyrate) P(3HB) by *Spirulina platensis. International Journal of Biological Macromolecules* **36**, 144–151.

Jiang, Y., Song, X., Gong, L., Li, P., Dai, C., and Shao, W. (2008). High poly(β-hydroxybutyrate) production by *Pseudomonas fluorescens* A2a5 from inexpensive substrates. *Enzyme and Microbial Technology* **42**, 167–172.

John, M. E. and Keller, G. (1996). Metabolic pathway engineering in cotton: Biosynthesis of polyhydroxybutyrate in fiber cells. *Proceeding of National Academy of Science. USA* **93**, 12768–12773.

Kahar, P., Tsuge, T., Taguchi, K., and Doi, Y. (2004). High yield production of polyhydroxyalkanoates from soybean oil by *Ralstonia eutropha* and its recombinant strain. *Polymer Degradation and Stability* **83**, 79–86.

Kawaguchi, Y. and Doi, Y. (1990). Structure of native poly(3- hydroxybutyrate) granules characterized by X-ray diffraction. *FEMS Mircobiology Letter* **70**, 151–156.

Kim, B. S. and Chang, H. N. (1998). Production of poly(3-hydroxybutyrate) from starch by Azotobacter chroococcum. *Biotechnology Letter* **20**, 109–112.

Kim, S. W., Kim, P., Lee, H. S., and Kim, J. H. (1996). High production of poly-β-hydroxybutyrate (PHB) from *Methylobacterium organophilum* under potassium limitation. *Biotechnology Letter* **18**, 25–30.

Koller, M., Bona, R., Chiellini, E., Fernandes, E. G., Horvat, P., Kutschera, C., Hesse, P., and Braunegg, G. (2008). Polyhydroxyalkanoate production from whey by *Pseudomonas hydrogenovora. Bioresource Technology* **99**, 4854–4863.

Kumar, M. S., Mudliar, S. N., Reddy, K. M. K., and Chakraborti, T. (2004). Production of Biodegradable Plastic from Activated Sludge Generated from the Food Processing Industrial Wastewater Treatment Plant. *Bioresource Technology* **95**, 327–330.

Kumar, T., Singh, M., Purohit, H. J., and Kalia, V. C. (2009). Potential of *Bacillus* sp. to produce polyhydroxybutyrate from biowaste. *Journal of Applied Microbiology* **106**, 2017–2023.

Kusaka, S., Abe, H., Lee, S. Y., and Doi, Y. (1997). Molecular mass of poly[(R)-3-hydroxybutyric acid] produced in a recombinant *Escherichia coli. Applied microbiology and Biotechnology* **47**(2), 140–143.

Leaf, T. A., Peterson, M. S., Stoup, S. K., Somers, D., and Srienc, F. (1996). *Saccharomyces cerevisiae* expressing bacterial polyhydroxybutyrate synthase produces poly-3-hydroxybutyrate. *Microbiology* **142**, 1169–1180.

Lee, S. Y. (1996). Bacterial polyhydroxy-alkonates. *Biotechnology and Bioengineering* **49**, 1–14.

Lee, S. Y., Lee, K. M., Chan, H. N., and Steinbuchel, A. (1994). Comparison of recombinant *Escherichia coli* strains for synthesis and accumulation of poly-(3-hydroxybutyric acid) and morphological changes. *Biotechnology and Bioengineering* **44**(11), 1337–1347.

Lee, S. Y., Lee, Y., and Wang, F. L. (1999). Chiral compounds from bacterial polyesters: sugars to plastics to fine chemicals. *Biotechnology and Bioengineering* **65**(3), 363–368.

Lee, W. H., Azizan, M. N. M., and Sudesh, K. (2004). Effects of culture conditions on the composition of poly(3-hydroxybutyrate-co-4-hydroxybutyrate) synthesized by *Comamonas acidovorans. Polymer Degradation and Stability* **84**, 129–134.

Lemoigne, M. (1925). Production of β-hydroxybutyric acid by certain bacteria of the *B. subtilis* group. *Ann Inst Pasteur* **39**, 144–156.

Lillo, J. G. and Rodriguez-Valera, F. (1998). Effect of culture conditions on poly(β-hydroxybutyric acid) production by *Haloferax mediterranei. Applied Microbiology* **56**, 2517–2521.

Liu, F., Li, W., Ridgway, D., Gu, T., and Shen, Z. (1998). Production of poly-β-hydroxybutyrate on molasses by recombinant *Escherichia coli. Biotechnology Letter* **20**, 345–348.

Macrae, R. M. and Wilkinson, J. R. (1958). Poly-β-hydroxybutyrate metabolism in washed suspensions of *Bacillus cereus* and *Bacillus megaterium*. *Journal of General Microbiology* **19**, 210–222.

Madison, L. L. and Huisman, G. W. (1999). Metabolic engineering of poly (3-hydroxyalkanoates): From DNA to plastic. *Microbiology and Molecular Biology Reviews* **63**, 21–53.

Md Din, M. F., Ujang, Z., van Loosdrecht, M. C. M., Razak, R., Wee, A., and Yunus, S. M. (2004). Accumulation of Sunflower Oil (SO) under Slowly Biosynthesis for Better Enhancement the Poly-β-hydroxybutyrate (PHB) Production Using Mixed Cultures Approach. Faculty of Civil Engineering (FKA), Universiti Teknologi Malaysia (UTM).

Merrick, J. M. and Doudoroff, M. (1964). Depolymerisation of poly-β-hydroxybutyrate by an intracellular enzyme system. *Journal of Bacteriology* **88**, 60–71.

Mittendorf, V. V., Robertson, E. J., Leech, R. M., Kruger, N., Steinbüchel, A., and Poirier, Y. (1998). Synthesis of medium-chain-length polyhydroxyalkanoates in *Arabidopsis thaliana* using intermediates of peroxisomal fatty acid β-oxidation. *Proceeding of National Academy of Science USA* **95**, 13397–13402.

Nakashita, H., Arai, Y., Yoshioka, K., Fukui, T., Doi, Y., Usami, R., Horikoshi, K., and Yamaguchi, I. (1999). Production of biodegradable polyester by a transgenic tobacco. *Bioscience Biotechnology and Biochemistry* **63**, 870–874.

Nawrath, C., Poirier, Y., and Somerville, C. (1994). Targeting of the polyhydroxybutyrate biosynthetic pathway to the plastids of Arabidopsis thaliana results in high levels of polymer accumulation. *Proceedings of the National Academy of Sciences of the USA* **91**(26), 12760–12764

Nikel, P. I., Pettinari, M. J., Galvagno, M. A., and Méndez, B. S. (2008). Poly(3-hydroxybutyrate) synthesis from glycerol by recombinant *Escherichia coli* arcA mutant in fed-batch microaerobic cultures. *Applied Journal of Microbiology and Biotechnology* **77**, 1337–1343.

Nikel, P. I., Pettinari, M. J., Galvagno, M. A., and Méndez, B. S. (2006). Poly(3-hydroxybutyrate) synthesis by recombinant *Escherichia coli* arcA mutants in microaerobiosis. *Applied and Environmental Microbiology* **72**, 2614–2620.

Nurbas, M. and Kutsal, T. (2004). Production of PHB and P(HB-co-HV) biopolymers by using *Alcaligenes eutrophus*. *Iranian Polymer Journal* **13**, 45–51.

Ostle, A. G. and Holt, J. G. (1982). Nile blue as a fluorescent stain for PHB. *Applied and Environmental Microbiology* **44**, 238–241.

Page, W. J. (1992). Production of Polyhydroxyalkanoates by *Azotobacter vinelandii* UWD in beet molasses culture . *FEMS Microbiological Letters* **103**(2–4), 149–157.

Page, W. J. and Cornish., A. (1993). Growth of *Azotobacter vinelandii* UWD in fish peptone medium and simplified extraction of poly-β-hydroxybutyrate. *Applied and Environmental Microbiology* **59**, 4236–4244.

Page, W. J., Bhanthumnavin, N., Manchalk, J., and Rumen, M. (1997). Production of Poly(Beta-Hydroxybutyrate-Beta-Hydroxyvalerate) Copolymer from Sugars by *Azotobacter-Salinestris*. *Applied Microbiology and Biotechnology* **48**(1), 88–93.

Poirier, Y., Dennis, D. E., Klomparens, K., and Somerville, C. (1992). Polyhydroxybutyrate, a biodegradable thermoplastic, produced in transgenic plants. *Science* **256**, 520–523.

Porwal, S., Kumar, T., Lal, S., Rani, A., Kumar, S., Cheema, S., Purohit, H. J., Sharma, R., Patel, S. K. S., and Kalia, V. C. (2008). Hydrogen and polyhydroxybutyrate producing abilities of microbes from diverse habitats by dark fermentative process. *Bioresource Technology* **99**, 5444–5451.

Povolo, S. and Casella, S. (2003). Bacterial production of PHA from lactose and cheese whey permeate. *Macromolecular Symposia* **197**, 1–9.

Punrattanasin, W. (2001). The Utilization of Activated Sludge Polyhidroxyalkanoates for the Production of Biodegradable Plastics. PhD Thesis, Virginia Polytechnic Institute.

Rapske, R. (1962). Nutritional requirements for *Hydrogenomonas eutropha*. *Journal of Bacteriology* **83**, 418–422.

Rawte, T. and Mavinkurve, S. (2002). A rapid hypochlorite method for the extraction of polyhydroxy alkonates from bacterial cells. *Indian Journal of Experimental Biology* **40**, 924–929.

Reusch, R. N. (1992). Biological complexes of poly-beta-hydroxybutyrate. *Bioscience Fems Microbiological Reviews* **9**(2–4), 119–129.

Reusch, R. N. (1995). Low molecular weight complexed poly(3-hydroxybutyrate): a dynamic and versatile molecule *in vivo*. *Canadian Journal of Microbiology* **41**, 50–54.

Reusch, R. N. and Sadoff, H. L. (1983). D-(-)-poly-beta-hydroxybutyrate in membranes of genetically competent bacteria. *Journal of Bacteriology* **156**(2), 778–788.

Reusch, R. N., Hiske, T. W., And Sadoff, H. L. (1986). Poly-beta-hydroxybutyrate membrane structure and its relationship to genetic transformability in Escherichia coli. *Journal of Bacteriology* **168**(2), 553–564.

Reusch, R. N., Hiske, T. W., Sadoff, H. L., Harris, R., and Beveridge, T. (1987) Cellular incorporation of poly-β-hydroxbutyrate into plasma membranes of *Escherichia coli* and *Azotobacter vinelandii*, alters native membrane structure. *Canadian Journal of Microbiology* **33**, 435–444.

Sayyed, R. Z. and Chincholkar, S. B. (2004). Production of Poly -b-hydroxy butyrate (PHB) from Alcaligenes faecalis. *Indian Journal of Microbiology* **44**(4), 269–272.

Sayyed, R. Z. and Gangurde, N. S. (2010). Poly-β-hydroxybutyrate production by Pseudomonas sp. RZS 1 under aerobic and semi-aerobic condition. *Indian Journal of Experimental Biology* **48**, 942–947.

Sayyed, R. Z., Gangurde, N. S., and Chincholkar, S. B. (2009). Hypochlorite digestion method for efficient recovery of PHB from *A. faecalis*. *Indian Journal of Microbiology* **49**(3), 230–232.

Schembri, M. A., Bayly, R. C., and Davies, J. K. (1994). Cloning and analysis of the poly hydroxyalkanoic acid synthase gene from an *Acinetobacter* sp.: Evidence that the gene is both plasmid and chromosomally located. *FEMS Microbiology Letter* **118**, 145–152.

Schlegel, H. G., Lafferty, R., and Krauss, I. (1970). The isolation of mutants not accumulating poly-beta-hydroxybutyric acid. *Archives of Microbiology* **71**, 283–294.

Serafim, L. S., Lemos, P. C., Oliveira, R., and Reis, M. A. (2004). Optimization of Polyhydroxybutyrate Production by Mixed Cultures Submitted to Aerobic Dynamic Feeding Conditions. *Biotechnology and Bioengineering* **87**(2), 145–160.

Shi, H. P., Lee, C. M., and Ma, W. H. (2007). Influence of electron acceptor, carbon, nitrogen, and phosphorus on polyhydroxyalkanoate (PHA) production by *Brachymonas* sp. P12. *World Journal of Microbiology and Biotechnology* **23**, 625–632.

Silva, L. F., Taciro, M. K., Ramos, M. E. M., Carter, J. M., Pradella, J. G., and Gomez, J. G. (2004). Poly-3-hydroxybutyrate (P3HB) production by bacteria from xylose, glucose and sugarcane bagasse hydrolysate. *Journal of Industrial Microbiology and Biotechnology* **31**, 245–254.

Slater, S., Houmiel, K. L., Tran, H., Mitsky, T. A., Taylor, N. B., Padgette, S. R., and Gruys, K. (1998). Multiple beta-ketothiolases mediate poly (beta-hydroxyalkanoate) copolymer synthesis in *Ralstonia eutropha*. *Journal of Bacteriology* **180**, 1979–1987.

Smet, M. J., Eggink, G., Witholt, B., Kingamma, J., and Wynberg, H., (1983). Characterization of intracellular inclusions formed by *Pseudomonas oleovorans* during growth on octane. *Journal of Bacteriology* **154**, 870–878.

Son, H., Park, G., and Lee, S. (2004). Growth-associated production of poly-β-hydroxybutyrate from glucose or alcoholic distillery wastewater by *Actinobacillus* sp. EL-9. *Biotechnology Letter* **18**, 1229–1234.

Spiekermann, P., Rehm, B. H. A., Kalscheuer, R., Baumeister, D., and Steinbüchel, A. (1999). A sensitive, viable-colony staining method using Nile red for direct screening of bacteria that accumulate polyhydroxyalkanoic acids and other lipid storage compounds. *Archives of Microbiology* **171**, 73–80.

Suzuki, T., Yamane, T., and Shimizu, S. (1986). Mass production of poly-β-hydroxybutyric acid by fed-batch culture with controlled carbon/nitrogen feeding. *Applied Journal of Microbiology and Biotechnology* **24**, 370–374.

Tuominen, J., Kylma, J., Kapanen, A., Venelampi, O., Itavaara, M., and Seppala, J. (2002). Biodegradation of lactic acid based polymers under controlled composting conditions and evaluation of the ecotoxicological impact. *Biomacromolecules* **3**, 445–455.

Ueno, T., Satoh, H., Mino, T. and Matsuo, T. (1993). Production of biodegradable plastics. *Polymer Preprints* **42**, 981–986.

Valappil, S. P., Boccaccini, A. R., Bucke, C., and Roy, I. (2007). Polyhydroxyalkanoates in Gram-positive bacteria: insight from the genera *Bacillus* and *Streptomyces*. *Antonie van Leeuwenhoek* **91**, 1–17.

Valappil, S. P., Misra, S. K., Boccaccini, A. R., Keshavarz, T., Bucke, C. and Roy, I. (2007). Large-scale production and efficient recovery of PHB with desirable material properties from the newly characterized *Bacillus cereus* SPV. *Journal of Biotechnology* **132**, 251–258.

Valentine, H. E., Broyles, D. L., Casagrande, L. A, Colburn, S. M., Creely, W. L., Delaquil, P. A., Felton, H. M., Gonzalez, K. A., Houmiel, K. L., Lutke, K., Mahadeo, D. A., Mitsky, T. A., Padgette, S. R., Reiser, S. E., Slater, S., Stark, D. M., Stock, R. T., Stone, D. A., Tylor, N. B., Thorne, G. M., Tran, M., and Gruys, K. J. (1999). PHA production, from bacteria to plants. *International Journal of Biological Macromolecules* **25**(1–3), 303–306.

Vijayendra, S. V. N., Rastogi, N. K., Shamala, T. R., Anil Kumar, P. K., Kshama, L., and Joshi, G. J. (2007). Optimization of polyhydroxybutyrate production by *Bacillus* sp CFR 256 with corn steep liquor as a nitrogen source. *Indian Journal of Microbiology* **47**, 170–175.

Wallen, L. L. and Rohwedder, W. K. (1974). Poly-β-hydroxyalkanoate from activated sludge. *Environmental Science and Technology* **8**, 576–597.

Wang, Y. J., Hua, F. L., Tseng, Y. F., Chan, S. Y., Sin, S. N., Chua, H., Yu, P. H. F., and Ren, N. Q. (2007). Synthesis of PHAs from waste water under various C:N ratios. *Bioresource Technology* **98**, 1690–1693.

Williams, M. D., Fieno, A. M., Grant, R. A., and Sherman, D. H. (1996). Expression and Analysis of a Bacterial Poly(hydroxyalkanoate) Synthase in Insect Cells Using a Baculovirus System. *Protein Expression and Purification* **7**, 203–211.

Williamson, D. H. and Wilkinson, J. F. (1958). The isolation and estimation of the poly-β-hydroxybutyrate inclusions of *Bacillus* species. *Journal of General Microbiology* **19**, 198–209.

Wong, H. H and Lee, S. Y. (1998). Poly-(3-hydroxybutyrate) production from whey by high-density cultivation of recombinant *Escherichia coli*. *Applied and Environmental Microbiology* **50**(1), 30–33

Wu, Q., Huang, H., Hu, G., Chen, J., Ho, K. P., and Chen, G. Q. (2001). Production of poly-3-hydroxybutyrate by *Bacillus* sp. JMa5 cultivated in molasses media. *Antonie van Leeuwenhoek* **80**, 111–118.

Yellore, V. and Desai, A. (1998). Production of poly-3-hydroxybutyrate from lactose and whey by *Methylobacterium* sp. ZP24. *Letter in Applied Microbiology* **26**, 391–397.

Young, F. K., Kastner, J. R., and May, S. W. (1994). Microbial production of poly-β-hydroxybutyric acid from D-xylose and lactose by *Pseudomonas cepacia*. *Applied and Environmental Microbiology* **60**, 4195–4198.

Yu, J. (2001). Production of PHA from starchy wastes via organic acids. *Journal of Biotechnology* **86**(2), 105–112.

Yuksekdag, Z. N. and Beyatli, Y. (2004). Production of Poly-beta-hydroxybutyrate (PHB) in different media by *Streptococcus thermophilus* Ba21S strain. *Journal of Applied Biological Sciences* **2**, 7–10.

Zakaria, M. R., Abd-Aziz, S., Ariffin, H., Rahman, N. A. A., Yee, P. L., and Hassan, M. A. (2008). *Comamonas* sp. EB172 isolated from digester treating palm oil mill effluent as potential polyhydroxyalkanoate (PHA) producer. *African Journal of Biotechnology* **7**, 4118–4121.

Zhang, H., Obias, V., Gonyer, K., and Dennis, D. (1994). Production of polyhydroxyalkanoates in sucrose-utilizing recombinant *Escherichia coli* and *Klebsiella* strains. *Applied Journal of Microbiology and Biotechnology* **60**, 1198–1205.

6

Albertsson, A-C. and Varma, I. K. (2002). Aliphatic polyesters: Synthesis, Properties and Applications. *Advances in Polymer Science* **157**, 1–40.

Alexandre, B., Langevin, D., Médéric, P., Aubry, T., Couderc, H., Nguyen, Q. T., Saiter, A., and Marais, S. (2009). Water barrier properties of polyamide 12/montmorillonite nanocomposite membranes: Structure and volume fraction effects. *Journal of Membrane Science* **328**, 186–204.

Averous, L. (2004). Biodegradable Multiphase Systems based on Plasticized Starch: A Review. *Journal of Macromolecular Science Part C – Polymer Reviews* **44**(3), 231–274.

Bastioli, C. (1998). Biodegradable materials— Present situation and future perspectives. *Macromolecular Symposia* **135**, 193–204.

Bhardwaj, R., Mohanty, A. K., Drzal, L. T., Pourboghrat, F., and Misra, M. (2006). Renewable Resource-based Green Composites from Recycled Cellulose Fiber and Poly(3-hydroxybutyrate-*co*-3-hydroxyvalerate) Bioplastic. *Biomacromolecules* **7**, 2044–2051.

Bordes, P., Pollet, E., and Averous, L. (2009). Nano-biocomposites: Biodegradable polyester/ nanoclay systems. *Progress in Polymer Science* **34**, 125–155.

Carothers, W. H., Dorough, G. L., and Van Natta, F. J. (1932). Studies of polymerization and ring formation. X. The reversible polymerization of six-membered cyclic esters. *Journals of the American Chemical Society* **54**, 761–772.

Chen, D. Z., Tang, C. Y., Chan, K. C., Tsui, C. P., Yu, P. H. F., Leung, M. C. P., and Uskokovic, P. S. (2007). Dynamic mechanical properties and *in vitro* bioactivity of PHBHV/HA nanocomposite. *Composites Science and Technology* **67**, 1617–1626.

De Smet, M. J., Eggink, G., Witholt, B., Kingma, J., and Wynberg, H. (1983). Characterization of intracellular inclusions formed by *Pseudomonas oleovorans* during growth on octane. *The Journal of Bacteriology* **154**, 870–878.

Doi, Y. (1990). In *Microbial Polyesters*. VCH Publishers, New York.

Duquesne, E., Moins, S., Alexandre, M., Dubois, P. (2007). How can Nanohybrids Enhance Polyester/Sepiolite Nanocomposite Properties? *Macromolecular Chemistry and Physics* **208**, 2542–2550.

Erceg, M., Kovačić, T., and Klarić, I. (2005). Dynamic thermogravimetric degradation of poly(3-hydroxybutyrate)/aliphatic-aromatic copolyester blends. *Polymer Degradation and Stability* **90**, 86–94.

Follain, N., Jouen, N., Dargent, E., Chivrac, F., Girard, F., and Marais, S. (2010). *Transport mechanism of small molecules through bacterial polyester films*. 2nd International Conference on Natural Polymers, Bio-Polymers, Bio-Materials, their Composites, Blends, IPNs, Polyelectrolytes and Gels: Macro to Nano Scales, Kottayam, Kerala, India.

Gouanve, F., Marais, S., Bessadok, A., Langevin, D., and Metayer, M. (2007). Kinetics of water sorption in flax and PET fibers. *European Polymer Journal* **43**, 586–598.

Gupta, B., Revagadea, N., and Hilborn, J. (2007). Poly(lactic acid) fiber: An overview. *Progress in Polymer Science* **32**, 455–482.

Haywood, G. W., Anderson, A. J., and Dawes, E. A. (1989). The importance of PHB-synthase substrate specificity in polyhydroxyalcanoate synthesis by Alcaligenes eutrophus. *FEMS Microbiology Letters* **57**, 1–6.

Holmes, P. A., Wright, L. F., Collins, S. H. (1981). *European Patent* **52**, 459.

Iordanskii, A. L., Kamaev, P. P., Ol'khov, A. A., and Wasserman, A. M. (1999). Water transport phenomena in "green" and "petrochemical" polymers. Differences and similarities. *Desalination* **126**, 139–145.

Iordanskii, A. L., Kamaev, P. P., and Zaikov, G. E. (1998). Water sorption and diffusion in poly(3-hydroxybutyrate) films. *International Journal of Polymeric Materials* **41**, 55–63.

Iordanskii, A. L., Krivandin, A. V., Startzev, O. V., Kamaev, P. P., and Hanggi, U. J. (1999) Transport phenomena in moderately hydrophilic polymers: poly(3-hydroxybutyrate). In: *Frontiers in biomedical polymer applications*. R. M. Ottenbrite (Ed.). Technomic Publishing, Lancaster, Basel, Vol 2, pp. 63–71.

Joly, C., Le Cerf, D., Chappey, C., Langevin, D., and Muller, G. (1999). Residual solvent effect on the permeation properties of fluorinated polyimide films. *Separation and Purification Technology* **16**, 47–54.

Kamaev, P. P., Aliev, I. I., Iordanskii, A. L., and Wasserman, A. M. (2001). Molecular dynamics of the spin probes in dry and wet poly(3-hydroxybutyrate) films with different morphology. *Polymer* **42**, 515–520.

Kamaev, P. P., Iordanskii, A. L., Aliev, I. I., Wasserman, A. M., and Hanggi, U. (1999). Transport water and molecular mobility in novel barrier membranes with different morphology features. *Desalination* **126**, 153–157.

Lehermeier, H. J., Dorgan, J. R., and Way, J. D. (2001). Gas permeation properties of poly(lactic acid). *Journal of Membrane Science* **190**, 243–251.

Lenz, R. W. and Marchessault, R. H. (2005). Bacterial polyesters: biosynthesis, biodegradable plastics and biotechnology. *Biomacromolecules* **6**(1), 1–8.

Lunt, J. (1998). Large-scale production, properties and commercial applications of polylactic acid polymers. *Polymer Degradation and Stability* **59**, 145–152.

Lüpke, T., Radusch, H. J., and Metzner, K. (1998). Solid state processing of PHB-powders. *Macromolecular Symposia* **127**, 227–240.

Macrae, R. M. and Wilkinson, J. F. (1958). Poly-β-hydroxybutyrate Metabolism in Washed Suspensions of Bacillus cereus and Bacillus megaterium. *Journal of General Microbiology* **19**, 210–222.

Marais, S., Métayer, M., and Labbé, M. (1999), Water diffusion and permeability in unsaturated polyester resin films characterized by measurements performed with a water-specific permeameter: Analysis of the transient permeation. *Journal of Applied Polymer Science* **74**(14), 3380–3395.

Miguel, O., Barbari, T. A., and Iruin, J. J. (1999b). Carbon Dioxide Sorption and Diffusion in Poly(3-hydroxybutyrate) and Poly(3-hydroxybutyrate-co-3-hydroxyvalerate). *Journal of Applied Polymer Science* **71**, 2391–2399.

Miguel, O., Fernandez-Berridi, M. J., and Iruin, J. J. (1997). Survey on Transport Properties of Liquids, Vapors, and Gases in Biodegradable Poly(3-hydroxybutyrate) PHB. *Journal of Applied Polymer Science* **64**, 1849–1859.

Miguel, O. and Iruin, J. J. (1999a). Water Transport Properties in Poly(3-hydroxybutyrate) and Poly(3-hydroxybutyrate-co-3-hydroxyvalerate) Biopolymers. *Journal of Applied Polymer Science* **73**, 455–468.

Müller, H. M. and Seebach, D. (1993). Poly(hydroxyfettsäureester), eine fünfte Klasse von physiologisch bedeutsamen organischen Biopolymeren? *Angewandte Chemie.* **105**, 483–509.

Oeding, V. and Schlegel, H. G. (1973). b-keto-thiolase from Hydrogenomonas eutropha H16 and its significance in the regulation of poly-b-hydroxybutyrate metabolism. *Biochemical Journal* **134**, 239–248.

Oliveira, N. S., Goncalves, C. M., Coutinho, J. A. P., Ferreira, A., Dorgan, J., and Marrucho, I. M. (2006). Carbon dioxide, ethylene and water vapor sorption in poly(lactic acid). *Fluid Phase Equilibria* **250**, 116–124.

Perego, G., Cella, G. D., and Bastioli, C. (1996). Effect of molecular weight and crystallinity on poly(lactic acid) mechanical properties. *Journal of Applied Polymer Science* **59**, 37–43.

Ramkumar, D. H. S. and Bhattacharia, M. (1998). Steady shear and dynamic properties of biodegradable polyesters. *Polymer Engineering and Science* **39**(9), 1426–1435.

Rogers, C. E. (1985). In: *Polymer Permeability*, J. Comyn (Ed.), Elsevier, London.

Sanchez-Garcia, M. D., Lagaron, J. M., and Hoa, S. V. (2010). Effect of addition of carbon nanofibers and carbon nanotubes on properties of thermoplastic biopolymers. *Composites Science and Technology* **70**(7), 1095–1105.

Senior, P. J. and Dawes, E. A. (1973). The regulation of poly-b-hydroxybutyrate metabolism in Azotobacter beijerinckii. *Biochemical Journal* **134**, 225–238.

Sodergard, A. and Stolt, M. (2002). Properties of lactic acid based polymers and their correlation with composition. *Progress in Polymer Science* **27**, 1123–1163.

Stanier, R. Y., Doudoroff, M., Kunisawa, R., and Contopoulou, R. (1959). The role of organic substrates in bacterial photosynthesis. *Proceedings of the National Academy of Sciences* **45**, 1246–1260.

Steinbüchel, A. and Lütke-Eversloh, T. (2003). Metabolic engineering and pathway construction for biotechnological production of relevant polyhydroxyalkanoates in microorganisms. *Biochemical Engineering Journal* **16**, 81–96.

Sudesh, K., Abe, H., and Doi, Y. (2000). Synthesis, structure and properties of polyhydroxyalkanoates: biological polyesters. *Progress in Polymer Science* **25**(10), 1503–1555.

Tokiwa, Y. and Suzuki, T. (1977). Hydrolysis of polyesters by lipases. *Nature (London)* **270**(5632), 76–78.

Tsujita, Y. (2003). Gas sorption and permeation of glassy polymers with microvoids. *Progress in Polymer Science* **28**, 1377–1401.

Tuominen, J., Kylma, J., Kapanen, A., Venelampi, O., Itavaara, M., and Seppala, J. (2002). Biodegradation of lactic acid based polymers under controlled composting conditions and evaluation of the ecotoxicological impact. *Biomacromolecules* **3**(3), 445–455.

Wang, S., Ma, P., Wang, R., Wang, S., Zhang, Y., and Zhang, Y. (2008). Mechanical, thermal and degradation properties of poly(d,l-lactide)/poly(hydroxybutyrate-co-hydroxyvalerate)/poly(ethylene glycol) blend. *Polymer Degradation and Stability* **93**, 1364–1369.

Williams, J. L., Hopfenberg, H. B., and Stannett, V. (1969). "Water Transport and Clustering in Poly-(VinylChloride), Poly(OxyMethylene) and Other Polymers". *Journal of Macromolecules Science: Physics: Part B* **3**(4), 711–725.

Williamson, D. H. and Wilkinson, J. F. (1958). The Isolation and Estimation of the Poly-β-hydroxybutyrate Inclusions of Bacillus Species. *Journal of General Microbiology* **19**, 198–209.

7

Aquino, E. M. F., Sarmento, L. P. S., and Oliveira, W. (2007). Moisture Effect on Degradation of Jute/Glass Hybrid Composites. *Journal of Reinforced Plastics and Composite* **26**(2), 219–233.

Bledzki, K. and Gassan, J. (1999). Composites Reinforced with Cellulose Based Fibers. *Progress in Polymer Science* **24**, 221–274.

Gay, D., Hoa, S. V., and Tsai, S. W. (2003). *Composite Material: Design and Applications*. CRC Press, 17–25.

Karmarkar, A., Chauhan, S. S., Modak, J. M., and Chanda, M. (2007). Mechanical Properties of Wood-Fiber Reinforced Polypropylene Composites: Effect of a Novel Compatibilizer with Isocyanate Functional Group. *Composites—Part A* **38**, 227–233.

Mitra, B. C., Basak, R. K., and Sarkar, M. (1998). Studies on Jute-Reinforced Composites, Its Limitation, and Some Solutions through Chemical Modifications of Fibers. *Journal of Applied Polymer Science* **67**, 1093–1100.

Singleton, A. C. N., Baillie, C. A., Beaumont, P. W. R., and Peijs, T., (2003). On the Mechanical Properties, Deformation and Fracture of a Natural Fiber/Recycled Polymer Composite. *Composites—Part B* **34**(5), 19–526.

Verpoest, I., Van Vuure, A. W., El Asmar, N., and Vanderbeke, J. (2010). *Silk Fiber Composites*. Patent application number: 20100040816, Washington DC, USA.

Wambua, P., Ivens, J. and Verpost, I. (2003). Natural Fibers: Can They Replace Glass in Fiber Reinforced Plastic? *Composite Science and Technology* **63**, 1259–1264.

8

Abbott, N. J. (2002). Astrocyte-endothelial interactions and blood-brain barrier permeability. *J. Anat.* **200**, 629638.

Baxendale, P. H. and Greenwood, P. E. (2010). Sustained oscillations for density dependent Markov processes. *J. Math. Biol.*

Bowman, P. D., Ennis, S. R., Rarey, K. E., Betz, A. L., and Goldstein, G. W. (1983). Brain microvessel endothelial cells in tissue culture: a model for study of blood-brain barrier permeability. *Ann. Neurol.* **14**, 396402.

Daniel, B. and DeCoster, M. A. (2004). Quantification of sPLA2-induced early and late apoptosis changes in neuronal cell cultures using combined TUNEL and DAPI staining. *Brain Res. Brain Res. Protoc.* **13**, 144150.

DeCoster, M. A., Lambeau, G., Lazdunski, M., and Bazan, N. G. (2002). Secreted phospholipase A2 potentiates glutamate-induced calcium increase and cell death in primary neuronal cultures. *J. Neurosci. Res.* **67**, 634645.

Falcke, M. (2004). Reading the patterns in living cellsThe physics of Ca^{2+} signaling. *Advances in Physics* 53.

Gordon, E. L., Pearson, J. D., and Slakey, L. L. (1986). The hydrolysis of extracellular adenine nucleotides by cultured endothelial cells from pig aorta. Feed-forward inhibition of adenosine production at the cell surface. *J. Biol. Chem.* **261**, 1549615507.

Guisoni, N. and de Oliveira, M. J. (2005). Lattice model for calcium dynamics. *Phys. Rev. E. Stat. Nonlin. Soft. Matter Phys.* **71**, 061910.

Janigro, D., West, G. A., Nguyen, T. S., and Winn, H. R. (1994). Regulation of blood-brain barrier endothelial cells by nitric oxide. *Circ. Res.* **75**, 528538.

Lyons, S. A., Chung, W. J., Weaver, A. K., Ogunrinu, T., and Sontheimer, H. (2007). Autocrine glutamate signaling promotes glioma cell invasion. *Cancer Res.* **67**, 94639471.

Mandal, S., Rouillard, J. M., Srivannavit, O., and Gulari, E. (2007). Cytophobic surface modification of microfluidic arrays for in situ parallel peptide synthesis and cell adhesion assays. *Biotechnol. Prog.* **23**, 972978.

Ma, S. H., Lepak, L. A., Hussain, R. J., Shain, W., and Shuler, M. L. (2005). An endothelial and astrocyte co-culture model of the blood-brain barrier utilizing an ultra-thin, nanofabricated silicon nitride membrane. *Lab Chip.* **5**, 7485.

Means, S., Smith, A. J., Shepherd, J., Shadid, J., Fowler, J., Wojcikiewicz, R. J., Mazel, T., Smith, G. D., and Wilson, B. S. (2006). Reaction diffusion modeling of calcium dynamics with realistic ER geometry. *Biophys. J.* **91**, 537557.

Mohammed, J. S., DeCoster, M. A., and McShane, M. J. (2004). Micropatterning of nanoengineered surfaces to study neuronal cell attachment in vitro. *Biomacromolecules* **5**, 17451755.

Patterson, M., Sneyd, J., and Friel, D. D. (2007). Depolarization-induced calcium responses in sympathetic neurons: relative contributions from Ca^{2+} entry, extrusion, ER/mitochondrial Ca^{2+} uptake and release, and Ca^{2+} buffering. *J. Gen. Physiol.* **129**, 2956.

Sehgal, A., Ricks, S., Warrick, J., Boynton, A. L., and Murphy, G. P. (1999). Antisense human neuroglia related cell adhesion molecule hNr-CAM, reduces the tumorigenic properties of human glioblastoma cells. *Anticancer Res.* **19**, 49474953.

Shaikh, M. J., DeCoster, M. A., and McShane, M. J. (2006). Fabrication of interdigitated micropatterns of self-assembled polymer nanofilms containing cell-adhesive materials. *Langmuir* **22**, 27382746.

Takano, T., Lin, J. H., Arcuino, G., Gao, Q., Yang, J., and Nedergaard, M. (2001). Glutamate release promotes growth of malignant gliomas. *Nat. Med.* **7**, 10101015.

Ventura, A. C., Bruno, L., and Dawson, S. P. (2006). Simple data-driven models of intracellular calcium dynamics with predictive power. *Phys. Rev. E. Stat. Nonlin. Soft. Matter Phys.* **74**, 011917.

Weiss, N., Miller, F., Cazaubon, S., and Couraud, P. O. (2009). The blood-brain barrier in brain homeostasis and neurological diseases. *Biochim. Biophys. Acta* **1788**, 842857.

Xing, Q., Zhao, F., Chen, S., McNamara, J., DeCoster, M. A., and Lvov, Y. M. (2010). Porous biocompatible three-dimensional scaffolds of cellulose microfiber/gelatin composites for cell culture. *Acta Biomater.* **6**, 21322139.

Ye, Z. C. and Sontheimer, H. (1999). Glioma cells release excitotoxic concentrations of glutamate. *Cancer Res.* **59**, 43834391.

9

Bakshi, S. R., Lahiri, D., and Agarwal, A. (2010). Carbon nanotube reinforced metal matrix composite—A review. *Int. Mater. Rev.* **55**(1), 41–64.

Borjesson, P., Gustavsson, L., Christersson, L., and Linder, S. (1997). Future production and utilisation of biomass in Sweden: Potentials and CO_2 mitigation. *Biomass Bioenerg.* **13**(6), 399–412.

Cheng, X. S., McEnaney, B., Mays, T. J, Alcaniz Monge, J., Cazorla Amoros, D., and Linares Solano, A. (1997). Theoretical and experimental studies of methane adsorption on microporous carbons. *Carbon* **35**(9), 1251–1258.

Corma, A. (1997). From microporous to mesoporous molecular sieve materials and their use in catalysis. *Chem. Rev.* **97**(6), 2373–2419.

Dillon, A. C. and Heben, M. J. (2001). Hydrogen storage using carbon adsorbents: Past, present and future. *Appl. Phys. A-Mater.* **72**(2), 133–142.

Emmett, P. H. (1948). Adsorption and pore-size measurements on charcoals and whetlerites. *Chem. Rev.* **43**(1), 69–148.

Gorska, O. (2009). Wybrane surowce naturalne jako potencjalne źródło do otrzymywania materiałów węglowych o właściwościach sito-wo-molekularnych. (Selected natural raw materials for fabrication of carbon moleculrs sieves). MSc thesis. Faculty of Chemistry, Nicholas Copernicus University, Torun, Poland.

Hassan, M. M., Ruthven, D. M., and Raghavan, N. S. (1986). Air separation by pressure swing adsorption on a carbon molecular sieve. *Chem. Eng. Sci.* **41**(5), 1333–1343.

Horvath, G. and Kawazoe, K. (1983). Method for the calculation of effective pore size distribution in molecular sieve carbon. *J. Chem. Eng. Jpn.* **16**(8), 470–475.

http//www.cheaptubes.com/ (accessed 20 February, 2011).

http//www.sigmaaldrich.com/ (accessed 20 February, 2011).

Hu, Y. H. and Ruckenstein, E. (2004). Pore size distribution of single-walled carbon nanotubes. *Ind. Eng. Chem. Res.* **43**(3), 708–711.

Ivanova, I. I., Kuznetsov, A. S., Yuschenko, V. V., and Knyazeva, E. E. (2004). Design of composite micro/mesoporous molecular sieve catalysts. *Pure Appl. Chem.* **76**(9), 1647–1658.

Labrecque, M., Teodorescu, T. I., and Daigle, S. (1997). Biomass productivity and wood energy of *Salix* species after 2 years growth in SRIC fertilized with wastewater sludge. *Biomass Bioenerg.* **12**(6), 409–417.

Lukaszewicz, J. P., Wesołowski, R. P., and Cyganiuk, A. (2009). Enrichment of *Salix viminalis* wood in metal Ions by means of phytoextraction. *Pol. J. Environ. Stud.* **18**(3), 507–511.

Mleczek, M., Magdziak, Z., Rissmann, I., and Golinski, P. (2009a). Effect of different soil conditions on selected heavy metals accumulation by *Salix viminalis* tissues. *J. Environ. Sci Heal. A: Tox. Hazard. Subst. Environ. Eng.* **44**(14), 1609–1616.

Mleczek, M., Łukaszewski, M., Kaczmarek, Z., Rissmann, I., and Goliński, P. (2009b). Efficiency of selected heavy metals accumulation by *Salix viminalis* roots. *Environ. Exp. Bot.* **65**(1), 48–53.

Ohata, M., Aoki, M., and Yoshizawa S. (2008). Effect of carbonization condition on microstructure of bamboo charcoal with practical furnace. Proceedings International Biochar Initiative

Conference, IBI, September 8–10, Newcastle, Great Britain.

Pilatos, G., Vermisoglou, E. C., Romanos, G. E., Karanikolos, G. N., Boukos, N., Likodimos, V., and Kanellopoulos, N. K. (2010). A closer look inside nanotubes: Pore structure evaluation of anodized alumina templated carbon nanotube membranes through adsorption and permeability studies. *Adv. Funt. Mater.* 20(15), 2500–2510.

Rege, S. U. and Yang, R. T. (2000). Kinetic separation of oxygen and argon using. molecular sieve carbon. *Adsorption* **6**(1), 15–22.

Rice-Evans, C. A., Miller, N. J., Bolwell, P. G., Bramley, P. M., and Pridham, J. B. (1995). The relative antioxidant activities of plant-derived polyphenolic flavonoids. *Free Radical Res.* **22**(4), 375–383.

Rosicka-Kaczmarek, J. (2004). Polifenole jako naturalne antyoksydanty w żywności (Polyphenols as natural food antioxidants). Przegląd Piekarski i Cukierniczy 6 (1), 12–16.

Ryoo, R., Joo, S. H., Kruk, M., and Jaroniec, M. (2001). Ordered mesoporous carbons. *Adv. Mater.* **13**(9), 677–681.

Schlottig, F., Textor, M., Georgi, U., and Roewer, G. (1999). Template synthesis of SiO_2 nanostructures. *J. Mater. Sci. Lett.* 18(8), 599–601.

Shirley, A. I. and Lemcoff, N. O. (2002). Air separation by carbon molecular sieves. *Adsorption* **8**(2), 147–155.

Upadhyayula, V. K., Deng, S., Mitchell, M. C., and Smith, G. B. (2009). Application of carbon nanotube technology for removal of contaminants in drinking water: A review. *Sci. Total Environ.* 408(1), 1–13.

10

Chen, L. F., Wong, B., and Baker, W. E. (1996). Melt-grafting of glycidyl methacrylate onto polypropylene and reactive compatibilization of rubber toughened polypropylene. *Polymer Engineering and Science* 36(12), 1594-1607.

Chen, X., Guo, Q., and Mi, Y. (1998). Bamboo-fiber-reinforced polypropylene composites: A study of the mechanical properties. *Journal of Applied Polymer Science* 69(10), 1891-1899.

Choi, S. H. and Nhö, Y. C. (1998). Introduction of phosphoric acid groups to polyethylene hollow fiber by radiation-induced graft copolymerization, and its adsorption characterization to Pb^{+2}, Cu^{+2}, and Co^{+2}. *Applied Chemistry* **2**(2), 664-667.

Deshpande, A. P., Rao, M. B., and Rao, C. L. (2000). Extraction of bamboo fibers and their use as reinforcement in polymeric composites. *Journal of Applied Polymer Science* **76**(1), 83-92.

Gaylor, N. G., Mehta, R., Kumar, V., and Tazi, M. (1989). High Density Polyethylene-g-Maleic Anhydride Preparation in Presence of Electron Donors. *Journal of applied Polymer Science* **38**(2), 359-371.

Heinen, W., Rosenmöller, C. H., Wenzel, C. B., Groot, H. J. M. de, Lugtenburg J., and Duin, M. van (1996). 13C NMR Study of the Grafting of Maleic Anhydride onto Polyethene, Polypropene, and Ethene-Propene Copolymers. *Macromolecules* **29**(4), 1151-1157

Ho, R. M., Su, A. C., Wu, C. H., and Chen, S. I. (1993). Functionalization of polypropylene via melt mixing. *Polymer* **34**(15), 3264-3269.

Jang, B. C., Huh, S. Y., Jang, J. G,. and Bae, Y. C. (2001). Mechanical Properties and Morphology of the Modified HDPE/Starch Reactive Blend. *Journal of Applied Polymer Science*, **82**(13), 3313–3320.

Kim, C. H., Cho, K.Y., and Park, J. K. (2001). Grafting of glycidyl methacrylate onto polycaprolactone: Preparation and characterization. *Polymer* **42**(12), 5135-5142.

Kim, J. P., Yoon, T. H., Mun, S. P., Rhee, J. M., and Lee, J. S. (2006). Wood-polypropylene composites using ethylene-vinyl alcohol copolymer as adhesion promoter. *Bioresource Technology* **97**(3), 494-499.

Kumar, S., Choudhary, V., and Kumar, R. (2010). Study on compatibility of unbleached and bleached bamboo-fiber with LLDPE matrix. *Journal of Thermal Analysis and Calorimetry* **102**(3), 751-761.

Liu, H., Wu, Q., Han, G., Yao, F., Kojima, Y., and Suzuki, S. (2008). Compatibilization and toughening bamboo flour-filled HDPE composites: mechanical properties and morphologies. *Composites Part A: Applied Science and Manufacturing* **39**(12), 1891-1900.

Machado, A.V., Covas, J. A., and Duin, M. van (2001). Effect of polyolefin structure on maleic anhydride grafting. *Polymer* **42**(8), 3649-3655.

Martínez, J. G., Benavides, R., Guerrero, C., and Reyes, B. E. (2004). UV sensitisation of polyethylenes for grafting of maleic anhydride. *Polymer Degradation and Stability* **86**(1), 129-134.

Moad, G. (1999). The synthesis of polyolefin graft copolymers by reactive extrusion. *Progress in Polymer Science* **24**(1), 81–142.

Nakason, C., Kaesaman, A., and Supasanthitikul, P. (2004). The grafting of maleic anhydride onto natural rubber. *Polymer Testing* **23**(1), 35-41.

Qi, R., Qian, J., Chen, Z., Jin, X., and Zhou, C. (2004). Modification of Acrylonitrile–Butadiene–Styrene Terpolymer by Graft Copolymerization with Maleic Anhydride in the Melt. II. Properties and Phase Behavior. *Journal of Applied Polymer Science* **91**(5), 2834-2839.

Qiu, W., Endo, T., and Hirotsu, T. (2005). A novel technique for preparing of maleic anhydride grafted polyolefins. *European Polymer Journal* **41**(9), 1979–1984.

Sclavons, M., Franquinet, P., Carlier, V., Verfaillie, G., Fallais, I., Legras, R., Laurent, M., and Thyrion F. C. (2000). Quantification of the maleic anhydride grafted onto polypropylene by chemical and viscosimetric titrations, and FTIR spectroscopy. *Polymer* **41**(6), 1989–1999.

Seema, J. and Kumar, R. (1994). Processing of bamboo-fiber reinforced plastic composites. *Materials and Manufacturing Processes* **9**(5), 813-828.

Shi, D., Yang, J., Yao, Z., Wang, Y., Huang, H., Jing, W., Yin, J., and Costa, G. (2001). Functionalization of isotactic polypropylene with maleic anhydride by reactive extrusion: Mechanism of melt grafting. *Polymer* **42**(13), 5549-5557.

Torres, N., Robin, J. J., and Boutevin, B. (2001). Functionalization of High-Density Polyethylene in the Molten State by Glycidyl Methacrylate Grafting. *Journal of Applied Polymer Science* **81**(3), 581–590.

11

de Albuquerque, A. C., Joseph, K., Hecker de Carvalho, L., and de Almeida, J. R. M. (2000).

Composites Science and Technology **60**, 833–844.

Bledzki, A. K. and Gassan, J. (1999). *J. Prog. Polym. Sci.* **24**, 221–274.

Gowda, T. N., Naidu, A. C. B., and Chhaya, R. (1999). *Composites* **30**, 277–284.

Milewski, J. V. and Katz, H. S. (1980). Handbook of filler and reinforcements for Plastics. *Reinhold* **3**, 118–140.

Uddin, M. K., Khan, M. A., and Ali, K. M. I. (1997). *Polym. Deg. Stab.* **55**, 1–10.

12

Aronson, N. N. Jr. and Davidson, E. A. (1967). Lysosomal hyaluronidase from rat liver: II Properties. *Journal of Biological Chemistry* **242**, 441-444.

Asteriou, T., Deschrevel, B., Delpech, B., Bertrand, P., Bultelle, F., Merai, C., and Vincent, J. C. (2001). An improved assay for the N-acetyl-D-glucosamine redusing ends of polysaccharides in the presence of proteins. *Analytical Biochemistry* **293**, 53-59.

Asteriou, T., Gouley, F., Deschrevel, B., and Vincent, J. C. (2002). In *Influence of Substrate and Enzyme Concentrations on Hyaluronan Hydrolysis Kinetics Catalyzed by Hyaluronidase*. J. F. Hyaluronan, G. O. Kennedy, and P. A. Phillips Williams (Eds.). Woodhead Publishing, Wrexham, Wales, vol. 1, pp. 249-252.

Auvinen, P., Tammi, R., Parkkinen, J., Tammi, M., Agren, U., Johansson, R., Hirvikoski, P., Eskelinen, M., and Kosma, V. M. (2000). Hyaluronan in peritumoral stroma and malignant cells associates with breast cancer spreading and predicts survival. *American Journal of Pathology* **156**, 529-536.

Asteriou, T., Vincent, J. C., Tranchepain, F., and Deschrevel, B. (2006). Inhibition of hyaluronan hydrolysis catalysed by hyaluronidase at high substrate concentration and low ionic strength. *Matrix Biology* **25**, 166-174.

Berriaud, N., Milas, M., and Rinaudo M. (1998). Characterization and properties of hyaluronic acid (hyaluronan). In *Polysaccharides: Structural Diversity, and Functional Versatility.* S. Dumitriu (Ed.). Marcel Decker Inc., New York, pp. 313-334.

Bertrand, P., Girard, B., Delpech, B., Duval, C., D'Anjou, J., and Dauce, J. P. (1992). Hyaluronan (hyaluronic acid) and hyaluronectin in the extracellular matrix of human breast carcinomas: Comparison between invasive and non-invasive areas. *International Journal of Cancer* **52**, 1-6.

Bertrand, P., Girard, N., Duval, C., D'Anjou, J., Chauzy, C., Ménard, J. F., and Delpech, B. (1997). Increased hyaluronidase levels in breast tumor metastases. *International Journal of Cancer* **73**, 327-331.

Bollet, A., Bonner, W., and Nance, J. (1963). The presence of hyaluronidase in various mammalian tissues. *Journal of Biological Chemistry* **238**, 3522-3527.

Chao, K. L., Muthukumar, L., and Herzberg, O. (2007). Structure of human hyaluronidase-1, a hyaluronan hydrolyzing enzyme involved in tumor growth and angiogenesis. *Biochemistry* **46**, 6911-6920.

Chichibu, K., Matsuura, T., Shichijo, S., and Yokoyama, M. (1989). Assay of serum hyaluronic acid in clinical application. *Clinica Chimica Acta* **181**, 317-323.

Cleland, R. L., Wang, J. L., and Detweiler, D. M. (1982). Polyelectrolyte properties of sodium hyaluronate. 2. Potentiometric titration of hyaluronic acid. *Macromolecules* **15**, 386-395.

Courel, M. N., Maingonnat, C., Tranchepain, F., Deschrevel, B., Vincent, J. C., Bertrand, P., and Delpech, B. (2002). Importance of hyaluronan length in a hyaladherin-based assay for hyaluronan. *Analytical Biochemistry* **302**, 285-290.

Cowman, M. K. and Matsuoka, S. (2005). Experimental approaches to hyaluronan structure. *Carbohydrate Research* **340**, 791-809.

David-Raoudi, M., Tranchepain, F., Deschrevel, B., Vincent, J. C., Bogdanowicz, P., Boumediene, K., and Pujol, J. P. (2008). Differential effects of hyaluronan and its fragments on fibroblasts: relation to wound healing. *Wound Repair and Regeneration* **16**, 274-287.

Day, A. J. (2001). Understanding hyaluronan-protein interactions. http://www.glycoforum.gr.jp/science/hyaluronan/HA16 (accessed 28 December, 2010)

Day, A. J. and Prestwich, G. D. (2002). Hyaluronan-binding proteins: tying up the giant. *The Journal of Biological Chemistry* **277**, 4585-4588.

Delmage, J. M., Powards, D. R., Jaynes, P. K., and Allerton, S. E. (1986). The selective suppression of immunogenicity by hyaluronic acid. *Annals of Clinical and Laboratory Science* **16**, 303-310.

Delpech, B., Bertrand, P., Maingonnat, C., Girard, N., and Chauzy, C. (1995). Enzyme-linked hyaluronectin: a unique reagent for hyaluronan assay and tissue location and for hyaluronidase activity detection. *Analytical Biochemistry* **229**, 35-41.

Delpech, B., Courel, M. N., Maingonnat, C., Chauzy, C., Sesboüé, R., and Pratesi, G. (2001). Hyaluronan digestion and synthesis in an experimental model of metastatic tumour. *Histochemical Journal* **33**, 553-558.

Delpech, B., Laquerrière, A., Maingonnat, C., Bertrand, P., and Freger, P. (2002). Hyaluronidase is more elevated in human brain metastases than in primary brain tumours. *Anticancer Research* **22**, 2423-2428.

Delpech, B., Maingonnat, C., Girard, N., Chauzy, C., Maunoury, R., Olivier, O., Tayot, J., and Creissard, P. (1993). Hyaluronan and hyaluronectin in the extracellular matrix of human brain tumor stroma. *European Journal of Cancer* **29A**, 1012-1017.

Deschrevel, B. (in press). Hyaluronan: Its unique chemical, physicochemical and biological properties make it an application-rich system. In *Medical Applications of Polymers*. M. Popa, R. M. Ottenbrite, and C. V. Uglea (Eds.). American Scientific Publishers, Los Angeles, U.S.

Deschrevel, B., Lenormand, H., Tranchepain, F., Levasseur, N., Asteriou, T., and Vincent, J. C. (2008a). Hyaluronidase activity is modulated by complexing with various polyelectrolytes including hyaluronan. *Matrix Biology* **27**, 242-253.

Deschrevel, B., Tranchepain, F., and Vincent, J. C. (2008b). Chain-length dependance of the kinetics of the hyaluronan hydrolysis catalyzed by bovine testicular hyaluronidase. *Matrix Biology* **27**, 475-486.

Elson, L. A. and Morgan, W. (1933). A colorimetric method for the determination of glucosamine and chondrosamine. *Biochemical Journal* **27**, 1824-1828.

Engstrom-Laurent, A., Laurent, U. B., Lilja, K., and Laurent, T. C. (1985). Concentration of sodium hyaluronate in serum. *Scandinavian Journal of Clinical and Laboratory Investigation* **45**, 497-504.

Evanko, S. P., Angello, J. C., and Wight, T. N. (1999). Formation of hyaluronan- and versican-rich pericellular matrix is required for proliferation and migration of vascular smooth muscle cells. *Arteriosclerosis Thrombosis and Vascular Biology* **19**, 1004-1013.

Folkman, J. (2002). Role of angiogenesis in tumor growth and metastasis. *Seminars in Oncology* **29**, 15-18.

Fouissac, E., Milas, M., and Rinaudo, M. (1993). Shear-rate, concentration, molecular weigth, and temperature viscosity dependences of hyaluronate, a wormlike polyelectrolyte. *Macromolecules* **26**, 6945-6951.

Franzmann, E. J., Schroeder, G. L., Goodwin, W. J., Weed, D. T., Fisher, P., and Lokeshwar, V. B. (2003). Expression of tumor markers hyaluronic acid and hyaluronidase (HYAL1) in head and neck tumors. *International Journal of Cancer* **106**, 438-445.

Gacesa, P., Savitsky, M. J., Dodgson, K. S., and Olavesen, A. H. (1981). A recommended procedure for the estimation of bovine testicular hyaluronidase in the presence of human serum. *Analytical Biochemistry* **118**, 76-84.

Gately, C. L., Muul, L. M., Greenwood, M. A., Papazoglou, S., Dick, S. J., Komblith, P. L., Smith, B. H., and Gately, M. K. (1984). *In vitro* studies on the cell-mediated immune response to human brain tumors. II. Leukocyte-induced coats of glycosaminoglycan increase the resistance of glioma cells to cellular immune attack. *Journal of Immunology* **133**, 3387-3395.

Girard, N., Courel, M. N., Véra, P., and Delpech, B. (2000). Therapeutic efficacy of intralesional 131I-labelled hyaluronectin in grafted human glioblastoma. *Acta Oncologica* **39**, 81-87.

Girish, K. S. and Kemparaju, K. (2005). Inhibition of Naja naja venom hyaluronidase by plant-derived bioactive components and polysaccharides. *Biochemistry (Moscow)* **70**, 948-952.

Girish, K. S. and Kemparaju, K. (2007). The magic glue hyaluronan and its eraser hyaluronidase: A biological overview. *Life Sciences* **80**, 1921-1943.

Gold, E. W. (1980). An interaction of albumin with hyaluronic acid and chondroitin sulfate: A study of affinity chromatography and circular dichroism. *Biopolymers* **19**, 1407-1414.

Gold, E. W. (1982). Purification and properties of hyaluronidase from human liver. *Biochemical Journal* **205**, 69-74.

Gribbon, P., Heng, B. C., and Hardingham, T. E. (1999). The molecular basis of the solution properties of hyaluronan investigated by confocal fluorescence recovery after photobleaching. *Biophysical Journal* **77**, 2210-2216.

Grymonpré, K. R., Staggemeier, A. B., Dubin, P. L., and Mattison, F. W. (2001). Identification by integrated computer modelingand light scattering studies of an electrostatic serum albumin-hyaluronic acid binding site. *Biomacromolecules* **2**, 422-429.

Hardingham, T. E. and Muir, H. (1972). The specific interaction of hyaluronic acid with cartilage proteoglycans. *Biochimica Biophysica Acta* **279**, 401-405.

Hascall, V. C. and Heinegard, D. (1974). Aggregation of cartilage proteoglycans: I. The role of hyaluronic acid. *Journal of Biological Chemistry* **249**, 4232-4241.

Hautmann, S. H., Lokeshwar, V. B., Schroeder, G. L., Civantos, F., Duncan, R. C., Gnann R., Friedrich, M. G., and Soloway, M. S. (2001). Elevated tissue expression of hyaluronic acid and hyaluronidase validates the HA-HAase urine test for bladder cancer. *Journal of Urology* **165**, 2068-2074.

Hofinger, E. S. A., Hoechstetter, J., Oettl, M., Bernhardt, G., and Buschauer, A. (2008). Isoenzyme-specific differences in the degradation of hyaluronic acid by mammalian-type hyaluronidases. *Glycoconjugates Journal* **25**, 101-109.

Hogde-Dufour, J., Noble, P. W., Horton, M. R., Bao, C., Wysoka, M., Burdick, M. D., Strieter, R. M., Trinchieri, G., and Pure, E. (1997). Induction of IL-12 and chemokines by hyaluronan requires adhesion-dependent priming of resident but not elicited macrophages. *Journal of Immunology* **159**, 2492-2500.

Hoon Han, J. and Lee, C. H. (1997). Determining isoelectric points of model proteins and Bacillus subtilis neutral protease by the cross partitioning using poly(ethylene glycol)/dextran aqueous two-phase systems. *Colloid Surface B: Biointerfaces* **9**, 131-137.

Isoyama, T., Thwaites, D., Selzer, M. G., Carrey, R. I., Barbucci, R., and Lokeshwar, V. B. (2006). Differential selectivity of hyaluronidase inhibitors towards acidic and basic hyaluronidases. *Glycobiology* **16**, 11-21.

Jedrzejas, M. J. and Stern, R. (2005). Structures of vertebrate hyaluronidases and their unique enzymatic mechanism of hydrolysis. *Proteins* **61**, 227-238.

Knudson, C. B. and Knudson, W. (1993). Hyaluronan-binding proteins in development, tissue homeostasis and disease. *FASEB Journal* **7**, 1233-1241.

Kovar, J. L., Johnson, M. A., Volcheck, W. M., Chen, J., and Simpson, M. A. (2006). Hyaluronidase expression induces prostate tumor metastasis in an orthotopic mouse model. *American Journal of Pathology* **69**, 1415-1426.

Lapčik, L. Jr., Lapčik, L., De Smedt, S., Demeester, J., and Chabreček, P. (1998). Hyaluronan: preparation, structure, properties, and applications. *Chemical Review* **98**, 2663-2684.

Laurent, T. C. (1970). Structure of hyaluronic acid. In *Chemistry and Molecular Biology of Intercellular Matrix*. E. A. Balazs (Ed.). Academic Press, London, pp. 703-732.

Laurent, T. C. (1987). Biochemistry of hyaluronan. *Acta Otolaryngologica Supplement (Stockholm)* **442**, 7-24.

Laurent, T. C. and Fraser, J. R. E. (1986). The properties and turnover of hyaluronan. In *Functions of the Proteoglycans*. D. Evered, J. Whelan (Eds.). Ciba Foundation Symposium, Wiley, Chichester, vol. 124, pp. 9-29.

Laurent, T. C. and Fraser, J. R. E. (1992). Hyaluronan. *FASEB Journal* **6**, 2397-2404.

Laurent, T. C. and Ogston, A. G. (1963). The interaction between polysaccharides and other macromolecules. 4. The osmotic pressure of mixtures of serum albumin and hyaluronic acid. *Biochemical Journal* **89**, 249-253.

Lenormand, H., Deschrevel, B., Vincent, J. C. (2008). Electrostatic interactions between hyaluronan and proteins at pH 4: how do they modulate hyaluronidase activity. *Biopolymers* **89**, 1088-1103.

Lenormand, H., Deschrevel, B., and Vincent, J. C. (2010a). pH effects on the hyaluronan hydrolysis catalysed by hyaluronidase in the presence of proteins: Part I. Dual aspect of the pH-dependence. *Matrix Biology* **29**, 330-337.

Lenormand, H., Deschrevel, B., Vincent, J. C. (2010b). Chain length effects on electrostatic interactions between hyaluronan fragments and albumin. *Carbohydrates Polymers* **82**, 887-894.

Lenormand, H., Tranchepain, F., Deschrevel, B., and Vincent, J. C. (2009). The hyaluronan-protein complexes at low ionic strength: how the hyaluronidase activity is controlled by the bovine serum albumin. *Matrix Biology* **28**, 365-372.

Liu, N., Lapcevich, R. K., Underhill, C. B., Han, Z., Gao, F., Swartz, G., Plum, S. M., Zhang, L., Green, S. J. (2001). Metastatin: a hyaluronan-binding complex from cartilage that inhibits tumor growth. *Cancer Research* **61**, 1022-1028.

Liu, D., Pearlman, E., Diaconu, E., Guo, K., Mori, H., Haqqi, T., Markowitz, S., Willson, J., and Sy, M. S. (1996). Expression of hyaluronidase by tumor cells induces angiogenesis *in vivo*. *Proceedings of the.National Academy of Science USA* **93**, 7832-7837.

Lokeshwar, V. B., Lokeshwar, B. L., Pham, H. T., and Block, N. L. (1996). Association of elevated levels of hyaluronidase, a matrix-degrading enzyme, with prostate cancer progression. *Cancer Research* **56**, 651-657.

Lokeshwar, V. B. and Selzer, M. G. (2008). Hyaluronidase: both a tumor promoter and suppressor. *Seminars in Cancer Biology* **18**, 281-287.

Maingonnat, C., Victor, R., Bertrand, P., Courel, M. N., Maunoury, R., and Delpech, B. (1999). Activation and inhibition of human cancer cell hyaluronidase by proteins. *Analytical Biochemistry* **268**, 30-34.

Malay, O., Bayraktar, O., and Batigun, A. (2007). Complex coacervation of silk fibroin and hyaluronic acid. *International Journal of Biological Macromolecules* **40**, 387-393.

Maksimenko, A. V., Petrova, M. L., Tischenko, E. G., and Schechilina, Y. V. (2001). Chemical modification of hyaluronidase regulates its inhibition by heparin. *European Journal of Pharmaceutics and Biopharmaceutics* **51**, 33-38.

Mathews, M. B. and Dorfman, A. (1955). Inhibition of hyaluronidases. *Physiological Reviews* **35**, 381-402.

McKee, C. M., Lowenstein, C. J., Horton, M. R., Wu, J., Bao, C., Chin, B. Y., Choi, A. M. K., and Noble, P. W. (1997). Hyaluronan fragments induce nitric-oxyde synthase in murine macrophages through a nuclear factor kappaB-dependent mechanism. *Journal of Biological Chemistry* **272**, 8013-8018.

McKee, C. M., Penno, M., Cowman, M., Burdick, M., Strieter, R., Bao, C., and Noble, P. (1996). Hyaluronan (HA) fragments induce chemokine gene expression in alveolar macrophages. The role of HA size and CD44. *Journal of Clinical Investigation* **98**, 2403-2413.

Meyer, K. (1971). Hyaluronidases. In *The Enzymes*. P. D. Boyer (Ed.). 3rd ed., Academic Press, New York, vol. 5, pp. 307-320.

Mio, K. and Stern, R. (2002). Inhibitors of the hyaluronidases. *Matrix Biology* **21**, 31-37.

Moss, J. M., Van Damme, P. M, Murphy, W. H., and Preston, B. N. (1997). Dependence of salt concentration on glycosaminoglycan-lysozyme interactions in cartilage. *Archives of Biochemistry and Biophysics* **348**, 49-55.

Noble, P. W. (2002). Hyaluronan and its catabolic products in tissue injury and repair. *Matrix Biology* **21**, 25-29.

Paiva, P., Van Damme, M. P., Tellbach, M., Jones, R. L., Jobling, T., and Salamonsen, L. A. (2005). Expression patterns of hyaluronan, hyaluronan synthases and hyaluronidases indicate a role for hyaluronan in the progression of endometrial cancer. *Gynecologic Oncology* **98**, 193-202.

Paris, S., Sesboüé, R., Chauzy, C., Maingonnat, C., and Delpech, B. (2006). Hyaluronectin modulation of lung metastasis in nude mice. *European Journal of Cancer* **42**, 3253-3259.

Pasquali-Ronchetti, I., Quaglino, D., Moris, G., Bachelli, B., and Ghosh, P. (1997). Hyaluronan-phospholipid interactions. *Journal of Structural Biology* **120**, 1-10.

Reissig, J., Strominger, J., and Leloir, L. (1955). A modified colorimetric method for the estimation of N-acetylamino sugars. *Journal of Biological Chemistry*. **217**, 959-966.

Rodriguez, E. and Roughley, P. (2006). Link protein can retard the degradation of hyaluronan in proteoglycan aggregates. *Osteoarthritis Cartilage.* **14**, 823-829.

Scheibner, K. A., Lutz, M. A., Boodoo, S., Fenton, M. J., Powell, J. D., and Horton, M. R. (2006). Hyaluronan fragments act as an endogenous danger signal by engaging TLR2. *Journal of Immunology.* **177**, 1272-1281.

Scott, D., Coleman, P. J., Mason, R. M., and Levick, J. R. (2000). Interaction of intraarticular hyaluronan and albumin in the attenuation of fluid drainage from joints. *Arthritis and Rheumatism* **43**, 1175-1182.

Simpson, M. A. (2006). Concurrent expression of hyaluronan biosynthetic and processing enzymes promotes growth and vascularization of prostate tumors in mice. *American Journal of Pathology* **169**, 247-257.

Stern, R. (2005). Hyaluronan metabolism: a major paradox in cancer biology. *Pathologie Biologie* **53**, 372-382.

Stern, R. (2008). Hyaluronidases in cancer biology. *Seminars in Cancer Biology* **18**, 275-280.

Stern, R., Asari, A. A., and Sugahara, K. N. (2006). Hyaluronan fragments: an information-rich system. *European Journal of Cell Biology* **85**, 699-715.

Stern, R., Kogan, G., Jedrzejas, M. J., and Šoltes, L. (2007). The many way to cleave hyaluronan. *Biotechnology Advances* **25**, 537-557.

Takahashi, Y., Li, L., Kamiryo, M., Asteriou, T., Moustakas, A., Yamashita, H., and Heldin, P. (2005). Hyaluronan fragments induce endothelial cell differentiation in a CD44- and CXCL1/GRO1-dependent manner. *Journal of Biological Chemistry* **280**, 24195-24204.

Taylor, K. R., Trowbridge, J. M., Rudisill, J. A., Termeer, C. C., Simon, J. C., and Gallo, R. L. (2004). Hyaluronan fragments stimulate endothelial recognition of injury through TLR4. *Journal of Biological Chemistry* **279**, 17079-17084.

Termeer, C. F., Benedix, J., Sleeman, J., Fieber, C., Voith, U., Ahrens, T., Miyake, K., Freudenberg, M., Galanos, C., and Simon, J. C. (2002). Oligosaccharides of hyaluronan activate dendritic cells via toll-like receptor 4. *Journal of Experimental Medicine* **195**, 99-111.

Toida, T., Ogita, Y., Suzuki, A., Toyoda, H., and Imanari, T. (1999). Inhibition of fully O-sulfonated glycosaminoglycans. *Archives of Biochemistry and Biophysics* **370**, 176-182.

Tolksdorf, S., McCready, M. H., McCullagh, D. R., and Schwenk, E. (1949). The turbidimetric assay of hyaluronidase. *Journal of Laboratory and Clinical Medicine* **34**, 74-89.

Toole, B. P. (2002). Hyaluronan promotes the malignant phenotype. *Glycobiology* **12**, 37R-42.

Tranchepain, F., Deschrevel, B., Courel, M. N., Levasseur, N., Le Cerf, D., Loutelier-Bourhis, C., and Vincent, J. C. (2006). A complete set of hyaluronan fragments obtained from hydrolysis catalyzed by hyaluronidase: application to studies of hyaluronan mass distribution by simple HPLC devices. *Analytical Biochemistry* **348**, 232-242.

Trochon, V., Mabilat-Pragnon, C., Bertrand, P., Legrand, Y., Soria, J., Soria, C., Delpech, B., and Lu, H. (1997). Hyaluronectin blocks the stimulatory effect of hyaluronan-derived fragments on endothelial cells during angiogenesis *in vitro*. *FEBS Letters* **418**, 6-10.

Underhill, C. B. and Toole, B. P. (1979). Binding of hyaluronate to the surface of cultured cells. *Journal of Cell Biology* **82**, 475-484.

Van Damme, M. P. I., Moss, J. M., Murphy, W. H., and Preston, B. N. (1994).Binding properties of glycosaminoglycans to lysozyme – effect of salt and molecular weight. *Archives of Biochemistry and Biophysics* **310**, 16-24.

Vincent, J. C., Asteriou, T., and Deschrevel, B. (2003). Kinetics of hyaluronan hydrolysis catalysed by hyaluronidase. Determination of the initial reaction rate and the kinetic parameters. *Journal of Biological Physics and Chemistry* **3**, 35-44.

Wang, Y. F., Gao, J. Y., and Dubin, P.L. (1996). Protein separation via polyelectrolyte coacervation: selectivity and efficiency. *Biotechnology Progress* **12**, 356-362.

West, D. C. and Kumar, S. (1989). The effect of hyaluronan and its oligosaccharides on endothelial cell proliferation and monolayer integrity. *Experimental Cell Research* **183**, 179-196.

Xu, S., Yamanaka, J., Sato, S., Miyama, I., and Yonese, M. (2000). Characteristics of complexes composed of sodium hyaluronate and bovine serum albumin. *Chemical and Pharmaceutical Bulletin* **48**, 779-783.

Yang, B., Yang, B. L., Savani, R. C., and Turley, E. A. (1994). Identification of a common hyaluronan binding motif in the hyaluronan binding proteins RHAMM, CD4 and link protein. *The EMBO Journal* **13**, 286-296.

13

Agapov, I. I., Pustovalova, O. L., Moisenovich, M. M., Bogush, V. G., Sokolova, O. S., Sevastyanov, V. I., Debabov, V. G., and Kirpichnikov, M. P. (2009). Three-Dimensional Scaffold Made from Recombinant Spider Silk Protein for Tissue Engineering. *Dokl. Biochem. Biophys.* **426**, 127-130.

Allmeling, C., Jokuszies, A., Reimers, K., Kall, S., and Vogt, P. M. (2006). Use of spider silk fibres as an innovative material in a biocompatible artificial nerve conduit. *J. Cell Mol. Med.* **10**, 770-777.

Allmeling, C., Jokuszies, A., Reimers, K., Kall, S., Choi, C. Y., Brandes, G., Kasper, C., Scheper, T., Guggenheim, M., and Vogt, P. M. (2008). Spider silk fibres in artificial nerve constructs promote peripheral nerve regeneration. *Cell Prolif.* **41**,408-420.

Arcidiacono, S., Mello, C. M., Butler, M., Welsh, E., Soares, J. W., Allen, A., Ziegler, D., Laue, T., and Chase, S. (2002). Aqueous processing and fiber spinning of recombinant spider silks. *Macromolecules* **35**, 1262-1266.

Arcidiacono, S., Mello, C., Kaplan, D., Cheley, S., and Bayley, H. (1998). Purification and characterization of recombinant spider silk expressed in *Escherichia coli*. *Applied Microbiology and Biotechnology* **49**,31-38.

Ayoub, N. A., Garb, J. E., Tinghitella, R. M., Collin, M. A., and Hayashi, C. Y. (2007). Blueprint for a high-performance biomaterial: Full-length spider dragline silk genes. *PLoS ONE* **2**,e514.

Baoyong, L., Jian, Z., Denglong, C., and Min, L. (2010). Evaluation of a new type of wound dressing made from recombinant spider silk protein using rat models. *Burns* **36**, 891-896.

Bini, E., Foo, C. W., Huang, J., Karageorgiou, V., Kitchel, B., and Kaplan, D. L. (2006). RGD-

functionalized bioengineered spider dragline silk biomaterial. *Biomacromolecules* **7**, 3139-3145.

Bogush, V. G., Sokolova, O. S., Davydova, L. I., Klinov, D. V., Sidoruk, K. V., Esipova, N. G., Neretina, T. V., Orchanskyi, I. A., Makeev, V. Y., Tumanyan, V. G., Shaitan, K. V., Debabov, V. G., and Kirpichnikov, M.P. (2008). A Novel Model System for Design of Biomaterials Based on Recombinant Analogs of Spider Silk Proteins. *J. Neuroimmune Pharmacol.*

Bon, M. (1710-1712). A Discourse upon the Usefulness of the Silk of Spiders. By Monsieur Bon, President of the Court of Accounts, Aydes and Finances, and President of the Royal Society of Sciences at Montpellier. Communicated by the Author. *Philosophical Transactions (1683-1775)* **27**, 2-16.

Brooks, A. E., Stricker, S. M., Joshi, S. B., Kamerzell, T. J., Middaugh, C. R., and Lewis, R. V. (2008). Properties of synthetic spider silk fibers based on Argiope aurantia MaSp2. *Biomacromolecules* **9**, 1506-1510.

Candelas, G. C. and Cintron, J. (1981). A spider fibroin and its synthesis. *J. Exp. Zool.* **216**, 1-6.

Chen, X., Knight, D. P., and Vollrath, F. (2002). Rheological characterization of nephila spidroin solution. *Biomacromolecules* **3**, 644-648.

Dicko, C., Vollrath, F., and Kenney, J. M. (2004). Spider silk protein refolding is controlled by changing pH. *Biomacromolecules* **5**, 704-710.

Doran, P. M. (2000). Foreign protein production in plant tissue cultures. *Curr. Opin. Biotechnol.* **11**, 199-204.

Exler, J. H., Hummerich, D., and Scheibel, T. (2007). The amphiphilic properties of spider silks are important for spinning. *Angew Chem. Int. Ed. Engl.* **46**, 3559-3562.

Fahnestock, S. R. and Bedzyk, L. A. (1997). Production of synthetic spider dragline silk protein in *Pichia pastoris*. *Appl. Microbiol. Biotechnol.* **47**, 33-39.

Fahnestock, S. R. and Irwin, S. L. (1997). Synthetic spider dragline silk proteins and their production in *Escherichia coli*. *Appl. Microbiol. Biotechnol.* **47**, 23-32.

Fredriksson, C., Hedhammar, M., Feinstein, R., Nordling, K., Kratz, G., Johansson, J., Huss, F., and Rising, A. (2009). Tissue Response to

Subcutaneously Implanted Recombinant Spider Silk: An *in Vivo* Study. *Materials* **2**. 1908-1922.

Fukushima, Y. (1998). Genetically engineered syntheses of tandem repetitive polypeptides consisting of glycine-rich sequence of spider dragline silk. *Biopolymers* **45**, 269-279.

Geisler, M., Pirzer, T., Ackerschott, C., Lud, S., Garrido, J., Scheibel, T., and Hugel, T. (2008). Hydrophobic and Hofmeister effects on the adhesion of spider silk proteins onto solid substrates: An AFM-based single-molecule study. *Langmuir* **24**, 1350-1355.

Gellynck, K., Verdonk, P., Forsyth, R., Almqvist, K. F., Van Nimmen, E., Gheysens, T., Mertens, J., Van Langenhove, L., Kiekens, P., and Verbruggen, G. (2008a). Biocompatibility and biodegradability of spider egg sac silk. *J. Mater. Sci. Mater. Med.* **19**, 2963-2970.

Gellynck, K., Verdonk, P. C., Van Nimmen, E., Almqvist, K. F., Gheysens, T., Schoukens, G., Van Langenhove, L., Kiekens, P., Mertens, J., and Verbruggen, G. (2008b). Silkworm and spider silk scaffolds for chondrocyte support. *J. Mater. Sci. Mater. Med.* **19**, 3399-3409.

Gosline, J. M., Guerette, P. A., Ortlepp, C. S., and Savage, K. N. (1999). The mechanical design of spider silks: From fibroin sequence to mechanical function. *Journal of Experimental Biology* **202**, 3295-3303.

Hagn, F., Eisoldt, L., Hardy, J., Vendrely, C., Coles, M., Scheibel, T., and Kessler, H. (2010). A conserved spider silk domain acts as a molecular switch that controls fibre assembly. *Nature* in press.

Hermanson, K. D., Harasim, M. B., Scheibel, T., and Bausch, A. R. (2007). Permeability of silk microcapsules made by the interfacial adsorption of protein. *Phys. Chem. Chem. Phys.* **9**, 6442-6446.

Hijirida, D. H., Do, K. G., Michal, C., Wong, S., Zax, D., and Jelinski, L. W. (1996). 13C NMR of Nephila clavipes major ampullate silk gland. *Biophys. J.* **71**, 3442-3447.

Hinman, M. B. and Lewis, R. V. (1992). Isolation of a clone encoding a second dragline silk fibroin. Nephila clavipes dragline silk is a two-protein fiber. *J. Biol. Chem.* **267**, 19320-19324.

Hu, X., Vasanthavada, K., Kohler, K., McNary, S., Moore, A. M., and Vierra, C. A. (2006).

Molecular mechanisms of spider silk. *Cell Mol. Life Sci.* **63**, 1986-1999.

Huang, J., Wong, C., George, A., and Kaplan, D. L. (2007). The effect of genetically engineered spider silk-dentin matrix protein 1 chimeric protein on hydroxyapatite nucleation. *Biomaterials* **28**, 2358-2367.

Huemmerich, D., Helsen, C. W., Quedzuweit, S., Oschmann, J., Rudolph, R., and Scheibel, T. (2004a). Primary structure elements of spider dragline silks and their contribution to protein solubility. *Biochemistry* **43**, 13604-13612.

Huemmerich, D., Scheibel, T., Vollrath, F., Cohen, S., Gat, U., and Ittah, S. (2004b). Novel assembly properties of recombinant spider dragline silk proteins. *Curr. Biol.* **14**, 2070-2074.

Huemmerich, D., Slotta, U., and Scheibel, T. (2006). Processing and modification of films made from recombinant spider silk proteins. *Applied Physics A* **82**,219-222.

Junghans, F. C. U., Scheibel, T., Heilmann, A., and Spohn, U. (2006). Preparation and mechanical properties of layers made of recombinant spider silk proteins and silk from silk worm. *Applied Physics A* **82**, 253-260.

Knight, D. P. and Vollrath, F. (2001). Changes in element composition along the spinning duct in a Nephila spider. *Naturwissenschaften* **88**, 179-182.

Kovoor, J. (1987). Comparative structure and histochemistry of silk-producing organs in arachnids. Berlin, Springer-Verlag.

Kuhbier, J. W., Allmeling, C., Reimers, K., Hillmer, A., Kasper, C., Menger, B., Brandes, G., Guggenheim, M., and Vogt, P. M. (2010). Interactions between spider silk and cells--NIH/3T3 fibroblasts seeded on miniature weaving frames. *PLoS One* **5**, e12032.

Lammel, A., Schwab, M., Slotta, U., Winter, G., and Scheibel, T. (2008). Processing conditions for spider silk microsphere formation. *ChemSusChem* **5**, 413-416.

Lazaris, A., Arcidiacono, S., Huang, Y., Zhou, J. F., Duguay, F., Chretien, N., Welsh, E. A., Soares, J. W., and Karatzas, C. N. (2002). Spider silk fibers spun from soluble recombinant silk produced in mammalian cells. *Science* **259**, 472-476.

Lewis, R. V., Hinman, M., Kothakota, S., and Fournier, M. J. (1996). Expression and purification of a spider silk protein: A new strategy for producing repetitive proteins. *Protein Expression and Purification* **7**, 400-406.

Liebmann, B., Hummerich, D., Scheibel, T., and Fehr, M. (2008). Formulation of poorly watersoluble substances using self-assembling spider silk protein. *Colloids and Surfaces A: Physicochem Eng Aspects* **331**, 126-132.

Lin, Z., Huang, W., Zhang, J., Fan, J. S., and Yang, D. (2009). Solution structure of eggcase silk protein and its implications for silk fiber formation. *Proc. Natl. Acad. Sci. USA* **106**, 8906-8911.

Ma, J. K., Drake, P. M., and Christou, P. (2003). The production of recombinant pharmaceutical proteins in plants. *Nat. Rev. Genet.* **4**, 794-805.

Mello, C. M., Soares, J. W., Arcidiacono, S., and Butler, M. M. (2004). Acid extraction and purification of recombinant spider silk proteins. *Biomacromolecules* **5**, 1849-1852.

Menassa, R., Zhu, H., Karatzas, C. N., Lazaris, A., Richman, A., and Brandle, J. (2004). Spider dragline silk proteins in transgenic tobacco leaves: Accumulation and field production. *Plant Biotechnol. J.* **2**, 431-438.

Metwalli, E. S. U., Darko, C., Roth, S., Scheibel, T., and Papadakis, C. (2007). Structural changes of thin films from recombinant spider silk proteins upon post treatment. *Applied Physics A* **89**, 655-661.

Michal, C. A., Simmons, A. H., Chew, B. G., Zax, D. B., and Jelinski, L. W. (1996). Presence of phosphorus in Nephila clavipes dragline silk. *Biophys. J.* **70**, 489-493.

Mieszawska, A. J., Nadkarni, L. D., Perry, C. C., and Kaplan, D. L. (2010). Nanoscale control of silica particle formation via silk-silica fusion proteins for bone regeneration. *Chem. Mater.* **22**, 5780-5785.

Morgan, A. W., Roskov, K. E., Lin-Gibson, S., Kaplan, D. L., Becker, M. L., Simon, C. G., Jr. (2008). Characterization and optimization of RGD-containing silk blends to support osteoblastic differentiation. *Biomaterials* **29**, 2556-2563.

Nemir, S. and West, J. L. (2010). Synthetic materials in the study of cell response to substrate rigidity. *Ann. Biomed. Eng.* **38**, 2-20.

Newman, J. and Newman, C. (1995). Oh what a tangled web: The medicinal uses of spider silk. *Dermatology* **34**, 290-292.

Patel, J., Zhu, H., Menassa, R., Gyenis, L., Richman, A., and Brandle, J. (2007). Elastin-like polypeptide fusions enhance the accumulation of recombinant proteins in tobacco leaves. *Transgenic Res.* **16**, 239-249.

Prince, J., Mcgrath, K., Digirolamo, C., and Kaplan, D. (1995). Construction, cloning, and expression of synthetic genes encoding spider dragline silk. *Biochemistry* **34**, 10879-10885.

Rammensee, S., Slotta, U., Scheibel, T., and Bausch, A. R. (2008). Assembly mechanism of recombinant spider silk proteins. *Proc. Natl. Acad. Sci. U S A* **105**, 6590-6595.

Rising, A., Hjalm, G., Engstrom, W., and Johansson, J. (2006). N-terminal nonrepetitive domain common to dragline, flagelliform, and cylindriform spider silk proteins. *Biomacromolecules* **7**, 3120-3124.

Rising, A., Johansson, J., Larson, G., Bongcam-Rudloff, E., Engstrom, W., and Hjalm, G. (2007). Major ampullate spidroins from Euprosthenops australis: Multiplicity at protein, mRNA and gene levels. *Insect Mol. Biol.* **16**, 551-561.

Rising, A., Widhe, M., Johansson, J., Hedhammar, M. Spider silk proteins: recent advances in recombinant production, structure-function relationships and biomedical applications. *Cell Mol. Life Sci.*

Sandkvist, M. and Bagdasarian, M. (1996). Secretion of recombinant proteins by Gram-negative bacteria. *Curr. Opin. Biotechnol.* **7**, 505-511.

Scheller, J., Gührs, K. H., Grosse, F., and Conrad, U. (2001). Production of spider silk proteins in tobacco and potato. *Nature Biotechnology* **19**, 573-577.

Scheller, J., Henggeler, D., Viviani, A., and Conrad, U. (2004). Purification of spider silk-elastin from transgenic plants and application for human chondrocyte proliferation. *Transgenic Res.* **13**, 51-57.

Slotta, U. T. M., Kremer, F., Koelsch, P., and Scheibel, T. (2006). Structural analysis of films cast from recombinant spider silk proteins. *Supramolecular Chemistry* **18**, 465-471.

Slotta, U. K., Rammensee, S., Gorb, S., and Scheibel, T. (2008). An engineered spider silk protein forms microspheres. *Angew Chem. Int. Ed. Engl.* **47**, 4592-4594.

Stark, M., Grip, S., Rising, A., Hedhammar, M., Engstrom, W., Hjalm, G., and Johansson, J. (2007). Macroscopic fibers self-assembled from recombinant miniature spider silk proteins. *Biomacromolecules* **8**, 1695-1701.

Stephens, J. S., Fahnestock, S. R., Farmer, R. S., Kiick, K. L., Chase, D. B., and Rabolt, J. F. (2005). Effects of electrospinning and solution casting protocols on the secondary structure of a genetically engineered dragline spider silk analogue investigated via Fourier transform Raman spectroscopy. *Biomacromolecules* **6**, 1405-1413.

Szela, S., Avtges, P., Valluzzi, R., Winkler, S., Wilson, D., Kirschner, D., and Kaplan, D. L. (2000). Reduction-oxidation control of beta-sheet assembly in genetically engineered silk. *Biomacromolecules* **1**, 534-542.

Teulé, F., Cooper, A. R., Furin, W. A., Bittencourt, D., Rech, E. L., Brooks, A., Lewis, R. V. (2009). A protocol for the production of recombinant spider silk-like proteins for artificial fiber spinning. *Nat. Protocols* **4**, 324-355.

Teulé, F., Furin, W. A., Cooper, A. R., Duncan, J. R., and Lewis, R. V. (2007). Modifications of spider silk sequences in an attempt to control the mechanical properties of the synthetic fibers. *Journal of Materials Science* **42**, 8974-8985.

Valluzzi, R., Szela, S., Avtges, P., Kirschner, D., and Kaplan, D. (1999). Methionine redox controlled crystallization of biosynthetic silk spidroin. *Journal of Physical Chemistry B* **103**, 11382-11392.

Wang, H., Wei, M., Zue, Z., and Li, M. (2009). [Cytocompatibility study of Arg-Gly-Asp-recombinant spider silk protein/poly vinyl alcohol scaffold] *Chinese* **23**, 747-750.

Werten, M. W., Moers, A. P., Vong, T., Zuilhof, H., van Hest, J. C., and de Wolf, F. A. (2008). Biosynthesis of an amphiphilic silk-like polymer. *Biomacromolecules* **9**, 1705-1711.

Widhe, M., Bysell, H., Nystedt, S., Schenning, I., Malmsten, M., Johansson, J., Rising, A., and Hedhammar, M. (2010). Recombinant spider silk as matrices for cell culture. *Biomaterials* **31**, 9575-9585.

Williams, D. (2003). Sows' ears, silk purses and goats' milk: New production methods and medical applications for silk. *Med. Device Technol.* **14**, 9-11.

Winkler, S., Szela, S., Avtges, P., Valluzzi, R., Kirschner, D. A., and Kaplan, D. (1999). Designing recombinant spider silk proteins to control assembly. *Int. J. Biol. Macromol.* **24**, 265-270.

Winkler, S., Wilson, D., and Kaplan, D. L. (2000). Controlling beta-sheet assembly in genetically engineered silk by enzymatic Phosphorylation/Dephosphorylation, by. *Biochemistry* **39**, 14002.

Vollrath, F., Barth, P., Basedow, A., Engstrom, W., and List, H. (2002). Local tolerance to spider silks and protein polymers *in vivo*. *In Vivo* **16**, 229-234.

Wong Po Foo, C., Patwardhan, S. V., Belton, D. J., Kitchel, B., Anastasiades, D., Huang, J., Naik, R. R., Perry, C. C., Kaplan, D. L. (2006). Novel nanocomposites from spider silk-silica fusion (chimeric) proteins. *Proc. Natl. Acad. Sci. U S A* **103**, 9428-9433.

Wong, S. L. (1995). Advances in the use of Bacillus subtilis for the expression and secretion of heterologous proteins. *Curr. Opin. Biotechnol.* **6**, 517-522.

Work, R. W. (1977). Mechanisms of major ampullate silk fiber formation by orb-web-spinning spiders. *Trans. Am. Microsc. Soc.* **96**, 170-189.

Xu, H. T., Fan, B. L., Yu, S. Y., Huang, Y. H., Zhao, Z. H., Lian, Z. X., Dai, Y. P., Wang, L. L., Liu, Z. L., Fei, J., and Li, N. (2007). Construct synthetic gene encoding artificial spider dragline silk protein and its expression in milk of transgenic mice. *Anim. Biotechnol.* **18**, 1-12.

Yang, J., Barr, L. A., Fahnestock, S. R., and Liu, Z. B. (2005). High yield recombinant silk-like protein production in transgenic plants through protein targeting. *Transgenic Res.* **14**, 313-324.

Zhou, S., Peng, H., Yu, X., Zheng, X., Cui, W., Zhang, Z., Li, X., Wang, J., Weng, J., Jia, W., and Li, F. (2008). Preparation and characterization of a novel electrospun spider silk fibroin/poly(D,L-lactide) composite fiber. *J. Phys. Chem. B* **112**, 11209-11216.

Zhou, Y., Wu, S., and Conticello, V. P. (2001). Genetically directed synthesis and spectroscopic analysis of a protein polymer derived from a flagelliform silk sequence. *Biomacromolecules* **2**, 111-125.

14

Alvarez, V., Vazquez, A., and Bernal, C. (2005). Fracture behaviour of sisal fibre-reinforced starch-based composites. *Polymer Composite* **26**(3), 316–323.

Belgacem, M. N. and Gandini, A. (2008). Surface modification of cellulose fibres. In: *Monomers, polymers and composites from renewable resources*. M. N. Belgacem and A. Gandini (Eds.), Elsevier, Amsterdam.

Bledzki, A. K. and Gassan, J. (1999). Composites reinforced with cellulose based fibres. *Progress in Polymer Science* **24**(2), 221–274.

Bledzki, A. K., Reihmane, S., and Gassan, J. (1998) Thermoplastics reinforced with wood fillers. A literature review. *Polymer-Plast Technology* **37**(4), 451–468.

Carvalho, A. J. F. (2008). Starch: Major Sources, Properties and Applications as Thermoplastic Materials. In: *Monomers, oligomers, polymers and composites from renewable resources*. M. N. Belgacem and A. Gandini (Eds.). Elsevier, Amsterdam.

Ciechanska, D. (2004). Multifunctional bacterial cellulose/chitosan composite materials for medical applications. *Fibres and Textile in Eastern Europe* **12**, 69–72.

Curvelo, A. A. S., de Carvalho, A. J. F., and Agnelli, J. A. M. (2001). Thermoplastic starch–cellulosic fibers composites: Preliminary results. *Carbohydrate Polymer* **5**(2), 183–188.

Derksen, J. T. P., Cuperus F. P., and Kolster P. (1996). Renewable resources in coatings technology: A review. *Progress in Organic Coatings* **27**, 45–53.

Dubey, V., Pandey, L. K., and Saxena, C. (2005). Pervaporative separation of ethanol/water azeotrope using a novel chitosan-impregnated bacterial cellulose membrane and chitosan-poly(vinyl alcohol) blends. *Journal of Membrane Science* **251**, 131–136.

Dufresne, A. (2008). Cellulose-Based Composites and Nanocomposites. In: *Monomers, oligomers, polymers and composites from renewable resources*. M. N. Belgacem and A. Gandini (Eds.). Elsevier, Amsterdam.

Dufresne, A., Dupeyre, D., and Vignon, M. R. (2000). Cellulose microfibrils from potato tuber cells: Processing and characterization of starch–cellulose microfibril composites. *J. Appl. Polym. Sci.* **76**(14), 2080–2092.

Fernandes, S. C. M., Freire, C. S. R., Silvestre, A. J. D., Pascoal Neto, C., Gandini, A., Berglund, L. A., and Salmén, L. (2010). Transparent chitosan films reinforced with a high content of nanofibrillated cellulose. *Carbohydrate Polymer* **81**(2), 394–401.

Fernandes, S. C. M., Oliveira, L., Freire, C. S. R., Silvestre, A. J. D., Neto, C. P., Gandini, A., and Desbrieres, J. (2009). Novel transparent nanocomposite films based on chitosan and bacterial cellulose. *Green Chemistry* **11**(12), 2023–2029.

Funke, U., Bergthaller, W., and Lindhauer, M. G. (1998). Processing and characterization of biodegradable products based on starch. In: *10th European conference and technology exhibition on biomass for energy and industry*. Wurzburg, Germany, pp. 454–457.

Grande, C. J., Torres, F. G., Gomez, C. M., Troncoso, O. P., Canet-Ferrer, J., and Martinez-Pastor, J. (2008). Morphological characterisation of bacterial cellulose–starch nanocomposites. *Polymer Composite* **16**(3), 181–185.

Hasegawa, H., Isogai, A., Onabe, F., Usuda, M., and Atalla, R. H. (1992). Characterization of cellulose-chitosan blend films. *Journal of Applied Polymer Science* **45**, 1873–1879.

Hosokawa, J., Nishiyama, M., Yoshihara, K., and Kubo, T. (1990). Biodegradable film derived from chitosan and homogenized cellulose. *Industrial and Engineering Chemistry Research* **29**, 800–805.

Hosokawa, J., Nishiyama, M., Yoshihara, K., Kubo, T., and Terabe, A. (1991). Reaction between chitosan and cellulose on biodegradable composite film formation. *Industrial and Engineering Chemistry Research* **30**, 788–792.

Klemm, D., Schumann, D., Kramer, F., Heßler, N., Koth, D., and Sultanova, B. (2009). Nanocellulose materials – different cellulose, different functionality. *Macromolecule Symposium* **280**: 60–71.

Lee, S. Y., Mohan, D. J., Kang, I. A., Doh, G. H., Lee, S., and Han, S. O. (2009). Nanocellulose reinforced PVA composite films: Effects of acid treatment and filler loading. *Fiber Polymer* **10**(1), 77–82.

Lima, I. S., Lazarin, A. M., and Airoldi, C. (2005). Favorable chitosan/cellulose film combinations for copper removal from aqueous solutions. *International Journal of Biological Macromolecules* **36**, 79–83.

Lu, Y. S., Weng, L. H., and Cao, X. D. (2005). Biocomposites of plasticized starch reinforced with cellulose crystallites from cottonseed linter. *Macromol. Biosci.* **5**(11), 1101–1107.

Lu, Y. S., Weng, L. H., and Cao, X. D. (2006). Morphological, thermal and mechanical properties of ramie crystallites—reinforced plasticized starch biocomposites. *Carbohydrate Polymers* **63**(2), 198–204.

Ma, X. F., Chang, P. R., and Yu, J. G. (2008). Properties of biodegradable thermoplastic pea starch/carboxymethyl cellulose and pea starch/microcrystalline cellulose composites. *Carbohydrate Polymers* **72**(3), 369–375.

Martins, I. M. G., Magina, S. P., Oliveira, L., Freire, C. S. R., Silvestre, A. J. D., Pascoal Neto C., and Gandini A. (2009). New biocomposites based on thermoplastic starch and bacterial cellulose. *Composite Sci. Technol.* **69**, 2163–2168.

Mucha, M. and Pawlak, A. (2003). Thermo gravimetric and FTIR studies of chitosan blends. *Thermochemical Acta* **396**, 153–166.

Mucha, M. and Pawlak, A. (2005). Themal analysis of chitosan and its blends. *Thermochimica Acta* **427**, 69–76.

Nakagaito, A. N. and Yano, H. (2004). The effect of morphological changes from pulp fiber towards nano-scale fibrillated cellulose on the mechanical properties of high-strength plant fiber based composites. *Appl. Phys. A* **78**(4), 547–552.

Nordqvist, D., Idermark, J., Hedenqvist M., Gällstedt, M., Ankerfors, M., and Lindström, T. (2007). Enhancement of the wet properties of transparent chitosan-acetic acid-salt films using microfibrillated cellulose. *Biomacromolecules* **8**, 2398–2403.

Orts, W. J., Shey, J., Imam, S. H., Glenn, G. M., Guttman, M. E., and Revol, J. F. (2005). Application of cellulose microfibrils in polymer nanocomposites. *J. Polym. Environ.* **13**(4), 301–306.

Pecoraro, E., Manzani, D., Messaddeq, Y., and Ribeiro, S. J. L. (2008). Bacterial Cellulose from Glucanacetobacter xylinus: Preparation, Properties and Applications. In: *Monomers, oligomers, polymers and composites from renewable resources*. M. N. Belgacem and A. Gandini (Eds.). Elsevier, Amsterdam.

Peniche, C., Argüelles-Monal, W., and Goycoolea, F. M. (2008). Chitin and Chitosan: Major Sources, Properties and Applications. In: *Monomers, oligomers, polymers and composites from renewable resources*. M. N. Belgacem and A. Gandini (Eds.). Elsevier, Amsterdam.

Pommet, M., Juntaro, J., Heng, J. Y. Y., Mantalaris, A., Lee, A. F., and Wilson, K., et al. (2008). Surface modification of natural fibers using bacteria: depositing bacterial cellulose onto natural fibers to create hierarchical fiber reinforced nanocomposites. *Biomacromolecules* **9**(6), 1643–1651.

Samir, M. A. S. A., Alloin, F., and Dufresne, A. (2005). Review of recent research into cellulosic whiskers, their properties and their application in nanocomposite field. *Biomacromolecules* **6**(2), 612–626.

Satyanarayana, K. G., Arizaga, G. G. C., and Wypych, F. (2009). Biodegradable composites based on lignocellulosic fibers: An overview. *Progress in Polymer Science*, **34**, 982–1021.

Shih, C.-M., Shield, Y.-T., and Twu, Y. K. (2009). Preparation and characterization of cellulose/chitosan blend films. *Carbohydrate Polymers* **78**, 169–174.

Twu, Y. K., Huang, H. I., Chang, S. Y., and Wang, S. L. (2003). Preparation and sorption activity of chitosan/cellulose blend beads. *Carbohydrate Polymers* **54**, 425–430.

15

ACD/ChemSketch (2009). *Version 12*, Advanced Chemistry Development, Inc., Toronto, ON, Canada.

Albarghouthi, M., Fara, D. A., Saleem, M., El-Thaher, T., Matalka, K., and Budwan, A. (2000). Immobilization of antibodies on alginate-chitosan beads. *International Journal of Pharmaceutics* **206**, 23–34.

Anson, S. I., Novikova, E. V., and Iozep, A. A. (2009). Intramolecular Esters of Alginic Acid. *Russian Journal of Applied Chemistry* **82**, 1095–1097.

Bubnis, W. A. and Ofner, C. M. (1992). The determination of e-amino groups in soluble and poorly soluble proteinaceous materials by a spectrophotometric method using trinitrobenzenesulfonic acid. *Analytical Biochemistry* **27**, 129–133.

Gombotz W. R. and We, S. F. (1998). Protein release from alginate matrices. *Advanced Drug Delivery Reviews* **31**, 267–285.

http://www.cgl.ucsf.edu./cgi-bin/gencollagen.py

Kadler, K. E., Baldock, C., Bella, J., and Raymond, P. (2007). Boot-Handford Collagens at a glance. *Journal of Cell Science* **120**, 1955–1958.

Kuga, T., Esato, K., Zempo, N., Fujioka, K., and Nakamurav, K. (1998). Detection of Type III Collagen Fragments in Specimens of Abdominal Aortic Aneurysms. *Surgery Today* **28**, 385–390.

Lin, Y. K. and Liu, D. C. (2006). Comparison of physical–chemical properties of Type I collagen from different species. *Food Chemistry* **99**, 244–251.

Madhan, B., Subramanian, V., Raghava Rao, J., Nair, B. U., Ramasami, T. (2005). Stabilization of collagen using plant polyphenol: Role of catechin. *International Journal of Biological Macromolecules* **37**, 47–53.

Mitra, T., Sailakshmi, G., Gnanamani, A., Raja, S. T. K.,Thiruselvi, T., Mangala Gowri, V., Selvaraj, N. V., Ramesh, G., and Mandal, A. B. (2010). Preparation and characterization of a thermostable and biodegradable biopolymers using natural cross-linker. *International Journal of Biological Macromolecules* (In Press).

Morris, G. M., Huey, R., Lindstrom, W., Sanner, M. F., Belew, R. K., Goodsell, D. S., and Olson, A. J. (2009). Autodock4 and AutoDockTools4: automated docking with selective receptor flexibility. *Journal of Computational Chemistry* **30**, 2785–2791.

Nomura, Y., Yamano, M., Hayakawa, C., Ishh, Y., and Shirai, K. (1997). Structural property and in vitro shark Type I collagen. *Bioscience Biotechnology and Biochemistry* **61**, 1919–1923.

Pavia, D. L., Lampman, G. M., and Kriz, G. S. (2001). *Introduction to Spectroscopy*, 3rd ed., Thomson Learning Inc., USA, pp. 13–101.

Ru, C. P., Hing, C. M., Liu, F. H., and Se, W. Y. (2005). Release characteristics and bioactivity of gelatin-tri calcium phosphate membrane covalently immobilized with new growth factors. *Biomaterials* **26**, 6579–6587.

Salome Machado, A. A., Martins, V. C. A., and Plepis, A. M. G. (2002). Thermal and rheological behavior of collagen chitosan blends. *Journal of Thermal Analysis and Calorimetry* **67**, 491–498.

Sung, H., Hsu, C., Wang, S., and Hsu, H. (1997). Degradation potential of biologic tissues fixed with various fixatives: An *in vitro* study. *Journal of Biomedical Materials Research* **35**, 147–155.

Van der Rest, M. and Garrone, R. (1991). Collagen family of proteins. *FASEB Journal* **51**, 2814–2823.

Wang, B., Jia, D. Y., Ruan, S. Q., and Qin, S. (2007). Structure and properties of collagen- Konjac glucomannan-sodium alginate blend films. *Journal of Applied Polymer Science* **106**, 327–332.

Wu, H. C., Wang, T. W., Kang, P. L., Tsuang, Y. H., Sun, J. S., and Lin, F. H. (2007). Coculture of endothelial and smooth muscle cells on a collagen membrane in the development of a small diameter vascular graft. *Biomaterials* **28**, 1385–1392.

Yang, S. R., Kwon, O. J., Kim, D. H., and Park, J. S. (2007). Characterization of the polyurethane foam using alginic acid as a polyol. *Fibers and Polymers* **8**, 257–262.

16

Albrecht, W., Reintjes, M., and Wulfhorst, B. (1997). Lyocell-Fasern (Alternative Zelluloseregeneratfasern). *Faserstoff-Tabellen nach*. P. A. Koch (Ed.). German.

Anonymous (2007). Bioplastics in Automotive Applications. *Bioplastics Magazine* **2**(1), 14–18.

Avella, M, Buzarovska, A, Errico, M. E, Gentile, G., and Grozdanov, A. (2009): Eco-Challenges of Bio-Based Polymer Composites. *Materials* **2**, 911–925.

Bax, B. and Müssig, J. (2008). Impact and tensile properties of PLA/Cordenka and PLA/flax composites. *Composites Science and Technology* **68**, 1601–1607.

Bhardwaj, R. and Mohanty, A. K. (2007). Advances in the Properties of Polylactides Based Materials A Review. *Journal of Biobased Materials and Bioenergy*. **1**, 191–209.

Bledzki, A. K., Jaszkiewicz, A., and Scherzer, D. (2009). Mechanical properties of PLA composites with man-made cellulose and abaca fibres. *Composites: Part A* **40**, 404–412.

Bodros, E., Pillin, I., Montrelay, N, and Baley, C (2007). Could biopolymers reinforced by randomly scattered flax fibre be used in structural applications? *Composites Science and Technology* **67**, 462–470.

Bourmaud, A. and Pimbert, S. (2008). Investigations on mechanical properties of poly(propylene) and poly(lactic acid) reinforced by miscanthus fibers. *Composites: Part A*, **39**, 1444–1454.

Bunsell, A. R., and Renard, J, (Eds.) (2005). *Fundamentals of Fibre Reinforced Composite Materials*. Institut of Physics Publishing, Bristol, Philadelphia, USA.

Cheung, H., Ho, M., Lau, K., Cardona, F.,and Hui, D. (2009). Natural fibre-reinforced composites for bioengineering and environmental engineering applications. *Composites Part B: Engineering* **40**, 655 –663.

Cho, D., Seo, J. M., Lee, H. S., Cho, C. W., Han, S. O., and Park, W. H. (2007). Property improvement of natural fiber-reinforced green composites by water treatment. *Advanced Composite Materials*, **16**, 299–314.

DIN EN ISO 61:1977: Glasfaserverstärkte Kunststoffe—Zugversuch (German).

DIN EN ISO 139:2005: Textilien—Normalklimate für die Probenvorbereitung und Prüfung (German).

DIN EN ISO 179:1996: Kunststoffe—Bestimmung der Charpy-Schlagzähigkeit (German).

Ganster, J. and Fink, H. P. (2006). Novel cellulose fibre reinforced thermoplastic materials. *Cellulose* **13**, 271–280.

Garlotta, D. (2001). A literature review of poly(lactic acid). *Journal of Polymers and the Environment* **9**(2), 63–84.

Gassan, J. and Bledzki, A. (1999). Möglichkeiten zur Kontrolle der Festigkeit und Impactzähigkeit von naturfaserverstärkten Kunststoffen. *Die Angewandte Makromolekulare Chemie* **272**(4745), 17–23.

Grashorn, C. (2007). Erstes Serienprodukt aus naturfaserverstärktem PLA im Spritzguss. 5. *N-FibreBase Kongress,* Mai 2007, Hürth, Germany, pp. 21–22.

Graupner, N. (2008). Application of lignin as natural adhesion promoter in cotton fibre-reinforced poly(lactic acid) (PLA) composites. *Journal of Materials Science* **43**, 5222–5229.

Graupner, N. (2009). Improvement of the Mechanical Properties of Biodegradable Hemp Fiber Reinforced Poly(lactic acid) (PLA) Composites by the Admixture of Man-made Cellulose Fibers. *Journal of Composite Materials* **43**(6), 689–702.

Graupner, N., Herrmann, A. S., and Müssig, J. (2009). Natural and man-made cellulose fibre-reinforced poly(lactic acid) (PLA) composites: An overview about mechanical characteristics and application areas. *Composites: Part A* **40**, 810–821.

Graupner, N. and Müssig, J. (2009). Man-made cellulose fibres as reinforcement for poly(lactic acid) (PLA) composites. *Journal of Biobased Materials and Bioenergy* **3**, 249–261.

Gupta, B., Revagade, N., and Hilborn, J. (2007). Poly(lactic acid) fiber: An overview. *Progress in Polymer Science,* **32**, 455–482.

Hu, R. and Lim, J. (2007). Fabrication and Mechanical Properties of Completely Biodegradable Hemp Fiber Reinforced Polylactic Acid Composites. *Journal of Composite Materials* **41**(13), 1655–1668.

Huber, T. and Müssig, J. (2008). Fibre matrix adhesion of natural fibres cotton, flax and hemp in polymeric matrices analyzed with the single fibre fragmentation test. *Composite Interfaces* **15**(2–3), 335–349.

Huda, M. S., Drzal, L. T., Mohanty, A. K., and Misra, M. (2008). Effect of fiber surface-treatments on the properties of laminated biocomposites from poly(lactic acid) (PLA) and kenaf fibers. Composites Science and Technology 68, 424–432.

Ibrahim, N., Yunus, W., Othman, M., Abdan, K., and Hadithon, K. (2010). Poly(Lactic Acid) (PLA)-reinforced Kenaf Bast Fiber Composites: The Effect of Triacetin. Journal of Reinforced Plastics and Composites, 29, 1099–1111.

Iji, M. (2008). Highly Functional Bioplastics Used for Durable Products. In The Netherlands Science and Technology (Organizer and Editor).

Innovative Technologies in Bio-Based Economy. The Netherlands, 8th April, Wageningen.

Jonoobi, M., Harun, J., Mathew, A. P., and Oksman, K. (2010). Mechanical properties of cellulose nanofiber (CNF) reinforced polylactic acid (PLA) prepared by twin screw extrusion. Composites Science and Technology 70(12), 1742–1747.

Karus, M. and Kaup, M. (2005). Natural fibres in the European automotive industry. Journal of Industrial Hemp 7, 131.

Karus, M., Ortmann, S., Gahle, C., and Pendarovski, C. (2006). Einsatz von Naturfasern in Verbundwerkstoffen für die Automobilproduktion in Deutschland von 1999 bis 2005. Nova Institut, Hürth, Germany.

Lee, B., Kim, H., Lee, S., Kim, H., and Dorgan, J. (2009). Bio-composites of kenaf fibers in polylactide: Role of improved interfacial adhesion in the carding process. Composites Science and Technology 69, 2573–2579.

Lee, S. H. and Ohkita, T. (2004). Bamboo fiber (BF)-filled poly(butylenes succinate) bio-composite—Effect of BF-e-MA on the properties and crystallization kinetics. Holzforschung 58, 7.

Lee, S. H. and Wong, S. (2006). Biodegradable polymers/bamboo fiber biocomposite with bio-based coupling agent. Composites: Part A 37, 80–91.

Lim, L., Auras, R., and Rubino, M. (2008). Processing technologies for poly(lactic acid). Progress in Polymer Science 33, 820–852.

Mieck, K. P., Reussmann, T., and Hauspurg, C. (2000). Correlations for the fracture work and falling weight impact properties of thermoplastic natural/long fibre composites. Materialwissenschaft und Werkstofftechnik 31, 169–174.

Moser, K. (Ed.) (1992). Faser-Kunststoff-Verbund. Entwurfs- und Berechnungsgrundlagen. Düsseldorf, Germany. VDI Verlag GmbH, (ISBN 3–18–401187–9).

Nampoothiri, K. M., Nair, N. R., and John, R. P. (2010). An overview of the recent developments in polylactide (PLA) research. Bioresource Technology 101, 8493–8501.

Plackett, D. (2004). Maleated Polylactide as an Interfacial Compatibilizer in Biocomposites. J. Polym. Environ. 12(3), 131–138.

Tokoro, R., Vu, D. M., Okubo, K., Tanaka, T., Fuji, T., and Fujura, T. (2008). How to improve mechanical properties of polylactic acid with bamboo fibers. J. Mater. Sci. 43, 775–785.

Yu, T., Ren, J., Li, S., Yuan, H., and Li, Y. (2010). Effect of fibre surface-treatments on the properties of poly(lactic acid)/ramie composites. Composites: Part A 41, 499–505.

17

Alemdar, A. and Sain, M. (2008). Isolation and characterization of nanofibers from agricultural residueswheatstraw and soy hulls. Bioresorce Technology 99, 16641671.

Ashori, A., Jalaluddin, H., Raverty, W. D., and Mohd Nor, M. Y. (2006). Chemical and morphological characteristics of Malaysia Cultivated Kenaf (Hibiscusecannabinus) fiber. Polym-Plast Technol. Eng. 45(1), 131134.

Azizi Samir, M. A. S., Alloin, F., Sanchez, J. Y., and Dufresne, A. (2004). Cellulose nanocrystal reinforced poly(oxythylene). Polymer 45, 41494157.

Bibin Mathew Cherian, Alcides Lopes Leão, Sivoney Ferreira de Souza, Sabu Thomas, Laly A. Pothan, M. Kottaisamy. (2010). Isolation of nanocellulose from pineapple leaf fibers by steam explosion. Carbohydrate Polymers 81, 720725.

Bondeson, D. and Oksman, K. (2007). Polylactic acid/cellulose whisker nanocomposites modified by polyvinyl alcohol, Composites Part A-Applied Science 38, 24862492.

Candanedo, S. B., Roman, M., and Gray, D. G. (2005). Effect of reaction conditions on the properties and behavior of wood cellulose nanocrystal suspensions. Biomacromolecule 6, 10481054.

De SouzeLima, M. M. and Borsali, R. (2004). Rodlike cellulose microcrystals: Structure, properties, and applications. Macromolecular Rapid Communications 25, 771787.

Fahmy, T. Y. A. and Mobarak, F. (2008). Nanocomposites from natural cellulose fibers filled with kaolin in presence of sucrose. Carbohydrate Polymers 72, 751755.

Franco, P. J. H. and Gonzalez, A. V. (2005). A study of the mechanical properties of short natural-fiber reinforced composite. *Composites Part B: Engineering* **67**, 597608.

Gardner, D. J., Oporto, G. S., Mills, R., and Azizi Samir, M. A. S. (2008). Adhesion and surface issues in cellulose and nanocellulose. *Journal of Adhesion Science and Technology* **22**, 545–567.

Kentaro, A., Shiniro, I., and Hiroyuki, Y. (2007). Obtaning cellulose nanofibers with a uniform width of 15 nm from wood. *Biomacromolecules* **8**, 32763278.

Kim, D. Y., Nishiyama, Y., Wada, M., and Kuga, S. (2001). High yield carbonization of cellulose by sulphoric acid impregnation. *Cellulose* **8**, 2933.

Klemm, D., Heublein, B., Fink, H. P. and Bohn, A. (2005). Fascinating biopolymer and sustainable raw material. Angewandte Chemie International Edition. *Cellulose* **44**, 33583393.

Mallick, P. K. (1988). *Fiber-Reinforced Composites*. Marcel Dekker Inc, America.

Maren, R. and William, T. W. (2004). Effect of sulfate groups from sulfuric acid hydrolysis on the thermal degradation behavior of bacterial cellulose. *Biomacromolecules* **5**, 16711677.

Mohd. Sapuan, S. (1999). Bahan Komposit Berasaskan Polimer. Serdang Penerbit Universiti, Putra Malaysia.

Morán, J. I., Alvarez, V. A., Cyras, V. P., and Vázquez, A. (2008). Extraction of cellulose and preparation of nanocellulosefrome sisal fibers. *Cellulose* **15**, 149159.

Mwaikambo, L. Y. and Ansell, M. P. (2002). Chemical modification of hemp, sisal, jute, and kapok fibers by alkalization. *Journal of Applied Polymer Science* **84**, 22222234.

Rongji, Li, Jianming, Fei, Yurong, Cai, Yufeng, Li, Jianqin, Feng, and Juming, Yao. (2009). Cellulose whiskers extracted from mulberry: Anovel biomass production. *Carbohydrate Polymers* **76**, 9499.

Roohani, M., Habibi, Y., Belgacem, N. M., Ebrahim, G., Karimi, A. N., and Dufresne, A. (2008). Cellulose whiskers reinforced polyvinyl alcohol copolymers nanocomposites. *J. Eur. Polym.* **44**, 24892498.

Sain, M. and Panthapulakkal, S. (2006). Bioprocess preparation of wheat straw fibers and their

characterization. *Industrial Crops and Products* **2**, 18.

Segal, L., Greely, L., Martin, A. E., and Conrad, C. M. (1959). An empirical method for estimating the degree of crystallinity of native cellulose using X-ray diffractometer. *Textille Res. J.* **29**, 786794.

Seydibeyoğlu, M. O. and Oksman, K. (2008). Novel nanocomposites based on polyurethane and micro fibrillated cellulose. *J. Compos. Sci. Technol.* **68**, 908914.

Shebani, A. N. and Van Reenen, A. J. (2008). The effect of wood extractives on the thermal stability of different wood species. *Thermochim. Acta.* **471**, 4350.

Sun-Young, Lee, Jagan, M. D., In-Aeh, K., Geum-Hyun, D., Soh, Lee, Seong, Ok Han. (2009). Nanocellulose reinforced PVE composite films: Effects of Acid Treatment and filler loading. *Fibers and Polymers* **10**, 7782.

Troedec, M., Sedan, D., Peyratout, C., Bonnet, J., Smith, A., Guinebretiere, R., Gloaguen, V., and Krausz, P. (2008). Influence of various chemical treatment on the composition and structure of hemp fibers. *Composites Part A* **39**(3), 514522.

Turbak, A. F., Snyder, F.W., and Sandberg, K. R. (1983). Microfibrillated cellulose, A new cellulose product. *Journal of Applied Polymer Science: ApplPolymSymp* **37**, 815827.

Xue, Li, Lope, G., and Tabil, S. P. (2007). Chemical treatments of natural fiber for use in natural fiber-reinforced composites: A review. *Journal Polymer Environment* **15**, 2533.

Yang, H. P., Yan, R., hen, H. P., Lee, D. H., and Zheng, C. G. (2007). Characteristics of hemicelluloses, cellulose and lignin pyrolysis. *Fuel* **86**, 17811788.

Youssef, H., Waleed, K., El-zawawy Maha, M., Ibrahim, and Alain, D. (2008). Processing and characterization of reinforced polyethylene composites made with lignocellulosic fibres from Egyptian agro-industrial residues. *Composites Science and Technology* **68**, 18771885.

18

Arora, A. and Padua, G. W. (2010). Review: Nanocomposites in food packaging. *Journal of Food Science* **75**, 43–48.

Blumstein, A. (1965). Polymerization of adsorbed monolayers. 1. Preparation of the clay-polymer complex. *Journal of Polymer Science Part A: Polymer Chemistry* **3**, 2653–2664.

Hansen, N. and Plackett, D. (2008). Sustainable films and coatings from hemicelluloses: A review. *Biomacromolecules* **9**, 1493–1505.

Hussain, F., Hojjati, M., Okamoto, M., and Gorga, R. E. (2006). Review article: Polymer-matrix nanocomposites, processing, manufacturing, and application: an overview. *Journal of Composite Materials* **40**, 1511–1575.

Lagarón, J. M. and Fendler, A. (2009). High water barrier nanobiocomposites of methyl cellulose and chitosan for film and coating applications. *Journal of Plastic Film and Sheeting* **25**, 47–59.

Lan, T., Kaviratna, P. D., and Pinnavaia, T. J. (1994). On the nature of polyimide-clay hybrid composites. *Chemistry of Materials* **6**, 573–575.

Messersmith, P. B. and Giannelis, E. P. (1995). Synthesis and barrier properties of poly(e-caprolactone)-layered silicate nanocomposites. *Journal of Polymer Science Part A: Polymer Chemistry* **33**, 1047–1057.

Pavlidou, S. and Papaspyrides, C. (2008). A review on polymer-layered silicate nanocomposites. *Progress in Polymer Science* **33**, 1119–1198.

Vartiainen, J., Tuominen, M., and Nättinen, K. (2010a). Biohybrid nanocomposite from sonicated chitosan and nanoclay. *Journal of Applied Polymer Science* **116**, 3638–3647.

Vartiainen, J., Tammelin, T., Pere, J., Tapper, U., and Harlin, A. (2010b). Biohybrid barrier films from fluidized pectin and nanoclay. *Carbohydrate Polymers* **82**, 989–996.

Yano, K., Usuki, A., and Okada, A. (1997). Synthesis and properties of polyimide-clay hybrid films. *Journal of Polymer Science Part A: Polymer Chemistry* **35**, 2289–2294.

19

Alluin, O., Wittmann, C., Marqueste, T., Chabas, J. F., Gracia, S., Lavaut, M. N., Guinard, D., Feron, F., and Decherchi P. (2009). Functional recovery after peripheral nerve injury and implantation of a collagen guide. *Biomaterials* **30**, 363–364.

Bodugoz-Senturk, H., Choi, J., Oral, E., Kung, J. H., Macias, C. E., Braithwaite, G., and Muratoglu, O. K. (2008). The effect of polyethylene glycol on the stability of pores in PVA hydrogels during annealing. *Biomaterials* **29**, 141–149.

Chilarski, A. and Pietrucha, K. (1996). Surgical treatment of gastroschisis: Primary fascial closure versus staged repair. Surgery in childhood. *International* **4**, 158–160.

Choi, J. S., Leong, K. W., and Yoo, H. S. (2008). *In vivo* wound healing of diabetic ulcers using electrospun nanofibers immobilized with human epidermal growth factor (EGF). *Biomaterials* **29**, 587–596.

Czaja, W. K., Young, D. J., Kawecki, M., and Brown, R. M. Jr. (2007). The future prospects of microbial cellulose in biomedical applications. *Biomacromolecules* **8**, 1–12.

Dai, W., Kawazoe, N., Lin, X., Dong, J., and Chen, G. (2010). The influence of structural design of PLGA/collagen hybrid scaffolds in cartilage tissue engineering. *Biomaterials* **31**, 2141–2152.

Doillon, C. J., Cote, M. F., Pietrucha, K., Laroche, G., and Gaudreault, R. G. (1994). Porosity and biological properties of polyethylene glycol-conjugated collagen materials. *Journal of Biomaterials Science-Polymer Edition* **6**, 715–728.

Doillon, C. J., Pietrucha, K., and Gaudreault, R. C. (1999). Biostable porous material comprising composite biopolymers. http://www.freepatentsonline.com/5863984.pdf.

Fatimi, A., Tassin, J. F., Quillard, S., Axelos, M. A. V., and Weiss P. (2008). The rheological properties of silated hydroxypropylmethylcellulose tissue engineering matrices. *Biomaterials* **29**, 533–543.

Gerard, C., Bordeleau, L. J., Barralet, J., and Doillon, C. J. (2010). The stimulation of angiogenesis and collagen deposition by copper. *Biomaterials* **31**, 824–831.

Helary, C., Bataille, I., Abed, A., Illoul, C., Anglo, A., Louedec, L., Letourneur, D., Meddahi-Pelle, A., and Giraud-Guille, M. M. (2010). Concentrated collagen hydrogels as dermal substitutes. *Biomaterials* **31**, 481–490.

Jagur-Grodzinski, J. (2006). Polymers for tissue engineering, medical devices, and regenerative medicine. Concise general review of recent studies. *Polymers for Advanced Technologies* 17, 395–418.

Kolacna, L., Bakesova, J., Varga, F., Kostakowa, E., Lukas, D., Amler, E., and Pelouch V. (2007). Biochemical and biophysical aspects of collagen nanostructure in the extracellular matrix. *Physiological Research* 11, 43–56.

Kuo, Y. C. and Yeh, C. F. (2011). Effect of surface-modified collagen on the adhesion, biocompatibility and differentiation of bone marrow stromal cells in poly(lactide-co-glicolide)/chitosan scaffolds. *Colloids and Surfaces B: Biointerfaces* 82, 624–631.

Lee, S. H. and Shin, H. (2007). Matrices and scaffolds for delivery of bioactive molecules in bone and cartilage tissue engineering. *Advanced Drug Delivery Reviews* 59, 339–359.

Ma, X. P. (2008). Biomimetic materials for tissue engineering. *Advanced Drug Delivery Reviews* 60, 184–198.

Madigan, N. N., McMahon, S., O'Brien, T., Yaszemski, M. J., and Windebank, A. J. (2009). Current engineering and novel therapeutic approaches to axonal regeneration following spinal cord injury using polymer scaffolds. *Respiratory Physiology and Neurobiology* 169, 183–199.

Marzec, E. and Pietrucha, K. (2008). The effect of different methods of cross-linking of collagen on its dielectric properties. *Biophysical Chemistry* 132, 89–96.

Matthews, J. M., Wnek, G. E., Simpson, D. G., and Bowlin, G. L. (2002). Electrspinning of collagen nanofibers. *Biomacromolecules* 3, 232–238.

Miron-Mendoza, M., Seemann, J., and Grinnel, F. (2010). The differential regulation of cell motile activity through matrix stiffness and porosity in three dimensional collagen matrices. *Biomaterials* 31, 6425–6435.

Morrison, W. A. (2009). Progress in tissue engineering of soft tissue and organs. *Surgery* 145, 127–130.

Murphy, C. M., Haugh, M. G., and O'Brien, F. J. (2010). The effect of mean pore size on cell attachment, proliferation and migration in collagen-glicosaminoglycan scaffolds for bone tissue engineering. *Biomaterials* 31, 461–466.

Nalinanon, S., Benjakul, S., Kishimura, H., and Osako, K. (2011). Type I collagen from the skin of ornate threadfin bream (*Nemipterus hexodon*): Characteristics end effect of pepsin hydrolysis. *Food Chemistry* 125, 500–507.

Narotam, P. K., van Dellen, J. R., O'Fee, R. P., McKinney, G. W., Archibald, S. J., and O'Grady, J. (1999). Dural/meningeal repair product using collagen matrix. http://www.freepatentsonline.com/5997895.pdf.

O'Brien, F. J., Harley, B. A., Yannas, I. V., and Gibson, L. J. (2005). The effect of pore size on cell adhesion in collagen-GAG scaffolds. *Biomaterials* 26, 433–441.

Okada, M., Miyamoto, O., Shibuya, S., Hang, X., Yamamoto, T., and Itano, T. (2007). Expression and role of Type I collagen in rat spinal cord contusion injury model. *Neuroscience Research* 58, 371–377.

Pang, Y., Wang, X., Ucuzian, A. A., Brey, E. M., Burgess, W. H., Jones, K. J., Alexander, T. D., and Greisler, H. P. (2010). Local delivery of a collagen-binding FGF-1 chimera to smooth muscle cells in collagen scaffolds for vascular engineering. *Biomaterials* 31, 878–885.

Parenteau-Bareil, R., Gauvin, R., and Berthod, F. (2010). Collagen-based Biomaterials for Tissue Engineering Applications. *Materials* 3, 1863–1887.

Park, S. N., Lee, H. J., Lee, K. H., and Sun, H. (2003). Biological characterization of EDC-cross-linked collagen-hyaluronic acid matrix in dermal tissue restoration. *Biomaterials* 24, 1631–1641.

Parlato, C., di Nuzzo, G., Luongo, M., Parlato, R. S., Accardo, M., Cuccurullo, L., and Moraci, A. (2011). Use of collagen biomatrix (Tissu-Dura®) for dura mater repair: a long-term neuroradiological and neuropathological evaluation. *Acta Neurochirurgica* 153, 142–147.

Peng, Y. Y., Stoichevska, V., Yoshizumi, A., Danon, S. J., Glattauer, V., Prokopenko, O., Mirochitchenko, O., Yu, Z., Inouye, M., Werkmeister, J. A., Brodsky, B., and Ramshaw, J. A. M. (2010). Evaluation of bacterial collagen-like protein constructs for biomedical applications. http://www.biopolymers.macromol.in.

Pieper, J. S., Hafmans, T., Veerkamp, J. H., and van Kuppevelt, T. H. (2000). Development of tailor-made collagen-glycosaminoglycan matrices: EDC/NHS cross-linking, and ultrastructural aspects. *Biomaterials* 21, 581–593.

Pieper, J. S., Oosterhol, A., Dijakstra, P. J., Veerkamp, J. H., and van Kuppevelt, T. H. (1999). Preparation and characterization of porous cross-linked collagenous matrices containing bioavailable chondroitin sulphate. *Biomaterials* **20**, 847–858.

Pietrucha, K. (1991). New collagen implant as dural substitute. *Biomaterials* **12**, 320–323.

Pietrucha, K. (2005). Changes in denaturation and rheological properties of collagen-hyaluronic acid scaffolds as a result of temperature dependencies. *International Journal of Biological Macromolecules* **36**, 299–304.

Pietrucha, K. (2007). Some biological and thermo-analytical studies of cross-linked collagen sponges. *Med. and Biol. Eng. and Comp. (JIFMBE)* **14**, 3369–3372.

Pietrucha, K. (2010a). Role of a new composite collagen-based material in the surgical treatment of dysraphic defects of the central nervous system in children. http://www.bit-ibio.com.

Pietrucha, K. (2010b). Biodegradable collagen-based 3-d structure as a potential scaffolds for tissue engineering. http://www.biopolymers.macromol.in.

Pietrucha, K. and Banas M. (2009). Diffusive properties of hybrid collagen derivative hydrogels for medical applications. www.hybridmaterialsconference.com.

Pietrucha, K. and Banas, M. (2010). The effect of different methods of cross-linking of fish collagen on some its properties. http://www.biopolymers.macromol.in.

Pietrucha, K. and Marzec, E. (2005). Dielectric properties of the collagen-glycosaminoglycans scaffolds in the temperature range of thermal decomposition. *Biophysical Chemistry* **118**, 51–56.

Pietrucha, K. and Marzec, E. (2007). Characterization of electrically conductive biodegradable materials for potential medical application. *Medical and Biological Engineering and Computing (JIFMBE)* **14**, 3376–3378.

Pietrucha, K. and Polis, L. (2009). New collagen implant as a dura mater of the brain and spinal cord. http://www.iiis.org/CDs2009/CD2009SCI/BMIC2009/PapersPdf/B085TK.pdf.

Pietrucha, K., Polis, L., and Bidziński, J. (2000). Long-term clinical experiences with a new dural substitute. *Molecular Crystals and Liquid Crystals* **354**, 243–248.

Pietrucha, K. and Verne, S. (2010). Synthesis and characterization of a new generation of hydrogels for biomedical applications. IFMBE Proceedings ISBN 978-3-642-03899-0, 25/10 World Congress on Medical Physics and Biological Engineering WC2009, 7–12 September, 2009, Munich, Germany. Biomaterials, Cellular and Tissue Engineering, Artificial Organs. O. Dossel and C. Schlegel (Eds.). Springer, Germany, pp. 1–4.

Polis, L., Pietrucha, K., Nowosławska, E., Szymański, W., and Zakrzewski, K. (1993). Use of a new collagen implant as a dural substitute in child neurosurgery with special regards to the treatment of dysraphic defects in the central nervous system. *Polish Journal of Neurology and Neurosurgery* **27**, 323–326.

Powell, H. M., Supp, D. M., and Boyce, S. T. (2008). Influence of electrospun collagen on wound contraction of engineered skin substitutes. *Biomaterials* **29**, 834–843.

Tanaka, E. M. and Ferretti, P. (2009). Considering the evolution of regeneration in the central nervous system. *Nature Reviews/Neuroscience* **10**, 713–723.

Thomas, V., Dean, D. R., Jose, M. V., Mathew, B., Chowdhury, S., and Vohra, Y. K. (2007). Nanostructured biocomposite scaffolds based on collagen coelectrospun with nanohydroxyapatite. *Biomacromolecules* **8**, 631–637.

Vin, F., Teot, L., and Meaume, S. (2002). The healing properties of Promogran in venous leg ulcers. *Journal of Wound Care* **11**, 335.

Wang, X., Zhang, J., Chen, H., and Wang, Q. (2009). Preparation and Characterization of Collagen-Based Composite Conduit for Peripheral Nerve Regeneration. *Journal of Applied Polymers Science* **112**, 3652–3662.

Wen, X., Shi, D., and Zhang, N. (2005). Applications of Nanotechnology in Tissue Engineering. In Handbook of *Nanostructured Biomaterials and Their Applications in Nanobiotechnology*. H. S. Nalwa (Ed.). American Scientific Publishers, California, pp. **1**, 1–23.

Wess, T. J. (2008). Collagen Fibrillar Structure and Hierarchies. In *Collagen Structure and Mechanics*. P. Fratzl (Ed.), Springer

Science+Business Media, LLC, New Jork, USA, pp. 49–74.

Williams, D. F. (2006). To engineer is to create: the link between engineering and regeneration. *Trends in Biotechnology* **24**, 4–8.

Yang, L., Fite, C. F. C., van der Werf, K. O., Bennink, M. L., Dijkstra, P. J., and Fejen J. (2008). Mechanical properties of single electrospun type I fibers. *Biomaterials* **29**, 955–962.

Yoshii, S., Ito, S., Sima, M., Taniguchi, A., and Akagi, M. (2009). Functional restoration of rabbit spinal cord using collagen-filament scaffold. *Journal of Tissue Engineering and Regenerative Medicine* **3**, 19–25.

Yunoki, S., Ikoma, T., and Tanaka J. (2010). Development of collagen condensation method to improve mechanical strength of tissue engineering scaffolds. *Materials Characterization* **61**, 907–911.

Yunoki, S., Mori, K., Suzuki, T., and Munekata, M. (2007). Novel elastic material from collagen for tissue engineering. *Journal of Materials Science: Materials in Medicine* **18**, 1369–1375.

Zhang, M., Liu, W., and Li, G. (2009). Isolation and characterisation of collagens from the skin of largefin longbarbel catfish (*Mystus macropterus*). *Food Chemistry* **115**, 826–831.

Zhao, H. P. and Feng, X. Q. (2007). Ultrasonic technique for extracting nanofibers from nature materials. *Applied Physics Letters* **90**, 073112.

Zheng, W., Zhang, W., and Jiang, X. (2010). Biomimetic collagen nanofibrous materials for bone tissue engineering. *Advanced Engineering Materials*, **12**(9), B451–465.

Zhong, S. P., Teo, W. E., Zhu, X., Beuerman, R., Ramakrishna, S., and Yung, L. Y. L. (2007). Development of a novel collagen-GAG nanofibrous scaffold via electrospinning. *Materials Science and Engineering C* **27**, 262–266.

Zhou, J., Cao, C., Ma, X., and Lin, J. (2010). Electrospinning of silk fibroin and collagen for vascular tissue engineering. *International Journal of Biological Macromolecules* **47**, 514–519.

of fruits of jujube (Zizyphus spinachristi L.) at different stages of maturity. *Journal of Horticultural Science* **63**, 337-339.

Albersheim, P. (1976). The primary cell wall. In *Plant Biochemistry*. J. Bonner and J. E. Varner (Eds.). 3rd ed.. Academic Press, New York, pp. 225-274.

Alexander, M. M. and Sulebele, G. A. (1980). Characterization of pectins from Indian citrus peels. *Journal of Food Science and Technology* **17**, 180-182.

Ali, Z. M., Armugam, S., and Lazan, H. (1995). Beat-galactosidase and its significance in ripening mango fruits. *Phytochemistry* **38**, 1109-1114.

Baig, M. M., Burgin, C. W., and Curda, J. J. (1980). Labelling and chemistry of grapefruit pectic substances. *Phytochemistry* **19**, 1425-1428.

Barnier, M. and Thibault, J. F. (1982). Pectic substances of cherry fruits. *Phytochemistry* **21**, 111-115.

Batisse, C., Fils-Lycaon, D., and Buret, M. (1994). Pectin changes in ripening cherry fruits. *Journal of Food Science* **59**, 389-393.

Biswas, A. B. and Rao C. V. N. (1969). Enzymic hydrolysis and berry degradation studies on pectic acid from the pulp of unripe papaya (*Carica papaya* L.) fruit. *Indian Journal of Chemistry* **1**, 588-591.

Boothby, D. (1980). The pectic components of plum fruits. *Phytochemistry* **19**, 1949-1953.

Boothby, D. (1983). Pectic substances in developing and ripening plum fruits. *Journal of the Science of Food and Agricultur* **34**, 1117-1122.

Bukhavev, V. T., Nour, B. A. A., and Nourt, V. F. (1987). Physical and chemical changes in dates during ripening with special reference to pectic substances. *Date Plum Journal* **5**, 199-207.

Cardoso, S. M., Ferreira, J. A., Mafra, I., Silva, A. M. S., and Coimbra, M. A. (2007). Structural ripening related changes of the arabinan rich pectic polysaccharides from olive pulp cell walls. *Journal of Agricultural and Food Chemistry* **55**, 7124-7130.

Carpita, N. C. and McCann, M. (2000). The cell wall. In *Biochemistry and Molecular Biology of Plants*. B. B. Buchanan, W. Gruissem, and R. L.

20

Abbas, M. F., Al-Niami, J. H., and Al-Ani, R. F. (1988). Some physiological characteristics

Jones (Eds.). American Society of Plant Biologists, Rockville, MU, USA, pp. 52-108.

Cohn, R. and Cohn, A. L. (1996). The by products of fruit processing. In *Fruit Processing.* D. Arthery and P. R. Ashurst (Eds.). Blackie Academic and Professional, USA, pp. 196-220.

Coimbra, M. A., Barros, A., Rutledge D. N., and Delgadillo I. (1999). FTIR spectroscopy as a tool for the analysis of olive pulp cell wall polysaccharide extracts. *Carbohydrate Research* **30**, 145-154.

Cruess, W. V. (1977). Pectin, Jellies and Marmalades. In *Commercial Fruit and Vegetables*. Allied Scientific Publishers, Bikaner, pp. 426-489.

Daito, H. and Sato, Y. (1984). Changes in the volatile flavor components of Satsuma orange fruits during maturation. *Journal of the Japanese Society for Horticultural Science* **53**, 141-149.

Das, A. and Majumder, K. (2010). Fractional changes of pectic polysaccharides in different tissue zones of developing guava (*Psidium guajava* L.) fruits. *Scientia Horticulturae* **125**, 406-410.

Dhingra, M. K., Gupta, O. P., and Chandawat, B. (1983). Studies on pectin yield and quality of some guava cultivars in relation to cropping season and fruit maturity. *Journal of Food Science and Technology (India)* **20**, 10-13.

Dourado, F., Cardoso, S. M., Silva, A. M. S., Gama, F. M., and Coimbra, M. A. (2006). NMR structural elucidation of the arabinan from Prunus dulcis immunobiological active pectic polysaccharides. *Carbohydrate Polymers* **66**, 27-33.

Efimova, K. F. (1981). Dynamics of accumulation of chemical substances in apples during growth and ripening. *Bulletin Vsesoyuznogo Ordena Lenina Institute Rastenievodstvaimemi Vavilova* **113**, 81-82.

El-Buluk, R. E., Babiker, E. E., and El-Tinay, A. H. (1995). Changes in chemical composition of guava fruits during growth and development. *Food Chemistry* **54**, 279-282.

El-Zeftawi, B. M. (1978). Factors affecting granulation and quality of late picked Valencia oranges. *Journal of Horticultural Science* **53**, 331-337.

El-Zoghbi, M. (1994). Biochemical changes in some tropical fruits during ripening. *Food Chemistry* **49**, 33-37.

Esteves, M. T., Da, C., Chitarra, M. I. E., Carvalho, V. D. De, Chitarra, A. B., and Paula, M. B. De (1984). Characteristics of fruits of six guava (*Psidium guajava* L.) cultivars during ripening III. Pectin cellulose and hemicellulose. *Empresa Catarinense de Pesquisa Agropecuaria* **2**, 501-513.

Femenia, A., Garcia-Conesa, M., Simal, S., and Rossello, C. (1997) Characterization of the cellwalls of loquat (*Eriobottrya japonica*) fruit tissues. *Annals of Botany* **79**, 695-701.

Fischer, M. and Amado, R. (1994). Changes in the pectic substances of apples during development and post-harvest ripening. Part-I. Analysis of the alcohol-insoluble residues. *Carbohydrate Polymers* **25**, 161-165.

Fisher, M., Arrigoni, E., and Amado, R. (1994). Changes in the pectic substances of apples during development and post-harvest ripening. Part-II. Analysis of the pectic fractions. *Carbohydrate polymers* **25**, 167-175.

Fishman, M. L., Levaj, B., Gillespie, D., and Sorza, R. (1993). Fruit pectin during on tree ripening and storage. *Journal of the American Society for Horticultural Science* **118**, 343-349.

Fuke, Y. and Matsuoka, H. (1984). Changes in content of pectic substances, ascorbic acids and polyphenols and activity of pectin esterase of kiwi fruit during growth and ripening after harvest, studies on constituents of kiwi fruit cultured in Japan. Part-II. *Journal of the Japanese Society of Food Science and Technology* **31**, 31-37.

Gangwar, B. M. and Tripathi, R. S. (1973). A study on physico-chemical changes during growth, maturity and ripening in mango. *Punjab Horticulture Journal* **13**, 230-236.

Goodwin, T. W. and Mercer, E. I. (1988). The plant cell wall. In *Introduction to Plant Biochemistry*, 2nd ed.. Pergamon Press, UK, pp. 55-91.

Goto, A. (1989). Relationship between pectic substances and calcium in healthy, gelated and granulated juice sacs of sanbokam (*Citrus sulcata*) fruits. *Plant and Cell Physiology* **30**, 801-806.

Gross, K. C. (1990). Recent developments on tomato fruit softening. *Post Harvest News Information* **1**, 109-112.

Hernandez-Unzon, H. Y. and Lakshminarayana, S. (1982). Biochemical changes during development and ripening of tamarind fruit (*Tamarindus indica* L.). *Hortscience* **17**, 940-942.

Hwang, Y. S., Huber, D. J., and Albrico, L. G. (1990). Composition of cell-wall components in normal and disordered juice vesicles of grapefruit. *Journal of the American Society for Horticultural Sciences* **115**, 281-287.

Inkyu, K., Juseop, K., and Jaekyun, B. (1996). The role of β-galactosidase on the modification of cell-wall components in persimmon fruit. *Journal of the Korean Society for Horticultural Science* **37**, 528-533.

Kashyap, D. R., Vohra, P. K., Chapra, S., and Tewari, R. (2001). Applications of pectinases in commercial sector: A review. *Bioresource Technology* **77**, 215-227.

Kertesz, Z. I. (1951). *The Pectic Substances*. Interscience Publishers Inc., New York.

Klein, J. D., Hanzon, J., Irwin, P. L., Shalom, N. B., and Lurie, S. (1995). Pectin esterase activity and pectin methyl esterification in heated Golden Delicious apples. *Phytochemistry* **39**, 491-494.

Kumar, S., Goswami, A. K., and Sharma, T. R. (1985). Changes in pectin content and polygalacturonase activity in developing apple fruits. *Journal of Food Science and Technology (India)* **22**, 282-283.

Liyama, K., Lam, T. B. T., and Stone B. A. (1994). Covalent cross-links in the cell wall. *Plant Physiology* **104**, 315-320.

Majumder, K. and Mazumdar, B. C. (2001). Effects of auxin and gibberellin on pectic substances and their degrading enzymes in developing fruits of cape-gooseberry (*Physalis peruviana* L.). *Journal of Horticultural Science and Biotechnology* **76**, 276-279.

Majumder, K. and Mazumdar, B. C. (2002). Changes of pectic substances in developing fruits of cape-gooseberry (*Physalis peruviana* L.) in relation to the enzyme activity and evolution of ethylene. *Scientia Horticulturae* **96**, 91-101.

Majumder, K. and Mazumdar, B. C. (2005). Ethephon-induced fractional changes of pectic polysaccharides in developing cape gooseberry (*Physalis peruviana* L.) fruits. *Journal of the Science of Food and Agriculture* **85**, 1222-1226.

Malis-Arad, S, Didi, S, Mizrahi, Y, and Kopeliovitch E. (1983). Pectic substances: Changes in

soft and firm tomato cultivars. *Journal of Horticultural Science* **58**, 111-116.

Maness, N. O., Chrz, D., Hedge, S., and Goffreda, J. C. (1993). Cell wall changes in ripening peach fruit from cultivars differing in softening rate. *Acta Horticulturae* **343**, 200-203.

Mangas, J. J., Dapena, E., Rodriguez, M. S., Moreno, J., Gutierrez, M. D., and Blanco, D. (1992). Changes in pectic fractions during ripening of cider apples. *Hort Science* **27**, 328-330.

Marcelin, O., Mourgues, J., and Talmann, A. (1990). The polysaccharides of guava (*Psidium guajava* L.): Development during growth and technological effects during puree and juice manufacture. *Fruits* **45**, 511-520.

Matsui, T. and Kitagawa, H. (1991). Seasonal changes in pectin methylesterase and polygalcturonase activities in persimmon fruits. *Technical Bulletin of Faculty of Agriculture, Kagawa University* **45**, 45-50.

Matsuura, Y. (1987). Limit to the de-esterification of citrus pectin by citrus pectinesterase. *Agricultural and Biological Chemistry* **51**, 1675-1677.

Miceli, A., Antonaci, C., Linsalata, V., and Leo, P. De (1995). Biochemical changes associated with ripening of kiwi fruits. *Agricultura Mediterranea* **125**, 121-127.

Mondal, S. K., Ray, B., Thibault, J. F., and Ghosal, P. K. (2002). Cell-wall polysaccharides from the fruits of Limonia acidissima: isolation, purification and chemical investigation. *Carbohydrate Polymers* **48**, 209-212.

Naggar, V. F., et al. (1992). Pectin, a possible matrix for oral sustained-release preparations of water-soluble drugs. *STP Pharma Sciences* **2**, 227-234.

Neukom, H, Amado, R., and Pfister, M. (1980). New insights into the structure of pectic substances. *Lebensmittet Weissenchaft and Technology* **13**, 1-6.

Ning, B., Kubo, Y., Inaba, A., and Nakamura, R. (1997). Softening characteristics of Chinese pear 'yali' fruit with special relation to changes in cell-wall polysaccharides and their degrading enzymes. *Scientific Reports of the Faculty of Agriculture, Okayama University* **86**, 71-78.

Nizamuddin, M., Hoffman, J., and Larm, O. (1982). Fractionation and characterization of carbohydrates from *Embilica officinalis. Swedish Journal of Agricultural Research* 12, 3-7.

Oakenfull, D. G. (1991). The chemistry of high-methoxyl pectins. In *The Chemistry and Technology of Pectin*. R. H. Walter (Ed.). Academic Press, New York.

O'beirne, D., Buren, J. P. van, and Mattick, L. R. (1982). Two distinct pectin fractions from senescent idovned apples extracted using non-degradative methods. *Journal of Food Science* 47, 173-176.

Oechslin, R., Lutz, M. V., and Amado, R. (2003) Pectic substances isolated from apple cellulosic residue: structural characterization of a new type of rhamnogalcturonan I. *Carbohydrate Polymers* 51, 301-310.

Owino, W. O., Nakano, R., Kubo, Y., and Inaba, A. (2004). Alterations in cell-wall polysaccharides during ripening in distinct anatomical tissue region of the fig (*Ficus carica* L.) fruits. *Postharvest Biology and Technology* 32, 67-77.

Pal, D. K. and Selvaraj Y. (1979). Changes in pectin and pectinesterase activity in developing guava fruits (*Psidium guajava* L.). *Journal of Food Science and Technology* 16, 115-116.

Pandey, N. C., Singh, A. R., Maurya, V. N., and Katiyar, R. K. (1986). Studies on the biochemical changes in bael (*Aegle marmelos* Correa fruit). *Progressive Hortculture* 18, 29-34.

Pilnik, W. and Veragen, A. G. J. (1970). Pectic substances and other uronides. In *The Biochemistry of Fruits and their Products Vol 1*. A. C. Hulme (Ed.). Academic Press Inc., London, pp. 55-80.

Platt-Alolia, K. A., Thomson, W. W., and Young, R. E. (1980). Ultra-structural changes in the wall of ripening ripening avocado: Transmission scanning and freeze fracture microscopy. *Botanical Gazette* 141, 366-373.

Proctor, A. and Peng, L. C. (1989). Pectic fractions during blueberry fruit development and ripening. *Journal of Food Science* 54, 385-387.

Procter, A. and Peng, L. C. (1990). Pectin changes during blueberry fruit development. *OARDC Special Circular Ohio Agricultural Research and Development Centre* 121, 11-14.

Ranawala, A. P., Suematsu, C. and Masuba, H. (1992). The use of β-galactosidase in the modifi-

cation of cell-wall components during muskmelon fruit ripening. *Plant Physiology* 100, 1318-1325.

Redgwell, R. J., Fischer, M., Kendale, E., and Macrae, E. A. (1997). Galactose loss and fruit ripening: High molecular weight arabinogalactans in the pectic polysaccharides of fruit cell walls. *Planta* 203, 174-181.

Redgwell, R. J., Melton, L. D., and Brasch, D. J. (2001). Cell-wall polysaccharides of kiwifruit (*Actinidia deliciosa*): Chemical features in different tissue zones of the fruit at harvest.

Ros, J. M., Schols, H. A., and Voragen, A. G. J. (1996). Extraction, characterization and enzymic degradation of lemon peel pectins. *Carbohydrate Researh* 282, 271-284.

Saeed, A. R., Tinay, A. H. E., and Khattab A.H. (1975). Viscosity of mango nectar as related to pectic substances. *Journal of Food Science* 46, 203-204.

Sathyanarayana, N. G. and Panda, T. (2003). Purification and Biochemical properties of microbial pectinases- a review. *Process Biochemistry* 38, 987-996.

Sato, Y., Yamakawa O., and Honda F. (1986). Varietal differences in fruit quality during ripening of strawberry. *Bulletin of the Vegetable and Ornamental Crops Research Station* 9, 23-30.

Saulnier, L. and Brillouet, J. M. (1998). Structural studies of pectic substances from the pulp of grape berries. *Carbohydrate Research* 182, 63-78.

Selli, R. and Sansavani, S. (1995). Sugar, acid and pectin content in relation to ripening and quality of peach and nectarine fruits. *Acta Horticulturae* 379, 345-358.

Seymour, G. B., Wainwright, H., and Tucker G. A. (1989). Cell wall changes in ripening mangoes. *Aspects of Applied Biology* 20, 93-94.

Singh, P. and Singh, I. S. (1995). Physico-chemical changes during fruit development in litchi (Litchi chinensis Sonn.) *Mysore Journal of Agricultural Science* 29, 252-255.

Slany, J., et al. (1981). Evaluation of tablets with pectin as a binding agent. *Farmaceuticky Obzor* 50, 491-498.

Soni, S. L. and Randhawa, G. S. (1969). Morphological and chemical changes in the developing fruits of lemon, Citrus limon L. *Indian Journal of Agricultural Science* **32**, 813-829.

Spayd, S. E. and Morris, J. R. (1981). Physical and chemical characteristics of puree from once over harvested strawberries. *Journal of the American Society for Horticultural Science* **106**, 101-105.

Sriamornsak, P. (2001). Pectin: The role in health. *Journal of Silpakorn University* **21-22**, 60-77.

Srirangarajan, A. N. and Shrikhande, A. J. (1977). Characterization of mango peel pectin. *Journal of Food Science* **42**, 279-280.

Tandon, D. K. and Kalra, S. K. (1984). Pectin changes during the development of mango fruit cv.Dashehari. *Journal of Horticultural Science* **59**, 283-286.

Tarchevsky, I. A. and Marchenko, G. N. (1991). Cell wall composition. In *Cellulose: Biosynthesis and Structure*. Springer-Verlag, New York, pp. 9-20.

Thakur, B. R., et al., (1997). Chemistry and uses of pectin—A review. *Critical Reviews in Food Science and Nutrition* **37**, 47-73.

Tischenko, V. P. (1973). Biochemical investigation of the diversity of soluble pectins in different apples cvs. *Referativny Zhurnal* **10**, 682.

Tripathi, V. K., Ram, H. B., Jain, S. P., and Singh, S. H. (1981). Changes in developing banana fruit. *Progressive Horticulture* **13**, 45-53.

Trukhiiya, S. R. and Shabel'skaya, E. F. (1983). Dynamics of pectins during ripening of Spanish red pineapple fruit. *Referativny Zhurnal* **4**, 679.

Tsantili, E. (1990). Changes during development of 'Tsapela' fig fruits. *Scientia Horticulturae* **44**, 227-234.

Vander Vlugt-Bergmans, C. J. B., Meeuwsen, P. J. A., Voragen, A. G. J., and van Oogen, A. J. J. (2000). Endo-xylogalacturonan hydrolase, a novel pectinolytic enzyme. *Applied Environmental Microbiology* **66**, 36-41.

Vargas, L., Lorente, F. A., San Chez, A., Valenzueia, J. L., and Romero, L. (1991). Phosphorus, calcium, pectin and carbohydrate fractions in varieties of watermelon. *Acta Horticulturae* **287**, 467-476.

Vierhuis, E., Schols, H. A., Beldman, G., and Voragen, A. G. J. (2000). Isolation and charac-

terization of cell wall material from olive fruit (*Olea europaea* cv koroneiki) at different ripening stages. *Carbohydrate Polymers* **43**, 195-203.

Wang, G., Han, Y., and Yu, L. (1995). Study on the activities of stage specific enzymes during the softening of Chinese gooseberry. *Acta Botanica Simica* **37**, 198-203.

Whitaker, J. R. (1970). Microbial pectinolytic enzymes. In *Microbial Enzymes and Biotechnology*. W. M. Fogarty and C. T. Kelly (Eds.). Elsevier Science, London, pp. 133-176.

Yamaki, S. and Matsuda, K. (1977). Changes in the activities of some cell-wall degrading enzymes during the development and ripening pear fruit (Prunus serotina var culta Rehder). *Plant and Cell Physiology* **18**, 81-93.

Yapo, B. M. and Koffi, K. L. (2008). The polysaccharide composition of yellow passion fruit cell wall: Chemical and macromolecular features of extractable pectins and hemicellulosic polysaccharides. *Journal of the Science of Food and Agriculture* **88**, 2125-2133.

Yapo, B. M., Lerouge, P.,Thibault, J. F., and Ralet, M. C. (2007). Pectins from citrus peel cell walls contain homogalacturonans homogenous with respect to molar mass, rhamnogalacturonan I and rhamnogalacturonan II. *Carbohydrate Polymers* **69**, 426-435.

21

Ashok Kumar, Stephenson, L. D., and Murray, J. N. (2006). Self-healing coatings for steel. *Progress in Organic Coatings* **55**, 244–253.

Blaiszik, B. J., Caruso, M. M., McIlroy, D. A., Moore, J. S., White, S. R., and Sottos, N. R. (2009). Microcapsules filled with reactive solutions for self-healing materials. *Progress in Polymer Science* **50**, 990–997.

Brown, E. N., Kessler, M. R., White, S. R., Sottos, N. R. (2003). *In situ* poly (urea-formaldehyde) microencapsulation of dicyclopentadiene. *Journal of Microencapsulation* **20**(6), 719–730.

Dong Yang Wu, Sam Meure, and David Solomon (2008). Self-healing polymeric materials: A review of recent developments. *Progress in Polymer Science* **33**, 479–522.

Erin, B. Murphy and Fred Wudl (2010). The world of smart healable materials. *Progress in Polymer Science* **35**, 223–251.

Suryanarayanaa C., Chowdoji K. Raob, and Dhirendra Kumara (2008). Preparation and characterization of microcapsules containing linseed oil and its use in self-healing coatings. *Progress in Organic Coatings* **63**, 72–78.

22

Atom (2010). "O tempo nas charges—Óleo Derramada Avança" (Time in the cartoons—Spilled Oil Forward) http://otemponaschargesefotos.blogspot.com/2010_06_01_archive.html (accessed on 13 December, 2010).

Barry, C. (2007). "Slick Death: Oil-spill treatment kills coral". Science News vol. 172, p. 67.

Bellamy, D. J., Clarke, P. H., John, D. M., Jones, D., Whittick, A., and Darke, T. (1967). "Effects of Pollution from the Torrey Canyon on Littoral and Sublittoral Ecosystems", *Nature* **216**, 1170–1173.

Biagini, D. (2010). "Vazamento no Golfo do México—Esse Mar É Nosso" (Leaking in the Gulf of Mexico—This is Our Sea) http://decimarbiagini.blogspot.com/2010_07_01_archive.html; (accessed on 13 December, 2010).

Bragg, J. R, Prince, R. C., Harner, E. J., and Atlas, R. M. (1994). "Effectiveness of bioremediation for the Exxon Valdez oil spill". *Nature* **368**, 413–418.

Carvalho, T. V., Craveiro, A. A.,and Craveiro, A. C. (2006). "Uso de quitina, quitosana e seus derivados na remoção de petróleo e seus resíduos de águas". Patent PI0404309-0.

Juraci H. Jr. (2010). "Castanha de Caju" (Cashew Nut) http://caju.t35.com/imagens/castanha%20de%20caju%201.html (accessed on 13 December, 2010).

Lopes, M. C., Oliveira, G. E., and Souza, F. G.. Jr. (2010). "Espumados magnetizáveis úteis em processos de recuperação ambiental". *Polímeros* **20**, 359–365.

Marcelino, A. (2010). "20 anos Depois as Praias do Alasca Ainda têm Crude do Exxon Valdez" (20 Years on the Beaches of Alaska are Still the Exxon Valdez Crude) http://terraeoxigenio.blogspot.com/2010/01/20-anos-depois-as-praias-do-alasca.html (accessed on 13 December, 2010).

Proquinor (2010). "Óleo de Mamona Padrão" (Castor Oil Standard) http://www.proquinor.com.br/oleodemamonapadrao.htm (accessed on 14 December, 2010).

Queiros, Y. G. C., Clarisse, M. D., Oliveira, R. S., Reis, B. D., Lucas, E. F., and Louvisse, A. M. T. (2006). "Materiais poliméricos para tratamento de água oleosa: utilização, saturação e regeneração". *Polímeros* **16**, 224–229.

Ramos, H. C.(2010). "Frutos do Brasil—Caju" (Fruits of Brazil—Cashew) http://frutosdobrasil.blogspot.com/2010/04/o-caju-e-muitas-vezes-tido-como-o-fruto.html (accessed in 13 December, 2010).

Schambeck, R. G. (2010). "Mancha aparece na Zimba…" (Oil slick appears in Zimba…) http://surf4ever.wordpress.com/category/meio-ambiente/ (accessed on 13 December, 2010).

Souza, F. G.. Jr., Marins, J. A., Pinto, J. C., Oliveira, G. E., and Rodrigues, C. H. M. (2009a). *J. Mat. Sci.* ICAM Special Issue, 4231.

Souza, F. G. Jr, Oliveira, G. E., Rodrigues, C. H. M., Soares, B.G., Nele, M., and Pinto, J. C. (2009b). *Macromol. Mater. Eng.* **294**, 484.

Souza, F. G. Jr., Marins, J. A., Rodrigues, C. H. M., and Pinto, J. C. (2010b). "A magnetic composite for cleaning of oil spills on water". *Macromolecular Materials and Engineering* **295**, 10, 942–948.

Souza, F. G. Jr., Oliveira,G. E., Marins, J. A., and Cosme, T. A. (2010a) "Processo para produção de biocompósitos magnéticos à base de cardanol e furfural, produtos resultantes desse processo e uso desses produtos para remoção de petróleo". Patent tranferred to UFRJ.

Souza, F. G. Jr., Soares, B. G., and Dahmouche, K. (2007). *J. Polym. Sci. B Polym. Phys.* **45**, 3069.

Tristão, G. (2010). "Linhaça Ajuda a Emagrecer e traz Benefícios a Saúde" (Flaxseed Helps Weight loss and health benefits) http://inhaarteartesanato.blogspot.com/2010_01_01_archive.html (accessed on 13 December, 2010).

23

Barry, C. (2007). "Slick Death: Oil-spill treatment kills coral". *Science News* **172**, 67.

Bellamy, D. J., Clarke, P. H., John, D. M., Jones, D., Whittick, A., and Darke, T. (1967). "Effects of Pollution from the Torrey Canyon on Littoral and Sublittoral Ecosystems". *Nature* **216**, 1170–1173.

Bouza, T. (2010). "EUA redobram esforços para conter mancha de óleo" (Redouble USA efforts to contain oil slick in 2010) http://www1.folha.uol.com.br/folha/ambiente/ult10007u732503.shtml (accessed on 12 December, 2010).

Bragg, J. R., Prince, R. C., Harner, E. J., and Atlas, R. M. (1994). "Effectiveness of bioremediation for the Exxon Valdez oil spill". *Nature* **368**, 413–418.

Caetano, J. (2010) "Perguntas Ardentes Sobre o Golfo do México de Derramamento de Óleo que Nós Merecemos Algumas Respostas" (Burning Questions About the Gulf of Mexico Oil Spill that we Deserve Some Answers in 2010) http://juniacaetano.blogspot.com/2010/07/16-perguntas-ardentes-sobre-o-golfo-do.html (accessed on 08 December, 2010).

Carvalho, T. V., Craveiro, A. A., and Craveiro, A. C. (2006). "Uso de quitina, quitosana e seus derivados na remoção de petróleo e seus resíduos de águas". Patent PI0404309-0.

CPDA NEWS (2010). "Igreja dos EUA Lança Guia De Oração Pela Tragédia No Golfo Do México" (USA Church Launches Guide Prayer For Tragedy In The Gulf Of Mexico in 2010) http://tvbrasa.wordpress.com/2010/06/21/igreja-dos-eua-lanca-guia-de-oracao-pela-tragedia-no-golfo-do-mexico (accessed on 08 December, 2010).

Deboni, H. (2010). "Biodiesel" http://www.deboni.he.com.br/empresa/biodiesel.htm (accessed on 08 December. 2010).

Flexor, G. (2010) "O Programa Nacional de Biodiesel: avanços e limites" (Biodiesel National Program: advances and limits in 2010) http://geografiaegeopolitica.blogspot.com/2010/06/o-programa-nacional-de-biodiesel.html (accessed on 08 December, 2010).

Lopes, M. C., Oliveira, G. E., and Souza, F. G. Jr. (2010). "Espumados magnetizáveis úteis em processos de recuperação ambiental". Polímeros (Accepted).

Luís, M. and Inês, M. (2011). "Marés Negras" (Alick in 2011) http://ambilog2007.blogspot.

com/2008_09_01_archive.html (accessed on 16 June, 2011).

Queiros, Y. G. C., Clarisse, M. D., Oliveira, R. S., Reis, B. D., Lucas, E. F., Louvisse, A. M. T. (2006). "Materiais poliméricos para tratamento de água oleosa: utilização, saturação e regeneração". *Polímeros* **16**, 224-229.

Souza, F. G. Jr., Marins, J. A., Rodrigues, C. H. M., and Pinto, J. C. (2010a). "A magnetic composite for cleaning of oil spills on water". *Macromolecular Materials and Engineering* **295**(10), 942-948.

Souza, F. G. Jr., Oliveira, G. E., Marins, J. A., and Cosme, T. A. (2010b). "Processo para produção de biocompósitos magnéticos à base de cardanol e furfural, produtos resultantes desse processo e uso desses produtos para remoção de petróleo". Patent transferred to UFRJ.

24

Alonso, B., Massiot, D., Florian, P., Paradies, H. H., Gaveau, P., and Mineva, T. (2009). [14]N and [81]Br. Quadrupolar nuclei as sensitive NMR probes of n-alkyltrimethylammonium bromide crystal structures. An experimental and theoretical study. *J. Phys. Chem. B.* **113**, 11906-11920.

Anderson, D. M., Gruner, S. M., and Leibler, S. (1988). Geometrical aspects of the frustration in the cubic phases of lyotropic liquid crystals. *Proc. Natl. Acad. Sci. USA* **85**, 5364.

Aschauer, H., Grob, A., Hildebrandt, J., Schuetze, E., and Stuetz, P. (1990). Highly purified lipid X is an LPS-antagonist only. I. Isolation and characterization of immunostimulating contaminants in a batch of synthetic lipid X. *J. Biol. Chem* **265**, 9159-9164.

Attard, G. S., Goltner, C. G., Corker, J. M., Henke, S., and Templer, R. H. (1997). Liquid crystals template for nanostructure metals. Angew. *Chem. Int. Ed. Engl.* **36**, 1315-1316.

Bates, F. S. and Fredrickson, G. H. (1999). Block copolymers-Designer soft materials. *Physics Today* **52**, 32.

Batista, V. M. and Miller, M. A. (2010). Crystallization of deformable spherical colloids. *Phys. Rev. Lett.* **105**, 088305-088309.

Bosse, J., Götze, W., and Lücke, M. (1978). Mode-coupling theory of simple classical liquids. *Phys. Rev. A* **17**, 434-446.

Bucci, S., Fagotti, C., Degiorgio, V., and Piazza, R. (1991). Small-angle neutron-scattering study of ionic-nonionic mixed micelles. *Langmuir* **7**, 824-826.

Christ, W. J., Hawkins, L. D., Lewis, M. D., and Kishi, Y. (2003). Synthetic lipid A antagonist for sepsis treatment. In *"Carbohydrate-based Drug Discovery"*, Vol. *1*, pp 341-355; C. H. Wong (Ed.), Wiley-VCH Verlag GmbH & Co, KGA.

Christ, W. J., McGuinness, P. D., Asano, O., Wang, Y., Mullarkey, M. A., Pere, M., Hawkins, L. D., Blythe, T. A., Dubuc, G. R., and Robidoux, A. L. (1994). *J. Amer. Chem. Soc.* **116**, 3637-3638.

Clancy, S. F. P. H., Steiger, D. A., Tanner, Thies, M., and Paradies, H. H. (1994). Micellar behavior of distearyldimethylammonium hydroxide and chloride in aqueous solutions. *J. Phys. Chem.* **98**, 11143-11162.

De Gennes, P. and Prost, J. (1993). *The Physics of Liquid Crystals.* International Series of Monographs on Physics. Oxford Science Publications (Clarendon Press, Oxford).

De Geyer, A., Guillermo, A., Rodriguez, V., and Molle, B. (2000). Evidence for spontaneous formation of three-dimensionally periodic cellular structures in a water/oil/surfactant/alcohol system. *J. Phys. Chem. B* **104**(28), 6610-6617.

Ernst, R. K., Yi, E. C., Guo, L., Lim, K. B., Burns, J. L., Hacket, M., and Miller, S. I. (1999). Specific lipopolysaccharide found in cystic fibrosis airway Pseudomonas aerugenosa. *Science* **286**, 1561.

Faunce, C. A. and Paradies, H. H. (2007). Density fluctuation in Coulombic colloid dispersion: Self-assembly of lipid A-phosphates. *Mater. Res. Soc. Symp. Proc.* **947**, 0947-A03-11.

Faunce, C. A. and Paradies, H. H. (2008). Observations of liquid like order of charged rod-like lipid A-diphosphate assemblies at pH 8.5. *J. Chem. Phys.* **128**, 065105-065113.

Faunce, C. A. and Paradies, H. H. (2009). Two new colloidal crystal phases of lipid A-monophosphate: Order-to-order transition in colloidal crystals. *J. Chem. Phys.* 244708-244728.

Faunce, C. A. and Paradies, H. H. (2010). Phase transition, structure, self-assembly and arrangements of lipid A-phosphates in aqueous dispersions. *Recent. Res. Devel. Physical Chem.* **10**, 77-141. Transworld Research Network, (37/661 92) Trivandrum-695 023 Kerala, India.

Faunce, C. A., Quitschau, P., and Paradies, H. H. (2003a). Solution and structural properties of colloidal charged lipid A (diphosphate) dispersions. *J. Phys. Chem. B.* **107**, 2214-2227.

Faunce, C. A., Reichelt, H., Paradies, H. H., Quitschau, P., Rusch, V., and Zimmermann, K. (2003b). The formation of colloidal crystals of lipid A-diphosphate: Evidence for the formation of nanocrystals at low ionic strength. *J. Phys. Chem. B.* **107**, 9943-9947.

Faunce, C. A., Reichelt, H., Quitschau, P., and Paradies, H. H. (2007). Ordering of lipid A-monophosphate clusters in aqueous solutions. *J. Chem. Phys.* **127**, 115103-11525.

Faunce, C. A., Reichelt, H., and Paradies, H. H. (2011). Studies on structures of lipid A-monophosphate clusters. *J. Chem. Phys.* **134**, 104902-104916.

Faunce, C. A., Reichelt, H., Quitschau, P., Zimmermann, K., and Paradies, H. H. (2005). The liquidlike ordering of lipid A-diphosphate colloidal crystals: The influence of Ca^{2+}, Mg^{2+}, Na^{+}, and K^{+} on the ordering of colloidal suspensions of lipid A-diphosphate in aqueous solutions. *J. Chem. Phys.* **122**, 214727-214750.

Frank, F. C. and Kasper, J. S. (1959) Complex alloy structures regarded as sphere packings. II. Analysis and classification of representative structures. *Acta Crystallograph* **12**, 483-499.

Gazit, E. (2010). Diversity for self-assembly. *Nature Chemistry* **2**, 1010-1011.

Hansen, J. P. and Verlet, l. (1969). Phase transitions of the Lennard-Jones systems. *Phys. Rev.* **184**, 151-161.

Hashmi, S. M., Wickman, H. H., and Weitz, D. A. (2005). "Tetrahedral calcite crystals facilitate self-assembly at the air-water interface." *Phys. Rev. E.* **72**, 041605-041612.

Koltover, I., Salditt, T., Rädler, J. O., and Safinya, C. R. (1998). An inverted hexagonal phase of cationic liposome-DNA complexes related to DNA release and delivery. *Science* **281**, 78-81.

Lee, S., Bluemle, M. J., and Bates, F. S. (2010). Discovery of a Frank-Kasper σ phase in sphere-forming block copolymer melts. *Science* **330**, 349-353.

Mackay, A. L. (1962). A dense non-crystallographic packing of equal spheres. *Acta Crystallograph* **15**, 916-918.

Paradies, H. H. (1992) Pharmaceutical Formulations. *US-Patent* **5,110**, 929.

Paradies, H. H. and Clancy, S. F. (1997) Uncoiling of DNA by double chained cationic surfactants. *Mat. Res. Soc. Symp. Proc.* **463**, 49-55.

Paradies, H. H. and Habben, F. (1993). Structure of N-hexadecylpyridinium chloride monohydrate. *Acta Crystallograph C* **49**, 744-747.

Raetz, C. R. and Whitefield, C. (2002). Lipopolysaccharide endotoxins. *Annu. Rev. Biochem.* **71**, 635-700.

Reichelt, H., Faunce, C. A., and Paradies, H. H. (2008). The phase transition of charged colloidal lipid A-diphosphate. *J. Phys. Chem.* **112**, 3290-3293.

Rivier, N. and Aste, T. (1996). "Organized packing". *Forma* **11**,223-231.

Rosen, J. B., Daniela, A., Wilson, D. A., Wilson, C. J., Peterca, M., Betty C., Won, B. C., Huang, C., Lipski, L. R., Zeng, X., Ungar, G., Heiney, P. A., and Percec, V. (2009). Predicting the structure of supramolecular dendrimers via the analysis of libraries of AB_3 and constitutional isomeric AB_2 biphenylpropyl ether self-assembling dendrons. *J. Amer. Chem. Soc.* **131**, 17500-17521.

Rusch, V., Ottendorfer, D., Zimmermann, K., Gebauer, F., Schrödl, W., Nowak, P., Skarabis, H., and Kunze, R. (2001) Results of an open, non-placebo controlled pilot study investigating the Immunomodulatory potential of autovaccine. *Drug Research* **51**, 690-697.

Sayers, T., Macher, I., Chung, J., and Kuhler, E. (1987). The production of tumor necrosis factor by mouse bone marrow-derived macrophages in response to bacterial lipopolysaccharide and a chemically-synthesized monosaccharide precursor. *J. Immunol.* **138**, 2935-2940.

Schneck, E., Oliveira, R., Rehfeldt, F., Deme, B., Brandenburg, K., Seydel, U., and Tanaka, M. (2009). Mechanical properties of interacting lipopolysaccharide membranes from bacteria mu-

tants studied by specular and off-specular neutron scattering. *Phys. Rev. E* **80**, 041929.

Seddon, J. M. (1996). Lyotropic phase behaviour of biological amphiphiles. *Ber. Bunsenges. Phys. Chem.* **100**, 380-393.

Thies, M, Clancy, S. F., and Paradies, H. H. (1996). Multicomponent diffusion of distearyldimethylammonium polyelectrolyte solution in the presence of salt and different pH: Coupled transport of sodium chloride. *J. Phys. Chem.* **100**, 9881-9891.

Torquato, S. and Stillinger, F. H. (2010). Jammed hard-particle packings: From Kepler to Bernal and beyond. *Reviews of Modern Physics* **82**, 2633-2672.

Tschierske, C. (2001). Micro-segregation, molecular shape and molecular topology partners for the design of liquid crystalline materials with complex mesophase morphologies. *J. Mater. Chem.* **11**, 2647-2671.

Ungar, G., Liu, Y., Parsec, V., and Cho, W. D. (2003). Giant supramolecular liquid crystal lattice. *Science* **229**, 1208-1211.

Weaire, D. and Phelan, R. (1994), Vertex instabilities in foams and emulsions. *J. Phys. Condens. Matter* **8** L37.

Wette, P., Klassen, I., Holland-Moritz, D., Herlach, D. M., Schöppe, Lorenz, N., Reiber, H., Palberg, T., and Vogt, S. V. (2010). Complete description of re-entrant phase behavior in a charge variable colloidal model system. *J. Chem. Phys.* **132**, 131102-131106.

Witten, T. A. and Pincus, P. (2004). *Structural Fluids*. Chapter 7.5.1. Oxford University Press, New York.

Yamanaka, J., Yoshida, H., Koga, T., Ise, N., and Hashimoto, T. (1999). Reentrant order-disorder transition in ionic colloidal dispersions by varying particle charge density. *Langmuir* **15**, 4198-4202.

Ziherl, P. and Kamien, R. D. (2001). Maximizing Entropy by Minimizing Area: Towards a new principle of self-organization. *J. Phys. Chem. B* **105**, 10147-10158.

Zimmermann, K., Rusch, V., and Paradies, H. H. (2003). Kyberdrug as autovaccines with immune-regulating effects. *European Patent (EP)* **1**, 341, 546.

25

Ahn, A. C. and Grodzinsky, A. J. (2009). Relevance of collagen piezoelectricity to "Wolff's Law": A critical review. *Medical Engineering and Physics* 31(7), 733–741.

Beaulé, P. E., Lee, J. L., Le Duff, M. J., Amstutz, H. C., and Ebramzadeh, E. (2004). Orientation of the femoral component in surface arthroplasty of the hip. A biomechanical and clinical analysis. *The Journal of Bone and Joint Surgery* 86(5), 2015–2021.

Dekhtyar, Y., Polyaka, N., and Sammons, R. (2008). Electrically Charged Hydroxyapatite Enhances Immobilization and Proliferation of Osteoblasts. In *14th Nordic-Baltic Conference on Biomedical Engineering and Medical Physics*, pp. 23–25.

ElMessiery, M. A. (1981). Physical basis for piezoelectricity of bone matrix. Physical Science, Measurement and Instrumentation, Management and Education, Reviews. *IEE Proceedings A* 128(5), 336–346.

Frias, C., Reis, J., Capela e Silva, F., Potes, J., Simões, J., and Marques, A. T. (2010). Polymeric piezoelectric actuator substrate for osteoblast mechanical stimulation. *Journal of Biomechanics* 43(6), 1061–1066.

Fukada, E. and Yasuda, I. (1957). On the piezoelectric effect of bone. *Journal of the Physical Society of Japan* 12(10), 1158–1162.

Fukada, E. and Yasuda, I. (1964). Piezoelectric effects in collagen. *Japanese Journal of Applied Physics* 3(2), 117–121.

Halperin, C., Mutchnik, S., Agronin, A., Molotskii, M., Urenski, P., Salai, M., and Rosenman G. (2004). Piezoelectric effect in human bones studied in nanometer scale. *Nano Letters*, 4(7), 1253–1256.

Harter, L. V., Hruska, K. A., and Duncan, R. L. (1995). Human osteoblast-like cells respond to mechanical strain with increased bone matrix protein production independent of hormonal regulation. *Endocrinology* 136(2), 528–535.

Huiskes, R. I. K., Weinans, H., and Rietbergen B. V. (1992). The relationship between stress shielding and bone resorption around total hip stems and the effects of flexible materials. *Clinical Orthopaedics and Related Research* 274, 124–134.

Hung, C. H., Lin, Y. L. and Young, T. H. (2006). The effect of chitosan and PVDF substrates on the behavior of embryonic rat cerebral cortical stem cells. *Biomaterials* 27(25), 4461–4469.

Kadow-Romacker, A., Hoffmann, J. E., Duda, G., Wildemann, B., and Schmidmaier, G. (2009). Effect of Mechanical Stimulation on Osteoblast- and Osteoclast Like Cells *in vitro Cells Tissues Organs* 190(2), 61–68.

Kumar, D., Gittings, J. P., Turner, I. G., Bowen, C. R., Bastida-Hidalgo, A., and Cartmell, S. H. (2010). Polarization of hydroxyapatite: Influence on osteoblast cell proliferation. *Acta Biomaterialia* 6(4), 1549–1554.

Marino, A. A. and Becker, R. O. (1974). Piezoelectricity in bone as a function of age. *Calcified Tissue International* 14(4), 327–331.

Mintzer, C. M., Robertson, D. D., Rackemann, S., Ewald, C., Scott, R. D., and Spector, M. (1990). Bone loss in the distal anterior femur after total knee arthroplasty. *Clinical Orthopaedics and Related Research* 260, 135–143.

Mosley, J. R., March, B. M., Lynch, J., and Lanyon, L. E. (1997). Strain magnitude related changes in whole bone architecture in growing rats. *Bone* 20(3), 191–198.

Mikuni-Takagaki, Y. (1999). Mechanical responses and signal transduction pathways in stretched osteocytes. *Journal of Bone and Mineral Metabolism* 17(1), 57–60.

Noris-Suárez, K., Lira-Olivares, J., Ferreira, A. M., Feijoo, J. L., Suárez, N., Hernández, M. C., and Barrios, E. (2007). *In vitro* deposition of hydroxyapatite on cortical bone collagen stimulated by deformation-induced piezoelectricity. *Biomacromolecules* 8(3), 941–948.

Pavlin, D., Zadro, R., and Gluhak-Heinrich, J. (2001): Temporal pattern of stimulation of osteoblast-associated genes during mechanically-induced osteogenesis. *In vivo*: Early Responses of Osteocalcin and Type I Collagen. *Connective Tissue Research* 42(2), 135–148.

Perrien, D. S., Brown, E. C., Aronson, J., Skinner, R. A., Montague, D. C., Badger, T. M., and Lumpkin C. K. Jr. (2002). Immunohistochemical study of osteopontin expression during distraction osteogenesis in the rat. *Journal of Histochemistry and Cytochemistry* **50**(4), 567–574.

Pienkowski, D. and Pollack, S. R. (1983). The origin of stress-generated potentials in fluid-saturated bone. *Journal of Orthopaedic Research* **1**(1), 30–41.

Ramtani, S. (2008). Electro-mechanics of bone remodeling. *International Journal of Engineering Science* **46**(11), 1173–1182.

Reinish, G. B. and Nowick A. S. (1975). Piezoelectric properties of bone as functions of moisture content. *Nature*, **253**(5493), 626–627.

Reis, J. C., Cabo-Verde, S., Capela-Silva, F., Frias, C., Potes, J., Melo, R., Nunes, I., Marcos, H., Silva, T., and Botelho M. L. (2010). Sterilization of an electronic medical device. In NIC 2010—NAARI International Conference. S. Sabharwal (Ed.). NAARI, Mumbai, India.

Smalt, R., Mitchell, F. T., Howard, R. L., and Chambers, T. J. (1997). Induction of NO and prostaglandin E2 in osteoblasts by wall-shear stress but not mechanical strain. *American Journal of Physiology—Endocrinology and Metabolism* **273**(4), E751–758.

Sumner, D. R. and Galante, J. O. (1992). Determinants of stress shielding: design versus materials versus interface. *Clinical Orthopaedics and Related Research* **274**, 202–212.

Tabary, N., Lepretre, S., Boschin, F., Blanchemain, N., Neut, C., Delcourt-Debruyne, E., Martel, B., Morcellet, M., and Hildebrand, H. F. (2007). Functionalization of PVDF membranes with carbohydrate derivates for the controlled delivery of chlorhexidin. *Biomolecular Engineering* **24**(5), 472–476.

Van't Hof, R. J. (2001). Nitric oxide and bone. *Immunology* **103**(3), 255–261.

26

Butler, M. F. and Cameron R. E. (2000). A study of the molecular relaxation in solid starch using dielectric spectroscopy. *Polymer* **41**, 2249–2263.

Einfiled, J., Meißner, D., and Kwasniewski, A. (2003). Contribution to the molecular origin of the dielectric relaxation processes in polysaccharides–The high temperature range. *Journal of Non Crystalline Solids* **320**, 40–55.

Einfiled J., Meißner, D., and Kwasniewski, A. (2001). Polymer dynamics of cellulose and other polysaccharides in solid state-secondary dielectric relaxation processes. *Prog. Polym. Sci.* **26**, 1419–1472.

Majumder, T. P., Meißner, D., and Schick, C. (2004). Dielectric process of wet and well-dried wheat starch. *Carbohydrate Polymers* **56**, 361–366.

Moates, G. K., Noel, T. R., Parker, R., and Ring, S. G. (2000). Dynamic mechanical and dielectric characterization of amylose-glycerol films. *Carbohydrate Polymers* **44**, 247–253.

Smits, A. L. M., Wubbenhorst, M., Kruiskamp, P. H., Van Soest, J. J. G., Vliegenthart, J. F. G., and Van Turnhout, J. (2001). Structure evolution in amylopectin/ethylene glycol mixtures by H-bond formation and phase separation studied with dielectric relaxation spectroscopy. *J. Phys. Chem. B* **105**, 5630–5636.

Viciosa, M. T., Dionisio, M., Silva, R. M., Reis, R. L., and Mano, J. F. (2004). Molecular motions in chitosan studied by deiectric relaxation spectroscopy. *Biomacromolecules* **5**, 2073–2078.

Wubbenhorst, M. and Turnhout, J. V. (2002). Analysis of complex dielectric spectra. One dimensional derivative techniques and three-dimensional modeling. *Journals of Non Crystalline Solids* **305**, 40–49.

27

Baiardo, M., Zini, E., and Scandola, M. (2004). Flax fiber-polyester composites. *Composites Part A: Applied Science and Manufacturing* **35**(6), 703–710.

Baley, C., Busnel, F., Grohens, Y., and Sire, O. (2006). Influence of chemical treatments on surface properties and adhesion of flax fiber-polyester resin. *Composites Part A: Applied Science and Manufacturing* **37**(10), 1626–1637.

Bijwe, J., Indumathi, J., and Ghosh, A. K. (2002). On the abrasive wear behavior of fabric-reinforced polyetherimide composites. *Wear* **253**(7–8), 768–777.

Borruto, A., Crivellone, G., and Marani, F. (1998). Influence of surface wettability on friction and wear tests. *Wear* **222**(1), 57–65.

Corbiere, N. T., Gfeller, L. B., Lundquist, L., Leterrier, Y., Manson, J. A. E., and Jolliet, O. (2001). Life cycle assessment of biofibers replacing glass fibers as reinforcement in plastics. *Resources,Conservation and Recycling* **33**(4), 267–287.

El-Tayeb, N. S. M. (2008a). A study on the potential of sugarcane fibers/polyester composite for tribological applications. *Wear* **265**(1–2), 223–235.

El-Tayeb, N. S. M. (2008b). Abrasive wear performance of untreated SCF reinforced polymer composite. *Journal of Materials Processing Technology* **206**(1–3), 305–314.

Gowda, T. M., Naidu, A. C. B., and Rajput, C. (1999). *Some mechanical properties of untreated jute fabric–reinforced polyester composite* **30**(3), 277–284.

Hashmi, S. A. R., Dwivedi, U. K., and Chand, N. (2007). Graphite modified cotton fiber reinforced polyester composites under sliding wear conditions. *Wear* **262**(11–12), 1426–1432.

Huda, M. S., Drzal, L. T., Mohanty, A. K., and Misra, M. (2008). Effect of fiber surface-treatments on the properties of laminated biocomposites from poly(lactic acid) (PLA) and kenaf fibers. *Composites Science and Technology* **68**(2), 424–432.

Joshi, S. V., Drzal, L. T., Mohanty, A. K., and Arora, S. (2004). Are natural fiber composites environmentally superior to glass fiber reinforced composites? *Composites Part A: Applied Science and Manufacturing* **35**(3), 371–376.

Liu, G., Chen, Y., and Li, H. (2004). A study on sliding wear mechanism of ultrahigh molecular weight polyethylene/polypropylene blends. *Wear* **256**(11–12), 1088–1094.

Navin, C. and Dwivedi, U. K. (2006). Effect of coupling agent on abrasive wear behavior of chopped jute fiber-reinforced polypropylene composites. *Wear* **261**, 1057–1063.

Nirmal, S. G. and Yousif, B. F. (2009). Wear and frictional performance of betelnut fiber-reinforced polyester composite. *Proceedings of the Institution of Mechanical Engineers, Part J: Journal of Engineering Tribology* **223**(2), 183–194.

Nishino, T., Hirao, K., Kotera, M., Nakamae, K., and Inagaki, H. (2003). Kenaf reinforced bio-degradable composite. *Composites Science and Technology* **63**(9), 1281–1286.

Pothan, L. A., Oommen, Z., and Thomas, S. (2003). Dynamic mechanical analysis of banana fiber reinforced polyester composites. *Composites Science and Technology* **63**(2), 283-293.

Sınmazcelik, T. and Yılmaz, T. (2007). Thermal aging effects on mechanical and tribological performance of PEEK and short fiber reinforced PEEK composites. *Materials & Design* **28**, 641–648.

Sumer, M., Unal, H., and Mimaroglu, A. (2008). Evaluation of tribological behaviour of PEEK and glass fiber reinforced PEEK composite under dry sliding and water lubricated conditions. *Wear* **265**(7–8), 1061–1065.

Thomas, B., Hadfield, M., and Austen, S. (2009). Wear observations applied to lifeboat slipway launches. *Wear* **267**(11), 2062–2069.

Tong, J., Arnell, R. D., and Ren, L. Q. (1998). Dry sliding wear behaviour of bamboo. *Wear* **221**(1), 37–46.

Tong, J., Ma, Y., Chen, D., Sun, J., and Ren, L. (2005). Effects of vascular fiber content on abrasive wear of bamboo. *Wear* **259**(1–6), 78–83.

Wambua, P., Ivens, J., and Verpoest, I. (2003). Natural fibers: can they replace glass in fibers reinforced plastics? *Composites Science and Technology* **63**, 1259–1264.

Wu, J. and Cheng, X. H. (2006). The tribological properties of Kevlar pulp reinforced epoxy composites under dry sliding and water lubricated condition. *Wear* **261**(11–12), 1293–1297.

Yamamoto, Y. and Takashima, T. (2002). Friction and wear of water lubricated PEEK and PPS sliding contacts. *Wear* **253**(7–8), 820–826.

Yousif, B. F. (2008). Frictional and wear performance of polyester composites based on coir fibers. *Proc. IMechE, Part J: J. Engineering Tribology* **223**(1), 51–59.

Yousif, B. F. and El-Tayeb, N. S. M. (2007a). The effect of oil palm fibers as reinforcement on tribological performance of polyester composite. *Surface Review and Letters* **14**(6), 1–8.

Yousif, B. F. and El-Tayeb, N. S. M. (2007b). Tribological evaluations of polyester composites considering three orientations of CSM glass fibers using BOR machine. *Applied Composite Materials* **14**, 105–116.

Yousif, B. F. and El-Tayeb, N. S. M. (2008a). High-stress three-body abrasive wear of treated and untreated oil palm fiber-reinforced polyester composites. *Proceedings of the Institution of Mechanical Engineers, Part J: Journal of Engineering Tribology* **222**(5), 637–646.

Yousif, B. F. and El-Tayeb, N. S. M. (2008b). Wear and friction characteristics of CGRP composite under wet contact condition using two different test techniques. *Wear* **265**(5–6), 856–864.

Yousif, B. F. and El-Tayeb, N. S. M. (2010). Wet adhesive wear characteristics of untreated oil palm fiber-reinforced polyester and treated oil palm fiber-reinforced polyester composites using the pin-on-disc and block-on-ring techniques. *Proceedings of the Institution of Mechanical Engineers, Part J: Journal of Engineering Tribology* **224**(2), 123–131.

Yousif, B. F., Nirmal, U., Lau, S. T. W., and Devadas, A. (2008). The Potential of Using Betelnut Fibers for Tribo-Polyester Composites Considering Three Different Orientations. *ASME Conference Proceedings* **2008**(48654), 79–84.

Yu, S., Hu, H., and Yin, J. (2008). Effect of rubber on tribological behaviors of polyamide 66 under dry and water lubricated sliding. *Wear* **265**(3–4), 361–366.

28

Akabori, K., Yamamoto, Y., Kawahara, S., Jinnai, H., and Nishioka, H. (2009). Field emission scanning electron microscopy combined with focused ion beam for rubbery material with nano-matrix structure. *Journal of Physics: Conference Series* **184**, 012–027.

Allen, P. W. and Bristow, G. M. (1963). *The gel phase in natural rubber* **7**(2), 603–615.

Beach, E., Keefe, M., Heeschen, W., and Rothe, D. (2005). Cross-sectional analysis of hollow latex particles by focused ion beam-scanning electron microscopy. *Polymer* **46**(25), 11195–11197.

Brostow, W., Gorman, B. P., and Olea-Mejia O. (2007). Focused ion beam milling and scanning electron microscopy characterization of polymer+metal hybrids. *Materials Letters* **61**(6), 1333–1336.

Creighton, T. E. (1984). *Proteins*. Freeman, New York, USA.

Eng, A. H., Ejiri, S., Kawahara, S., and Tanaka, Y. (1994). Structural characteristics of natural rubberrole of ester groups. *J. Appl. Polym. Sci.: Appl. Polym. Symp.* **53**, 5–14.

Eng, A. H., Kawahara, S., and Tanaka, Y. (1993). Determination of low nitrogen content of purified natural rubber. *J. Nat. Rubb. Res.* **8**(2), 109–113.

Eng, A. H., Tanaka, Y., and Gan, S. N. (1992). FTIR studies on amino groups in purified Hevea rubber. *J. Nat. Rubb.Res.* **7**(2), 152–155.

Fernando, W. S. E.,Yapa, P. A. J., and Jayasinghe, P. P. (1985). Preparation and properties of low-nitrogen natural rubber. *KautschukGummi-Kunststoffe* **38**(11), 1010–1011.

Fukushima, Y., Kawahara, S., and Tanaka, Y. (1998). Synthesis of graft copolymers from highly deproteinized natural rubber. *J. Rubb. Res.* **1**(3), 154–166.

Kaneko, T., Nishioka, H., Nishi, T., and Jinnai, H. (2005). Reduction of anisotropic image resolution in transmission electron microtomography by use of quadrangular prism-shaped section. *J. Electron Micros.* **54**(5), 437–444.

Kato, M., Ito, T., Aoyama, Y., Sawa, K., Kaneko, T., Kawase, N., and Jinnai, H. (2007). Three-dimensional structural analysis of a block copolymer by scanning electron microscopy combined with a focused ion beam. *J. Polym. Sci., Part B: Polym. Phys.* **45**(6), 677–683.

Kawahara, S., Isono, Y., Washino, K., Morita, T., and Tanaka, Y. (2001). High-resolution latex state 13C-NMR spectroscopy: Part II. Effect of particle size and temperature. *Rubb. Chem. Technol.* **74**(2), 295–302.

Kawahara, S., Kawazura, T., Sawada, T., and Isono, Y. (2003). Preparation and characterization of natural rubber dispersed in nano-matrix. *Polymer* **44**(16), 4527–4531.

Kawahara, S. and Yamamoto, Y. (2007). Nano-matrix structure formed for NR. *Rubber World* **237**(2), 26–31.

Kawahara, S., Yamamoto, Y., Fujii, S., Isono, Y., Niihara, K., Jinnai, H., Nishioka, H., and Takaoka, A. (2008). FIB-SEM and TEMT observation of highly elastic rubbery material with nanomatrix structure. *Macromolecules* **41**(12), 4510–4513.

Kawase, N., Kato, M., Nishioka, H., and Jinnai, H. (2007). Transmission electron microtomography without the "missing wedge" for quantitative structural analysis. *Ultramicroscopy* **107**(1), 8–15.

Patterson, D. J. and Koenig, J. L. (1987). A Fourier-transform infrared and nuclear magnetic resonance study of cyclized natural rubber. *Makromol. Chem.* **188**(10), 2325–2337.

Pukkate, N., Kitai, T., Yamamoto, Y., Kawazura, T., Sakdapipanich, J., and Kawahara, S. (2007). Nano-matrix structure formed by graft-copolymerization of styrene onto natural rubber. *Eur. Polym. J.* **43**(8), 3208–3214.

Roberts, A. D. (1988). *Natural Rubber Science and Technology*. Oxford University Press, Oxford, USA.

Steinbüchel, A. and Koyama, T. (2001). *Biopolymer, Vol 2, Polyisoprenoides*. Wiley-VCH Verlag GmbH, Weinheim, Germany.

Suksawad, P., Kawahara, S., Yamamoto, Y., and Kuroda, H. (2009). Nanomatrix channel for ionic molecular transportation. *Macromolecules* **42**(21), 8557–8560.

Suksawad, P., Kosugi, K., Yamamoto, Y., Akabori, K., Kuroda, H., and Kawahara, S. (2011). Polymer electrolyte membrane with nanomatrix channel prepared by sulfonation of natural rubber grafted with polystyrene. *J. Appl. Polym. Sci.* **122**, 2403–2414.

Takayanagi, M., Imada, K., and Kajiyama, T. (1966). Mechanical properties and fine structure of drawn polymers. *J. Polym. Sci., Polym. Symp.* **15**, 26380.

Tanaka, Y. (1989). Structural characterization of naturally occurring *cis*- and *trans*-polyisoprenes by carbon-13NMR spectroscopy. *J. Appl.Polym. Sci.: Appl. Polym. Symp.* **44**, 19.

Tanaka, Y. (1989). Structure and biosynthesis mechanism of natural polyisoprene. *Prog. Polym. Sci.* **14**(3), 339–371.

Tanaka, Y., Kawahara, S., and Tangpakdee, J. (1997). Structural characterization of natural rubber. *KautschukGummiKunststoffe* **50**(1), 6–11.

Tanaka, Y., Sato, H., and Kageyu, A. (1983). Structure and biosynthesis mechanism of natural *cis*-polyisoprene from goldenrod. *Rubb. Chem. Technol.* **56**(2), 299–303.

Tangpakdee, J. and Tanaka, Y. (1997). Purification of natural rubber. *J. Nat. Rubb. Res.* **12**(2), 112–119.

Tangpakdee, J. and Tanaka, Y. (1997). Characterization of sol and gel in Hevea natural rubber. *Rubb. Chem. Technol.* **70**(5), 707–713.

Tuampoemsab, S. and Sakdapipanich, J. (2007). *Kautsch. Gummi Kunst.***57**, 678–684.

Yusof, N. H., Kawahara, S., and Said, M. M. (2008). Modification of deproteinised natural rubber by graft-copolymerization of methyl methacrylate. *J. Rubb. Res.* **11**(2), 97–110.

Index

Milton Keynes UK
Ingram Content Group UK Ltd.
UKHW031139141024
449569UK00024B/1217